# Lecture Notes in Computer Science

*Commenced Publication in 1973*
Founding and Former Series Editors:
Gerhard Goos, Juris Hartmanis, and Jan van Leeuwen

## Editorial Board

Mariano Consens   Gonzalo Navarro (Eds.)

# String Processing and Information Retrieval

12th International Conference, SPIRE 2005
Buenos Aires, Argentina, November 2-4, 2005
Proceedings

 Springer

Volume Editors

Mariano Consens
University of Toronto
Department of Mechanical and Industrial Engineering
Department of Computer Science
Toronto, Canada
E-mail: consens@cs.toronto.edu

Gonzalo Navarro
University of Chile
Center for Web Research, Dept. of Computer Science, Chile
E-mail: gnavarro@dcc.uchile.cl

Library of Congress Control Number: 2005934415

CR Subject Classification (1998): H.3, H.2.8, I.2, E.1, E.5, F.2.2

ISSN        0302-9743
ISBN-10     3-540-29740-5 Springer Berlin Heidelberg New York
ISBN-13     978-3-540-29740-6 Springer Berlin Heidelberg New York

Springer is a part of Springer Science+Business Media

springeronline.com

© Springer-Verlag Berlin Heidelberg 2005
Printed in Germany

Typesetting: Camera-ready by author, data conversion by Scientific Publishing Services, Chennai, India
Printed on acid-free paper        SPIN: 11575832        06/3142        5 4 3 2 1 0

# Preface

The papers contained in this volume were presented at the 12th edition of the International Symposium on String Processing and Information Retrieval (SPIRE), held November 2–4, 2005, in Buenos Aires, Argentina. They were selected from 102 papers submitted from 25 countries in response to the Call for Papers. A total of 27 submissions were accepted as full papers, yielding an acceptance rate of about 26%. In view of the large number of good-quality submissions the conference program also included 17 short papers that also appear in the proceedings. In addition, the Steering Committee invited the following speakers: Prabhakar Raghavan (Yahoo! Research, USA), Paolo Ferragina (University of Pisa, Italy), and Gonzalo Navarro (University of Chile, Chile).

Papers solicited for SPIRE 2005 were meant to constitute original contributions to areas such as string processing (dictionary algorithms, text searching, pattern matching, text compression, text mining, natural language processing, and automata-based string processing); information retrieval languages, applications, and evaluation (IR modeling, indexing, ranking and filtering, interface design, visualization, cross-lingual IR systems, multimedia IR, digital libraries, collaborative retrieval, Web-related applications, XML, information retrieval from semi-structured data, text mining, and generation of structured data from text); and interaction of biology and computation (sequencing and applications in molecular biology, evolution and phylogenetics, recognition of genes and regulatory elements, and sequence-driven protein structure prediction).

SPIRE has its origins in the South American Workshop on String Processing (WSP). Since 1998 the focus of the conference was broadened to include information retrieval. Starting in 2000, Europe has been the conference venue on even years. The first 11 meetings were held in Belo Horizonte (Brazil, 1993), Valparaíso (Chile, 1995), Recife (Brazil, 1996), Valparaíso (Chile, 1997), Santa Cruz (Bolivia, 1998), Cancún (Mexico, 1999), A Coruña (Spain, 2000), Laguna San Rafael (Chile, 2001), Lisboa (Portugal, 2002), Manaus (Brazil, 2003), and Padova (Italy, 2004).

SPIRE 2005 was held in tandem with LA-WEB 2005, the 3rd Latin American Web Congress, with both conferences sharing a common day in Web Retrieval.

SPIRE 2005 was sponsored by Centro Latinoamericano de Estudios en Informática (CLEI), Programa Iberoamericano de Ciencia y Tecnología para el Desarrollo (CYTED), Center for Web Research (CWR, University of Chile), and Sociedad Argentina de Informática e Investigación Operativa (SADIO).

We thank the local organizers for their support in the organization of SPIRE and the members of the Program Committee and the additional reviewers for providing timely and detailed reviews of the submitted papers and for their active participation in the email discussions that took place before we could assemble the final program. Finally, we would like to thank Ricardo Baeza-Yates, who, on behalf of the Steering Committee, invited us to chair the Program Committee.

November 2005                                                    Mariano P. Consens,
                                                                Gonzalo Navarro

# SPIRE 2005 Organization

## Steering Committee

| | |
|---|---|
| Ricardo Baeza-Yates (Chair) | ICREA-Universitat Pompeu Fabra (Spain) and Universidad de Chile (Chile) |
| Alberto Apostolico | Università di Padova (Italy) and Georgia Tech (USA) |
| Alberto Laender | Universidade Federal de Minas Gerais (Brazil) |
| Massimo Melucci | Università di Padova (Italy) |
| Edleno de Moura | Universidade Federal do Amazonas (Brazil) |
| Mario Nascimento | University of Alberta (Canada) |
| Arlindo Oliveira | INESC (Portugal) |
| Berthier Ribeiro-Neto | Universidade Federal de Minas Gerais (Brazil) |
| Nivio Ziviani | Universidade Federal de Minas Gerais (Brazil) |

## Program Committee Chairs

| | |
|---|---|
| Mariano Consens | Dept. of Mechanical and Industrial Engineering Dept. of Computer Science University of Toronto, Canada |
| Gonzalo Navarro | Center for Web Research Dept. of Computer Science Universidad de Chile, Chile |

## Program Committee Members

| | |
|---|---|
| Amihood Amir | Bar-Ilan University (Israel) |
| Alberto Apostolico | Università di Padova (Italy) and Georgia Tech (USA) |
| Ricardo Baeza-Yates | ICREA-Universitat Pompeu Fabra (Spain) and Universidad de Chile (Chile) |
| Nieves R. Brisaboa | Universidade da Coruña (Spain) |
| Edgar Chávez | Universidad Michoacana (Mexico) |
| Charles Clarke | University of Waterloo (Canada) |
| Bruce Croft | University of Massachussetts (USA) |
| Paolo Ferragina | Università di Pisa (Italy) |
| Norbert Fuhr | Universität Duisburg-Essen (Germany) |
| Raffaele Giancarlo | Università di Palermo (Italy) |
| Roberto Grossi | Università di Pisa (Italy) |
| Carlos Heuser | Universidade Federal de Rio Grande do Sul (Brazil) |

| Carlos Hurtado | Universidad de Chile (Chile) |
| Lucian Ilie | University of Western Ontario (Canada) |
| Panagiotis Ipeirotis | New York University (USA) |
| Juha Kärkkäinen | University of Helsinki (Finland) |
| Nick Koudas | University of Toronto (Canada) |
| Mounia Lalmas | Queen Mary University of London (UK) |
| Gad Landau | University of Haifa (Israel) and Polytechnic University (NY, USA) |
| Stefano Lonardi | University of California at Riverside (USA) |
| Yoelle Maarek | IBM Haifa Research Lab (Israel) |
| Veli Mäkinen | Bielefeld University (Germany) |
| Mauricio Marín | Universidad de Magallanes (Chile) |
| João Meidanis | UNICAMP (Brazil) |
| Massimo Melucci | Università di Padova (Italy) |
| Edleno de Moura | Universidade Federal do Amazonas (Brazil) |
| Ian Munro | University of Waterloo (Canada) |
| Arlindo Oliveira | INESC (Portugal) |
| Kunsoo Park | Seoul National University (Korea) |
| Prabhakar Raghavan | Yahoo Inc. (USA) |
| Berthier Ribeiro-Neto | Universidade Federal de Minas Gerais (Brazil) |
| Kunihiko Sadakane | Kyushu University (Japan) |
| Marie-France Sagot | INRIA (France) |
| João Setubal | Virginia Tech (USA) |
| Jayavel Shanmugasundaram | Cornell University (USA) |
| Ayumi Shinohara | Tohoku University (Japan) |
| Jorma Tarhio | Helsinki University of Technology (Finland) |
| Jeffrey Vitter | Purdue University (USA) |
| Hugh Williams | Microsoft Corporation (USA) |
| Hugo Zaragoza | Microsoft Research (UK) |
| Nivio Ziviani | Universidade Federal de Minas Gerais (Brazil) |
| Justin Zobel | RMIT (Australia) |

## External Reviewers

| | |
|---|---|
| Jussara Almeida | Michela Bacchin |
| Ramurti Barbosa | Bodo Billerbeck |
| Sebastian Böcker | Michael Cameron |
| David Carmel | Luis Coelho |
| Marco Cristo | Giorgio Maria Di Nunzio |
| Alair Pereira do Lago | Shiri Dori |
| Celia Francisca dos Santos | Fan Yang |
| Feng Shao | Nicola Ferro |
| Kimmo Fredriksson | Gudrun Fisher |
| Paulo B. Golgher | Alejandro Hevia |
| Jie Zheng | Carmel Kent |

Shahar Keret
Sascha Kriewel
Nicholas Lester
Julia Mixtacki
Henrik Nottelmann
Rodrigo Paredes
Hannu Peltola
Nadia Pisanti
Bruno Possas
Davi de Castro Reis
Luis Russo
Marinella Sciortino
Darren Shakib
S.M.M. (Saied) Tahaghoghi
Andrew Turpin
Ying Zhang

Tsvi Kopelowitz
Michael Laszlo
Saadia Malik
Viviane Moreira Orengo
Nicola Orio
Laxmi Parida
Patrícia Peres
Benjamin Piwowarski
Jussi Rautio
Nora Reyes
Klaus-Bernd Schürmann
Rahul Shah
Riva Shalom
Eric Tannier
Rodrigo Verschae

## Local Organization

SADIO (Argentine Society for Informatics and Operations Research)

| | |
|---|---|
| SADIO President | Gabriel Baum |
| Local Arrangements Chair | Héctor Monteverde |
| Steering Committee Liaison | Ricardo Baeza-Yates |
| Administrative Manager | Alejandra Villa |

# Table of Contents

## String Processing and Information Retrieval 2005

Enhanced Byte Codes with Restricted Prefix Properties
*J. Shane Culpepper, Alistair Moffat* .............................. 1

Experimental Analysis of a Fast Intersection Algorithm for Sorted
Sequences
*Ricardo Baeza-Yates, Alejandro Salinger* ......................... 13

Compressed Perfect Embedded Skip Lists for Quick Inverted-Index
Lookups
*Paolo Boldi, Sebastiano Vigna* .................................. 25

XML Retrieval with a Natural Language Interface
*Xavier Tannier, Shlomo Geva* .................................. 29

Recommending Better Queries from Click-Through Data
*Georges Dupret, Marcelo Mendoza* ............................. 41

A Bilingual Linking Service for the Web
*Alessandra Alaniz Macedo, José Antonio Camacho-Guerrero,*
*Maria da Graça Campos Pimentel* ............................. 45

Evaluating Hierarchical Clustering of Search Results
*Juan M. Cigarran, Anselmo Peñas, Julio Gonzalo, Felisa Verdejo* .... 49

Counting Suffix Arrays and Strings
*Klaus-Bernd Schürmann, Jens Stoye* ............................ 55

Towards Real-Time Suffix Tree Construction
*Amihood Amir, Tsvi Kopelowitz, Moshe Lewenstein,*
*Noa Lewenstein* ................................................ 67

Rank-Sensitive Data Structures
*Iwona Bialynicka-Birula, Roberto Grossi* ......................... 79

Cache-Conscious Collision Resolution in String Hash Tables
*Nikolas Askitis, Justin Zobel* .................................. 91

Measuring the Difficulty of Distance-Based Indexing
*Matthew Skala* ................................................. 103

$N$-Gram Similarity and Distance
*Grzegorz Kondrak* .............................................. 115

Using the $k$-Nearest Neighbor Graph for Proximity Searching in Metric
Spaces
*Rodrigo Paredes, Edgar Chávez* ................................. 127

Classifying Sentences Using Induced Structure
*Menno van Zaanen, Luiz Augusto Pizzato, Diego Mollá* ............. 139

Counting Lumps in Word Space: Density as a Measure of Corpus
Homogeneity
*Magnus Sahlgren, Jussi Karlgren* ............................... 151

Multi-label Text Categorization Using K-Nearest Neighbor Approach
with M-Similarity
*Yi Feng, Zhaohui Wu, Zhongmei Zhou* ........................... 155

Lydia: A System for Large-Scale News Analysis
*Levon Lloyd, Dimitrios Kechagias, Steven Skiena* ................. 161

Composite Pattern Discovery for PCR Application
*Stanislav Angelov, Shunsuke Inenaga* ........................... 167

Lossless Filter for Finding Long Multiple Approximate Repetitions
Using a New Data Structure, the Bi-factor Array
*Pierre Peterlongo, Nadia Pisanti, Frederic Boyer,*
*Marie-France Sagot* ............................................ 179

Linear Time Algorithm for the Generalised Longest Common Repeat
Problem
*Inbok Lee, Yoan José Pinzón Ardila* ............................. 191

Application of Clustering Technique in Multiple Sequence Alignment
*Patrícia Silva Peres, Edleno Silva de Moura* ..................... 202

Stemming Arabic Conjunctions and Prepositions
*Abdusalam F.A. Nwesri, S.M.M. Tahaghoghi, Falk Scholer* .......... 206

XML Multimedia Retrieval
*Zhigang Kong, Mounia Lalmas* .................................. 218

Retrieval Status Values in Information Retrieval Evaluation
*Amélie Imafouo, Xavier Tannier* ................................ 224

A Generalization of the Method for Evaluation of Stemming Algorithms
Based on Error Counting
   *Ricardo Sánchez de Madariaga, José Raúl Fernández del Castillo,*
   *José Ramón Hilera* . . . . . . . . . . . . . . . . . . . . . . . . . . . . . . . . . . . . . . . . . . . . .    228

Necklace Swap Problem for Rhythmic Similarity Measures
   *Yoan José Pinzón Ardila, Raphaël Clifford, Manal Mohamed* . . . . . . . .    234

Faster Generation of Super Condensed Neighbourhoods Using Finite
Automata
   *Luís M.S. Russo, Arlindo L. Oliveira* . . . . . . . . . . . . . . . . . . . . . . . . . . . . .    246

Restricted Transposition Invariant Approximate String Matching
Under Edit Distance
   *Heikki Hyyrö* . . . . . . . . . . . . . . . . . . . . . . . . . . . . . . . . . . . . . . . . . . . . . . . . .    256

Fast Plagiarism Detection System
   *Maxim Mozgovoy, Kimmo Fredriksson, Daniel White, Mike Joy,*
   *Erkki Sutinen* . . . . . . . . . . . . . . . . . . . . . . . . . . . . . . . . . . . . . . . . . . . . . . . . .    267

A Model for Information Retrieval Based on Possibilistic Networks
   *Asma H. Brini, Mohand Boughanem, Didier Dubois* . . . . . . . . . . . . . . . .    271

Comparison of Representations of Multiple Evidence Using a Functional
Framework for IR
   *Ilmério R. Silva, João N. Souza, Luciene C. Oliveira* . . . . . . . . . . . . . .    283

Deriving TF-IDF as a Fisher Kernel
   *Charles Elkan* . . . . . . . . . . . . . . . . . . . . . . . . . . . . . . . . . . . . . . . . . . . . . . . .    295

Utilizing Dynamically Updated Estimates in Solving the Longest
Common Subsequence Problem
   *Lasse Bergroth* . . . . . . . . . . . . . . . . . . . . . . . . . . . . . . . . . . . . . . . . . . . . . . . .    301

Computing Similarity of Run-Length Encoded Strings with Affine Gap
Penalty
   *Jin Wook Kim, Amihood Amir, Gad M. Landau, Kunsoo Park* . . . . . .    315

$L_1$ Pattern Matching Lower Bound
   *Ohad Lipsky, Ely Porat* . . . . . . . . . . . . . . . . . . . . . . . . . . . . . . . . . . . . . . . .    327

Approximate Matching in the $L_\infty$ Metric
   *Ohad Lipsky, Ely Porat* . . . . . . . . . . . . . . . . . . . . . . . . . . . . . . . . . . . . . . . .    331

An Edit Distance Between RNA Stem-Loops
   *Valentin Guignon, Cedric Chauve, Sylvie Hamel* . . . . . . . . . . . . . . . . . . .    335

A Multiple Graph Layers Model with Application to RNA Secondary
Structures Comparison
     *Julien Allali, Marie-France Sagot* .................................. 348

Normalized Similarity of RNA Sequences
     *Rolf Backofen, Danny Hermelin, Gad M. Landau,*
     *Oren Weimann* ...................................................... 360

A Fast Algorithmic Technique for Comparing Large Phylogenetic Trees
     *Gabriel Valiente* ................................................... 370

Practical and Optimal String Matching
     *Kimmo Fredriksson, Szymon Grabowski* ............................ 376

A Bit-Parallel Tree Matching Algorithm for Patterns with Horizontal
VLDC's
     *Hisashi Tsuji, Akira Ishino, Masayuki Takeda* ..................... 388

A Partition-Based Efficient Algorithm for Large Scale Multiple-Strings
Matching
     *Ping Liu, Yan-bing Liu, Jian-long Tan* ............................ 399

**Author Index** ................................................... 405

# Enhanced Byte Codes with Restricted Prefix Properties

J. Shane Culpepper and Alistair Moffat

NICTA Victoria Laboratory,
Department of Computer Science and Software Engineering,
The University of Melbourne, Victoria 3010, Australia

**Abstract.** Byte codes have a number of properties that make them attractive for practical compression systems: they are relatively easy to construct; they decode quickly; and they can be searched using standard byte-aligned string matching techniques. In this paper we describe a new type of byte code in which the first byte of each codeword completely specifies the number of bytes that comprise the suffix of the codeword. Our mechanism gives more flexible coding than previous constrained byte codes, and hence better compression. The structure of the code also suggests a heuristic approximation that allows savings to be made in the prelude that describes the code. We present experimental results that compare our new method with previous approaches to byte coding, in terms of both compression effectiveness and decoding throughput speeds.

## 1 Introduction

While most compression systems are designed to emit a stream of bits that represent the input message, it is also possible to use bytes as the basic output unit. For example, Scholer et al. [2002] describe the application of standard byte codes – called vbyte encoding in their paper – to inverted file compression; and de Moura et al. [2000] consider their use in a compression system based around a word-based model of text.

In this paper we describe a new type of byte code in which the first byte of each codeword completely specifies the number of bytes that comprise the suffix of the codeword. The new structure provides a compromise between the rigidity of the static byte codes employed by Scholer et al., and the full power of a radix-256 Huffman code of the kind considered by de Moura et al. The structure of the code also suggests a heuristic approximation that allows savings to be made in the prelude that describes the code. Rather than specify the codeword length of every symbol that appears in the message, we partition the alphabet into two sets – the symbols that it is worth taking care with, and a second set of symbols that are treated in a more generic manner.

Our presentation includes experimental results that compare the new methods with previous approaches to byte coding, in terms of both compression effectiveness and decoding throughput speeds.

## 2 Byte-Aligned Codes

In the basic byte coding method, denoted in this paper as bc, a stream of integers $x \geq 0$ is converted into a uniquely decodeable stream of bytes as follows: for each integer $x$, if $x < 128$, then $x$ is coded as itself in a single byte; otherwise, $(x \text{ div } 128) - 1$

M. Consens and G. Navarro (Eds.): SPIRE 2005, LNCS 3772, pp. 1–12, 2005.

is recursively coded, and then $x \bmod 128$ is appended as a single byte. Each output byte contains seven data bits. To force the code to be prefix-free, the last output byte of every codeword is tagged with a leading zero bit, and the non-final bytes are tagged with a leading one bit. The following examples show the simple byte code in action – bytes with a decimal value greater than 127 are *continuers* and are always followed by another byte; bytes with a decimal value less than 128 are *stoppers* and are terminal.

$$
\begin{array}{lll}
0 \to 000 & 1,000 \to 134\text{-}104 & 1,000,000 \to 188\text{-}131\text{-}064 \\
1 \to 001 & 1,001 \to 134\text{-}105 & 1,000,001 \to 188\text{-}131\text{-}065 \\
2 \to 002 & 1,002 \to 134\text{-}106 & 1,000,002 \to 188\text{-}131\text{-}066
\end{array}
$$

To decode, a radix-128 value is constructed. For example, $188\text{-}131\text{-}066$ is decoded as $((188 - 127) \times 128 + (131 - 127)) \times 128 + 66 = 1,000,002$.

The exact origins of the basic method are unclear, but it has been in use in applications for more than a decade, including both research and commercial text retrieval systems to represent the document identifiers in inverted indexes. One great advantage of it is that each codeword finishes with a byte in which the top (most significant) bit is zero. This identifies it as the last byte before the start of a new codeword, and means that compressed sequences can be searched using standard pattern matching algorithms. For example, if the three-element source sequence "$2; 1,001; 1,000,000$" is required, a byte-wise scan for the pattern $002\text{-}134\text{-}105\text{-}188\text{-}131\text{-}064$ in the compressed representation will find all locations at which the source pattern occurs, without any possibility of false matches caused by codeword misalignments. In the terminology of Brisaboa et al. [2003b], the code is "end tagged", since the last byte of each codeword is distinguished. de Moura et al. [2000] consider byte codes that are not naturally end-tagged.

The simple byte code is most naturally coupled to applications in which the symbol probabilities are non-increasing, and in which there are no gaps in the alphabet caused by symbols that do not occur in the message. In situations where the distribution is not monotonic, it is appropriate to introduce an *alphabet mapping* that permutes the sparse or non-ordered symbol ordering into a ranked equivalent, in which all mapped symbols appear in the message (or message block, if the message is handled as a sequence of fixed-length blocks), and each symbol is represented by its rank.

Brisaboa et al. [2003b] refer to this mapping process as generating a *dense* code. For example, consider the set of symbol frequencies:

$$20, 0, 1, 8, 11, 1, 0, 5, 1, 0, 0, 1, 2, 1, 2$$

that might have arisen from the analysis of a message block containing 53 symbols over the alphabet $0 \ldots 14$. The corresponding dense frequency distribution over the alphabet $0 \ldots 10$ is generated by the alphabet mapping

$$[0, 4, 3, 7, 12, 14, 2, 5, 8, 11, 13] \to [0, 1, 2, 3, 4, 5, 6, 7, 8, 9, 10],$$

that both extracts the $n = 11$ subset of alphabet symbols that occur in the message, and also indicates their rank in the sorted frequency list. Using dense codes, Brisaboa et al. were able to obtain improved compression when the underlying frequency distribution was not monotonically decreasing, with compressed searching still possible by mapping the pattern's symbols in the same manner. Our experimentation below includes a permuted alphabet dense byte coder, denoted dbc. The only difference between it and bc is that each message block must have a *prelude* attached to it, describing the alphabet mapping in use in that block. Section 4 considers in more detail the implications of including a prelude in each block of the compressed message.

In followup work, Brisaboa et al. [2003a] (see also Rautio et al. [2002]) observe that there is nothing sacred about the splitting point of 128 used to separate the stoppers and the continuers in the simple byte coder, and suggest that using values $S$ and $C$, with $S + C = 256$, gives a more flexible code, at the very small cost of a single additional parameter in the prelude. One way of looking at this revised scheme is that the tag bit that identifies each byte is being arithmetically coded, so that a little more of each byte is available for actual "data" bits.

The codewords generated by a $(S, C)$-dense coder retain the end-tagged property, and are still directly searchable using standard character-based pattern matching algorithms. The same per-block prelude requirements as for the dbc implementation apply to scdbc, our implementation of $(S, C)$-dense coding.

Brisaboa et al. describe several mechanisms for determining an appropriate value of $S$ (and hence $C$) for a given frequency distribution, of which the simplest is brute-force – simply evaluating the cost of each alternative $S$, and choosing the $S$ that yields the least overall cost. Pre-calculating an array of cumulative frequencies for the mapped alphabet allows the cost of any proposed set of codeword lengths to be evaluated quickly, without further looping. Brute-force techniques based on a cumulative array of frequencies also play a role in the new mechanism described in Section 3.

Finally in this section, we note that Brisaboa et al. [2005] have recently described an adaptive variant of the $(S, C)$-dense mechanism, in which the prelude is avoided and explicit "rearrange alphabet mapping now" codes are sent as needed.

## 3  Restricted Prefix Byte Codes

The $(S, C)$-dense code is a byte-level version of the Golomb code [Golomb, 1966], in that it matches best with the type of self-similar frequency sets that arise with a geometric probability distribution. For example, once a particular value of $S$ has been chosen, the fraction of the available code-space used for one byte codewords is $S/(S + C)$; of the code-space allocated to multi-byte codewords, the fraction used for two byte codes is $S/(S + C)$; and so on, always in the same ratio.

On the other hand, a byte-level Huffman code of the kind examined by de Moura et al. [2000] exactly matches the probability distribution, and is minimum-redundancy over all byte codes. At face value, the Huffman code is much more versatile, and can assign any codeword length to any symbol. In reality, however, a byte-level Huffman code on any plausible probability distribution and input message block uses just four different codeword lengths: one byte, two bytes, three bytes, and four bytes. On an $n$-symbol decreasing probability distribution, this observation implies that the set of dense symbol identifiers $0 \ldots (n - 1)$ can be broken into four contiguous subsets – the symbols that are assigned one-byte codes, those given two-byte codes, those given three-byte codes, and those given four-byte codes. If the sizes of the sets are given by $h_1$, $h_2$, $h_3$, and $h_4$ respectively, then for all practical purposes a tuple $(h_1, h_2, h_3, h_4)$ completely defines a dense-alphabet byte-level Huffman code, with $n = h_1 + h_2 + h_3 + h_4$.

In the $(S, C)$-dense code, the equivalent tuple is infinite, $(S, CS, C^2S, \ldots)$, and it is impossible, for example, for there to be more of the total codespace allocated to two-

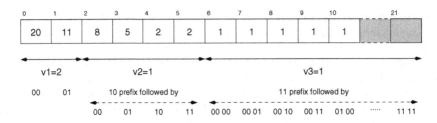

**Fig. 1.** Example of a restricted prefix code with $R = 4$ and $n = 11$, and $(v_1, v_2, v_3) = (2, 1, 1)$. The codewords for symbols 11 to 21 inclusive are unused. The 53 symbols are coded into 160 bits, compared to 144 bits if a bitwise Huffman code is calculated, and 148 bits per symbol if a radix-4 Huffman code is calculated. Prelude costs are additional.

byte codewords than to one-byte codewords. On a input message that consists primarily of low probability symbols, compression effectiveness must suffer.

Our proposal here adds more flexibility. Like the radix-256 Huffman code, we categorize an arrangement using a 4-tuple of numbers $(v_1, v_2, v_3, v_4)$, and require that the Kraft inequality be satisfied. But the numbers in the tuple now refer to initial digit ranges in the radix-$R$ code, and are set so that $v_1 + v_2 + v_3 + v_4 \leq R$. The code itself has $v_1$ one-byte codewords; $Rv_2$ two-byte codewords; $R^2 v_3$ three-byte codewords; and $R^3 v_4$ four-byte ones. To be feasible, we thus also require $v_1 + v_2 R + v_3 R^2 + v_4 R^3 \geq n$, where $R$ is the radix, typically 256. We will denote as *restricted prefix* a code that meets these criteria. The codeword lengths are not as freely variable as in an unrestricted radix-256 Huffman code, but the loss in compression effectiveness compared to a Huffman code is slight.

Figure 1 shows an example code that has the restricted prefix property, calculated with a radix $R = 4$ for a dense alphabet covering $n = 11$ symbols. In this code, the first two-bit unit in each codeword uniquely identifies the number of two-bit units in the suffix. Two symbols have codes that are one unit long ($v_1 = 2$); four symbols have codes that are two units long, prefixed by 10; and five symbols have codes that are two units long, prefixed by 11. There are eleven unused codewords.

The great benefit of the additional constraint is that the first unit (byte) in each codeword unambiguously identifies the length of that codeword, in the same way that in the $K$-flat code of Liddell and Moffat [2004] each codeword commences with a $k$-bit binary prefix that determines the length of the suffix part for that codeword, for some fixed value $k$. In particular, for the code described by $(v_1, v_2, v_3, v_4)$, the first byte of any one-byte codeword will be in the range $0 \ldots (v_1 - 1)$; the first byte of any two-byte codeword in the range $v_1 \ldots (v_1 + v_2 - 1)$; and the first byte of any three-byte codeword will lie between $(v_1 + v_2) \ldots (v_1 + v_2 + v_3 - 1)$. With this structure, it is possible to create an $R$-element array *suffix* that is indexed by the first byte of each codeword and exactly indicates the total length of that codeword.

Algorithm 1 shows how the *suffix* array, and a second array called *first*, are initialized, and then used during the decoding process. Once the codeword length is known, the mapped symbol identifier is easily computed by concatenating suffix bytes together, and adding a pre-computed value from the *first* array.

**Algorithm 1.** Decoding a message block

input: a block-length $m$, a radix $R$ (typically 256), and control parameters $v_1$, $v_2$, $v_3$, and $v_4$, with $v_1 + v_2 + v_3 + v_4 \leq R$.

1: $create\_tables(v_1, v_2, v_3, v_4, R)$
2: **for** $i \leftarrow 0$ **to** $m - 1$ **do**
3:     assign $b \leftarrow get\_byte()$ and $offset \leftarrow 0$
4:     **for** $i \leftarrow 1$ **to** $suffix[b]$ **do**
5:         assign $offset \leftarrow offset \times R + get\_byte()$
6:     assign $output\_block[i] \leftarrow first[b] + offset$

output: the $m$ symbols coded into the message block are available in the array $output\_block$

**function** $create\_tables(v_1, v_2, v_3, v_4, R)$

1: assign $start \leftarrow 0$
2: **for** $i \leftarrow 0$ **to** $v_1 - 1$ **do**
3:     assign $suffix[i] \leftarrow 0$ and $first[i] \leftarrow start$ and $start \leftarrow start + 1$
4: **for** $i \leftarrow v_1$ **to** $v_1 + v_2 - 1$ **do**
5:     assign $suffix[i] \leftarrow 1$ and $first[i] \leftarrow start$ and $start \leftarrow start + R$
6: **for** $i \leftarrow v_1 + v_2$ **to** $v_1 + v_2 + v_3 - 1$ **do**
7:     assign $suffix[i] \leftarrow 2$ and $first[i] \leftarrow start$ and $start \leftarrow start + R^2$
8: **for** $i \leftarrow v_1 + v_2 + v_3$ **to** $v_1 + v_2 + v_3 - v_4 - 1$ **do**
9:     assign $suffix[i] \leftarrow 3$ and $first[i] \leftarrow start$ and $start \leftarrow start + R^3$

---

**Algorithm 2.** Seeking forward a specified number of codewords

input: the tables created by the function $create\_tables()$, and a seek offset $s$.

1: **for** $i \leftarrow 0$ **to** $s - 1$ **do**
2:     assign $b \leftarrow get\_byte()$
3:     adjust the input file pointer forwards by $suffix[b]$ bytes

output: a total of $s - 1$ codewords have been skipped over.

---

The new code is not end tagged in the way the $(S, C)$-dense method is, a change that opens up the possibility of false matches caused by byte misalignments during pattern matching. Algorithm 2 shows the process that is used to seek forward a fixed number of symbols in the compressed byte stream and avoid that possibility. Because the suffix length of each codeword is specified by the first byte, it is only necessary to touch one byte per codeword to step forward a given number $s$ of symbols. By building this mechanism into a pattern matching system, fast compressed searching is possible, since standard pattern matching techniques make use of "shift" mechanisms, whereby a pattern is stepped along the string by a specified number of symbols.

We have explored several methods for determining a minimum-cost reduced prefix code. Dynamic programming mechanisms, like those described by Liddell and Moffat [2004] for the $K$-flat binary case, can be used, and have asymptotically low execution costs. On the other hand, the space requirement is non-trivial, and in this preliminary study we have instead made use of a generate-and-test approach, described in Algorithm 3, that evaluates each viable combination of $(v_1, v_2, v_3, v_4)$, and chooses the one with the least cost. Even when $n > 10^5$, Algorithm 3 executes in just a few hundredths or tenths of a second, and requires no additional space. In particular, once the cumulative frequency array $C$ has been constructed, on average just a few hundred thousand

---

**Algorithm 3.** Calculating the code split points using a brute force approach

---

input: a set of $n$ frequencies, $f[0 \ldots (n-1)]$, and a radix $R$, with $n \leq R^4$.

1: assign $C[0] \leftarrow 0$
2: **for** $i \leftarrow 0$ **to** $n - 1$ **do**
3:   assign $C[i+1] \leftarrow C[i] + f[i]$
4: assign $mincost \leftarrow partial\_sum(0, n) \times 4$
5: **for** $i_1 \leftarrow 0$ **to** $R$ **do**
6:   **for** $i_2 \leftarrow 0$ **to** $R - i_1$ **do**
7:     **for** $i_3 \leftarrow 0$ **to** $R - i_1 - i_2$ **do**
8:       assign $i_4 \leftarrow \lceil (n - i_1 - i_2 R - i_3 R^2)/R^3 \rceil$
9:       **if** $i_1 + i_2 + i_3 + i_4 \leq R$ **and** $cost(i_1, i_2, i_3, i_4) < mincost$ **then**
10:         assign $(v_1, v_2, v_3, v_4) \leftarrow (i_1, i_2, i_3, i_4)$ and $mincost \leftarrow cost(i_1, i_2, i_3, i_4)$
11:       **if** $i_1 + i_2 R + i_3 R^2 \geq n$ **then**
12:         **break**
13:     **if** $i_1 + i_2 R \geq n$ **then**
14:       **break**
15:   **if** $i_1 \geq n$ **then**
16:     **break**
output: the four partition sizes $v_1$, $v_2$, $v_3$, and $v_4$.

**function** $partial\_sum(lo, hi)$:
1: **if** $lo > n$ **then**
2:   assign $lo \leftarrow n$
3: **if** $hi > n$ **then**
4:   assign $hi \leftarrow n$
5: **return** $C[hi] - C[lo]$

**function** $cost(i_1, i_2, i_3, i_4)$
1: **return** $partial\_sum(0, i_1) \times 1 +$
      $partial\_sum(i_1, i_1 + i_2 R) \times 2 +$
      $partial\_sum(i_1 + i_2 R, i_1 + i_2 R + i_3 R^2) \times 3 +$
      $partial\_sum(i_1 + i_2 R + i_3 R^2, i_1 + i_2 R + i_3 R^2 + i_4 R^3) \times 4$

---

combinations of $(i_1, i_2, i_3, i_4)$ are evaluated at step 9, and there is little practical gain in efficiency possible through the use of a more principled approach.

## 4   Handling the Prelude

One of the great attractions of the simple bc byte coding regime is that it is completely static, with no parameters. To encode a message, nothing more is required than to transmit the first message symbol, then the second, and so on through to the last. In this sense it is completely *on-line*, and no input buffering is necessary. On the other hand, all of the dense codes are *off-line* mechanisms – they require that the input message be buffered into *message blocks* before any processing can be started. They also require that a *prelude* be transmitted to the decoder prior to any of the codewords in that block.

As well as a small number of scalar values (the size of the block; and the code parameters $v_1$, $v_2$, $v_3$, and $v_4$ in our case) the prelude needs to describe an ordering of the codewords. For concreteness, suppose that a message block contains $m$ symbols in total; that there are $n$ distinct symbols in the block; and that the largest symbol identifier in the block is $n_{max}$.

The obvious way of coding the prelude is to transmit a permutation of the alphabet [Brisaboa et al., 2003a,b]. Each of the $n$ symbol identifiers requires approximately $\log n_{max}$ bits, so to transmit the decreasing-frequency permutation requires a total of $n \log n_{max}$ bits, or an overhead of $(n \log n_{max})/m$ bits per message symbol. When $n$ and $n_{max}$ are small, and $m$ is large, the extra cost is negligible. For character-level coding applications, for example with $n \approx 100$ and $n_{max} \approx 256$, the overhead is less than $0.001$ bits per symbol on a block of $m = 2^{20}$ symbols. But in more general applications, the cost can be non-trivial. When $n \approx 10^5$ and $n_{max} \approx 10^6$, the overhead cost on the same-sized message block is $1.9$ bits per symbol.

In fact, an exact permutation of the alphabet is not required – all that is needed is to know, for each alphabet symbol, whether or not it appears in this message block, and how many bytes there are in its codeword. This realization leads to a better way of describing the prelude: first of all, indicate which $n$-element subset of the symbols $0 \ldots n_{max}$ appears in the message block; and then, for each symbol that appears, indicate its codeword length. For example, one obvious tactic is to use a bit-vector of $n_{max}$ bits, with a zero in the $k$th position indicating "$k$ does not appear in this message block", and a one in the $k$th position indicating that it does. That bit-vector is then followed by a set of $n$ two-bit values indicating codeword lengths between 1 and 4 bytes. Using the values $n \approx 10^5$ and $n_{max} \approx 10^6$ bits, the space required would thus be $n_{max} + 2n \approx 1.2 \times 10^6$, or $1.14$ bits per symbol overhead on a message block of $m = 2^{20}$ symbols.

Another way in which an ordered subset of the natural numbers can be efficiently represented is as a sequence of *gaps*, taking differences between consecutive items in the set. Coding a bit-vector is tantamount to using a unary code for the gaps, and more principled codes can give better compression when the alphabet density differs significantly from one half, either globally, or in locally homogeneous sections.

In a byte coder, where the emphasis is on easily decodeable data streams, it is natural to use a simple byte code for the gaps. The sets of gaps for the symbols with one-byte codes can be encoded; then the set of gaps of all symbols with two-byte codes; and so on. To estimate the cost of this prelude arrangement, we suppose that all but a small minority of the gaps between consecutive symbols are less than 127, the largest value that is coded in a single byte. This is a plausible assumption unless, for example, the sub-alphabet density drops below around 5%. Using this arrangement, the prelude costs approximately $8n$ bits, and when $n \approx 10^5$ corresponds to $0.76$ bit per symbol overhead on a message block of $m = 2^{20}$ symbols.

The challenge is to further reduce this cost. One obvious possibility is to use a code based on half-byte nibbles rather than bytes, so as to halve the minimum cost of coding each gap. But there is also another way of improving compression effectiveness, and that is to be precise only about high-frequency symbols, and to let low-frequency ones be assigned default codewords without their needing to be specified in the prelude. The motivation for this approach is that spending prelude space on rare symbols may, in the long run, be more expensive than simply letting them be represented with their "natural" sparse codes.

Algorithm 4 gives details of this *semi-dense* method, and Figure 2 gives an example. A threshold $t$ is used to determine the number of high-frequency symbols for which prelude information is supplied in a dense part to the code; and all symbols (including

**Algorithm 4.** Determining the code structure with a semi-dense prelude

input: an integer $n_{max}$, and an unsorted array of symbol frequency counts, with $c[s]$ recording the frequency of $s$ in the message block, $0 \leq s \leq n_{max}$; together with a threshold $t$.

1: assign $n \leftarrow 0$
2: **for** $s \leftarrow 0$ to $n_{max}$ **do**
3:     assign $f[t + s].sym \leftarrow s$ and $f[t + s].freq \leftarrow c[s]$
4: identify the $t$ largest $freq$ components in $f[t \ldots (t + n_{max})]$, and copy them and their corresponding symbol numbers into $f[0 \ldots (t - 1)]$
5: **for** $s \leftarrow 0$ to $t - 1$ **do**
6:     assign $f[f[s].sym].freq \leftarrow 0$
7: assign $shift \leftarrow 0$
8: **while** $f[t + shift] = 0$ **do**
9:     assign $shift \leftarrow shift + 1$
10: **for** $s \leftarrow t + shift$ to $n_{max}$ **do**
11:     assign $f[s - shift] \leftarrow f[s]$
12: use Algorithm 3 to compute $v_1, v_2, v_3$, and $v_4$ using the $t + n_{max} + 1 - shift$ elements now in $f[i].freq$
13: sort array $f[0 \ldots (t - 1)]$ into increasing order of the $sym$ component, keeping track of the corresponding codeword lengths as elements are exchanged
14: transmit $v_1, v_2, v_3$, and $v_4$ and the first $t$ values $f[0 \ldots (t - 1)].sym$ as a prelude, together with the matching codeword lengths for those $t$ symbols
15: sort array $f[0 \ldots (t - 1)]$ into increasing order of codeword length, with ties broken using the $sym$ component
16: **for** each symbol $s$ in the message block **do**
17:     **if** $\exists x < t : f[x].sym = s$ **then**
18:         code $s$ as the integer $x$, using $v_1, v_2, v_3$, and $v_4$
19:     **else**
20:         code $s$ as the integer $t + s - shift$, using $v_1, v_2, v_3$, and $v_4$

those in the dense code) are allocated sparse codewords. A minimum-redundancy restricted prefix code for the augmented symbol set is calculated as before; because the highest frequency symbols are in the dense set, and allocated the shortest codewords, compression effectiveness can be traded against prelude size by adjusting the control knob represented by $t$. For example, $t$ might be set to a fixed value such as 1,000, or might be varied so as to ensure that the symbols that would originally be assigned one-byte and two-byte codewords are all in the dense set.

Chen et al. [2003] describe a related mechanism in which symbols over a sparse alphabet are coded as binary offsets within a bucket, and a Huffman code is used to specify bucket identifiers, based on the aggregate frequency of the symbols in the bucket. In their method, each bucket code is sparse and self-describing, and the primary code is a dense one over buckets. In contrast, we partially permute the alphabet to create a dense region of "interesting" symbols, and leave the uninteresting ones in a sparse zone of the alphabet.

## 5   Experiments

Table 1 describes the four test files used to validate the new approach to byte coding. They are all derived from the same source, a 267 MB file of SGML-tagged newspaper

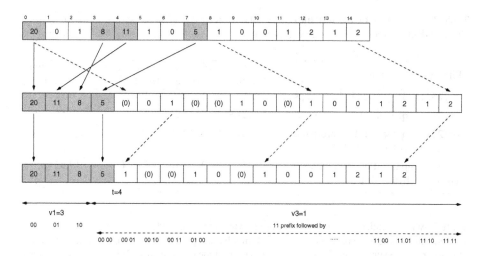

**Fig. 2.** Example of a semi-dense restricted prefix code with $R = 4$, $n_{\max} = 14$, and a threshold of $t = 4$. The largest four frequencies are extracted, and the rest of the frequency array shifted right by four positions, with zeros inserted where elements have been extracted. In the third row, $shift = 2$ leading zeros are suppressed. The end array has $t + n_{\max} + 1 - shift = 13$ elements, and is minimally represented as a $(v_1, v_2, v_3) = (3, 0, 1)$ code, with a cost of 162 bits. The modified prelude contains only four symbols, seven less than is required when the code is dense.

text, but processed in different ways to generate streams of integer symbol identifiers. The first two files are of particular interest, and can be regarded as respectively representing the index of a mid-sized document retrieval system, and the original text of it. In this example the index is stored as a sequence of $d$-gaps (see Witten et al. [1999] for a description of inverted index structures), and the text using a word-based model.

Table 2 shows the compression effectiveness achieved by the experimental methods for the four test files described in Table 1, when processed as a sequence of message blocks (except at the end of the file) of $m = 2^{20}$ symbols. Table 2 does not include any prelude costs, and hence only those for the basic byte coder bc represent actual achievable compression. The file wsj267.repair shows the marked improvement possible with the rpbc approach compared to the scdbc method – on this file there are almost no one-byte codewords required, and a large number of two-byte codewords.

The first three columns of Table 3 show the additional cost of representing a dense prelude, again when using blocks of $m = 2^{20}$ symbols. Storing a complete permutation of the alphabet is never effective, and not an approach that can be recommended. Use of a bit-vector is appropriate when the sub-alphabet density is high, but as expected, the gap-based approach is more economical when the sub-alphabet density is low.

The fourth column of Table 3 shows the cost of the semi-dense prelude approach described in Section 4. It is expressed in two parts – the cost of a partial prelude describing the dense subset of the alphabet, plus a value that indicates the extent to which compression effectiveness of the rpbc method is reduced because the code is no longer dense. In these experiments, in each message block the threshold $t$ was set to the sum $v_1 + v_2 R$ generated by a preliminary fully-dense evaluation of Algorithm 3, so that all

**Table 1.** Parameters of the test files. The column headed "$n/n_{max}$" shows the average sub-alphabet density when the message is broken into blocks each containing $m = 2^{20}$ symbols.

| File name and origin | Total symbols | Maximum value | $n/n_{max}$ $(m = 2^{20})$ | Self-information (bits/sym) |
|---|---|---|---|---|
| wsj267.ind: Inverted index $d$-gaps | 41,389,467 | 173,252 | 10.4% | 6.76 |
| wsj267.seq: Word-parsed sequence | 58,421,983 | 222,577 | 22.5% | 10.58 |
| wsj267.seq.bwt.mtf: Word-parsed sequence BWT'ed and MTF'ed | 58,421,996 | 222,578 | 20.8% | 7.61 |
| wsj267.repair: Phrase numbers from a recursive byte-pair parser | 19,254,349 | 320,016 | 75.3% | 17.63 |

**Table 2.** Average codeword length for different byte coding methods. Each input file is processed as a sequence of message blocks of $m = 2^{20}$ symbols, except at the end. Values listed are in terms of bits per source symbol, excluding any necessary prelude components. Only the column headed bc represents attainable compression, since it is the only one that does not require a prelude.

| File | Method | | | |
|---|---|---|---|---|
| | bc | dbc | scdbc | rpbc |
| wsj267.ind | 9.35 | 9.28 | 9.00 | 8.99 |
| wsj267.seq | 16.29 | 12.13 | 11.88 | 11.76 |
| wsj267.seq.bwt.mtf | 10.37 | 10.32 | 10.17 | 10.09 |
| wsj267.repair | 22.97 | 19.91 | 19.90 | 18.27 |

symbols that would have been assigned one-byte and two-byte codes were protected into the prelude, and symbols with longer codes were left in the sparse section.

Overall compression is the sum of the message cost and the prelude cost. Comparing Tables 2 and 3, it is apparent that on the files wsj267.ind and wsj267.seq.bwt.mtf with naturally decreasing probability distributions, use of a dense code is of no overall benefit, and the bc coder is the most effective. On the other hand, the combination of semi-dense prelude and rpbc codes result in compression gains on all four test files.

Figure 3 shows the extent to which the threshold $t$ affects the compression achieved by the rpbc method on the file wsj267.seq. The steady decline through to about $t = 200$ corresponds to all of the symbols requiring one-byte codes being allocated space in the prelude; and then the slower decline through to 5,000 corresponds to symbols warranting two-byte codewords being promoted into the dense region.

Table 4 shows measured decoding rates for four byte coders. The bc coder is the fastest, and the dbc and scdbc implementations require around twice as long to decode each of the four test files. However the rpbc code recovers some of the lost speed, and even with a dense prelude, outperforms the scdbc and dbc methods. Part of the bc coder's speed advantage arises from not having to decode a prelude in each block. But the greater benefit arises from the absence of the mapping table, and the removal of the per-symbol array access incurred in the symbol translation process. In particular, when the mapping table is large, a cache miss per symbol generates a considerable speed penalty. The benefit of avoiding the cache misses is demonstrated in the final column

**Table 3.** Average prelude cost for four different representations. In all cases the input file is processed as a sequence of message blocks of $m = 2^{20}$ symbols, except for the last. Values listed represent the total cost of all of the block preludes, expressed in terms of bits per source symbol. In the column headed "semi-dense", the use of a partial prelude causes an increase in the cost of the message, the amount of which is shown (for the `rpbc` method) as a secondary component.

| File | Prelude representation | | | |
|------|-------------|------------|------|------------|
|      | permutation | bit-vector | gaps | semi-dense |
| `wsj267.ind`         | 0.31 | 0.20 | 0.14 | 0.08+0.00 |
| `wsj267.seq`         | 0.59 | 0.22 | 0.27 | 0.13+0.01 |
| `wsj267.seq.bwt.mtf` | 0.65 | 0.25 | 0.29 | 0.15+0.01 |
| `wsj267.repair`      | 4.44 | 0.78 | 1.87 | 0.49+0.02 |

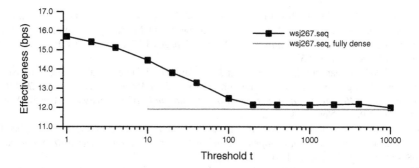

**Fig. 3.** Different semi-dense prelude thresholds $t$ used with `wsj267.seq`, and the `rpbc` method

**Table 4.** Decoding speed on a 2.8 Ghz Intel Xeon with 2 GB of RAM, in millions of symbols per second, for complete compressed messages including a prelude in each message block, and with blocks of length $m = 2^{20}$. The bc method has no prelude requirement.

| File | bc | dbc | scdbc | rpbc | |
|------|--------|-------|-------|-------|------------|
|      | (none) | dense | dense | dense | semi-dense |
| `wsj267.ind`         | 68 | 30 | 30 | 47 | 59 |
| `wsj267.seq`         | 59 | 24 | 24 | 36 | 43 |
| `wsj267.seq.bwt.mtf` | 60 | 26 | 26 | 39 | 50 |
| `wsj267.repair`      | 49 | 9  | 9  | 12 | 30 |

of Table 4 – the `rpbc` method with a semi-dense prelude operates with a relatively small decoder mapping, and symbols in the sparse region of the alphabet are translated without an array access being required. Fast decoding is the result.

## 6   Conclusion

We have described a restricted prefix code that obtains better compression effectiveness than the $(S, C)$-dense mechanism, but offers many of the same features. In addition, we have described a semi-dense approach to prelude representation that offers a useful pragmatic compromise, and also improves compression effectiveness. On the file wsj267.repair, for example, overall compression improves from $19.90 + 0.78 = 20.68$ bits per symbol to $18.27 + (0.49 + 0.02) = 18.78$ bits per symbol, a gain of close to 10%. In combination, the new methods also provide significantly enhanced decoding throughput rates compared to the $(S, C)$-dense mechanism.

*Acknowledgment.* The second author was funded by the Australian Research Council, and by the ARC Center for Perceptive and Intelligent Machines in Complex Environments. National ICT Australia (NICTA) is funded by the Australian Government's Backing Australia's Ability initiative, in part through the Australian Research Council.

# References

N. R. Brisaboa, A. Fariña, G. Navarro, and M. F. Esteller. $(S, C)$-dense coding: An optimized compression code for natural language text databases. In M. A. Nascimento, editor, *Proc. Symp. String Processing and Information Retrieval*, pages 122–136, Manaus, Brazil, October 2003a. LNCS Volume 2857.

N. R. Brisaboa, A. Fariña, G. Navarro, and J. R. Paramá. Efficiently decodable and searchable natural language adaptive compression. In *Proc. 28th Annual International ACM SIGIR Conference on Research and Development in Information Retrieval*, Salvador, Brazil, August 2005. ACM Press, New York. To appear.

N. R. Brisaboa, E. L. Iglesias, G. Navarro, and J. R. Paramá. An efficient compression code for text databases. In *Proc. 25th European Conference on Information Retrieval Research*, pages 468–481, Pisa, Italy, 2003b. LNCS Volume 2633.

D. Chen, Y.-J. Chiang, N. Memon, and X. Wu. Optimal alphabet partitioning for semi-adaptive coding of sources of unknown sparse distributions. In J. A. Storer and M. Cohn, editors, *Proc. 2003 IEEE Data Compression Conference*, pages 372–381. IEEE Computer Society Press, Los Alamitos, California, March 2003.

E. S. de Moura, G. Navarro, N. Ziviani, and R. Baeza-Yates. Fast and flexible word searching on compressed text. *ACM Transactions on Information Systems*, 18(2):113–139, 2000.

S. W. Golomb. Run-length encodings. *IEEE Transactions on Information Theory*, IT–12(3): 399–401, July 1966.

M. Liddell and A. Moffat. Decoding prefix codes. December 2004. Submitted. Preliminary version published in *Proc. IEEE Data Compression Conference*, 2003, pages 392–401.

J. Rautio, J. Tanninen, and J. Tarhio. String matching with stopper encoding and code splitting. In A. Apostolico and M. Takeda, editors, *Proc. 13th Ann. Symp. Combinatorial Pattern Matching*, pages 42–51, Fukuoka, Japan, July 2002. Springer. LNCS Volume 2373.

F. Scholer, H. E. Williams, J. Yiannis, and J. Zobel. Compression of inverted indexes for fast query evaluation. In M. Beaulieu, R. Baeza-Yates, S. H. Myaeng, and K. Järvelin, editors, *Proc. 25th Annual International ACM SIGIR Conference on Research and Development in Information Retrieval*, pages 222–229, Tampere, Finland, August 2002. ACM Press, New York.

I. H. Witten, A. Moffat, and T. C. Bell. *Managing Gigabytes: Compressing and Indexing Documents and Images*. Morgan Kaufmann, San Francisco, second edition, 1999.

# Experimental Analysis of a Fast
# Intersection Algorithm for Sorted Sequences

Ricardo Baeza-Yates and Alejandro Salinger

Center for Web Research, Department of Computer Science,
University of Chile, Blanco Encalada 2120,
Santiago, Chile

**Abstract.** This work presents an experimental comparison of intersection algorithms for sorted sequences, including the recent algorithm of Baeza-Yates. This algorithm performs on average less comparisons than the total number of elements of both inputs ($n$ and $m$ respectively) when $n = \alpha m$ ($\alpha > 1$). We can find applications of this algorithm on query processing in Web search engines, where large intersections, or differences, must be performed fast. In this work we concentrate in studying the behavior of the algorithm in practice, using for the experiments test data that is close to the actual conditions of its applications. We compare the efficiency of the algorithm with other intersection algorithm and we study different optimizations, showing that the algorithm is more efficient than the alternatives in most cases, especially when one of the sequences is much larger than the other.

## 1   Introduction

In this work we study different algorithms to compute the intersection of sorted sequences. This problem is a particular case of a generic problem called multiple searching [3] (see also [15], research problem 5, page 156), which consists of, given an $n$-element data multiset, $D$, drawn from an ordered universe, search $D$ for each element of an $m$-element query multiset, $Q$, drawn from the same universe. The found elements form, exactly, the intersection of both multisets.

The sorted sequences intersection problem finds its motivation in Web search engines, since most of them use inverted indices, where for each different word, we have a list of positions or documents where it appears. Generally, these lists are ordered by some criterion, like position, a global precomputed ranking, frequency of occurrence in a document, etc. To compute the result of a query, in most cases we need to intersect these lists. In practice these lists can have hundreds of millions of elements, hence, it is useful to have an algorithm that is fast and efficient on average.

In the case when $D$ and $Q$ are sets (and not multisets) already ordered, multiple search can be solved by merging both sets. However, this is not optimal for all possible cases. In fact, if $m$ is small (say if $m = o(n/\log n)$), it is better to do $m$ binary searches obtaining an $O(m \log n)$ algorithm [2], where the complexity metric is the number of comparisons between any pair of elements. Baeza-Yates'

M. Consens and G. Navarro (Eds.): SPIRE 2005, LNCS 3772, pp. 13–24, 2005.

algorithm matches both complexities depending on the value of $m$. On average, it performs less than $m + n$ comparisons when both sets are ordered under some pessimistic assumptions.

This work focuses on the experimental study of this algorithm, as well as different optimizations to it. The experiments consisted on measuring the running time of the original algorithms and its optimizations with sequences of ordered random integer numbers and comparing it to an algorithm based on merging and to the *Adaptive* algorithm, proposed by Demaine et al. [10], which seems to be the most used in practice. Our results show that Baeza-Yates' algorithm is slightly better than Adaptive and much better than Merge when the length of the sequences differ considerably.

In Section 2 we present related work. Section 3 presents the motivation for our problem and some practical issues. Section 4 presents the algorithms used for the comparison, including the algorithm of Baeza-Yates and a proposed optimization. Section 5 presents the experimental results obtained. Throughout this paper $n \geq m$ and logarithms are base two unless explicitly stated otherwise.

## 2   Related Work

In order to solve the problem of determining whether any elements of a set of $n + m$ elements are equal, we require at least $\Theta((n + m) \log(n + m))$ comparisons in the worst case (see [13]). However, this lower bound does not apply to the multiple search problem nor, equivalently, to the set intersection problem. Conversely, the lower bounds of the search problem do apply to the element uniqueness problem [12]. This idea was exploited by Demaine *et al.* to define an adaptive multiple set intersection algorithm [10,11] that finds the common elements by searching in an unbounded domain. They also define the difficulty of a problem instance, which was refined later by Barbay and Kenyon [8].

For the ordered case, lower bounds on set intersection are also lower bounds for merging both sets. However, the converse is not true, as in set intersection we do not need to find the actual position of each element in the union of both sets, just if one element is in the other set or not. Although there has been a lot of work on minimum comparison merging in the worst case, almost no research has been done on the average case because it does not make much of a difference. However, this is not true for multiple search, and hence for set intersection [3].

The algorithm of Baeza-Yates [1] adapts to the input values. In the best case, the algorithm performs $\lceil \log(m + 1) \rceil \lceil \log(n + 1) \rceil$ comparisons, which for $m = O(n)$, is $O(\log^2(n))$. In the worst case, the number of comparisons performed by the algorithm is

$$W(m, n) = 2(m + 1) \log((n + 1)/(m + 1)) + 2m + O(\log(n))$$

Therefore, for small $m$, the algorithm is $O(m \log(n))$, while for $n = \alpha m$ it is $O(n)$. In this case, the ratio between this algorithm and merging is $2(1 + \log(\alpha))/(1 + \alpha)$ asymptotically, being 1 when $\alpha = 1$. The worst case is worse than merging for $1 < \alpha < 6.3197$ having its maximum at $\alpha = 2.1596$, where it is

1.336 times slower than merging. Hence the worst case of the algorithm matches the complexity of both, the merging and the multiple binary search, approaches, adapting nicely to the size of $m$. For the average case, under pessimistic assumptions, the number of comparisons is:

$$A(m,n) = (m+1)(\ln((n+1)/(m+1)) + 3 - 1/\ln(2)) + O(\log n)$$

For $n = \alpha m$, the ratio between this algorithm and merging is $(\ln(\alpha) + 3 - 1/\ln(2))/(1 + \alpha)$ which is at most 0.7913 when $\alpha = 1.2637$ and 0.7787 when $\alpha = 1$. The details of the algorithm are presented on section 4.2.

# 3 Motivation: Query Processing in Inverted Indices

Inverted indices are used in most text retrieval systems [4]. Logically, they are a vocabulary (set of unique words found in the text) and a list of references per word to its occurrences (typically a document identifier and a list of word positions in each document). In simple systems (Boolean model), the lists are sorted by document identifier, and there is no ranking (that is, there is no notion of relevance of a document). In that setting, an intersection algorithm applies directly to compute Boolean operations on document identifiers: union (OR) is equivalent to merging, intersection (AND) is the operation on study (we only keep the repeated elements), and subtraction implies deleting the repeated elements, which is again similar to an intersection. In practice, long lists are not stored sequentially, but in blocks. Nevertheless, these blocks are large, and the set operations can be performed in a block-by-block basis.

In complex systems ranking is used. Ranking is typically based in word statistics (number of word occurrences per document and the inverse of the number of documents having it). Both values can be precomputed and the reference lists are then stored by decreasing intra-document word frequency order to have first the most relevant documents. Lists are then processed by decreasing inverse extra-document word frequency order (that is, we process the shorter lists first), to obtain first the most relevant documents. However, in this case we cannot always have a document identifier mapping such that lists are sorted by that order. Nevertheless, they are partially ordered by identifier for all documents of equal word frequency.

The previous scheme was used initially on the Web, but as the Web grew, the ranking deteriorated because word statistics do not always represent the content and quality of a Web page and also can be "spammed" by repeating and adding (almost) invisible words. In 1998, Page and Brin [9] described a search engine (which was the starting point of Google) that used links to rate the quality of a page, a scheme called PageRank. This is called a global ranking based in popularity, and is independent of the query posed. It is out of the scope of this paper to explain PageRank, but it models a random Web surfer and the ranking of a page is the probability of the Web surfer visiting it. This probability induces a total order that can be used as document identifier. Hence, in a pure link based search engine we can use the intersection algorithm as before. However, nowadays hybrid ranking schemes that combine link and word evidence are used. In spite of this, a link based mapping still gives good results as it approximates well the true ranking (which can be corrected while is computed).

Another important type of query is sentence search. In this case we use the word position to know if a word follows or precedes a word. Hence, as usually sentences are small, after we find the Web pages that have all of them, we can process the first two words[1] to find adjacent pairs and then those with the third word and so on. This is like to compute a particular intersection where instead of finding repeated elements we try to find correlative elements ($i$ and $i + 1$), and therefore we can use again the intersection algorithm as word positions are sorted. The same is true for proximity search. In this case, we can have a range $k$ of possible valid positions (that is $i \pm k$) or to use a different ranking weight depending on the proximity.

Finally, in the context of the Web, an adaptive algorithm is in practice much faster because the uniform distribution assumption is pessimistic. In the Web, the distribution of word occurrences is quite biased. The same is true with query frequencies. Both distributions follow a power law (a generalized Zipf distribution) [4,6]. However, the correlation of both distributions is not high [7] and even low [5]. That implies that the average length of the lists involved in the query are not that biased. That means that the average lengths of the lists, $n$ and $m$, when sampled, will satisfy $n = \Theta(m)$ (uniform), rather than $n = m + O(1)$ (power law). Nevertheless, in both cases our algorithm makes an improvement.

## 4   The Algorithms

Suppose that $D$ is sorted. In this case, obviously, if $Q$ is small, will be faster to search every element of $Q$ in $D$ by using binary search. Now, when $Q$ is also sorted, set intersection can be solved by merging. In the worst or average case, straight merging requires $m + n - 1$ comparisons. However, we can do better for set intersection. Next, we describe here the different algorithms compared. We do not include the merging algorithm as it is well known.

### 4.1   Adaptive

This algorithm [10,11] works as follows: we take one of the sets, and we choose its first element, which we call *elim*. We search *elim* in the other set, making exponential jumps, this is, looking at positions $1, 2, 4, \ldots, 2^i$. If we overshoot, that is, the element in the position $2^i$ is larger than *elim*, we binary search *elim* between positions $2^{i-1}$ and $2^i$[2] If we find it, we add it to the result. Then, we remember the position where *elim* was (or the position where it should have been) so we know that from that position backwards we already processed the

---

[1] Actually, it is more efficient to use the two words with shorter lists, and so on until we get to the largest list if the intersection is still non empty.

[2] This is the classical result of Bentley and Yao for searching an element in an unbounded set which is $O(\log n)$.

set. Now we chose *elim* as the smallest element of the set that is greater than the former *elim* and we exchange roles, making jumps from the position that signals the processed part of the set. We finish when there is no element greater than the one we are searching.

## 4.2 Baeza-Yates

Baeza-Yates' algorithm is based on a double binary search, improving on average under some pessimistic assumptions. The algorithm introduced in [1] can be seen as a balanced version of Hwang and Lin's [14] algorithm adapted to our problem.

The algorithm works as follows. We first binary search the median (middle element) of $Q$ in $D$. If found, we add that element to the result. Found or not, we have divided the problem in searching the elements smaller than the median of $Q$ to the left of the position found on $D$, or the position the element should be if not found, and the elements bigger than the median to the right of that position. We then solve recursively both parts (left sides and right sides) using the same algorithm. If in any case, the size of the subset of $Q$ to be considered is larger than the subset of $D$, we exchange the roles of $Q$ and $D$. Note that set intersection is symmetric in this sense. If any of the subsets is empty, we do nothing.

A simple way to improve this algorithm is to apply the original algorithm not over the complete sets $D$ and $Q$, but over a subset of both sets where they actually overlap, and hence, where we can really find elements that are part of the intersection.

We start by comparing the smallest element of $Q$ with the largest of $D$, and the largest of $Q$ with the smallest of $D$. If both sets do not overlap, the intersection is empty. Otherwise, we search the smallest and largest element of $D$ in $Q$, to find the overlap, using just $O(\log m)$ time. Then we apply the previous algorithm just to the subsets that actually overlaps. This improves both, the worst and the average case. The dual case is also valid, but then finding the overlap is $O(\log n)$, which is not good for small $m$. This optimization is mentioned in [1], but it is effectiveness is not studied.

## 5  Experimental Results

We compared the efficiency of the algorithm, which we call *Intersect* in this section, with an intersection algorithm based on merging, and with an adaptation of the *Adaptive* algorithm [10,11] for the intersection of two sequences. In addition, we show the results obtained with the optimizations of the algorithm.

We used sequences of integer random numbers, uniformly distributed in the range $[1, 10^9]$. We varied the length of one of the lists $(n)$ from 1,000 to 22,000 with a step of 3,000. For each of these lengths we intersected those sequences with sequences of four different lengths $(m)$, from 100 to 400. We use twenty random instances per case and ten thousand runs (to eliminate the variations due to the operating system given the small resulting times).

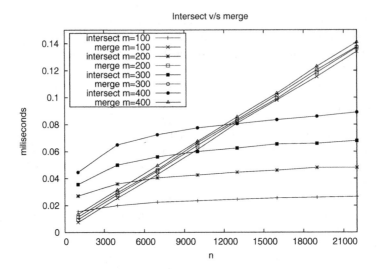

**Fig. 1.** Experimental results for Intersect and Merge for different values of $n$ and $m$

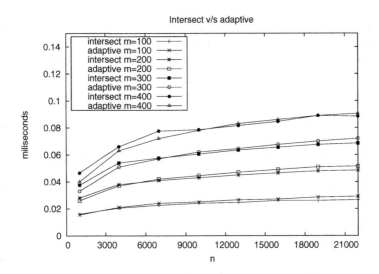

**Fig. 2.** Experimental results for Intersect and Adaptive, for different values of $n$ and $m$

The programs were implemented in $C$ using the Gcc 3.3.3 compiler in a Linux platform running an Intel(R) Xeon(TM) CPU 3.06GHz with 512 Kb cache and 2Gb RAM.

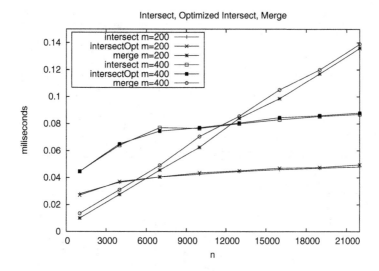

**Fig. 3.** Experimental results for Intersect, optimized Intersect and Merge, for different values of $n$ and $m = 200$ y $m = 400$

Figure 1 shows a comparison between Intersect and Merge. We can see that Intersect is better than Merge when $n$ increases and that the time increases for larger values of $m$.

Figure 2 shows a comparison between the times of Intersect and Adaptive. We can see that the times of both algorithms follow the same tendency and that Intersect is better than Adaptive.

Figure 3 shows the results obtained with the Intersect algorithm and the optimization described at the end of the last section. For this comparison, we also added the computation of the overlap of both sequences to Merge.

We can see that there is no big difference between the original and the optimized algorithm, and moreover, the original algorithm was a bit faster than the optimized one. The reason why the optimization did not result in an improvement can be the uniform distribution of the test data. As the random numbers are uniformly distributed, in most cases the overlap of both sets covers a big part of $Q$. Then, the optimization does not produce any improvement and it only results in a time overhead due to the overlap search.

### 5.1 Hybrid Algorithms

We can see from the experimental results obtained that there is a section of values of $n$ where Merge is better than Intersect. Hence, a natural idea is to combine both algorithms in one hybrid algorithm that runs each of them when convenient.

In order to know where is the cutting point to use one algorithm instead of the other, we measured for each value of $n$ the time of both algorithms with

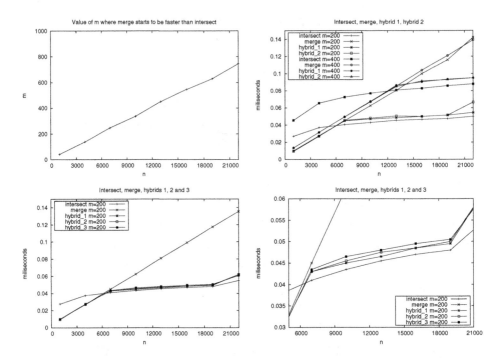

**Fig. 4.** Up: on the left, value of $m$ from which Merge is faster than Intersect. On the right, a comparison between the original algorithm, Merge and the hybrids 1 and 2 for $m = 200$ y $m = 400$. Down: comparison between Intersect, Merge and the three hybrids for $m = 200$. The plot on the right is a zoom of the one on the left.

different values of $m$ until we identified the value of $m$ where Merge was faster than Intersect. These values of $m$ form a straight line as a function of $n$, which we can observe in Fig. 4. This straight line is approximated by $m = 0.033n + 8.884$, with a correlation of $r^2 = 0.999$.

The hybrid algorithm works by running Merge whenever $m > 0.033n + 8.884$, and running Intersect otherwise. The condition is evaluated on each step of the recursion.

When we modify the algorithm, the cutting point changes. We would like to find the optimal hybrid algorithm. Using the same idea again, we found the straight line that defines the values where Merge is better than the hybrid algorithm. This straight line can be approximated by $m = 0.028n + 32.5$, with $r^2 = 0.992$. Hence, we define the algorithm Hybrid2, which runs Merge whenever $m > 0.028n + 32.5$ and runs Intersect otherwise. Finally, we combined both hybrids, creating a third version where the cutting line between Merge and Intersect is the average between the lines of the hybrids 1 and 2. The resulting straight line is $m = 0.031n + 20.696$. Figure 4 shows the cutting line between the original algorithm and Merge, and the results obtained with the hybrid algorithms. The optimal algorithm would be on theory the Hybrid.$i$ when $i$ tends to infinity, as we are looking for a fixed point algorithm.

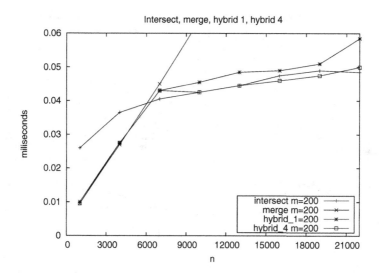

**Fig. 5.** Experimental results for Intersect, Merge and the hybrids 1 and 4 for different values of $n$ and for $m = 200$

We can observe that the hybrid algorithms registered lower times than the original algorithm in the section where the latter is slower than Merge. However, in the other section the original algorithm is faster than the hybrids, due to the fact that in practice we have to evaluate the cutting point in each step of the recursion. Among the hybrid algorithms, we can see that the first one is slightly faster than the second one, and that this one is faster than the third one. An idea to reduce the time in the section that the original algorithm is faster than the hybrids is to create a new hybrid algorithm that runs Merge when it is convenient and that then runs the original algorithm, without evaluating the relation between $m$ and $n$ in order to run Merge. This algorithm shows the same times than Intersect in the section where the latter is better than Merge, combining the advantages of both algorithms in the best way. Figure 5 show the results obtained with this new hybrid algorithm.

## 5.2   Sequence Lengths with Zipf Distribution

As we said before, one of the applications of the algorithm is the search of Web documents, where the number of documents in which a word appears follows a Zipf distribution.

It is interesting to study the behavior of the Intersect algorithm depending of the ratio between the lengths of the two sequences when these lengths follow a Zipf distribution and the correlation between both sets is zero (ideal case). For this experiment, we took two random numbers, $a$ and $b$, uniformly distributed between 0 and 1,000. With these numbers we computed the lengths of the sequences $D$ and $Q$ as $n = K/a^\alpha$ y $m = K/b^\alpha$, respectively, with $K = 10^9$ and $\alpha = 1.8$ (a typical value for word occurrence distribution in English), making

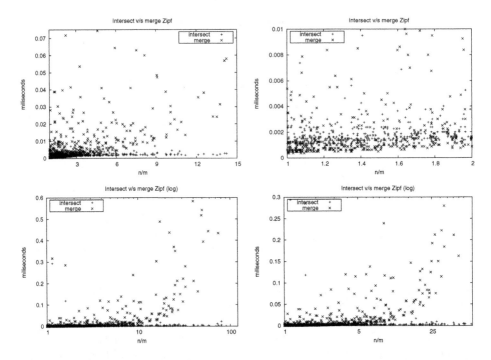

**Fig. 6.** Up: times for Intersect and Merge as a function of the ratio between the lengths of the sequences when they follow a Zipf distribution. The plot on the right is a zoom of the one on the left. Down: times for Intersect and Merge in logarithmic scale. The plot on the right is a zoom of the one on the left.

sure that $n > m$. We did 1,000 measurements, using 80 different sequences for each of them, and repeating 1,000 times each run.

Figure 6 shows the times obtained with both algorithms as a function of $n/m$, in normal scale and logarithmic scale.

We can see that the times of Intersect are lower than the times of Merge when $n$ is much greater than $m$. When we decrease the ratio between $n$ and $m$, it is not so clear anymore which of the algorithms is faster. When $n/m < 2$, in most cases the times of Merge are better.

## 6   Conclusions

In this work we have experimentally studied a simple sorted set intersection algorithm that performs quite well in average and does not inspect all the elements involved. Our experiments showed that Baeza-Yates' algorithm is faster than Merge when one of the sequences is much larger than the other one. This improvement is more evident when $n$ increases. In addition, Baeza-Yates' algorithm surpasses Adaptive [10,11] for every relation between the sizes of the sequences. The hybrid algorithm that combines Merge and Baeza-Yates' algorithm according to the empiric information obtained, takes advantage of both algorithms and became the most efficient one.

In practice, we do not need to compute the complete result of the intersection of two lists, as most people only look at less than two result pages [6]. Moreover, computing the complete result is too costly if one or more words occur several millions of times as happens in the Web and that is why most search engines use an intersection query as default. Hence, lazy evaluation strategies are used and the results is completed at the user's request.

If we use the straight classical merging algorithm, this naturally obtains first the most relevant Web pages. The same is true for the Adaptive algorithm. For Baeza-Yates' algorithm, it is not so simple, because although we have to process first the left side of the recursive problem, the Web pages obtained do not necessarily appear in the correct order. A simple solution is to process the smaller set from left to right doing binary search in the larger set. However this variant is efficient only for small $m$, achieving a complexity of $O(m \log n)$ comparisons. An optimistic variant can use a prediction on the number of pages in the result and use an intermediate adaptive scheme that divides the smaller sets in non-symmetric parts with a bias to the left side. Hence, it is interesting to study the best way to compute partial results efficiently.

As the correlation between both sets in practice is between 0.2 and 0.6, depending on the Web text used (Zipf distribution with $\alpha$ between 1.6 y 2.0) and the queries (Zipf distribution with a lower value of $\alpha$, for example 1.4), we would like to extend our experimental results to this case. However, we already saw that in both extremes (correlation 0 or 1), the algorithm on study is competitive.

## Acknowledgements

We thank the support of Millennium Nucleus Grant P04-067-F.

## References

1. R. Baeza-Yates. A Fast Set Intersection Algorithm for Sorted Sequences. In *Proceedings of the 15th Annual Symposium on Combinatorial Pattern Matching (CPM 2004)*, Springer LNCS 3109, pp 400-408, Istanbul, Turkey, July 2004.
2. R.A. Baeza-Yates. *Efficient Text Serching*. PhD thesis, Dept. of Computer Science, University of Waterloo, May 1989. Also as Research Report CS-89-17.
3. Ricardo Baeza-Yates, Phillip G. Bradford, Joseph C. Culberson, and Gregory J.E. Rawlins. The Complexity of Multiple Searching, unpublished manuscript, 1993.
4. R. Baeza-Yates and B. Ribeiro-Neto, *Modern Information Retrieval*, ACM Press/Addison-Wesley, England, 513 pages, 1999.
5. R. Baeza-Yates, and Felipe Sainte-Jean. A Three Level Search Engine Index bases in Query Log Distribution. SPIRE 2003, Springer LNCS, Manaus, Brazil, October 2003.
6. Ricardo Baeza-Yates. Query Usage Mining in Search Engines. In Web Mining: Applications and Techniques, Anthony Scime, editor. Idea Group, 2004.

7. Ricardo Baeza-Yates, Carlos Hurtado, Marcelo Mendoza and Georges Dupret. Modeling User Search Behavior, LA-WEB 2005, IEEE CS Press, October 2005.

8. Jérémy Barbay and Claire Kenyon. Adaptive Intersection and $t$-Threshold Problems. In *Proceedings of the 13th Annual ACM-SIAM Symposium on Discrete Algorithms*, pp. 390-399, San Francisco, CA, January 2002.

9. S. Brin and L. Page. The anatomy of a large-scale hypertextual Web search engine. In *7th WWW Conference*, Brisbane, Australia, April 1998.

10. Erik D. Demaine, Alejandro López-Ortiz, and J. Ian Munro. Adaptive set intersections, unions, and differences. In *Proceedings of the 11th Annual ACM-SIAM Symposium, on Discrete Algorithms*, pages 743-752, San Francisco, California, January 2000.

11. Erik D. Demaine, Alejandro López-Ortiz, and J. Ian Munro. Experiments on Adaptive Set Intersections for Text Retrieval Systems. In *Proceedings of the 3rd Workshop on Algorithm Engineering and Experiments*, LNCS, Springer, Washington, DC, January 2001.

12. Dietz, Paul, Mehlhorn, Kurt, Raman, Rajeev, and Uhrig, Christian; "Lower Bounds for Set Intersection Queries", *Proceedings of the 4th Annual Symposium on Discrete Algorithms*, 194-201, 1993.

13. Dobkin, David and Lipton, Richard; "On the Complexity of Computations Under Varying Sets of Primitives", *Journal of Computer and Systems Sciences*, 18, 86-91, 1979.

14. F.K. Hwang and S. Lin. A Simple algorithm for merging two disjoint linearly ordered lists, *SIAM J. on Computing* 1, pp. 31-39, 1972.

15. Rawlins, Gregory J. E.; *Compared to What?: An Introduction to the Analysis of Algorithms*, Computer Science Press/W. H. Freeman, 1992.

# Compressed Perfect Embedded Skip Lists for Quick Inverted-Index Lookups

Paolo Boldi and Sebastiano Vigna

Dipartimento di Scienze dell'Informazione, Università degli Studi di Milano

**Abstract.** Large inverted indices are by now common in the construction of web-scale search engines. For faster access, inverted indices are indexed internally so that it is possible to skip quickly over unnecessary documents. To this purpose, we describe how to embed efficiently a *compressed perfect skip list* in an inverted list.

## 1 Introduction

The birth of web search engines has brought new challenges to traditional inverted index techniques. In particular, *eager* (or *term-at-a-time*) query evaluation has been replaced by *lazy* (or *document-at-a-time*) query evaluation. In the first case, the inverted list of one of the terms of the query is computed first (usually, choosing the rarest term [4]), and then, it is merged or filtered with the other lists. When evaluation is lazy, instead, inverted lists are scanned in parallel, retrieving in sequence each document satisfying the query.

Lazy evaluation requires keeping constantly in sync several inverted lists. To perform this operation efficiently, it is essential that a *skip* method is available that allows the caller to quickly reach the first document pointer larger than or equal to a given one. The classical solution to this problem [1] is that of embedding skipping information in the inverted list itself: at regular intervals, some additional information describe a *skip*, that is, a pair given by a document pointer and the number of bits that must be skipped to reach that pointer. The analysis of skipping given in [1] concludes that skips should be spaced as the square root of the term frequency, and that one level of skip is sufficient for most purposes.

Nonetheless, the abovementioned analysis has two important limitations. First of all, it does not contemplate the presence of *positions*, that is, of a description of the exact position of each occurrence of a term in a document, or of application-dependent additional data, thus underestimating the cost of *not* skipping a document; second, it is fundamentally based on eager evaluation, and its conclusions cannot be extended to lazy evaluation. Motivated by these reasons, we are going to present a generic method to self-index inverted lists with a very fine level of granularity. Our method does not make any assumption on the structure of a document record, or on the usage pattern of the inverted index. Skips to a given pointer (or by a given amount of pointers) can always be performed with a logarithmic number of low-level reads and bit-level skips: nonetheless, the size of the index grows just by a few percents.

M. Consens and G. Navarro (Eds.): SPIRE 2005, LNCS 3772, pp. 25–28, 2005.

Our techniques are particularly useful for *in-memory indices*, that is, for indices kept in core memory (as it happens, for instance, in Google), where most of the computational cost of retrieving document is scanning and decoding inverted lists (as opposed to disk access), and at the same time a good compression ratio is essential. All results described in this paper have been implemented in MG4J, available at `http://mg4j.dsi.unimi.it/`.

## 2   Perfect Embedded Skip Lists for Inverted Indices

**Perfect skip lists.** Skip lists [3] are a data structure in which elements are organised as in an ordered list, but with additional references that allow one to skip forward in the list as needed. More precisely, a *skip list* is a singly linked list of increasingly ordered items $x_0$, $x_1$, ... , $x_{n-1}$ such that every item $x_i$, besides a reference to the next item, contains a certain number $h_i \geq 0$ of extra references, that are called the *skip tower* of the item; the $t$-th reference in this tower addresses the first item $j > i$ such that $h_j \geq t$.

We are now going to describe *perfect skip lists*, a deterministic version of skip lists (which were originally formulated in randomised terms) that is suitable for inverted lists. We fix two limiting parameters: the number of items in the list, and the maximum height of a tower.

For sake of simplicity, we start by describing an ideal, infinite version of a perfect skip list. Let $\mathrm{LSB}(x)$ be defined as the least significant bit of $x$ if $x$ is a positive integer, and $\infty$ if $x = 0$. Then, define the height of the skip tower of item $x_i$ in an infinite perfect skip list as $h_i = \mathrm{LSB}(i)$. We say that a finite skip list is *perfect* w.r.t a given height $h$ and size $T$ when no tower contains more than $h + 1$ references, and all references that would exist in an infinite perfect skip list are present, provided that they refer to an item with index smaller than or equal to $T$, and that they do not violate the first requirement.

**Theorem 1.** *In a perfect skip list with $T$ items and maximum height $h$, the height of a tower at element $i$ is $\min(h, \mathrm{LSB}(k), \mathrm{MSB}(T - i)) + 1$, where $k = i \bmod 2^h$, and $\mathrm{MSB}(x)$ is the most significant bit of $x > 0$, or $-1$ if $x = 0$. In particular, if $i < T - T \bmod 2^h$ the tower has height $\min(h, \mathrm{LSB}(k)) + 1$, whereas if $i \geq T - T \bmod 2^h$ the tower has height $\min(\mathrm{LSB}(k), \mathrm{MSB}(T \bmod 2^h - k)) + 1$.*

Addressing directly all pointers in an inverted list would create unmanageable indices. Thus, we shall index only one item out of $q$, where $q$ is a fixed *quantum* that represents the minimally addressable block of items. Figure 1 shows a perfect skip list.

### Embedding Skip Lists into Inverted Indices

The first problem that we have to deal with when trying to embed skip lists into an inverted index is that we want to access data in a strictly sequential manner, so the search algorithm we described cannot be adopted directly: we must store not only the bit offset of the referenced item, but also the pointer contained therein.

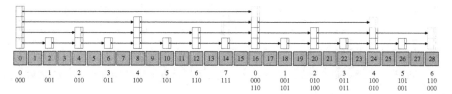

**Fig. 1.** A perfect skip list with $T = 31$ items, $q = 2$ and $h = 3$. The first line show the values of $k$; the second line their $h$-bit binary expansion. The list of made of two *blocks*, and for the latter (a *defective* block of $L = 15$ items), also the binary expansion of $\lfloor L/q \rfloor - k$ is shown. The ghosted references do not exist (they are truncated by minimisation with MSB($\lfloor L/q \rfloor - k$)).

**Pointer Skips.** Let us consider a document collection of $N$ documents, where each term $t$ appears with relative frequency $p_t$; according to the *Bernoulli model* [4], every term $t$ is considered to appear independently in each document with probability $p_t$. As a result, the random variable describing a pointer skip to $\ell$ documents farther is approximated by a normal distribution with mean $\ell/p_t$ and standard deviation $\sqrt{\ell(1 - p_t)}/p_t$ (details appear in the full paper). We also suggest to predict the pointer skips that are not a tower top using a *halving model*, in which a pointer skip of level $s$ is stored as the difference from the skip of level $s + 1$ that contains it, divided by 2 (standard deviation drops to $\sqrt{\ell(1 - p_t)/2}/p_t$).

**Gaussian Golomb Codes.** We are now left with the problem of coding integers normally distributed around 0. We compute approximately the best Golomb code for a given normal distribution. This is not, of course, an optimal code for the distribution, but for $\sigma \gg 1$ it is excellent, and the Golomb modulus can be approximated easily in closed form: integers distributed with standard deviation $\sigma$ are Golomb-coded optimally with modulus $2 \ln 2 \sqrt{\frac{2}{\pi}} \sigma \approx 1.106\,\sigma$.

**Bit Skips.** The strategy we follow for bit skips is absolutely analogous to that of pointers, but with an important difference: it is very difficult to model correctly the distribution of bit skips. We suggest a prediction scheme based on the average bit length of a quantum, coupled with a universal code such that $\gamma$ or $\delta$.

## 3   Inherited Towers

As remarked in the previous sections, the part of a tower that has greater variance (and thus is more difficult to compress) is the tower top. However this might seem strange, our next goal is to avoid writing tower tops at all. When scanning sequentially an inverted list, it is possible to maintain an *inherited tower* that represent all "skip knowledge" gathered so far (see Figure 2), similarly to *search fingers* [2].

Note that inherited entries might not reach height $h$ if the block is defective (see the right half of Figure 1). Supposing without loss of generality that $q = 1$, we have:

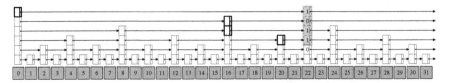

**Fig. 2.** Scanning the list ($q = 1$, $h = 5$), we are currently positioned on element $x = 22$, with a tower of height 2; its inherited tower is represented in grey, and the items it is inheriting from are in bold

**Theorem 2.** *The highest valid entry in an inherited tower for a defective block of length $L$ is* $\mathrm{MSB}(L \oplus k)$, *where* $\oplus$ *denotes bit-by-bit exclusive or.*

The above computation leads us the following, fundamental observation: *a non-truncated tower with highest entry* $\bar{h}$ *inherits an entry of level* $\bar{h} + 1$ *that is identical to its top entry.* As a consequence, if lists are traversed from their beginning, top entries of non-truncated towers can be omitted. The omission of top entries halves the number of entries written, and, as we observed at the start of the paragraph, reduces even further the skip structure size.

### Experimental Data and Conclusions

We gathered statistics indexing a partial snapshot of 13 Mpages taken from the .uk domain containing about 50GiB of parsed text (the index contained counts and occurrences). The document distribution in the snapshot was highly skewed, as documents appeared in crawl order. Adding an embedded perfect skip list structure with arbitrary tall towers caused an increase in size of 2.3% (317 MiB) with $q = 32$ and 1.23% when $q = 64$; indexing using the square-root spaced approach caused an increase of 0.85%. Compressing the same skip structures using a $\gamma$ or $\delta$ code instead of Gaussian Golomb codes for pointer skips caused an increase in pointer-skip size of 42% and 18.2%, respectively.

Speed is, of course, at the core of our interests. The bookkeeping overhead of skip lists increases by no more than 5% (and by .5% on the average) the time required to perform a linear scan. On the contrary, tests performed on synthetically generated queries in disjunctive normal form show an increase in speed between 20 and 300% w.r.t. the square-root spaced approach.

## References

1. Alistair Moffat and Justin Zobel. Self-indexing inverted files for fast text retrieval. *ACM Trans. Inf. Syst.*, 14(4):349–379, 1996.
2. William Pugh. A skip list cookbook. Technical report UMIACS-TR-89-72.1, Univ. of Maryland Institute for Advanced Computer Studies, College Park, College Park, MD, USA, 1990.
3. William Pugh. Skip lists: a probabilistic alternative to balanced trees. *Commun. ACM*, 33(6):668–676, 1990.
4. Ian H. Witten, Alistair Moffat, and Timothy C. Bell. *Managing Gigabytes: Compressing and Indexing Documents and Images.* Morgan Kaufmann Publishers, Los Altos, CA 94022, USA, second edition, 1999.

# XML Retrieval with a Natural Language Interface

Xavier Tannier[1] and Shlomo Geva[2]

[1] École Nationale Supérieure des Mines de Saint-Etienne, 158 Cours Fauriel,
F-42023 Saint-Etienne, France
`tannier@emse.fr`
[2] Centre for Information Technology Innovation, Faculty of Information Technology,
Queensland University of Technology,
GPO Box 2434, Brisbane Q 4001, Australia
`s.geva@qut.edu.au`

**Abstract.** Effective information retrieval in XML documents requires the user to have good knowledge of document structure and of some formal query language. XML query languages like XPath and XQuery are too complex to be considered for use by end users. We present an approach to XML query processing that supports the specification of both textual and structural constraints in natural language. We implemented a system that supports the evaluation of both formal XPath-like queries and natural language XML queries. We present comparative test results that were performed with the INEX 2004 topics and XML collection. Our results quantify the trade-off in performance of natural language XML queries vs formal queries with favourable results.

## 1  Introduction and Motivation

Applications of Natural Language Processing to Information Retrieval have been extensively studied in the case of textual (flat) collections (see overviews [1, 2, 3, 4]). Among other techniques, linguistic analysis of queries was meant to bring about decisive improvements in retrieval processes and in ergonomy. However, only few linguistic methods, such as phrasal term extraction or some kinds of query expansion, are now commonly used in information retrieval systems.

The rapidly growing spread of XML document collections brings new motivating factors to the use of natural language techniques:

- Benefits that can be gained from the use of natural language queries are probably much higher in XML retrieval than in traditional IR. In the later, a query is generally a list of keywords which is quite easy to write. In XML retrieval, such a list is not sufficient to specify queries on both content and structure; for this reason, advanced structured query languages have been devised.

  XML is now widely used, particularly on the Internet, and that implies that novice and casual users ought to be able to query any XML corpus. From that perspective, two major difficulties arise, because we cannot expect such users to:

M. Consens and G. Navarro (Eds.): SPIRE 2005, LNCS 3772, pp. 29–40, 2005.
© Springer-Verlag Berlin Heidelberg 2005

- learn a complex structured formal query language (a language with formalized semantics and grammar, as opposed to natural language);
- have full knowledge of the DTD and its semantics.
- In structured documents, a well-thought and semantically strong structure formally marks up the meaning of the text; this can make easier query "understanding", at least when this query refers (partly) to the structure.
- Finally, formal queries do not permit information retrieval in heteregenous collections (with different and unknown DTDs). A natural language interface could resolve this problem, since users can express their information need conceptually.

Note that these comments could be made about the database domain too, and that the issues seem quite similar. Many natural language interfaces for databases have been developed, most of them transforming natural language into Structured Query Language (SQL) (see [5, 6, 7] for overviews). But the problem is different for the following reasons:

- Unlike databases, XML format looks set to be used and accessed by the general public, notably through the Internet. Although unambiguous, machine-readable, structured and formal query languages are necessary to support the retrieval process (in order to actually extract the answers), the need for simpler interfaces will become more and more important in the future.
- Database querying is a strict interrogation; it is different to Information Retrieval. The user knows what kind of information is contained in the database, her information need is precise, and the result she gets is either right or wrong. This means that the natural language analysis must interpret the query perfectly and unambiguously, failing which the final answer is incorrect and the user disatisfied.

  In XML IR, as well as in traditional IR, the information need is loosely defined and often there is no perfect answer to a query. A natural language interface is a part of the retrieval process, and thus it can interpret some queries imperfectly, and still return useful results. The problem is then made "easier" to solve...

## 2   INEX, NLPX Track and NEXI

### 2.1   INEX

The Initiative for Evaluation of XML Retrieval, INEX [8], provides a test collection consisting of over 500 Mbytes of XML documents, topics and relevance assessments. The document set is made up of 12,107 articles of the IEEE Computer Society's publications. Topics are divided into two categories:

- *Content-and-Structure* (CAS) queries, which contain structural constraints. *e.g.: Find* paragraphs *or* figure-captions *containing the definition of Godel, Lukasiewicz or other fuzzy-logic implications.* (Topic 127)

 – *Content-Only* (CO) queries that ignore the document structure.
   *e.g.: Any type of coding algorithm for text and index compression.* (Topic 162)

This article focuses on CAS topics. Figure 1 shows an example of CAS topic. The `description` element is a natural language (English) description of the user's information need; The `title` is a faithful translation of this need into a formal XPath-like language called Narrowed Extended XPath I (NEXI) [9]. The `narrative` part is a more detailed explanation of the information need.

```
<inex_topic topic_id="130" query_type="CAS">

    <title>
        //article[about(.//p,object
        database)]//p[about(.,version management)]
    </title>
    <description>
        We are searching paragraphs dealing with version management
        in articles containing a paragraph about object databases.
    </description>
    <narrative>
        The elements to be considered relevant are ...
    </narrative>
    <keywords>object database version management</keywords>

</inex_topic>
```

**Fig. 1.** An example of CAS topic

The participants in the *ad-hoc* INEX task use only NEXI titles in order to retrieve relevant elements. We adapted our system so that it takes the topic description (natural language expression depicting the same query) as input, and returns a well-formed NEXI title. Going through this pivot language presents many advantages: it allows the use of an existing NEXI search engine in the retrieval process. Furthermore a user can still specify her query in this formal language if she prefers to. Finally, we can evaluate the translation by comparing the translated queries with the original hand-crafted NEXI titles. On the other hand, the transformation to a pre-existing restrictive language may result in loss of information.

## 2.2   NEXI

NEXI CAS queries have the form $//\mathbf{A[B]}//\mathbf{C[D]}$ where A and C are paths and B and D are filters. We can read this query as *"Return C descendants of A where A is about B and C is about D"*. B and D correspond to disjunctions or conjunctions of 'about' clauses $\mathbf{about(//E, F)}$, where E is a path and F a list of terms. The `'title'` part of Fig. 1 gives a good example of a query formulated in NEXI. More information about NEXI can be found in [9].

## 3    Description of Our Approach

Requests are analysed through several steps:

1. A part-of-speech (POS) tagging is performed on the query. Each word is labeled by its word class (*e.g.*: noun, verb, adjective...).
2. A POS-dependant semantic representation is attributed to each word. For example the noun *'information'* will be represented by the predicate *information(x)*, or the verb *'identify'* by *evt($e_1$, identify)*.
3. Context-free syntactic rules describe the most current grammatical constructions in queries and questions. Low-level semantic actions are combined with each syntactic rule. Two examples of such operations, applied to the description of topic 130 (Fig. 1), are given in Fig. 2. The final result is a logical

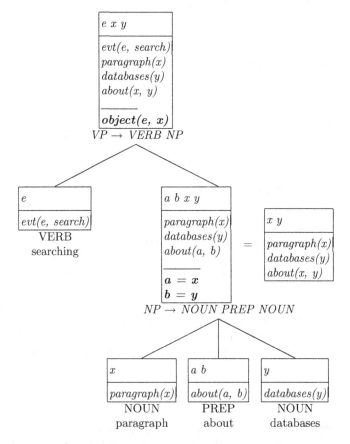

**Fig. 2.** Example of rule application for the verbal phrase *"searching paragraphs about databases"* (rules *NP → NOUN PREP NOUN* and *VP → VERB NP*). Basic semantic representations are attributed to part-of-speeches (leaf components). When applying syntactic rules, components are merged and semantic actions are added (here identity relations and verbal relation predicate – bold predicates).

representation shown in the left part of Fig. 3. This representation is totally independant from the queried corpus, it is obtained by general linguistic operations.

4. The semantic representation is then reduced with the help of specific rules:
   - a recognition of some typical constructions of a query (*e.g.: Retrieve + object*) or of the corpus (*e.g.: "an article written by [...]"* refers to the tag *au – author*);
   - and a distinction between semantic elements mapping on the structure and, respectively, mapping on the content;

   This part is the only one that uses corpus-specific information, among which the DTD, a dictionary of specific tag name synonyms (*e.g.: paper=`article`*), some simple ontologic structures (*"a article citing somebody"* refers to bibliography in INEX collection). Figure 3 shows the specific rules that apply to the example.
5. A treatment of relations existing between different elements;
6. The construction of a well-formed NEXI query.

Steps 1 to 5 are explained in more details in [10], as well as necessary corpus knowledge and the effect of topic complexity on the analysis.

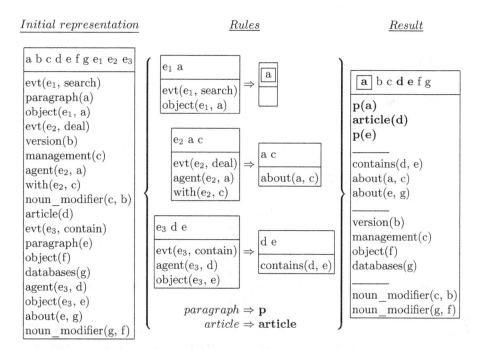

| *Initial representation* | *Rules* | *Result* |

**Fig. 3.** The semantic analysis of topic 130 (left), is reduced by some generic rules (center), leading to a new representation (right). Bold predicates emphasize words representing XML tag names and the framed letter stands for the element that should be returned to the user. The first three rules deal with verbal phrases *"to search sth"*, *"to deal with sth"* and *"to contain sth"*.

The representation obtained at the end of step 5 does not depend on any retrieval system or query language. It could be transformed (with more or less information loss) into any existing formal language.

### 3.1 Getting to NEXI

Transformation process from our representation to NEXI is not straightforward. Remember that a NEXI query has the form //**A[B]**//**C[D]**.

- At structural level, a set of several tag identifiers (that can be DTD tag names or wildcards) has to be distributed into parts A, B, C and D, that we respectively call support requests, support elements, return requests and return elements.
- At content level, linguistic features (like `noun_modifier` in the example) cannot be kept and must be transformed in an appropriate manner.

**Structural Level.** These four parts A, B, C and D are built from our representation (Fig. 3) in the following way:

- C is the 'framed' (selected) element name (see Fig. 3 and its caption);
- D is composed of all C children (relation *contains*) and their textual content (relation *about*);
- A is the highest element name in the DTD tree, that is not C or one of its children;
- B is composed of all other elements and their textual content.

Wildcard-identified tags of the same part are merged and are considered to be the same element. See an example in Sect. 4.

**Table 1.** Examples of linguistic features and of their NEXI equivalents

| Predicate | Initial text | Representation | NEXI content |
|---|---|---|---|
| `noun_property` | "definition of a theorem" | *definition(a)* *theorem(b)* *noun_property(a, b, of)* | "defition of theorem" |
| `noun_modifier` | "version management" | *version(a)* *management(b)* *noun_modifier(b, )* | "version management" |
| `adjective` | "digital library" | *digital(a)* *library(b)* *adjective(b, a)* | "digital library" |
| `disjunction` | "definition of Godel or Lukasiewicz" | *definition(a)* *noun_property(a, b, of)* $b = c \lor d$ Godel(c) Lukasiewicz(d) | "definition of Godel definition of Lukasiewicz" |

**Content Level.** The main linguistic predicates generated by our system are `np_property`, `noun_modifier`, `adjective` and disjunction or conjunction relations. NEXI format requires 'about' clauses to contain only textual content. In most cases we chose to reflect as far as possible the initial text, because the search engine can deal with noun phrases. In the case of disjunctions and conjunctions, for the same reason, we built separated noun phrases. Examples of each operation are given in Tab. 1.

The transformation of the semantic representation of Fig. 3 results in:

```
//article[(about(.//p, object databases))]//p[(about(.,
                version management))]
```

## 4   Example

We give here a significant example, with the analysis of topic 127 (INEX 2004). Several syntactic parsings could be possible for the same sentence. In practice a "score" is attributed to each rule release, depending on several parameters (among which distance between words that are linked, length of phrases, type of relations... Unfortunately we lack space to explain more precisely this process). In our sample topic only the best scored result is given.

| $c1\ c2\ c3\ c4\ c5\ c6\ c7\ c8\ c9$ |
|---|
| $c10\ c11\ c12\ c13$ |
| *1. event(c1, find)* |
| *2. object(c1, c2)* |
| *3. c2 = c3 ∨ c4* |
| *4. paragraph(c3)* |
| *5. figure(c5)* |
| *6. caption(c4)* |
| *7. rel_ noun_ modifier(c4, c5)* |
| *8. event(c6, contain)* |
| *9. agent(c6, c2)* |
| *10. object(c6, c7)* |
| *11. definition(c7)* |
| *12. rel_ np_ relation(c7, c8, of)* |
| *13. c8 = c9 ∨ c10, c10 = c11 ∨ c12* |
| *14. c9 = Godel, c11 = Lukasiewicz* |
| *15. 'fuzzy-logic'(c13)* |
| *16. implication(c12)* |
| *17. rel_ noun_ modifier(c12, c13)* |
| *18. rel_ adjective(c12, other)* |

$\xrightarrow[\text{rules}]{\text{reduction}}$

| $c3\ c4\ c5\ \mathbf{c6}\ \mathbf{c8}\ c9\ c10\ c11$ |
|---|
| $\boxed{c12}\ c13$ |
| $\mathbf{p(c3)}$ |
| $\mathbf{fgc(c4)}$ |
| ——— |
| *about(c2, c7)* |
| ——— |
| *c2 = c3 ∨ c4* |
| *definition(c7)* |
| *c8 = c9 ∨ c10, c10 = c11 ∨ c12* |
| *c9 = Godel, c11 = Lukasiewicz* |
| *'fuzzy-logic'(c13)* |
| *implication(c12)* |
| ——— |
| *rel_ np_ relation(c7, c8, of)* |
| *rel_ noun_ modifier(c12, c13)* |
| *rel_ adjective(c12, other)* |

```
//article//(p|fgc)[(about(., "definition of Godel" "definition of
    Lukasiewicz" "definition of fuzzy-logic implications"))]
```

**Fig. 4.** Semantic representations of topic 127, and automatic conversion into NEXI

(127) Find paragraphs or figure captions containing the definition of Godel, Lukasiewicz or other fuzzy-logic implications.

Figure 4 shows the three major steps of the analysis of topic 127. The left frame represents the result of step 3 (see Sect. 3). Some IR- and corpus-specific reduction rules are then applied and lead to right frame: terms *paragraph* and *figure-captions* are recognized as tag names **p** and **fgc** (lines 4 to 7); the construction "*c2 contains c7*" is changed into **about(c2, c7)** (lines 8 to 11). The other relations are kept.

Translation into NEXI is performed as explained above, disjunctions *c8* and *c10* result in the repetition of the term "*definition*" with preposition "*of*" and three distinct terms.

## 5   Processing NEXI Queries

### 5.1   XML File Inversion

In our scheme each term in an XML document is identified by three elements: its filename, its absolute XPath context, and its ordinal position within the XPath context. An inverted list for a given term is depicted in Tab. 2.

**Table 2.** Inverted file

| Document | XPath | Position |
|---|---|---|
| e1303.xml | article[1]/bdy[1]/sec[6]/p[6] | 23 |
| e1303.xml | article[1]/bdy[1]/sec[7]/p[1] | 12 |
| e2404.xml | article[1]/bdy[1]/sec[2]/p[1]/ref[1] | 1 |
| f4576.xml | article[1]/bm[1]/bib[1]/bibl[1]/bb[13]/pp[1] | 3 |
| f4576.xml | article[1]/bm[1]/bib[1]/bibl[1]/bb[14]/pp[1] | 2 |
| g5742.xml | article[1]/fm[1]/abs[1] | 7 |

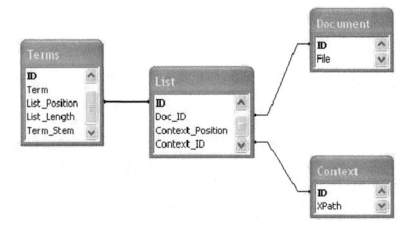

**Fig. 5.** Schema for XML Inverted File

In principle at least, a single table can hold the entire cross reference list (our inverted file). Suitable indexing of terms can support fast retrieval of term inverted lists. However, it is evident that there is extreme redundancy in the specification of partial absolute XPath expressions (substrings). There is also extreme redundancy in full absolute XPath expressions where multiple terms in the same document share the same leaf context (e.g. all terms in a paragraph). Furthermore, many XPath leaf contexts exist in almost every document (e.g. /article[1]/fm[1]/abs[1] in INEX collection). For these reasons we chose to normalize the inverted list table to reduce redundancy.

The structure of the database used to store the inverted lists is depicted in Fig. 5. It consists of four tables. The `Terms` table is the starting point of a query on a given term. The `Term_Stem` column holds the Porter stem of the original term. The `List_Position` is a foreign key from the `Terms` table into the `List` Table. It identifies the starting position in the inverted list for the corresponding term. The `List_Length` is the number of list entries corresponding to that term. The `List` table is (transparently) sorted by `Term` so that the inverted list for any given term is contiguous.

## 5.2   Ranking Scheme

Elements are ranked according to a relevance judgment score. Leaf and branch elements need to be treated differently. Data usually occur at leaf elements, and thus, our inverted list mostly stores information about leaf elements. A leaf element is considered candidate for retrieval if it contains at least one query term. A branch node is candidate if it contains a relevant child element. Once an element (either leaf or branch) is deemed to be a candidate for retrieval its relevancy judgment score is calculated. A heuristically derived formula (Equation (1)) is used to calculate the relevance judgment score of leaf elements. The same equation is used for both return and support elements. The score is determined from query terms contained in the element. It penalises elements with frequently occurring query terms (frequent in the collection), and it rewards elements with evenly distributed query term frequencies within the elements.

$$L = K^{n-1} \sum_{i=1}^{n} \frac{t_i}{f_i} \qquad (1)$$

Here $n$ is the number of unique query terms contained within the leaf element, $K$ is a small integer (we used $K = 5$). The term $K^{n-1}$ scales up the score of elements having multiple distinct query terms. We experimented with $K = 3$ to 10 with little difference in results. The sum is over all terms where $t_i$ is the frequency of the $i^{th}$ query term in the leaf element and $f_i$ is the frequency of the $i^{th}$ query term in the collection. This sum rewards the repeat occurrence of query terms, but uncommon terms contribute more than common terms.

Once the relevance judgment scores of leaf elements have been calculated, they can be used to calculate the relevance judgment score of branch elements. A naïve solution would be to just sum the relevance judgment score of each branch relevant children. However, this would ultimately result in root elements accumulating at the top of the ranked list, a scenario that offers no advantage over document-level retrieval. Therefore, the relevance judgment score of children elements should be somehow decreased while being propagated up the XML tree.

A heuristically derived formula (Equation (2)) is used to calculate the scores of intermediate branch elements:

$$R = D(n) \sum_{i=1}^{n} L_i \qquad (2)$$

Where:

- $n$ = the number of children elements
- $D(n)$ = 0.49 if $n = 1$
        0.99 otherwise
- $L_i$ = the $i^{th}$ return child element

The value of the decay factor $D$ depends on the number of relevant children that the branch has. If the branch has one relevant child then the decay constant is 0.49. A branch with only one relevant child will be ranked lower than its child. If the branch has multiple relevant children the decay factor is 0.99. A branch with many relevant children will be ranked higher than its descendants. Thus, a section with a single relevant paragraph would be judged less relevant than the paragraph itself, but a section with several relevant paragraphs will be ranked higher than any of the paragraphs.

Having computed scores for all result and support elements, the scores of support elements are added to the scores of the corresponding result elements that they support. For instance, consider the query:

//A[about(.//B,C)]//X[about(.//Y,Z)]

The score of a support element //A//B will be added to all result elements //A//X//Y where the element A is the ancestor of both X and Y.

Finally, structural constraints are only loosely interpreted. Elements are collected regardless of the structural stipulations of the topic. Ancestors or descendants of Y may be returned, depending on their score and final rank.

More information about this system can be found in [11].

## 6   Results

We tested our system using the INEX 2004 collection (set of topics and evaluation metrics). Recall/precision graphs have been calculated by the official INEX evaluation program. In the following we call $S_{adhoc}$ the system using official, hand-crafted NEXI titles. $S_{NLP}$ is the same system, but using natural language queries, automatically translated into NEXI.

$S_{adhoc}$ is ranked $1^{st}$ from 52 submitted runs in the task, with an average precision of 0.13. $S_{NLP}$ is ranked $5^{th}$ with an average precision of 0.10, and $1^{st}$ among the systems using natural language queries.

The Recall/Precision curves are presented in Fig. 6. The top bold dashed curve represents results for $S_{adhoc}$, the lower bold one is $S_{NLP}$ curve, and the other curves are all the official runs submitted at INEX 2004.

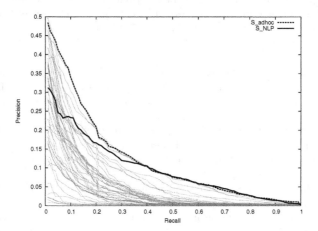

**Fig. 6.** INEX'04 VCAS Recall/Precision curve averaged over all topics and all metrics

The precision loss caused by the natural language interface is limited. $S_{NLP}$ looses only four ranks with this interface, and still outperforms most of ad-hoc systems. We think that this trade-off is very interesting; indeed, as we pointed out in the introduction, the benefits brought by a natural language interface compensate for the precision loss, at least for non-expert users[1].

## 7   Conclusion

In this paper we presented an XML retrieval system that allows the user to express a query over an XML collection, using both structural and content retrieval cues, in natural English expression. An NLP module analyses this expression syntactically and semantically, applies some specific rules and translates the result into a formal query language. This intermediate language is then processed by a backend system.

This system had been tested with INEX 2004 collection, topics, and relevance assessments and with good results. This study shows that natural language queries over XML collections can offer promising prospects for deploying in general public applications.

---

[1] We can note that before an online NEXI parser became available for INEX topic developers, the majority of submitted topics were not well formed (depicting the wrong meaning) and/or syntactically incorrect. However INEX participants are XML retrieval professionals that have at least a good knowledge of XPath and NEXI. The task would have been much more difficult for casual users.

# References

[1] Smeaton, A.F.: Information Retrieval: Still Butting Heads with Natural Language Processing? In Pazienza, M., ed.: Information Extraction – A Multidisciplinary Approach to an Emerging Information Technology. Volume 1299 of Lecture Notes in Computer Science. Springer-Verlag (1997) 115–138

[2] Smeaton, A.F.: Using NLP or NLP Resources for Information Retrieval Tasks. [12] 99–111

[3] Arampatzis, A., van der Weide, T., Koster, C., van Bommel, P.: Linguistically-motivated Information Retrieval. In Kent, A., ed.: Encyclopedia of Library and Information Science. Volume 69. Marcel Dekker, Inc., New York, Basel (2000) 201–222

[4] Sparck Jones, K.: What is the role of NLP in text retrieval? [12] 1–24

[5] Androutsopoulos, I., G.D.Ritchie, P.Thanisch: Natural Language Interfaces to Databases – An Introduction. Journal of Natural Language Engineering **1** (1995) 29–81

[6] A.Copestake, Jones, K.S.: Natural Language Interfaces to Databases. The Knowledge Engineering Review **5** (1990) 225–249

[7] Perrault, C., Grosz, B.: Natural Language Interfaces. Exploring Articial Intelligence (1988) 133–172

[8] Fuhr, N., Lalmas, M., Malik, S., Szlàvik, Z., eds.: Advances in XML Information Retrieval. Third Workshop of the Initiative for the Evaluation of XML retrieval (INEX). Volume 3493 of Lecture Notes in Computer Science., Schloss Dagstuhl, Germany, Springer-Verlag (2005)

[9] Trotman, A., Sigurbjörnsson, B.: Narrowed Extended XPath I (NEXI). [8]

[10] Tannier, X., Girardot, J.J., Mathieu, M.: Analysing Natural Language Queries at INEX 2004. [8] 395–409

[11] Geva, S.: GPX - Gardens Point XML Information Retrieval at INEX 2004. [8]

[12] Strzalkowski, T., ed.: Natural Language Information Retrieval. Kluwer Academic Publisher, Dordrecht, NL (1999)

# Recommending Better Queries from Click-Through Data[*]

Georges Dupret[1] and Marcelo Mendoza[2]

[1] Center for Web Research, Department of Computer Science,
Universidad de Chile
gdupret@dcc.uchile.cl
[2] Department of Computer Science, Universidad de Valparaiso
marcelo.mendoza@uv.cl

**Abstract.** We present a method to help a user redefine a query based on past users experience, namely the click-through data as recorded by a search engine. Unlike most previous works, the method we propose attempts to recommend better queries rather than related queries. It is effective at identifying query specialization or sub-topics because it take into account the co-occurrence of documents in individual query sessions. It is also particularly simple to implement.

The scientific literature follows essentially two research lines in order to recommend related queries: query clustering algorithms [1,4,5] and query expansion [2,3]. Most of these works use terms appearing in the queries and/or the documents. By contrast, the simple method we propose here aims at discovering alternate queries that improve the search engine ranking of documents: We order the documents selected during past sessions of a query according to the ranking of other past queries. If the resulting ranking is better than the original one, we recommend the associated query.

Before describing the algorithm, we need first to introduce a definition for consistency. A document is consistent with a query if it has been selected a significant number of times during the sessions of the query. Consistency ensures that a query and a document bear a natural relation in the opinion of users and discards documents that have been selected by mistake once or a few time. Similarly, we say that a query and a set of documents are consistent if each document in the set is consistent with the query.

Consider the set $D(s_q)$ of documents selected during a session $s_q$ of a query $q$. If we make the assumption that this set represents the information need of the user who generated the session, we might wonder if alternate queries exist in the logs that 1) are consistent with $D(s_q)$ and 2) better rank the documents of $D(s_q)$. If these alternate queries exist, they return the same information $D(s_q)$ as the original query, but with a better ranking and are potential query recommendations. We then repeat the procedure for each session of the original query,

---

[*] This research was supported by Millennium Nucleus, Center for Web Research (P04-067-F), Mideplan, Chile.

M. Consens and G. Navarro (Eds.): SPIRE 2005, LNCS 3772, pp. 41–44, 2005.

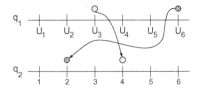

**Fig. 1.** Comparison of the ranking of two queries. A session of the original query $q_1$ contains selections of documents $U_3$ and $U_6$ appearing at position 3 and 6 respectively. The rank of this set of document is 6 by virtue of Def. 1. By contrast, query $q_2$ achieves rank 4 for the same set of documents and therefore qualifies as a candidate recommendation.

select the alternative queries that appear in a significant number of sessions and propose them as recommendations to the user interested in $q$.

To apply the above algorithm, we need a criteria to compare the ranking of a set of documents for two different queries. We first define the **rank** of document $u$ in query $q$, denoted $r(u, q)$, as the position of document $u$ in the listing returned by the search engine. We say that a document has a high or large rank if the search engine estimates that its relevance to the query is comparatively low. We extend this definition to sets of documents:

**Definition 1 (Rank of a Set of Documents).** *The rank of a set $U$ of documents in a query $q$ is defined as*

$$r(U, q) = \max_{u \in U} r(u, q) \ .$$

In other words, the documents with the worse ranking determines the rank of the set. If a set of documents achieves a lower rank in a query $q_a$ than in a query $q_b$, then we say that $q_a$ ranks the documents better than $q_b$. This criteria is illustrated in Fig. 1 for a session containing two documents.

**Definition 2 (Ranking Comparison).** *A query $q_a$ ranks better a set $U$ of documents than a query $q_b$ if $r(U, q_a) < r(U, q_b)$.*

One might be reluctant to reject an alternative query that orders well most documents of a session and poorly only one or a few of them. The argument against recommending such a query is the following: If the documents with a large rank in the alternate query appear in a significant number of sessions, they are important to the users. Presenting the alternate query as a recommendation would mislead users into following recommendations that conceal part of their information need. We can formalize the method as follows:

**Definition 3 (Recommendation).** *A query $q_a$ is a recommendation for a query $q_b$ if a significant number of sessions of $q_a$ are consistent with $q_b$ and are ranked better by $q_a$ than by $q_b$.*

The recommendation algorithm induces a directed graph between queries. The original query is the root of a tree with the recommendations as leaves.

**Fig. 2.** Queries `Valparaiso`(a), `Naval battle in Iquique`(b), `Fiat`(c) and their associated recommendations. Numbers inside the nodes represent the number of query sessions for the associated query, and number on the edges represented the number of query sessions improved by the pointed query.

Each branch of the tree represents a different specialization or sub-topic of the original query. The depth between a root and its leaves is always one, because we require the recommendations to improve the associated document set ranking.

The algorithm was implemented using the logs of the `TodoCL` search engine for a period of three months (`www.todocl.cl`). `TodoCL` is a search engine that mainly covers the `.cl` domain (Chilean web pages). Over three months the logs gathered 20,563,567 requests, most of them with no selections: Meta search engines issue queries and re-use the answer of *TodoCL* but do not return information on user selections. A total of 213,540 distinct queries lead to 892,425 registered selections corresponding to 387,427 different URLs.

We intend to illustrate that the recommendation algorithm has the ability to identify sub-topics and suggest query refinement. For example, Fig. 2(a) shows the recommendation graph for the query `valparaiso`. The number inside nodes refer to the number of sessions registered in the logs for the query. The edge numbers count the sessions improved by the pointed query. Valparaiso is an important touristic and harbor city, with various universities. It also recommends some queries that are typical of any city of some importance like city hall, municipality, a local newspaper and so on. The more potentially beneficial recommendations have a higher link number.

For another example (see figure 2(b)), a recommendation concerns the ''naval battle in Iquique'' in May 21, 1879 between Peru and Chile. The point we want to illustrate here is the ability of the algorithm to extract from the logs and to suggest to users alternative search strategies. Out of the 47 sessions, 2 sessions were better ranked by armada of Chile and sea of Chile. In turn, out of the 5 sessions for armada of Chile, 3 would have been better answered by sea of Chile. An interesting recommendation is for biography of Arturo Prat, who was captain of the "Esmeralda", a Chilean ship involved in the battle.

Finally, another recommendation related to the query ''fiat'', suggests to specify the car model he is interested in, if he wants spare parts, or if he is interested in selling or buying a fiat. It also suggests to a user interested in – say – the history or the profitability of the company to issue a query more specific to his needs. The graph is shown in fig 2(c).

Contrary to methods based on clustering where related queries are proposed to the user, the recommendation method we propose are made only if they are expected to improve. Among the limitations of the method, one is inherent to query logs: Only queries present in the logs can be recommended and can give rise to recommendations. Problems might also arise if the search engine presents some weakness or if the logs are not large enough. Currently, we are working on query recommendation systems that overcome these limitations.

## References

1. R. Baeza-Yates, C. Hurtado, and M. Mendoza. Query recommendation using query logs in search engines. In *LNCS 3268*, pages 588–596, 2004.
2. B. Billerbeck, F. Scholer, H. E. Williams, and J. Zobel. Query expansion using associated queries. In *CIKM03*, pages 2–9, New York, NY, USA, 2003. ACM Press.
3. F. Scholer and H. E. Williams. Query association for effective retrieval. In *CIKM02*, pages 324–331, New York, NY, USA, 2002. ACM Press.
4. J. Wen, J. Nie, and H. Zhang. Clustering user queries of a search engine. In *10th WWW Conference*, 2001.
5. O. R. Zaiane and A. Strilets. Finding similar queries to satisfy searches based on query traces. In *EWIS02*, Montpellier, France, 2002.

# A Bilingual Linking Service for the Web

Alessandra Alaniz Macedo[1], José Antonio Camacho-Guerrero[2],
and Maria da Graça Campos Pimentel[3]

[1] FFCLRP, Sao Paulo University, Ribeirão Preto, Brazil
[2] 3WT, São Carlos, Brazil
[3] ICMC, Sao Paulo University, São Carlos, Brazil

## 1   Introduction

The aim of Cross-Language Information Retrieval (CLIR) area is to address situations where a query is made in one language and the application is able to return documents in another. Many CLIR techniques attempt to translate the user's query to the language of the target documents using translation dictionaries. However, these techniques have limitations in terms of lexical coverage of the dictionary adopted. For some applications, the dictionaries are manually edited towards improving the results — but this may require much effort to represent a large collection of information.

In this article we propose an infrastructure for defining automatically relationships between Web documents written in different languages. Our approach is based on the Latent Semantic Indexing Technique, which tries to overcome the problems common to the lexical approach due to words with multiple meanings and multiple words with the same meaning. LSI automatically organizes text objects into a semantic structure appropriate for matching [3]. To support the identification of relationships among documents in different languages, the proposed infrastructure manipulates the stem portion of each word in order to index the corresponding Web documents when building the information space manipulated by LSI. To experiment this proposal, we studied the creation of links among news documents in English and Spanish in three different categories: entertainment, technology and world. The results were positive.

## 2   A Bilingual Linking Approach

**LinkDigger** is a Web application allowing monolingual repositories to be related according to their set of words [5]. We extended LinkDigger to support the creation of links between bilingual Web repositories by extracting stems from their textual information in order to compose the information space manipulated by LSI. The current Bilingual LinkDigger relates Spanish and English repositories.

To activate LinkDigger, a user (i) selects the number of sites to be linked and feeds their URLs, (ii) informs an email address and (iii) activates the service. The service runs in the background in order to support large sites to be indexed.

The windows shown on the right portion of Fig. 1 correspond to the selection of the title "CNN.com - Robot ship braces for death by Jove - Sep. 19,

M. Consens and G. Navarro (Eds.): SPIRE 2005, LNCS 3772, pp. 45–48, 2005.

**Fig. 1.** The page in English (left) automatically related to the pages in Spanish (right)

2003" in the user interface of LinkDigger, and the related two titles to the first: "CNNenEspañol.com - Sonda Galileo concluye misión estrellándose en Júpiter - Septiembre 21, 2003" and "CNNenEspañol.com - Con un impacto contra Júpiter concluye la misión de la sonda Galileo - Septiembre". Those hyperlinks — relating the page in English with the two pages in Spanish — are a good example of the relevant links created in the experiment described next.

**The LinkDigger Infrastructure** supporting the processing carried out by the Bilingual LinkDigger is built with several components. Towards facilitating the presentation of those components, we grouped the processes in three levels: Linking Level, Storage Level and Anchoring Level. The *Linking Level* defines links between the bilingual Web documents, and its processing is as follows:

- *Crawling Web Repositories.* Initially the repositories are indexed. Significant words, excluding stop words, are extracted from the Web pages.
- *Stemming.* The relevant words extracted from the pages are analyzed by a stemmer, specific to the language of repositories manipulated [4] [7].
- *Weighting the stems.* To differentiate the stems from words, appropriate weights must be given to them. Weights are composed by the number of times a radical appears in a given document relative to the number of times that it appears in each document in the whole repository of documents [8].
- *Generating the Term by Document Matrix.* The index resulting from the previous steps is used to generate a term by document matrix $X$, which is used in the process of Latent Semantic Indexing [3].
- *Computing SVD.* The matrix $X$ is decomposed into the product of three other component matrices $T$, $S$ and $D'$ using the Single Value Decomposition (SVD) exploited by LSI. Following SVD, the $k$ most important dimensions are selected, and all others are omitted. The choice of $k$ is detailed in [2].

- *Defining the Semantic Matrix.* A semantic matrix is generated by the inner-product among each column of the matrix generated on the last step.
- *Computing Similarities.* Given the semantic matrix generated in the previous step, relationships between Web documents are identified by considering the cells that have the higher values of similarity. A threshold level of similarity is used to filter the relationships created towards generating a relevance semantic matrix which is used to define bilingual links between documents.

In the *Anchoring Level*, the links created can be stored in databases or open hypermedia linkbases. The communication between the Linking Level and the Storage Level is supported by procedures in the Anchoring Level.

Our current implementation uses as *Storage Level* the Web Linkbase Service (WLS) [1] to store the links created. WLS is an XML-based open linkbase, developed as an API, that aims at providing hypermedia functionalities to applications in general. By using WLS, hyperlinks generated by Bilingual LinkDigger can be exploited by other Web applications — such characteristic makes Bilingual LinkDigger an example of applying open hypermedia on the Web.

On top of the Linking Level, the *Request Interface* supports user-interaction functionalities. For further details about the Storage Level and the Linking Level, respectively, see [5] and [6].

## 3   Experimentation

We run an experiment to relate documents from two news services in two different languages: CNN in English (CNN-E: http://cnn.com) and CNN in Spanish (CNN-S: http://cnnenespanol.com). In September 2003, our service scanned, in each language, 3 sections: Technology, Entertainment and World. The overall data for the experiment is summarized in Table 1. A total of 86 pages were collected from CNN-S and a total of 126 pages from CNN-E. Pages containing only images or that could not be read at the time of the harvesting were removed.

Considering that the documents in the repositories were in different languages, we used a threshold of 20% on the level of similarity that defines when pages are related. We had a total of 7 links relating pages on Technology, 27 links relating the pages on Entertainment and 10 links relating the pages in the World sections. A native Spanish speaking user, highly fluent in English, analyzed each link to rate whether the link was of high, medium, low or no relevance. For the links relating the sections on Technology, all 7 links were considered highly relevant. For the links relating the Entertainment sections, 12 links were rated highly relevant, 1 link was rated with medium relevance, 1 link was rated with low relevance and 13 links were considered not relevant. For the links relating the World section, from the 10 links generated, 5 were rated highly relevant, 3 were rated with medium relevance and 2 were rated not relevant.

The fact that most links (66%) were considered relevant — indeed, almost 55% were rated highly relevant — is a good result. It is also important to observe that the total number of links generated is not a small one considered the total number of pages in the collections.

**Table 1.** Characteristics of the Experiments with the Bilingual Linking Service

|  | Technology | | Entertainment | | World | |
|---|---|---|---|---|---|---|
|  | Spanish | English | Spanish | English | Spanish | English |
| Web pages in collections | 83 | 48 | 38 | 89 | 33 | 92 |
| Web pages after filtering | 48 | 25 | 21 | 45 | 17 | 56 |
| Total of terms | 12327 | 11996 | 10410 | 23213 | 8207 | 26490 |
| Stop Words | 351 | 570 | 351 | 570 | 351 | 570 |
| Relevance(High,Medium,Low,None) | (7,0,0,0) | | (12,1,1,13) | | (5,3,0,2) | |
| Precision Average | 60 | | 48.15 | | 100 | |
| Recall Average | 60 | | 36.12 | | 22.58 | |

# 4   Conclusion

Bilingual LinkDigger is a service that can be exploited towards the internation-alization of the contents of the Web. The results obtained in the experiment motivate the investigation of other techniques towards improving LinkDigger. Possible improvements could be achieved exploiting implicit and explicit user-feedback and linguistic analysis. Another alternative is to exploit the structure of the document contained in the repositories. We also intend to investigate the benefits that a thesaurus would bring to our service.

**Acknowledgments.** The authors are supported by FAPESP.

# References

1. R. F. Bulcão Neto, C. A. Izeki, M. G. C. Pimentel, R. P. M. Pontin, and K. N. Truong. An open linking service supporting the authoring of web documents. In *Proc. of the ACM DocEng*, pages 66–73, 2002.
2. S. Deerwester, S. T. Dumais, T. Landauer, G. Furnas, and R. Harshman. Indexing by latent semantic analysis. *J. of the Society for Inf. Science*, 41(6):391 – 407, 1990.
3. G. W. Furnas, S. Deerwester, S. T. Dumais, T. K. Landauer, R. A. Harshman, L. A. Streeter, and K. E. Lochbaum. Information retrieval using a singular value decomposition model of latent semantic structure. In *Proc. of the ACM SIGIR*, pages 465–480, 1988.
4. S. Group. Spanish stemming algorithm. Internet: snowball.tartarus.org, 2002.
5. A. A. Macedo, M. G. C. Pimentel, and J. A. Camacho-Guerrero. An infrastructure for open latent semantic linking. In *Proc. of ACM Hypertext*, pages 107–116, 2002.
6. A. A. Macedo, M. G. C. Pimentel, and J. A. C. Guerrero. Latent semantic linking over homogeneous repositories. In *Proc. of the ACM DocEng*, pages 144–151, 2001.
7. M. Porter. An algorithm for suffix stripping. *Program*, 14(3):130 – 137, 1980.
8. G. Salton and C. Buckley. Term-weighting approaches in automatic text retrieval. *Information Processing and Management*, 24(5):513 – 523, 1988.

# Evaluating Hierarchical Clustering
# of Search Results

Juan M. Cigarran, Anselmo Peñas, Julio Gonzalo, and Felisa Verdejo

Dept. Lenguajes y Sistemas Informáticos, E.T.S.I. Informática UNED[**]
{juanci, anselmo, julio, felisa}@lsi.uned.es

**Abstract.** We propose a goal-oriented evaluation measure, *Hierarchy Quality*, for hierarchical clustering algorithms applied to the task of organizing search results -such as the clusters generated by *Vivisimo* search engine-. Our metric considers the content of the clusters, their hierarchical arrangement, and the effort required to find relevant information by traversing the hierarchy starting from the top node. It compares the effort required to browse documents in a baseline ranked list with the minimum effort required to find the same amount of relevant information by browsing the hierarchy (which involves examining both documents and node descriptors).

## 1 Motivation

Clustering search results is an increasingly popular feature of web search and meta-search engines; examples include many industrial systems like Vivisimo, Kartoo, Mooter, Copernic, IBoogie, Groxis, Dogpile and Clusty and research prototypes such as Credo [1] or SnakeT [4].

For navigational search goals (finding a specific web site), clustering is probably useless. But for complex information needs, where there is a need to compile information from various sources, hierarchical clustering has clear advantages over plain ranked lists. According to [9], around 60% of web searches are informational, suggesting that techniques to organize and visualize search results can play an increasingly important role in search engines.

Although evidences show that this kind of systems seems to work fine and may be helpful, there is not a consensus about what kind of metrics are suitable to evaluate and to compare them in a quantitative way.

The research described in this paper attempts to design task-oriented evaluation metrics for hierarchical clusters that organize search results, according to their ability of minimizing the amount of irrelevant documents that has to be browsed by the user to access all the relevant information. Our metrics compare generated clusters with the original ranked list considering the navigational properties of the cluster, counting all clusters with relevant information and also

---

[**] This work has been partially supported by the Spanish Ministry of Science and Technology within the project(TIC-2003-07158-C04) Answer Retrieval From Digital Documents (R2D2).

M. Consens and G. Navarro (Eds.): SPIRE 2005, LNCS 3772, pp. 49–54, 2005.

how they are interconnected. Although our metrics compare the original ranked list with the clustering created, we have to remark that we work in an scenario where recall is maximum.

Regarding task-oriented clustering, there has been some attempts to evaluate the quality of clusters for Information Retrieval tasks. Scatter/Gather system [5], for instance, generates a hierarchical clustering but the evaluation issues are only focused on the highest scoring cluster to be compared with the original ranked list of documents and without considering the effort required to reach it. [8] also assume that a good clustering algorithm should put relevant and nonrelevant documents into separate classes but, again, in a flat approach that ignores the hierarchical and navigational properties of the cluster. More recently, [7] and [6] consider the hierarchical properties of the clusters in a task-oriented evaluation methodology and propose metrics to measure the time spent to find all the relevant information. [7] scores hierarchies estimating the time it takes to find all relevant documents by calculating the total number of nodes that must be traversed and the number of documents that must be read. In this case, the measure is used to compare the structure of hierarchies built using different approaches. The algorithm calculates the optimal path to each relevant document and then averages the results. No differences between the cognitive cost of reading a document and a cluster description are made in this approach. [6] present a similar strategy that compares the retrieval improvements of different clustering strategies versus the original ranked list. They consider differences between the cognitive cost of reading a document and a cluster description but, again, the algorithm operates over each relevant document separately and then averages the results. Our metrics do not calculate an optimal path to each relevant document averaging the results. Instead, a Minimal Browsing Area (MBA) (i.e. an optimal area within the hierarchy containing all the relevant documents) is calculated and then the measures (i.e. Distillation Factor and Hierarchical Quality) are applied. MBA is a way to reflect the power of the clustering to isolate relevant information allowing to work with it as a whole (i.e. we suppose the user is going to access all the relevant documents but taking the advantages of the set of nodes previously traversed). This approach relies on the idea of working at maximum recall.

The paper is organized as follows. First we present some preliminaries about how should be a good clustering organization and the basic assumptions in which our metrics are based. Then we explain the metrics, Distillation Factor and Hierarchical Quality and finally we present the conclusions and the future work.

## 2   Basic Assumptions

At least, four features of a hierarchical clustering have to be considered: a) *the content of the clusters*. A clustering that effectively groups relevant information is better than a clustering that mixes relevant and non-relevant documents; b) *the hierarchical arrangement of the clusters*. For instance, if the user has to browse a chain of clusters with irrelevant information before reaching a cluster with several relevant documents, the clustering is non-optimal for the task;

c) *the number of clusters*. To be effective, the number of clusters that have to be considered to find all relevant information should be substantially lower than the actual number of documents being clustered. Otherwise, the cognitive cost of examining a plain ranked list would be smaller than examining the cluster hierarchy, and; d) *how clusters are described*. A good cluster description should help the user predicting the relevance of the information that it contains. Optimal cluster descriptions will discourage users from exploring branches that only contain irrelevant information. We think that this last feature cannot be measured objectively without user studies. But before experimenting with users - which is costly and does not produce reusable test beds - we can optimize hierarchical clustering algorithms according to the first three features: contents of the clusters, number of clusters and hierarchical arrangement. Once the clustering is optimal according to these features, we can compare cluster descriptions by performing user studies. In this paper we will not discuss the quality of the cluster descriptions, focusing on those aspects that can be evaluated objectively and automatically (given a suitable test bed with relevance judgments).

In order to make our metrics to work, we propose the following two main assumptions: a) the first assumption considers that a hierarchical clustering should build each cluster only with those documents fully described by its descriptors. For instance, a cluster about *physics* with sub-clusters such as *nuclear physics* *astrophysics* should only contain those generic documents about physics, but not those about nuclear physics or astrophysics. Moreover, specific documents about any of the physics subtopics should be place in lower clusters (i.e. otherwise, we do not have a hierarchical clustering but only a hierarchical description of clusters). From the evaluation point of view, it has no sense to have a top high level cluster containing all the documents retrieved because it will will force the user to read the whole list at the very first time. This approach is the same as used in web directories and is considered the natural way of browsing hierarchical descriptions; b) the second assumption considers that clustering is made with an 'open-world' view. This means that, if a document is about very different topics, it should be placed (i.e. repeated) in its corresponding topic clusters and, as a consequence, it should appear in different parts of the hierarchy. For instance, if we have a document about *physics* which also includes some *jokes about physics* and a clustering hierarchy with clusters about *physics* and *jokes* without any connection between them, clustering should place the document anywhere in the physics hierarchy and it should repeat it in the jokes hierarchy. This 'open-world' view is more realistic than the classical 'closed-world' view, applied by some hierarchical clustering algorithms, and where a document only belongs to one part of the hierarchy. As a drawback of this assumption, it is very difficult to deal with evaluation issues when repeated documents appear in different parts of the hierarchy. As a possible solution, we propose the use of lattices instead of hierarchies as the data models used to represent the clustering. From the modeling and browsing point of view there are no differences between lattices and hierarchies and it is always possible to unfold a lattice into a hierarchy and viceversa. [2], [3] and [1] show how to deal with concept lattices in a document clustering framework.

# 3   Evaluation Measures

Let us start with a ranked list obtained as a result of a search which is going to be clustered using a lattice. Let $N$ be the set of nodes of the lattice $L$, where each node is described by a pair $(DOCS, DESC)$, with $DOCS$ the set of documents associated to the node (but not to its subnodes) and $DESC$ the description that characterizes the cluster (if any).

Our proposal is to measure the quality of a lattice, for the purpose of browsing search results, as the potential reduction of the cognitive load required to find all the relevant information as compared to the original ranked list. This can be expressed as a gain factor:

$$\text{Quality}(\text{lattice}) \equiv \frac{\text{cognitive load}(\text{ranked list})}{\text{cognitive load}(lattice)}$$

The effort required to browse a ranked list is roughly proportional to the number of documents in the list. Of course, the non-trivial issue is how to estimate the effort required to browse the lattice. In the remainder of this section, we will discuss two approaches to this problem: the first one is an initial approach that only considers the cost of examining documents. The second approach, in addition, also considers the cost of taking decisions (which nodes to explore, which nodes to discard) when traversing the lattice.

## 3.1   Distillation Factor

Let us assume that the user begins browsing the lattice at the top node; Let us also assume that, in our evaluation testbed, we have manual assessments indicating whether each document is relevant or not for the query that produced the ranked list of documents.

We can then define the *Minimal Browsing Area* (MBA) as the smallest part of the lattice that has to be explored by the user to find all relevant information contained in the lattice (i.e. a complete description of how to build MBA can be found in [2]). If we compute the cognitive cost of exploring the lattice as the cost of examining the documents contained in the MBA, then the quality of a lattice $L$ would be given by the ratio between the cost of examining the list of documents and the cost of examining the documents in the MBA.

Then, if $k_d$ is the cognitive cost of examining a document, $D_{RankedList}$ is the number of documents in the ranked list, and $D_{MBA}$ is the number of documents in the minimal browsing area, then the quality of the lattice, according to the definition above, would be:

$$\text{DF}(L) = \frac{k_d * D_{RankedList}}{k_d * D_{MBA}} = \frac{D_{RankedList}}{D_{MBA}}$$

We call this measure *Distillation Factor* (DF), because it describes the ability of the lattice to 'distill' or filter relevant documents in a way that minimizes user effort. Its minimum value is 1 (when all nodes in the lattice have to be examined to retrieve all relevant documents), which means that there is no improvement over the ranked list.

Notice that $DF$ can also be seen as the factor between the *precision* of the set of documents in the MBA and the *precision* of the original ranked list:

$$DF(L) = \frac{D_{Relevant}/D_{MBA}}{D_{Relevant}/D_{RankedList}} = \frac{D_{RankedList}}{D_{MBA}}$$

## 3.2 Hierarchy Quality

The DF measure is only concerned with the cost of reading documents, but browsing a conceptual structure has the additional cost (i.e. as compared to a ranked list of documents) of examining node descriptors and deciding whether each node is worth exploring. For instance, a lattice may lead us to $n$ relevant documents and save us from reading another $m$ irrelevant ones, ... but force us to traverse a thousand nodes to find the relevant information! Obviously, the number of nodes has to be considered when computing the cost of using the lattice to find relevant information.

To compute the cost of browsing the lattice we need to count all node descriptions that have to be considered to explore all relevant nodes. Let us call this set of nodes $N_{view}$. It can be computed starting with the nodes in the MBA and adding the lower neighbors of every node in the MBA.

Then, if the cognitive cost of examining a node description is $k_n$, the quality of the lattice can be defined as:

$$HQ(L) = \frac{k_d * D_{RankedList}}{k_d * D_{MBA} + k_n * |N_{view}|} = \frac{D_{RankedList}}{D_{MBA} + \frac{k_n}{k_d}|N_{view}|}$$

We call this measure *Hierarchy Quality* (HQ). This is the improved measure that we propose to evaluate the quality of a lattice for the task of organizing and visualizing search results. Note that it depends on a parameter $k \equiv \frac{k_n}{k_d}$, which estimates the ratio between the effort needed to examine a document (i.e. its title, its snippet or another kind of description) and the effort required to examine a node description. This value has to be settled according to the retrieval scenario and the type of node descriptions being considered.

Unlike the Distillation Factor, the HQ measure can have values below 1, if the number of nodes to be considered is too large. In this case, the HQ measure would indicate that the lattice is worse than the original ranked list for browsing purposes.

Of course, this formula implies to fix (i.e. for each retrieval scenario) a value for $k$. This value has a strong influence on the final HQ values and should be estimated conducting user studies.

# 4    Conclusions

We have introduced task-oriented evaluation metrics for hierarchical clustering algorithms that organize search results. These metrics consider the features of a good clustering for browsing search results. Our main evaluation measure, HQ (*Hierarchy Quality*), compares the cognitive effort required to browse the lattice, with the effort required to browse the original ranked list of results. Our measure is computed using the concept of *Minimal Browsing Area* and the related concept $N_{view}$ of minimal set of node descriptions which have to be considered in order to traverse the minimal browsing area. Our future work aims at trying to estimate the value of $k$ over different kinds of documents and cluster descriptions using different clustering strategies.

# References

1. C. Carpineto and G. Romano. *Concept Data Analysis. Data and Applications.* Wiley, 2004.
2. J. Cigarran, J. Gonzalo, A. Peñas, and F. Verdejo. Browsing search results via formal concept analysis: Automatic selection of attributes. In *Concept Lattices.* Second International Conference on Formal Concept Analysis, ICFCA 2004, Springer.
3. J. Cigarran, A. Peñas, J. Gonzalo, and F. Verdejo. Automatic selection of noun phrases as document descriptors in an fca-based information retrieval system. In *Formal Concept Analysis.* Third International Conference, ICFCA 2005, Springer, 2005.
4. P. Ferragina and A. Gulli. A personalized search engine based on web-snippet hierarchical clustering. In *WWW '05: Special interest tracks and posters of the 14th international conference on World Wide Web*, pages 801–810, New York, NY, USA, 2005. ACM Press.
5. M. Hearst and J. Pedersen. Reexamining the cluster hypothesis: Scatter/gather on retrieval results. In *Proceedings of SIGIR-96, 19th ACM International Conference on Research and Development in Information Retrieval*, pages 76–84, Zurich, CH, 1996.
6. K. Kummamuru, R. Lotlikar, S. Roy, K. Singal, and R. Krishnapuram. A hierarchical monothetic document clustering algorithm for summarization and browsing search results. In *WWW '04: Proceedings of the 13th international conference on World Wide Web*, pages 658–665, New York, NY, USA, 2004. ACM Press.
7. D. Lawrie and W. Croft. Discovering and comparing topic hierarchies. In *Proceedings of RIAO 2000.*, 2000.
8. A. Leouski and W. Croft. An evaluation of techniques for clustering search results, 1996.
9. D. E. Rose and D. Levinson. Understanding user goals in web search. In *WWW '04: Proceedings of the 13th international conference on World Wide Web*, pages 13–19. ACM Press, 2004.

# Counting Suffix Arrays and Strings

Klaus-Bernd Schürmann and Jens Stoye

AG Genominformatik, Technische Fakultät, Universität Bielefeld, Germany
Klaus-Bernd.Schuermann@CeBiTec.Uni-Bielefeld.DE

**Abstract.** Suffix arrays are used in various application and research areas like data compression or computational biology. In this work, our goal is to characterize the combinatorial properties of suffix arrays and their enumeration. For fixed alphabet size and string length we count the number of strings sharing the same suffix array and the number of such suffix arrays. Our methods have applications to succinct suffix arrays and build the foundation for the efficient generation of appropriate test data sets for suffix array based algorithms. We also show that summing up the strings for all suffix arrays builds a particular instance for some summation identities of Eulerian numbers.

## 1 Introduction

In the early 1990s, Manber and Myers [1] and Gonnet *et al.* [2] introduced the suffix array as an alternative data structure to suffix trees. Since then the application of and the research on suffix arrays advanced over the years [3,4,5,6].

In bioinformatics and text mining applications suffix arrays with some further annotations are often used as an indexing structure for fast string querying [3], and also in the data compression community suffix arrays received more and more attention over the last decade. At first, this interest has arisen from the close relation with the Burrows-Wheeler-Transform [7] which is mainly based on the fact that computing the Burrows-Wheeler-Transform by block-sorting the input string is equivalent to suffix array construction.

Moreover, in the last years, the task of full-text index compression emerged after Grossi and Vitter introduced the compressed suffix array [8] that reduces the space requirements to a linear number of bits. Other compressed indices of that type are Ferragina and Manzini's FM-index [9] based on the Burrows-Wheeler-Transform, a compressed suffix array based index of Sadakane [10] that does not use the text itself, and a suffix array based succinct index developed by He *et al.* [11], just to mention a few. Also lower bounds for the size of such indices are known. Demaine and López-Ortiz [12] proved a lower bound for indices providing substring search, and Miltersen [13] showed lower bounds for selection and rank indices.

All these developments on compressed indices, however, restrict themselves to certain queries. Therefore, information may get lost when compressing an original base index, like the suffix array. We believe that a profound knowledge of the algebraic and combinatorial properties of suffix arrays is essential to develop suffix array based, succinct indices preserving their original functionality.

M. Consens and G. Navarro (Eds.): SPIRE 2005, LNCS 3772, pp. 55–66, 2005.

Duval and Lefebvre [6] already characterized strings for the same suffix array. Further on, Crochemore *et al.* [14] recently showed combinatorial properties of the related Burrows-Wheeler transformation, but these properties are unassignable for suffix arrays. They rely on the fact that the Burrows-Wheeler transform is based on the order of cyclic shifts of the input sequence, whereas the suffix array is based on suffixes cut at the end of the string which destroys that nice group structure.

Most suffix array applications face strings with small, fixed alphabets like the DNA, amino acid, or ASCII alphabet. The possible suffix arrays for such strings are just a small fraction of all possible permutations. Therefore, besides discovering their combinatorial structure, our goal is to enumerate the different suffix arrays for strings over fixed size alphabets.

In Section 2 we give the basic definitions and notations concerning alphabets, strings, permutations, and suffix arrays. Strings sharing the same suffix array are counted in Section 3 and distinct suffix arrays in Section 4. Section 5 proves identities by summing up over suffix arrays and their strings, and Section 6 concludes.

An extend version of this article including all proofs is available as [15].

## 2   Strings, Permutations, and Suffix Arrays – Definitions and Terminology

The interval $[g, h] = \{z \in \mathbf{Z} \mid g \leq z \leq h\}$ denotes the set of all integers greater than or equal to $g$, and less than or equal to $h$.

*Alphabet and Strings.* Let $\Sigma$ be a finite set of size $|\Sigma|$, the *alphabet*, and $t = t_1 t_2 \ldots t_n \in \Sigma^n$ a string over $\Sigma$ of length $n$, the *text*. $\Sigma(t) = \{c \in \Sigma \mid \exists i \in [1, |t|] : t_i = c\}$ is the subset of characters actually occurring in $t$ and is called the *character set* of $t$. We usually use $\sigma$ for the alphabet size but, if the strings are required to use all characters such that their character set equals the alphabet, we use $k$.

For $i \in [1, n]$, $t[i]$ denotes the $i^{th}$ character of $t$, and for all pairs of indices $(i, j)$, $1 \leq i \leq j \leq n$, $t[i, j] = t[i], t[i + 1], \ldots, t[j]$ denotes the substring of $t$ starting at position $i$ and ending at position $j$. Substrings $t[i, n]$ ending at position $n$ are *suffixes* of $t$. The starting position $i$ of a suffix $t[i, n]$ is called its *suffix number*.

We deal with different kinds of equivalences of strings. The natural definition is that strings are *equivalent* if they are equal, and *distinct* otherwise.

In order to define the other two equivalences, we first introduce a bijective mapping $m$ of the characters of a string $t$ to the first $|\Sigma(t)|$ integers, $m : \Sigma(t) \longrightarrow [1, |\Sigma(t)|]$ such that $m(t) = m(t[1])m(t[2]) \ldots m(t[n])$. We call $m$ *order-preserving* if $c_1 < c_2 \Leftrightarrow m(c_1) < m(c_2)$ for all pairs of characters $(c_1, c_2) \in \Sigma^2(t)$.

We call two strings $t_1$ and $t_2$ *order-equivalent*, if there exists an order-preserving bijection $m_1$ for $t_1$ and another such bijection $m_2$ for $t_2$ such that $m_1(t_1) = m_2(t_2)$; otherwise the strings are *order-distinct*. If there exist not necessarily order-preserving mappings $m_1$ and $m_2$ such that $m_1(t_1) = m_2(t_2)$, we call $t_1$ and $t_2$ *pattern-equivalent*; otherwise the strings are *pattern-distinct*.

Equivalent strings are also order-equivalent and order-equivalence implies pattern-equivalence. The strings AT and AG, for example, are distinct but order-equivalent, and the strings AG and GA are order-distinct but pattern-equivalent.

*Permutations and Suffix Arrays.* Let $P$ be a permutation of $[1, n]$. Then $i \in [1, n-1]$ is a *permutation descent* if $P[i] > P[i+1]$. Conversely, a non-extendable ascending segment $P[i] < P[i+1] < \ldots < P[j]$ of $P$ is called a *permutation run*, denoted by the index pair $(i, j)$. Each permutation run of $P$ is bordered by permutation descents, or the permutation boundaries 1 or $n$. Hence, the permutation runs define the permutation descents and vice versa.

The *suffix array* $sa(t)$ of $t$ is a permutation of the suffix numbers $\{1, \ldots, n\}$ according to the lexicographic ordering of the $n$ suffixes of $t$. More precisely, a permutation $P$ of $[1, n]$ is the suffix array for string $t$ of length $n$ if for all pairs of indices $(i, j)$, $1 \leq i < j \leq n$, the suffix $t[P[i]], t[P[i] + 1], \ldots, t[n]$ at position $i$ in the permutation is lexicographically smaller than the suffix $t[P[j]], t[P[j] + 1], \ldots, t[n]$ at position $j$ in the permutation.

The *rank array* $R_P$, further on simply denoted by $R$, and sometimes called the inverse suffix array, for the permutation $P$, is defined as follows. For all indices $i \in [1, n]$ the rank of $i$ is $j$, $R[i] = j$, if $i$ occurs at position $j$ in the permutation, $P[j] = i$. We extend the rank array by $R[n+1] = 0$, indicating that the empty suffix, not contained in the suffix array, is always the lexicographically smallest.

Further on, we define the $R_+$-array to be $R_+[i] = R[P[i] + 1]$ for all $i \in [1, n]$. We define the $R_+$-descents and $R_+$-runs of $P$ similar to the permutation descents and permutation runs, respectively: A position $i \in [1, n-1]$ is called an $R_+$-*descent* if $R_+[i] > R_+[i+1]$. A non-extendable ascending segment $R[P[i]+1] < R[P[i+1]+1] < \ldots < R[P[j]+1]$, denoted by the index pair $(i, j)$, $i < j$, is called an $R_+$-*run*. Moreover, the set of $R_+$-descents $\{i \in [1, n-1] \mid R[P[i]+1] > R[P[i+1]+1]\}$ is denoted by $R_+$-$desc(P)$, or shortly $desc(P)$, and the set of $R_+$-runs $\{(i, j) \in [1, n]^2 \mid i < j \wedge (i = 1 \vee i - 1 \in desc(P)) \wedge (j = n \vee j \in desc(P)) \wedge \forall h \in [i, j-1] : h \notin desc(P)\}$ is denoted by $R_+$-$runs(P)$, or shortly $runs(P)$.

*Further Definitions.* Besides the binomial coefficient $\binom{x}{y} = \frac{x!}{y!(x-y)!}$, combinatorial objects related to permutations that are important for this work are the Stirling numbers and the Eulerian numbers. Although these numbers have a venerable history, their notation is less standard. We will follow the notation of [16] where the Stirling number of the second kind $\left\{{n \atop k}\right\}$ stands for the number of ways to partition a set of $n$ elements into $k$ nonempty subsets, and the Eulerian number $\left\langle {n \atop d} \right\rangle$ gives the number of permutations of $[1, n]$ having exactly $d$ permutation descents, also defined through the recursion (i) $\left\langle {n \atop 0} \right\rangle = 1$, (ii) $\left\langle {n \atop d} \right\rangle = 0$ for $d \geq n$, and (iii) $\left\langle {n \atop d} \right\rangle = (d+1) \left\langle {n-1 \atop d} \right\rangle + (n-d) \left\langle {n-1 \atop d-1} \right\rangle$ for $0 < d < n$.

## 3 Counting the Strings Per Suffix Array

In this section, we count the number of strings over a fixed size alphabet all sharing the same suffix array.

### 3.1   Characterizing Strings Sharing the Same Suffix Array

We repeat a characterization of the set of strings $sa^{-1}(P)$ sharing the same suffix array $P$ that states that the order of consecutive suffixes in the suffix array is determined by their first character and by the order of suffixes with respect to offset one. This result was already given by Burkhardt and Kärkkainen [5], and equivalent characterizations were given by Duval and Lefebvre [6].

**Theorem 1.** *Let $P$ be a permutation of $[1, n]$ and $t$ a string of length $n$. $P$ is the suffix array of $t$ if and only if for all $i \in [1, n]$ the following two conditions hold:*

*(a)   $t[P[i]] \leq t[P[i+1]]$ and*
*(b)   $R[P[i] + 1] > R[P[i+1] + 1] \Rightarrow t[P[i]] < t[P[i+1]]$.*

Theorem 1 characterizes the strings in the preimage $sa^{-1}(P)$ of $P$, and it also suggests criteria to divide the strings in equivalence classes according to their suffix array that will be counted in Section 4.

For a permutation $P$ with $d$ $R_+$-descents, Bannai *et al.* [4] already showed that the number of different characters in a string $t$ with suffix array $P$ is at least the number of $R_+$-descents plus one, $|\Sigma(t)| \geq d + 1$. They also presented an algorithm to construct a unique string $b_P$ consisting of exactly $d + 1$ characters, $|\Sigma(b_P)| = d + 1$.

W.l.o.g., we assume the character set of $b_P$ contains the smallest natural numbers, $\Sigma(b_P) = [1, d+1]$, and call $b_P$ the *base string* of the suffix array $P$.

The algorithm suggested in [4] works as follows. It starts with the initial character $c = 1$. For each index position $i \in [1, n]$ in ascending order, the algorithm proceeds through all suffix numbers from $P[1]$ to $P[n]$ by setting $P[i]$ to $c$. If $i$ is an $R_+$-descent, $c$ is incremented by one to satisfy condition (2) of Theorem 1, such that $b_P[i] = d_i + 1$ where $d_i$ is the number of descents in the prefix $P[1, \ldots, i]$ of the suffix array $P$.

*Remark 1.* Let $P$ be a permutation with $d$ descents, then the base string $b_P$ has the properties

(a)  $b_P[P[1]] = 1$ and $b_P[P[n]] = d + 1$,
(b)  for $i \in ([1, n-1] \setminus R_+\text{-desc}(P)) : b_P[P[i]] = b_P[P[i+1]]$,
(c)  for $i \in R_+\text{-desc}(P) : b_P[P[i]] + 1 = b_P[P[i+1]]$.

### 3.2   Counting Strings Composed of Up to $\sigma$ Distinct Characters

Strings sharing the same suffix array $P$ of length $n$ can be derived from the base string for the suffix array by applying a certain sequence of rewrite-operations to the base string, after which the order of suffixes remains untouched. The modification sequence starts with the largest suffix. Increasing the first character of the largest suffix by $r$ does not change the order of suffixes. Then, the first character of the second largest suffix can be increased by at most $r$ without changing the order of suffixes, and so on.

**Definition 1.** *Let $P$ be a permutation of $[1, n]$ with base string $b_P$. Moreover, let $m$ be a sequence of length $n$ of numbers from $[0, \psi]$, for some $\psi \in \mathbf{N}$. The $m$-incremented sequence $s_{P,m}$ of $P$ is defined as*

$$s_{P,m}[i] = b_P[P[i]] + m[i] \quad \text{for all } i \in [1, n].$$

We show a relationship between the sequences sharing the same suffix array and non-decreasing sequences.

**Theorem 2.** *Let $P$ be a permutation of $[1, n]$ with $d$ $R_+$-descents and $\mathcal{S}_{P,\Sigma}$ the set of sequences over the ordered alphabet $\Sigma$, $\sigma = |\Sigma|$, with suffix array $P$. Moreover, let $\mathcal{M}$ be the set of non-decreasing sequences of length $n$ over the ordered alphabet $[0, \sigma - d - 1]$.*

*There exists an isomorphism between $\mathcal{S}_{P,\Sigma}$ and $\mathcal{M}$.*

*Proof.* Let $b_P$ be the base string for permutation $P$. W.l.o.g., we assume $\Sigma = [1, \sigma]$. For each $m \in \mathcal{M}$, the corresponding $m$-incremented string $s_{P,m}$ fulfills the conditions of Theorem 1. Each other sequence $o$ of length $n$, $o \notin \mathcal{M}$, is either decreasing at one position, or it is not a sequence over $[0, \sigma - d - 1]$. If it is decreasing at one position, then $s_{P,o}$ contradicts Theorem 1. Otherwise, if $o$ is not a sequence over $[0, \sigma - d - 1]$, then the character set of $s_{P,o}$ is not covered by $\Sigma$. $\quad\square$

To count the number of non-decreasing sequences of length $n$ over $k + 1$ elements, we observe the following.

**Lemma 1.** *Let $M(n, a)$ be the number of non-decreasing sequences of length $n$ of elements in $[0, a - 1]$. For any positive integers $n$ and $a$*

$$M(n, a) = \binom{n + a - 1}{a - 1}.$$

*Proof.* The non-decreasing sequences of length $n$ on $a$ symbols can be modeled as a sequence of two different operations. Initially, the current symbol is set to 0. Then, we apply a sequence of operations to generate non-decreasing sequences of length $n$. One possible operation is to write the current symbol behind the so far written symbols, and the other one is to increment the symbol by 1. To generate a non-decreasing sequence, we apply $n + a - 1$ operations, $n$ to write down the non-decreasing sequence and $a - 1$ to increment the current symbol until $a - 1$ is reached. For this sequence of length $n + a - 1$, we have $\binom{n+a-1}{a-1}$ possibilities to choose the $a - 1$ positions of the increment operations. $\quad\square$

Applying this observation to Theorem 2, we get the number of strings sharing the same suffix array.

**Theorem 3.** *Let $P$ be a permutation of length $n$ with $d$ $R_+$-descents and $\Sigma$ an alphabet of $\sigma = |\Sigma|$ ordered symbols. The number of strings over $\Sigma$ with suffix array $P$ is $|S_{P,\Sigma}| = \binom{n+\sigma-d-1}{\sigma-d-1}$.*

The non-decreasing sequences of length $n$ over $[0, \sigma - d - 1]$ can simply be enumerated in-place by applying one change operation at a time, beginning with the sequence $0^n$. The bijection described through Definition 1 suggests to apply these enumeration steps directly to the base string of a certain suffix array. In this way, we can enumerate all $|\mathcal{S}_{P,\Sigma}|$ strings over a given alphabet $\Sigma$ for a certain suffix array $P$ in optimal $O(n + |\mathcal{S}_{P,\Sigma}|)$ time, where $n$ steps are used to construct the base string.

### 3.3   Counting Strings Composed of Exactly $k$ Distinct Characters

So far, we have considered the strings over a fixed alphabet all sharing the same suffix array. Now, we characterize the subset of such strings all composed of exactly $k$ different characters.

**Theorem 4.** *Let $P$ be a permutation of length $n$ with $d$ $R_+$-descents. The number of strings with suffix array $P$ composed of exactly $k$ different characters is $\binom{n-d-1}{k-d-1}$.*

*Proof.* The proof works similar as for Theorem 3. Obviously, we have to count each non-decreasing sequence $m$ in $M$ for which $s_{P,m}$ consists of exactly $k$ letters. To assure that none of the $k$ characters $[1, k]$ is left out, it is sufficient to count all $m$ such that $s_{P,m}[P[1]] = 0$, $s_{P,m}[P[n]] = k$, and consecutive characters are not differing by more than one. This property is realized by a sequence $m$, if and only if, (a) $m[1] = 0$ and $m[n] = k - d - 1$, (b) $m[i] = m[i+1]$ or $m[i] + 1 = m[i+1]$ if $i$ is not a descent position, and (c) $m[i] = m[i+1]$ if $i$ is a descent position.

We again represent these kind of non-decreasing sequences as $n$ write operations and $a - 1$ increment operations, as it has been modeled above. Here, for the placement of the $k - d - 1$ increment operations, we are restricted by the mentioned conditions.

In order not to hurt these conditions, (a) an increment operation must not appear before or after the first or last write operation, (b) at most one increment operation must appear between two write operations, and (c) the $d$ descent positions are blocked for the increments. We are thus left with $n - 1 - d$ mutually exclusive positions from which we choose $k - d - 1$ increment operations.   □

### 3.4   Filling the Gaps

For a given permutation $P$ of length $n$ with $d$ $R_+$-descents, we have already counted the strings over an alphabet of size $\sigma$ and the strings composed of exactly $k$ distinct characters, respectively.

Table 1 summarizes the results. For different conditions, it shows the number of distinct, order-distinct, and pattern-distinct strings of length $n$. The first row shows the number of strings composed of exactly $k$ different characters, the second row the number of strings over a certain alphabet of size $\sigma$, and the third and fourth rows the number of such strings sharing the same suffix array.

Some of the numbers were proven by other authors or in the previous sections, but there are yet some gaps to be filled. We start with the first row. Moore *et*

**Table 1.** Number of distinct, order-distinct and pattern-distinct strings of length $n$ in general, and those mapped to the same suffix array. In the analyses, $d$ is always the number of $R_+$-descents for the respective suffix array.

| number of | distinct | order-distinct | pattern-distinct |
|---|---|---|---|
| strings with exactly $k$ letters | $\left\{ {n \atop k} \right\} \cdot k!$ | $\left\{ {n \atop k} \right\} \cdot k!$ | $\left\{ {n \atop k} \right\}$ [17] |
| strings for alphabet size $\sigma$ | $\sigma^n$ | $\sum_{k=1}^{\sigma} \left\{ {n \atop k} \right\} \cdot k!$ | $\sum_{k=1}^{\sigma} \left\{ {n \atop k} \right\}$ |
| string with exactly $k$ letters sharing same suffix array | $\binom{n-d-1}{k-d-1}$ [Thm. 4] | $\binom{n-d-1}{k-d-1}$ | – |
| strings for alphabet size $\sigma$ sharing same suffix array | $\binom{n+\sigma-d-1}{\sigma-d-1}$ [Thm. 3] | $\sum_{k=d+1}^{\sigma} \binom{n-d-1}{k-d-1}$ | – |

*al.* [17] already showed that the number of pattern-distinct strings composed of exactly $k$ different characters is $\left\{ {n \atop k} \right\}$. For each pattern-distinct string, we permute the alphabet in $k!$ different ways to get a total of $\left\{ {n \atop k} \right\} k!$ order-distinct strings. These are already all the distinct strings since we have no flexibility to choose different characters to produce distinct strings yet order-equivalent.

The numbers of strings over a given alphabet of size $\sigma$ are shown in the second row. Needless to say, we have $\sigma^n$ distinct strings. For the order- and pattern-distinct strings, we just sum up the number of strings for all possible $k$.

The number of distinct strings composed of exactly $k$ different characters sharing a suffix array $P$ with $d$ $R_+$-descents was given in Theorem 4. All these strings are again order-distinct. For a pattern-distinct string, we cannot necessarily determine a unique suffix array. For example, $ab$ and $ba$ are pattern-equivalent, but have different suffix arrays. This is indicated by a dash in the table.

The number of distinct and order-distinct strings over an alphabet of size $\sigma$ sharing the same suffix array are given in the fourth row. Theorem 3 gave the number of distinct strings, and for the order-distinct strings we just sum up over all possible $k$.

# 4    Counting Suffix Arrays for Strings with Fixed Alphabet

In this section, the distinct suffix arrays for strings over a fixed size alphabet are counted. This also yields a tight lower bound for the compressibility of suffix arrays.

We first confine ourselves to the equivalent problem of counting the number of suffix arrays with a certain number of $R_+$-descents.

Bannai *et al.* [4] already stated that the number of suffix arrays of length $n$ with exactly $d$ $R_+$-descents is equal to the Eulerian number $\left\langle {n \atop d} \right\rangle$. In their explanation, they interpret Eulerian numbers as the number of permutations of length $n$ with $d$ permutation descents, and explain how their algorithm checks for these permutation descents. In fact, their algorithm counts the number of

$R_+$-descents, but the $R_+$-array is not a permutation. Nevertheless, as we show in this section, their proposition is true.

For a permutation $P$ of length $n - 1$, we map $P$ to a set $\mathcal{P}'$ of successor permutations, each of length $n$. We show some relations between $P$ and $\mathcal{P}'$, finally leading to the recursive definition of the Eulerian numbers.

First of all, we define the mapping from $P$ to $\mathcal{P}'$.

**Definition 2.** *Let $P$ be a permutation of length $n - 1$. A set of successor permutations $\mathcal{P}'$ of $P$ is defined as $\mathcal{P}' = \{P'_i \mid i \in [1, n]\}$ where $P'_i$ evolves from $P$ by incrementing each element of $P$ by one and inserting the missing 1 at position $i$, such that each position $j$ in $P$ corresponds to a position $j'$ in $P'_i$:*

$$j' = j, \qquad\qquad\qquad \text{if } j < i.$$
$$\text{and} \quad j' = j + 1, \qquad\qquad \text{if } j \geq i,$$

*and*

$$P'_i[j'] = P[j] + 1, \qquad\qquad \text{if } j' \neq i$$
$$\text{and} \quad P'_i[j'] = 1, \qquad\qquad\qquad \text{if } j' = i.$$

The insertion at position $i$ shifts the elements at positions $j$, $j \geq i$, to the right resulting in an increased rank for the respective elements of $P'_i$.

**Lemma 2.** *Let $P$ be a permutation of length $n - 1$ and $P' = P'_i$ a successor of $P$ with insertion position $i$, then we have for all $e \in [1, n - 1]$ that*

(a)    $R'[e + 1] = R[e]$        *if $R[e] < i$,*
(b)    $R'[e + 1] = R[e] + 1$    *if $R[e] \geq i$,*    *and*
(c)    $R'[1] = i$.

Furthermore, mapping $P$ to $P'$ basically preserves the $R_+$-order:

**Lemma 3.** *Let $P$ be a permutation of length $n - 1$ with successor $P'$. For all indices $g$ and $h$, $g, h \in [1, n - 1]$,*

$$R_+[g] < R_+[h] \implies R'_+[g'] < R'_+[h'].$$

Lemma 3 considers the $R_+$-order of $P'$, but leaves out the insertion position $i$. The next lemma states that the $R_+$-order at position $i$ just depends on the position $R[1]$ of element 1 in the permutation $P$.

**Lemma 4.** *Let $P'$ be a successor of $P$ with insertion position $i$ and $g$ an index of $P$, then*

$$R_+[g] < R[1] \iff R'_+[g'] < R'_+[i] \qquad \text{for all } g \in [1, n - 1].$$

After characterizing the $R_+$-order of successor permutations, we now prove that through the mapping from $P$ to an arbitrary successor permutation the number of $R_+$-descents is preserved or increased by one.

**Lemma 5.** *Let $P$ be a permutation of length $n - 1$ with $d$ $R_+$-descents and $\mathcal{P}'$ the set of successor permutations for $P$, then for all successor permutations $P'_i \in \mathcal{P}'$, we have $|\mathrm{desc}(P)| \leq |\mathrm{desc}(P'_i)| \leq |\mathrm{desc}(P)| + 1$.*

*Proof.* Due to Lemma 3, the mapping with respect to a certain insertion position $i$ does only touch the $R_+$-order of positions adjacent to $i$. Lemma 4 then adds that, through the mapping, the number of descents is preserved or increased by one, depending on whether $i$ is a descent, or not.    □

**Lemma 6.** *Let $P$ be a permutation with $d$ $R_+$-descents and $\mathcal{P}'$ the set of successor permutations for $P$, then the number of successor permutations with $d$ $R_+$-descents is $d + 1$.*

*Proof.* By considering several distinct cases it can be shown that for each $R_+$-run of $P$ there exists exactly one insertion position that does not add a descent through the mapping, and $P$ has $d + 1$ $R_+$-runs. We omit the details here.    □

**Theorem 5.** *Let $A(n, d)$ be the number of permutations of length $n$ with $d$ $R_+$-descents, then $A(n, d) = \left\langle {n \atop d} \right\rangle$.*

*Proof.*

(i) Since the permutation $(n, n - 1, \ldots, 1)$ is the only one without any $R_+$-descent, $A(n, 0) = 1$.

(ii) Obviously, the number of potential $R_+$-descents is limited by $n - 1$. Hence, there is no permutation of length $n$ with more than $n - 1$ $R_+$-descents, and thus $A(n, d) = 0$ for $d \geq n$.

(iii) As mentioned before, mapping each permutation $P$ of length $n$ to $P'_i$ with all possible insertion positions $i$ leads to $n$ successor permutations each of length $n$. If $P$ contains $d$ descents, then Lemma 6 implies: there exist exactly $d + 1$ successor permutations with $d$ descents and, according to Lemma 5, the other $n - d$ successors permutations contain $d + 1$ descents. Combining these observations leads to the recursion $A(n, d) = (d + 1)A(n - 1, d) + (n - d)A(n - 1, d - 1)$ for $0 < d < n$.

The propositions (i),(ii), and (iii) yield the same recursion as for the Eulerian numbers. Hence, $A(n, d) = \left\langle {n \atop d} \right\rangle$.    □

Bannai *et al.* [4] showed that each suffix array with $d$ $R_+$-descents can be associated with a string of at least $d + 1$ different characters. Therefore, we sum up the appropriate suffix arrays to obtain the number of suffix arrays for strings over a fixed size alphabet.

**Corollary 1.** *Let $\Sigma$ be a fixed size alphabet, $\sigma = |\Sigma|$. The number of distinct suffix arrays of length $n$ for strings over $\Sigma$ is $\sum_{d=0}^{\sigma-1} \left\langle {n \atop d} \right\rangle$.*

*Proof.* After Bannai *et al.* [4], all suffix arrays with up to $\sigma - 1$ descents have at least one string with no more than $\sigma$ characters.    □

Many application areas for suffix arrays handle small alphabets like the DNA, amino acid, or ASCII alphabet. Corollary 1 thus limits the number of distinct suffix arrays for such applications. For a DNA alphabet of size 4, for example, the number of distinct suffix arrays of length 15 is $861,948,404 = \sum_{d=0}^{3} \left\langle {15 \atop d} \right\rangle$, whereas the number of possible permutations of length 15 is $1,307,674,368,000 = 15!$ which is about $1,517$ times larger, and this difference rapidly increases for larger $n$.

Moreover, we achieve a lower bound on the compressibility of the whole information content of suffix arrays.

**Corollary 2.** *For strings of length $n$ over an alphabet of size $\sigma$, the lower bound for the compressibility of their suffix arrays in the Kolmogorov sense is* $\log \sum_{d=0}^{\sigma-1} \left\langle {n \atop d} \right\rangle$.

*Proof.* There are $\sum_{d=0}^{\sigma-1} \left\langle {n \atop d} \right\rangle$ distinct suffix arrays. Among them, there exists at least one binary representation with Kolmogorov complexity not less than $\log \sum_{d=0}^{\sigma-1} \left\langle {n \atop d} \right\rangle$. □

## 5   Summation Identities

We present constructive proofs for two long known summation identities of Eulerian numbers deduced by summing up the number of different suffix arrays for fixed alphabet size and string length. We believe that our constructive proofs are simpler than previous ones.

The identity $\sigma^n = \sum_i \left\langle {n \atop i} \right\rangle \binom{\sigma+i}{n}$, as given in [16–eq. 6.37], was proved by J. Worpitzki, already in 1883. We prove it by summing up the number of distinct strings of length $n$ over a given alphabet of size $\sigma$ for each suffix array:

$$\sigma^n = \sum_{d=0}^{\sigma-1} \left\langle {n \atop d} \right\rangle \binom{n+\sigma-d-1}{\sigma-d-1} \tag{1}$$

$$= \sum_{d=0}^{\sigma-1} \left\langle {n \atop n-1-d} \right\rangle \binom{n+\sigma-d-1}{n} \tag{2}$$

$$= \sum_{i=n-\sigma}^{n-1} \left\langle {n \atop i} \right\rangle \binom{\sigma+i}{n} \tag{3}$$

$$= \sum_i \left\langle {n \atop i} \right\rangle \binom{\sigma+i}{n}. \tag{4}$$

Equality (2) follows from the symmetry rule for Eulerian and binomial numbers, equality (3) from substituting $i = n - d - 1$, and equality (4) from $\left\langle {n \atop i} \right\rangle = 0$ for all $i \geq n$ and $\binom{a+i}{n} = 0$ for all $i < n - \sigma$, respectively.

The second summation identity, which we are concerned with, is the summation rule for Eulerian numbers to generate the Stirling numbers of the second kind [16–Eq. 6.39]: $k! \left\{ {n \atop k} \right\} = \sum_i \left\langle {n \atop i} \right\rangle \binom{i}{n-k}$. To prove this identity, we count the $k! \left\{ {n \atop k} \right\}$ strings composed of exactly $k$ different characters. Summing up these strings for each suffix array gives

$$k! \begin{Bmatrix} n \\ k \end{Bmatrix} = \sum_{d=0}^{k-1} \left\langle {n \atop d} \right\rangle \binom{n-d-1}{k-d-1} \tag{5}$$

$$= \sum_{d} \left\langle {n \atop d} \right\rangle \binom{(n-k)+(k-d-1)}{k-d-1} \tag{6}$$

$$= \sum_{d} \left\langle {n \atop n-1-d} \right\rangle \binom{n-d-1}{n-k} \tag{7}$$

$$= \sum_{i} \left\langle {n \atop i} \right\rangle \binom{i}{n-k}. \tag{8}$$

Equality (6) holds since $\left\langle {n \atop d} \right\rangle = 0$ for $d \geq k$, equality (7) follows from the symmetry rule for Eulerian and binomial numbers, and equality (8) from substituting $d = n - 1 - i$.

## 6    Conclusion

We have presented constructive proofs to count the strings sharing the same suffix array as well as the distinct suffix arrays for fixed size alphabets. For alphabets of size $\sigma$, $\binom{n+\sigma-d-1}{\sigma-d-1}$ strings share the same suffix array (with $d$ $R_+$-descents) among which $\binom{n-d-1}{\sigma-d-1}$ are composed of exactly $\sigma$ distinct characters. For these strings we have given a bijection into the set of non-decreasing sequences over $\sigma - d$ integers. The number of distinct suffix arrays is $\sum_{d=0}^{\sigma-1} \left\langle {n \atop d} \right\rangle$. This has yielded a $\log \sum_{d=0}^{\sigma-1} \left\langle {n \atop d} \right\rangle$ lower bound for the compressibility of such suffix arrays.

Moreover, summing up the number of strings for each suffix array yields constructive proofs for Worpitzki's identity and for the summation rule of Eulerian numbers to generate the Stirling numbers of the second kind, respectively. One could also say the number of suffix arrays and its strings form a particular instance of these identities.

Of further interest will be the development of efficient enumeration algorithms for which our constructive proofs have already suggested suitable methods. For the enumeration of strings sharing the same suffix array, we have proved the equivalence to the enumeration of non-decreasing sequences which can be easily performed in optimal time, whereas the enumeration of distinct suffix arrays in optimal time requires further development.

*Acknowledgments.* We thank Veli Mäkinen, Hans-Michael Kaltenbach, and Constantin Bannert for helpful discussions.

## References

1. Manber, U., Myers, E.W.: Suffix arrays: A new method for on-line string searches. SIAM Journal on Computing **22** (1993) 935–948
2. Gonnet, G.H., Baeza-Yates, R.A., Snider, T.: New indices for text: Pat trees and pat arrays. In Frakes, W.B., Baeza-Yates, R.A., eds.: Information retrieval: data structures and algorithms. Prentice-Hall (1992) 66–82

3. Abouelhoda, M.I., Kurtz, S., Ohlebusch, E.: Replacing suffix trees with enhanced suffix arrays. Journal of Discrete Algorithms **2** (2004) 53–86
4. Bannai, H., Inenaga, S., Shinohara, A., Takeda, M.: Inferring strings from graphs and arrays. In: Proceedings of the 28th International Symposium on Mathematical Foundations of Computer Science (MFCS 2003). Volume 2747 of LNCS., Springer Verlag (2003) 208–217
5. Burkhardt, S., Kärkkäinen, J.: Fast lightweight suffix array construction and checking. In: Proceedings of the 14th Annual Symposium on Combinatorial Pattern Matching (CPM 2003). Volume 2676 of LNCS., Springer Verlag (2003) 55–69
6. Duval, J.P., Lefebvre, A.: Words over an ordered alphabet and suffix permutations. RAIRO – Theoretical Informatics and Applications **36** (2002) 249–259
7. Burrows, M., Wheeler, D.J.: A block-sorting lossless data compression algorithm. Technical Report 124, Digital System Research Center (1994)
8. Grossi, R., Vitter, J.S.: Compressed suffix arrays and suffix trees with applications to text indexing and string matching. In: Proceedings of the 32nd Annual ACM Symposium on Theory of Computing (STOC 2000). (2000) 397–406
9. Ferragina, P., Manzini, G.: Opportunistic data structures with applications. In: Proceedings of the 41st Annual Symposium on Foundations of Computer Science (FOCS 2000), IEEE Computer Society (2000) 390–398
10. Sadakane, K.: Compressed text databases with efficient query algorithms based on the compressed suffix array. In: Proceedings of the 11th International Symposium on Algorithms and Computation (ISAAC 2000). Volume 1969 of LNCS., Springer Verlag (2000) 410–421
11. He, M., Munro, J.I., Rao, S.S.: A categorization theorem on suffix arrays with applications to space efficient text indexes. In: Proceedings of the 16th Annual ACM-SIAM Symposium on Discrete Algorithms (SODA 2005), SIAM (2005) 23–32
12. Demaine, E.D., López-Ortiz, A.: A linear lower bound on index size for text retrieval. Journal of Algorithms **48** (2003) 2–15
13. Miltersen, P.B.: Lower bounds on the size of selection and rank indexes. In: Proceedings of the 16th Annual ACM-SIAM Symposium on Discrete Algorithms (SODA 2005), SIAM (2005) 11–12
14. Crochemore, M., Désarménien, J., Perrin, D.: A note on the Burrows-Wheeler transformation. Theoretical Computer Science **332** (2005) 567–572
15. Schürmann, K.B., Stoye, J.: Counting suffix arrays and strings. Technical Report 2005-04, Technische Fakultät, Universität Bielefeld, Germany (2005)
16. Graham, R.L., Knuth, D.E., Patashnik, O.: Concrete Mathematics. Second edn. Addison-Wesley (1994)
17. Moore, D., Smyth, W.F., Miller, D.: Counting distinct strings. Algorithmica **23** (1999) 1–13

# Towards Real-Time Suffix Tree Construction

Amihood Amir[1,2,*], Tsvi Kopelowitz[1],
Moshe Lewenstein[1,**], and Noa Lewenstein[3]

[1] Department of Computer Science, Bar-Ilan University, Ramat-Gan 52900, Israel
[2] College of Computing, Georgia Tech, Atlanta GA 30332-0280
[3] Department of Computer Science, Netanya College, Israel

**Abstract.** The quest for a real-time suffix tree construction algorithm is over three decades old. To date there is no convincing understandable solution to this problem. This paper makes a step in this direction by constructing a suffix tree online in time $O(\log n)$ per every single input symbol. Clearly, it is impossible to achieve better than $O(\log n)$ time per symbol in the comparison model, therefore no true real time algorithm can exist for infinite alphabets. Nevertheless, the best that can be hoped for is that the construction time for every symbol does not exceed $O(\log n)$ (as opposed to an amortized $O(\log n)$ time per symbol, achieved by current known algorithms). To our knowledge, our algorithm is the first that spends in the *worst case* $O(\log n)$ per every *single* input symbol.

We also provide a simple algorithm that constructs online an indexing structure (the *BIS*) in time $O(\log n)$ per input symbol, where $n$ is the number of text symbols input thus far. This structure and fast LCP (Longest Common Prefix) queries on it, provide the backbone for the suffix tree construction. Together, our two data structures provide a searching algorithm for a pattern of length $m$ whose time is $O(\min(m \log |\Sigma|, m + \log n) + tocc)$, where $tocc$ is the number of occurrences of the pattern.

## 1   Introduction

*Indexing* is one of the most important paradigms in searching. The idea is to preprocess the text and construct a mechanism that will later provide answer to queries of the the the form "does a pattern $P$ occur in the text" in time proportional to the size of the *pattern* rather than the text. The suffix tree [7,13,16,17] and suffix array [11,12] have proven to be invaluable data structures for indexing.

One of the intriguing questions that has been plaguing the algorithms community is whether there exists a *real time* indexing algorithm. An algorithm is *online* if it accomplishes its task for the $i$th input without needing the $i + 1$st input. It is *real-time* if, in addition, the time it operates between inputs is a constant. While not all suffix trees algorithms are online (e.g. McCreight, Farach)

---

\* Partly supported by NSF grant CCR-01-04494 and ISF grant 282/01.
\*\* Partially supported by an IBM Faculty Fellowship.

M. Consens and G. Navarro (Eds.): SPIRE 2005, LNCS 3772, pp. 67–78, 2005.
© Springer-Verlag Berlin Heidelberg 2005

some certainly are (e.g. Weiner, Ukkonen). Nevertheless, the quest for a real-time indexing algorithm is over 30 years old. It should be remarked that Weiner basically constructs an online reverse prefix tree. In other words, to use Weiner's algorithm for online indexing queries, one would need to reverse the pattern. For real-time construction there is some intuition for constructing prefix, rather than suffix trees, since the addition of a single symbol in a suffix tree may cause $\Omega(n)$ changes, whereas this is never the case in a prefix tree.

It should be remarked that for infinite alphabets, no real-time algorithm is possible since the suffix tree can be used for sorting. All known comparison-based online suffix tree construction algorithms have amortized time $O(n \log n)$, for suffix tree or suffix array construction, and $O(m \log n + tocc)$ (in suffix trees), or $O(m + \log n + tocc)$ (in suffix arrays) search time. The latter uses non-trivial pre-processing for LCP (longest common prefix) queries. However, the best that can be hoped for (but not hitherto achieved) is an algorithm that pays at least $\Omega(\log n)$ time for every *single* input symbol.

The problem of *dynamic indexing*, where changes can be made anywhere in the text has been addressed as well. Real-time indexing can be viewed as a special case of dynamic indexing, where the change is insertions and deletions at the end (or, symmetrically, the beginning) of the text. If dynamic indexing could be done in constant time per update, then it would result in a real-time indexing algorithm. Sahinalp and Vishkin [15] provide a dynamic indexing where updates are done in time $O(\log^3 n)$. This result was improved by Alstrup, Brodal and Rauhe [14] to an $O(\log^2 n \log \log n \log^* n)$ update time and $O(m + \log n \log \log n + tocc)$ search time. To date, the pattern matching community still seeks a simple algorithm for the *real-time indexing* problem. The motivation for real-time indexing is the case where the data arrives in a constant stream and indexing queries are asked while the data stream is still arriving. Clearly a real-time suffix tree construction answers this need.

Grossi and Italiano [9] provide a dynamic data structure for online handling of multidimensional keys. Their algorithm needed to be further tuned to allow indexing of a single text. Such improvement was made by Franceschini and Grossi [8]. However, their algorithm uses complex data structures for handling LCP and does not provide a suffix tree.

Our contribution is the first algorithm for online suffix tree construction over unbounded alphabets where the worst case processing per input symbol is $O(\log n)$, where $n$ is the length of the text input so far. Furthermore, the search time for a pattern of length $m$ is $O(\min(m \log |\Sigma|, m + \log n))$, where $\Sigma$ is the alphabet. This matches the best times for amortized algorithms in the comparison model.

Similar complexity results for searches on a suffix list can also be achieved by the techniques of Franceschini and Grossi [8]. They give a full constant-time LCP query mechanism. We present a data-structure that avoids their heavy machinery and solves the problem more simply and directly. In some sense we use the simpler techniques of [9] to solve the indexing problem. In addition, this is the first known algorithm that builds a suffix tree on-line with time $O(\log n)$ per input symbol.

Our paper is divided into two main thrusts. In the first we construct a *balanced indexing data structure* which enables insertions (and deletions) in time $O(\log n)$. For the $O(m+\log n+tocc)$ search it was necessary to devise an efficient dynamic LCP mechanism.

The more innovative part is the second thrust of the paper, where we maintain and insert incoming symbols into a suffix tree in time $O(\log n)$ per symbol. We employ some interesting observations that enable a binary search on the path when a new node needs to be added to the suffix tree.

We note that deletions of letters from the beginning of the text can also be handled within the same bounds. This is left for the journal version. Also, all of the omitted proofs are left for the journal version.

## 2   Preliminaries and Definitions

Consider a text $S$ of length $n$ and let $S_1, \cdots, S_n$ be the suffixes of $S$. We describe a new data structure for indexing, which we call the *Balanced Indexing Structure*, *BIS* for short. The Balanced Indexing Structure has as its basic elements, $n$ nodes, one for each suffix $S_i$. The node corresponding to the suffix $S_i$ is denoted by $node(S_i)$. The suffix associated with a given node $u$ (or the suffix that $u$ represents) is denoted by $suffix(u)$. However, when clear from the context, we allow ourselves to abuse notation and to refer to node $u$ as the suffix it represents.

The BIS data structure will incorporate three data structures on these nodes. (1) An ordered list of the suffixes in lexicographic order, as in a suffix array or the leaves of a suffix tree, (2) a balanced search tree on the suffixes, where the node ordering is lexicographic (we specifically use AVL trees [1] - although other binary search trees can also be used such as red-black trees [3] and B-trees [2]) and (3) a textual ordering of the suffixes.

Within the BIS data structure we maintain pointers on the nodes to capture the three data structures we just mentioned, namely:

1. Suffix order pointers: next(u) and previous(u) representing the node $v$ that is the lexicographic successor and predecessor of $suffix(u)$, respectively.
2. Balanced tree pointers: left(u), right(u), and parent(u).
3. Text pointers textlink(u) - if node $u$ represents suffix $ax$ then textlink(u) is the node $v$ representing suffix $x$.

A node $u$ is called a *leaf* if it is a leaf with respect to the balanced tree. A node $u$ is called the *root* if it is the root with respect to the balanced tree. Likewise, when we refer to a subtree of a node $u$ we mean the subtree of node $u$ with respect the balanced tree. A node $u$ is said to be a *descendant* of $v$ if it appears in $v$'s subtree. A leaf $v$ is said to be the *rightmost*, or *leftmost*, leaf of a node $u$ if for all nodes $x$ such that $x$ is in $u$'s subtree and $v$ is in $x$'s subtree, $v$ is in the subtree of $right(x)$, respectively $left(x)$. A node $u$ is said to be between nodes $x$ and $y$ if $x$ is $u$'s lexicographic predecessor and $y$ is $u$'s lexicographic successor.

We now state a couple of connections between the lexicographic ordered list and the balanced tree that will be useful in handling the BIS data structure. The following observations can be easily proven and hence we omit their proofs.

**Observation 1.** *Let $u$ be a node that is not a leaf. Then the rightmost leaf in the subtree of left(u) is previous(u) and the leftmost leaf in the subtree of right(u) is next(u).*

**Observation 2.** *Let $r$ be a node. Then (1) if $u = left(r)$, then $v$ is between $u$ and $r$ iff $v$ is in $u$'s right subtree, and (2) if $u = right(r)$, then $v$ is between $u$ and $r$ iff $v$ is in $u$'s left subtree.*

We also maintain a copy of the text and point out that although the text changes dynamically by insertions to its head it can be dynamically maintained with $O(1)$ time per operation by using standard doubling and halving de-amortization techniques. Likewise we directly index into the text taking into account that the insertions are done at the head of the text.

We note that auxiliary information will still be needed within the BIS data structure for the queries. We show how this information is stored and maintained in Section 4.

## 3   BIS Operations

Obviously, using a suffix-array the suffix order pointers can be constructed in linear time, from which the balanced tree pointers can be created. The text pointers are easy to create. However, the purpose of the BIS is to support indexing when the text is online. So, our goal is to support $Addition(a, x)$ (updating the data structure for $x$ into the data structure for $ax$, and $Query(p, x)$ (finding all of the occurrences of $p$ in $x$). In this section we show how to support the addition operation and in the next we show how to handle queries.

### 3.1   Addition Operation

Assume we have the BIS for string $\bar{T} = x$ of size $n$, and we want to update the BIS to be for $T = a\bar{T} = ax$ where $a \in \Sigma$ by inserting the new suffix $ax$ into the data structure. To do so, we traverse the BIS on its balanced tree pointers, with $ax$, starting from the root. We make a lexicographic comparison between $ax$ and the suffix represented by each node and continue left or right accordingly. We continue comparing until we reach a leaf (we must reach a leaf as any node $u$ in the BIS, and in particular the nodes on the traversal path, have $u \neq ax$, as $ax$ is the longest suffix). Assume the suffix associated with the leaf is $S$. So from the properties of binary search trees we know that either $ax$ is between $S$ and $next(S)$, or between $previous(S)$ and $S$. By comparing $ax$ with $S$ we can know which of the two options is correct, and then we can insert $ax$ into the list and into the balanced tree (maintaining balancing).

In order for us to compare $ax$ with a suffix of $\bar{T}$ we can note that the order is determined either by the first letters of the strings, and if those letter are the same, then using the textlinks and the data structure presented by Dietz and Sleator in [6] (or the simplified data structures of Bender, Cole, Demaine, Farach-Colton and Zito [4]) we can compare the two strings in constant time. Of course, when inserting a new suffix, we must also insert it into the Dietz and Sleator data structure, and balance the BIS. Full details will be prvided in the journal version. This gives us the following:

**Theorem 3.** *We can maintain the BIS of an online text under additions in* $O(\log n)$ *worst-case time per update, where $n$ is the size of the current text.*

## 4    Navigating and Querying Balanced Indexing Structures

In the previous section we saw how to construct an online BIS. In this section we show how to answer indexing queries on the BIS. An indexing *query* is a pattern $P = p_1 p_2 \cdots p_m$ for which we wish to find all occurrences of $P$ in a text $T$.

We use the BIS of $T$ in a similar way to that of the query on a suffix array, which takes $O(m + \log n + tocc)$ time. When answering a query using the suffix array we preprocess information in order to be able to answer *longest common prefix (LCP)* queries between any two suffixes in constant time. One seeks the pattern in a binary search of the suffix array, where the LCP is used to speed up the comparison and decide the direction of the binary search. The following lemma can be easily proven and is crucial for answering queries in a suffix array.

**Lemma 1.** *For string $T = t_1 t_2 ... t_n$ let $\pi(i)$ denote the rank of the $i^{th}$ suffix of $T$ in the lexicographic ordering of the suffixes of $T$. Then for $\pi(i) < \pi(j)$:*

$$lcp(T_{\pi(i)}, T_{\pi(j)}) = min_{\pi(i) \leq k < \pi(j)} lcp(T_k, T_{k+1})$$

**Corollary 1.** *For string $T = t_1 t_2 ... t_n$ let $\pi(i)$ denote the rank of the $i^{th}$ suffix of $T$ in the lexicographic ordering of the suffixes of $T$. Then for any $\pi(i) < \ell < \pi(j)$:*

$$lcp(T_{\pi(i)}, T_{\pi(j)}) = min(lcp(T_{\pi(i)}, T_\ell), lcp(T_\ell, T_{\pi(j)}))$$

However, there are two issues that need to be considered in our online setting. The first is how to perform a binary-like search on the ordered list of suffixes, and the second is how to implement the LCP queries (being that we cannot preprocess in an online setting). For the first we use the induced balanced tree of the BIS to perform a binary-like search on the ordered list of suffixes (we elaborate on this later). For the second we maintain the following information for each node $u$ in the BIS: the LCP between the leftmost leaf and the rightmost leaf of the subtree of $u$(denoted by $lcpsub(u)$), and the LCP between $u$ and $previous(u)$. We first show how this information helps us answer indexing queries in $O(m + \log n + tocc)$ time, and then, how to maintain this information through additions.

### 4.1    Answering Indexing Queries

We are given a pattern $P = p_1 p_2 \cdots p_m$ and wish to find all of the occurrences of $P$ in $T$ using the ordered list of suffixes of $T$. As we mentioned before, the idea is to perform a binary-like search on the ordered list of suffixes using the induced balanced tree of the BIS. We begin with the following lemmas that will lead to our query algorithm.

**Lemma 2.** *Let $r$ be a node in the BIS of $T$, $u$ a node in $left(r)$'s subtree ($right(r)$'s subtree, respectively), and $k = lcp(r, P)$. If all of the occurrences of $P$ in $T$ correspond to nodes which are in the subtree of $u$, then: (1) if $lcp(r, u) < k$ then all occurrences of $P$ in $T$ correspond to nodes in the subtree of $right(u)$ (subtree of $left(u)$, respectively), and (2) if $lcp(r, u) > k$, then all occurrences of $P$ in $T$ correspond to nodes in the subtree of $left(u)$ (subtree of $right(u)$, respectively).*

We now present a recursive procedure which answers an indexing query.

**IndexQuery(node $root$, string $P$)**

1. for($i = 1; i < m; i + +$)
   (a) if $p_i < t_{location(root)+i-1}$ then return IQR($u$,$left(root)$,P,$i − 1$)
   (b) if $p_i > t_{locatin(root)+i-1}$ then return IQR($u$,$right(root)$,P,$i − 1$)
2. return $root$

**IQR(node $r$, node $u$, string $P$, int $k$)**

1. $\ell \leftarrow lcp(r, u)$.
2. if $\ell = k$ then
   (a) for($i = k + 1; i < m; i + +$)
       i. if $p_i < t_{location(u)+i-1}$ then return IQR($u$,$left(u)$,P,$i − 1$)
       ii. if $p_i > t_{location(u)+i-1}$ then return IQR($u$,$right(u)$,P,$i − 1$)
   (b) return $u$
3. if $u$ is a leaf then return NIL
4. if $u$ is in $r$'s left subtree then
   (a) if $\ell < k$ then return IQR($r$,right($u$),P,k)
   (b) else then return IQR($r$,left($u$),P,k)
5. if $u$ is in $r$'s right subtree then
   (a) if $\ell < k$ then return IQR($r$,left($u$),P,k)
   (b) else then return IQR($r$,right($u$),P,k)

**Theorem 4. (Correctness:)** *The algorithm IndexQuery(root, P), where root is the root of the BIS of $T$ and $P$ is the query pattern, returns NIL if $P$ does not occur in $T$, and otherwise, it returns a node $v$ such that $P$ matches $T$ at location($v$), and all other occurrences of $P$ in $T$ correspond to suffixes with nodes in the subtree of $v$.*

Now, in order to analyze the running time of IndexQuery, we need the following lemma regarding the time it takes to calculate the LCP in step 1 of the IQR procedure.

**Lemma 3.** *Step 1 in IQR can be implemented to take constant time.*

**Lemma 4. (Running Time:)** *The procedure IndexQuery takes has running time $O(m + \log n)$.*

**Theorem 5.** *Let $T$ be a text of size $n$, $P$ be a pattern of size $m$, $G$ be the BIS for $T$, and root be the root of $G$. Denote by occ the number of occurrences of $P$ in $T$. Then there exists an algorithm which runs in $O(m + \log n + occ)$ time, returns NIL if $occ = 0$, and otherwise, it returns all occ nodes in $G$ such that $P$ matches $T$ at the locations of those nodes.*

*Proof.* The IndexQuery and IQR can easily be manipulated to find all occurrences in the desired upper bound.                                                   □

### 4.2  Answering LCP Queries and Maintaining the LCP Data

As we will soon see, in order to be able to maintain the LCP data, we need to be able to answer LCP queries between any two suffixes $u$ and $v$ in the BIS in $O(\log n)$ time. This can be done by finding the least common ancestor of $u$ and $v$ in the balanced tree (denote this node by $R_{u,v}$), and then use Corollary 1 and Lemma 3 to calculate the LCp in $O(\log n)$ time. The full details are left for the journal version.

Now, when inserting the suffix $ax$ into the BIS, ordered after $S = s_1 s_2 ... s_r$ and before $S' = s'_1 s'_2 ... s'_{r'}$ we need to update $lcp(node(ax), previous(node(ax)))$, and $lcp(node(S'), previous(node(S')))$. If $S = ay$ then $lcp(S, ax) = 1 + lcp(x, y)$ which we showed can be calculated in $O(\log n)$ time. If $s_1 \neq a$ then $lcp(S, ax) = 0$. The computation of $lcp(node(S'), previous(node(S')))$ is done in a symmetric manner. Regarding $lcpsub$ for the nodes in the BIS, we note that the nodes affected due to the new addition are only the nodes that are ancestors of the new node in the BIS. Thus, we can traverse up the BIS from the new node, and update the LCP data in $O(\log n)$ time. Of course, we must also balance the BIS. Again, the full details are left for the journal version.

## 5  Online Construction of Suffix Trees

We now present an online procedure which builds a suffix tree in $O(\log n)$ time per insertion of a letter to the beginning of the text. Our solution uses the BIS for the construction. We already showed that the BIS can be constructed in $O(\log n)$ time per addition. This is now extended to the suffix tree construction.

### 5.1  Suffix Tree Data

We first describe the relevant information maintained within each inner node in the suffix tree. Later we will show how to maintain it through insertions. For a static text, each node in the suffix tree has a maximum outdegree of $|\Sigma|$ where $\Sigma$ is the size of the alphabet of the string  We use a balanced search tree for each node in the suffix tree (not to be confused with the balanced search tree from the previous section). Each such balanced search tree contains only nodes corresponding to letters of outgoing edges. This gives linear space, but we spend $O(\log |\Sigma|)$ time locating the outgoing edge at a node. We can use this solution in the online scenario because balanced search trees allow us to insert and delete new edges dynamically. Therefore, we assume that the cost of adding or deleting an outgoing edge is $O(\log |\Sigma|)$. The time for locating an outgoing edge is also $O(\log |\Sigma|)$. Note that we always have $|\Sigma| \leq n$. Hence, if during the process of an addition we insert or remove a constant number edges in the suffix tree, the total cost is bounded by $O(\log n)$ as needed. In addition, for each node in the suffix tree we maintain the length of the string corresponding to the path from the root to that node. For a node $u$ in the suffix tree we denote this length by $length(u)$. We also note that it is possible to maintain suffix links through insertions, which have proven useful in several applications [10]. The details are left for the journal version.

For the sake of completeness we note that many of the operations done on suffix tree (assuming we want linear space) use various pointers to the text in order to save space for labelling the edges. We leave such issues for the journal paper. We also note that a copy of the text can be saved in array format may be necessary for various operations, requiring direct addressing. This can be done with constant time update by standard deamortization techniques.

## 5.2    Some Simple Cases

We now proceed to the online construction of the suffix tree. We begin by looking at some special cases of additions which will give us a better understanding towards the more general scenarios. It is a known fact that a depth first search (DFS) on the suffix tree encounters the leaves, which correspond to suffixes, in lexicographic order of the suffixes. Hence, the leaves of the suffix tree in the order encountered by the DFS form the lexicographic ordering of suffixes, a part of the BIS. So, upon inserting $ax$ into the tree, $node(ax)$ is inserted as a leaf, and according to the lexicographic ordering of the suffixes, we know between (as defined by the DFS) which two leaves of the suffix tree $node(ax)$ will be inserted. If the LCP between $ax$ and each of its neighboring suffixes in the lexicographic order of the suffixes equals zero, then the letter $a$ at the beginning of the text appears *only* at the beginning of the text. In such a case we add an outgoing edge from the root of the suffix tree to $node(ax)$ and we are done. So we assume that the LCP between $ax$ and at least one of its neighbors is larger than zero. If the LCP between $ax$ and only one of its neighboring suffixes in the lexicographic order of the suffixes is more than zero, we know that in the suffix tree for the string $ax$, that neighbor and $node(ax)$ share a path from the root till their least common ancestor (LCA) in the suffix tree. This is because the LCA is equivalent to the LCP between them. If the LCP between $ax$ and both of its neighboring suffixes in the lexicographic order of the suffixes is more than zero, the neighbor whose suffix has the larger LCP with $ax$ has a longer common path with $node(ax)$ in the suffix tree for the string $ax$, and what we need to do is find the location on that neighbor's path (from the root) that splits towards $node(ax)$. If we can find this location efficiently we will be done. This is discussed in the next subsection.

## 5.3    A More General Scenario - Looking for the Entry Node

Continuing the more general case from the previous subsection, we assume w.l.o.g., that $node(ax)$ is inserted between $node(ay)$ and $node(z)$, where $x$ is lexicographically bigger than $y$ (hence $node(ax)$ follows $node(ay)$ in the lexico-graphic order of the suffixes), and $ax$ is lexicographically smaller than $z$ (hence $node(z)$ follows $node(ax)$ in the lexicographic order of the suffixes). Also, we assume ,w.l.o.g., that $node(ax)$ will be inserted into the path from the root of the suffix tree to $node(ay)$, (meaning that $lcp(ax, ay) \geq lcp(ax, z)$). All of the lemmas and definitions in the rest of this section are under these assumptions. The other possible scenarios use a similar method.

**Definition 1.** *Let $x$ be a string, and $a$ be a character. Then the entry node of $node(ax)$ in $ST(x)$ is the node $v$ (in $ST(x)$) of maximal length such that $label(v)$ is a prefix of $ax$.*

We denote by $v$ the entry node for $node(ax)$. First, note that $v$ is on the path from the root of $ST(x)$ to $node(ay)$. Also, note that the LCA of $node(ay)$ and $node(z)$ is either $v$ or an ancestor $v$ (due to the maximality of $v$). Finally, note that when we add $node(ax)$ to $ST(x)$ (in order to get $ST(ax)$), all of the changes made in the suffix tree are in close vicinity of $v$. Specifically, the only nodes or edges which might change are $v$ and its outgoing edges. This is because $node(ax)$ will enter into the subtree of $v$, and for each son of $v$ in $ST(x)$, $node(ax)$ cannot be in its subtree because the label of the son is not a prefix of $ax$. So our first task will be to find $v$ which will be the place of entry for $node(ax)$ (we explain how to do this in the next subsection). After finding $v$, the two options at hand will be either to add $node(ax)$ as one of $v$'s sons, or break one of $v$'s outgoing edges into two, creating a new node that will be the parent of $node(ax)$.

In order for us to find $v$ we note that if $k = lcp(ay, ax)$ (we can calculate this in $O(\log n)$ time, as shown above), then $v$ is the node with maximum length on the path from the root of the suffix tree to $node(ay)$ for which $length(v) \leq k$. Specifically, if we have equality, then $v$ will be the parent of $node(ax)$, and if there isn't equality, then we must break an edge for the parent of $node(ax)$. This observation leads us to the following lemma, which will help us find $v$ efficiently.

**Lemma 5.** *Let $x$ be a string, $a$ be a character, and $v$ be the entry node for $node(ax)$ in $ST(x)$. If $v \neq LCA(node(ay), node(z)$ in $ST(x)$, then there exists a leaf $u$ in the suffix tree such that $v$ is the LCA $u$ and $node(ay)$, and $u$ is to the left of $node(ay)$ in the lexicographic order of the suffixes. We call the rightmost such node in the lexicographic order of the suffixes the beacon of $node(ax)$.*

*Proof.* Note that if $v$ is not the LCA of $node(ay)$ and $node(z)$, then $node(ay)$ is the right most leaf in $v$'s subtree. This is because, as mentioned above, the LCA of $node(ay)$ and $node(z)$ is either $v$ (which we assume is not the case) or a parent of $v$. Hence, $node(z)$ is not in $v$'s subtree. There obviously exists another leaf $u$ in $v$'s subtree for which the LCA of $u$ and $node(ay)$ is $v$, because the outdegree of $v$ is at least two (by the definition of the suffix tree). □

## 5.4   Searching for the Entry Node

We now consider the task of finding the entry node $v$. We first find the LCA of $node(ay)$ and $node(z)$ in $ST(x)$. Being that the LCP of any two suffixes $S$ and

$S'$ corresponds to the length in the suffix tree of the LCA of those two suffixes, the maintenance of the data needed to answer LCA queries is done in a similar method to that of maintaining the lcp-data. Thus, we can locate the LCA of any two nodes already in the suffix tree in $O(\log n)$. We note that other more complicated solutions exist (Cole and Hariharan [5]), but they are not needed in order to maintain our $O(\log n)$ upper bound.

After finding the LCA of $node(ay)$ and $node(z)$ in $ST(x)$, we can quickly check whether this is the entry node $v$ that we are looking for in the following manner. We simply check whether the length of the LCA is exactly $k$, and if not (then it must be less - it cannot be more), in $O(\log(\Sigma))$ we can find the outgoing edge from the LCA leading towards $node(ay)$, and in constant time we can check the length of the node on the other side of that edge - we check if the length of this node is more than $k$. If both fail, $v$ is not the LCA of $node(ax)$ and $node(z)$ in $ST(x)$, and we continue as follows in the next paragraph.

If $v$ is not the LCA of $node(ay)$ and $node(z)$ in $ST(x)$, then from Lemma 5 there exists a node $u$ which is the beacon for $node(ax)$. Once we find $u$, we know that $v$ is the LCA of $u$ and $node(ay)$ in $ST(x)$, and $u$ is a leaf. So, as we previously mentioned, we can find $v$ in $O(\log n)$ time. Observe that for any node $w$ between $u$ and $node(ay)$, $lcp(w, ay) > k$. This is because $lcp(next(u), ay) > k$ and Lemma 1. With this observation in hand we can use the balanced tree pointers in order to find the beacon $u$ using the following method.

Recall that a subtree refers to the subtree of the node in the $BIS$, unless stated otherwise. We separate the task of finding $u$ into two. If $u$ is in the subtree rooted by $node(ay)$, then $u$ must be in the subtree of $left(node(ay))$. If $u$ is not in the subtree rooted by $node(ay)$, then we observe that there exists some subtree that has both $node(ay)$ and $u$ in it. Specifically, if we traverse the path from $node(ay)$ to the root of the subtree, until we reach the first node $w$ that has $u$ in its subtree, then $u$ is in the subtree of $left(w)$, and $node(ay)$ is in the subtree of $right(w)$. This is because if both $u$ and $node(ay)$ are in the subtree of $left(w)$ or $right(w)$, then $w$ would not be the first node that has $u$ in its subtree when traversing the path from $node(ay)$ to the root. So, now we observe that $w$ is lexicographically between $u$ and $node(ay)$ and if $w \neq u$, then we know that $lcp(u, ay) = lcp(w, u)$. This is because by definition of the beacon, $LCA(w, node(ay))$ in $ST(x)$ is an internal node on the path from $w$ $LCA(u, node(ay))$ in $ST(x)$, thus $lcp(w, ay) > k$, and being that $k = lcp(u, ay) = min(lcp(u, w), lcp(w, ay))$ we have $lcp(u, ay) = lcp(w, u) = k$. Using the last two observations, we can traverse up the tree, spending constant time at each node (We look for the first node $w'$ on the path from $node(ay)$ to the root for which $lcpsub(left(w')) \leq k$), until we reach $w$.

Now we are left with the situation where we have a node $w$, the beacon $u$ is in the subtree of $left(w)$, and $lcp(u, ay) = lcp(w, u)$. Specifically, if $u$ is in the subtree of $node(ay)$, then $w = node(ay)$, and otherwise, $w$ is the first node that has $u$ in its subtree when traversing the path from $node(ay)$ to the root. Now, we begin searching downwards in the subtree rooted by $left(w)$, looking for $u$. We use the following recursive procedure in order to achieve this task.

The recursive procedure uses the node-variables $r$ and $w$. First, we assume the following invariants in our recursive procedure: $u$ is in the subtree rooted by $r$, the right most descendant in the subtree rooted by $right(r)$ is $previous(w)$, and the LCA in the suffix tree of $w$ and $u$ is $v$ (hence $lcp(u, ay) = lcp(w, u)$). We initialize $w = w$ and $r = left(w)$ - the invariants obviously hold. The recursion is as follows.

**FindBeacon(node $r$, node $w$)**

1. if $min(lcpsub(root(right(r))), lcp(w, previous(w)), lcp(r, next(r))) > k$ then return FindBeacon($left(r)$,$r$)
2. if $min(lcpsub(root(B')), lcp(w, previous(w))) \leq k$ then return FindBeacon ($right(r)$,w)
3. return $r$

In the first step we check whether $u$ is in the subtree rooted by $left(r)$. We do this by looking at $lcp(r, w)$. If it is more than $k$ then we know that $u$ is to the right of $r$. So we continue recursively to the subtree of $left(r)$ (the invariants obviously hold). In the second step (assuming $u$ is not in the subtree rooted by $left(r)$) we check whether $u$ is in the subtree rooted by $right(r)$. we do this by looking at $lcp(next(r), w)$. If it is not greater than $k$, this means that $r$ is not $u$, because it is not the rightmost in the lexicographic order of the suffixes which is to the left of $node(ay)$ for which the lcp is less than or equal to $k$. So, we continue searching in the subtree of $right(r)$. finally, if we reach the third step, we know from the invariant that $u$ is in the subtree rooted by $r$, but due to steps one and two, $u$ is not in the subtree rooted by either of $r$'s sons. So it must be that $u = r$. To complete the correctness of the procedure we note that in the case we reach a leaf (during the recursion), then due to the invariant, that leaf must be $u$ - so the algorithm will always return our beacon.

## 5.5    Final Touch

After finding $u$, we can find the LCA of $u$ and $node(ay)$ in $O(\log n)$ (as mentioned before), giving us the entry node $v$. After we have found $v$, we wish to insert $node(ax)$ into the suffix tree. If $length(v) = k$, then we simply add an outgoing edge to $v$ entering $node(ax)$. If $length(v) < k$ then we must break the edge leading from $v$ to $node(ay)$ to two, creating a new node with length $k$, and this new node has a new outgoing edge which connects to $node(ax)$.

The time required to update the suffix tree due to an addition operation comes from performing a constant number of LCA queries, traversing up and down the balanced search tree, and performing a constant number of searches, additions, and removal of outgoing edges. This takes a total of $O(\log n)$ time.

Searching for a pattern on the suffix tree can be done in the usual manner in time $O(m \log |\Sigma|)$ by following the path from the suffix tree root. It is then possible to find all occurrences by DFS on the subtree rooted at the suffix tree node corresponding to the pattern, or, if the pattern ends in an edge, the tail of that edge. The size of this subtree is $O(number\ of\ leaves)$ because the suffix tree has no degree-one nodes.

# References

1. G.M. Adelson-Velskii and E.M. Landis. An algorithm for the organizaton of information. *Soviet Math. Doklady*, 3:1259–1263, 1962.
2. R. Bayer. Symetric Binary B-trees: Data structure and maintenance algorithms. In *Acta Informatica*, 1:290–306, 1972.
3. R. Bayer, E. M. McCreight. Organization and maintenance of large ordered indexes, In *Acta Informatica*, 1(3):173–189, 1972.
4. M. Bender, R. Cole, E. Demaine, M. Farach-Colton, and J. Zito. Two simplified algorithms for maintaining order in a list. In *Proc. 10th Annual European Symposium on Algorithms (ESA 2002)*, pages 152–164, 2002.
5. R. Cole and R. Hariharan. Dynamic lca queries in trees. In *Proc. 10th ACM-SIAM Symposium on Discrete Algorithms (SODA)*, pages 235–244, 1999.
6. P.F. Dietz and D.D. Sleator. Two algorithms for maintaining order in a list. In *Proc. 19th ACM Symposium on Theory of Computing (STOC)*, pages 365–372, 1987.
7. M. Farach. Optimal suffix tree construction with large alphabets. *Proc. 38th IEEE Symposium on Foundations of Computer Science*, pages 137–143, 1997.
8. G. Franceschini and R. Grossi. A general technique for managing strings in comparison-driven data structures. In *Proc. 31 Intl. Col. on Automata, Languages and Programming (ICALP)*, pages 606–617, 2004.
9. R. Grossi and G. F. Italiano. Efficient techniques for maintaining multidimensional keys in linked data structures. In *Proc. 26th Intl. Col. on Automata, Languages and Programming (ICALP)*, pages 372–381, 1999.
10. Dan Gusfield. *Algorithms on Strings, Trees, and Sequences: Computer Science and Computational Biology.* Cambridge University Press, 1997.
11. Juha Kärkkäinen and Peter Sanders. Simple linear work suffix array construction. In *Proc. 30th International Colloquium on Automata, Languages and Programming (ICALP 03)*, pages 943–955, 2003. LNCS 2719.
12. U. Manber and G. Myers. Suffix arrays: A new method for on-line string searches. In *Proc. 1st ACM-SIAM Symp. on Discrete Algorithms (SODA)*, pages 319–327, 1990.
13. E. M. McCreight. A space-economical suffix tree construction algorithm. *J. of the ACM*, 23:262–272, 1976.
14. T. Rauhe S. Alstrup, G. S. Brodal. Pattern matching in dynamic texts. In *Proc. 11th ACM-SIAM Symposium on Discrete algorithms(SODA)*, pages 819–828, 2000.
15. S. C. Sahinalp and U. Vishkin. Efficient approximate and dynamic matching of patterns using a labeling paradigm. *Proc. 37th FOCS*, pages 320–328, 1996.
16. E. Ukkonen. On-line construction of suffix trees. *Algorithmica*, 14:249–260, 1995.
17. P. Weiner. Linear pattern matching algorithm. *Proc. 14 IEEE Symposium on Switching and Automata Theory*, pages 1–11, 1973.

# Rank-Sensitive Data Structures

Iwona Bialynicka-Birula and Roberto Grossi

Dipartimento di Informatica, Università di Pisa,
Largo Bruno Pontecorvo 3, 56127 Pisa, Italy
{iwona, grossi}@di.unipi.it

**Abstract.** Output-sensitive data structures result from preprocessing $n$ items and are capable of reporting the items satisfying an on-line query in $O(t(n) + \ell)$ time, where $t(n)$ is the cost of traversing the structure and $\ell \leq n$ is the number of reported items satisfying the query. In this paper we focus on *rank-sensitive* data structures, which are additionally given a *ranking* of the $n$ items, so that just the top $k$ best-ranking items should be reported at query time, *sorted in rank order*, at a cost of $O(t(n) + k)$ time. Note that $k$ is part of the query as a parameter under the control of the user (as opposed to $\ell$ which is query-dependent). We explore the problem of adding rank-sensitivity to data structures such as suffix trees or range trees, where the $\ell$ items satisfying the query form $O(\text{polylog}(n))$ intervals of consecutive entries from which we choose the top $k$ best-ranking ones. Letting $s(n)$ be the number of items (including their copies) stored in the original data structures, we increase the space by an additional term of $O(s(n) \lg^\epsilon n)$ memory words of space, each of $O(\lg n)$ bits, for any positive constant $\epsilon < 1$. We allow for changing the ranking on the fly during the lifetime of the data structures, with ranking values in $0 \ldots O(n)$. In this case, query time becomes $O(t(n) + k)$ plus $O(\lg n / \lg \lg n)$ per interval; each change in the ranking and each insertion/deletion of an item takes $O(\lg n)$ time; the additional term in space occupancy increases to $O(s(n) \lg n / \lg \lg n)$.

## 1 Introduction

Output-sensitive data structures are at the heart of text searching [13], geometric searching [5], database searching [28], and information retrieval in general [3,31]. They are the result of preprocessing $n$ items (these can be textual data, geometric data, database records, multimedia, or any other kind of data) into $O(n \, \text{polylog}(n))$ space in such a way, as to allow quickly answering on-line queries in $O(t(n) + \ell)$ time, where $t(n) = o(n)$ is the cost of querying the data structure (typically $t(n) = \text{polylog}(n)$). The term output-sensitive means that the query cost is proportional to $\ell$, the number of reported items satisfying the query, assuming that $\ell \leq n$ can be much smaller than $n$. In literature, a lot of effort has been devoted to minimizing $t(n)$, while the dependency on the *variable* cost $\ell$ has been considered unavoidable because it depends on the items satisfying the given query and cannot be predicted before querying.

In recent years we have been literally overwhelmed by the electronic data available in fields ranging from information retrieval, through text processing and

M. Consens and G. Navarro (Eds.): SPIRE 2005, LNCS 3772, pp. 79–90, 2005.

computational geometry to computational biology. For instance, the number $\ell$ of items reported by search engines can be so huge as to hinder any reasonable attempt at their post-processing. In other words, $n$ is very large but $\ell$ is very large too (even if $\ell$ is much smaller than $n$). Output-sensitive data structures are too optimistic in a case such as this, and returning all the $\ell$ items is not the solution to the torrent of information.

**Motivation.** Search engines are just one example; many situations arising in large scale searching share a similar problem. But what if we have some preference regarding the items stored in the output-sensitive data structures? The solution in this case involves assigning an application-dependent *ranking* to the items, so that the top $k$ best-ranking items among the $\ell$ ones satisfying an on-line query can be returned *sorted in rank order*. (We assume that $k \leq \ell$ although the general bound is indeed for $\min\{k, \ell\}$.) Note that the overload is significantly reduced when $k \ll \ell$. For example, PageRank [24] is at the heart of the Google engine, but many other rankings are available for other types of data. Z-order is useful in graphics, since it is the order in which geometrical objects are displayed on the screen [14]. Records in databases can be returned in the order of their physical location (to minimize disk seek time) or according to a time order (e.g. press news). Positions in biological sequences can be ranked according to their biological function and relevance [13]. These are just basic examples, but more can be found in statistics, geographic information systems, etc.

**Our Results for Rank-Sensitive Data Structures.** In this paper, we study the theoretical framework for a class of *rank-sensitive* data structures. They are obtained from output-sensitive data structures such as suffix trees [30,27] or range trees [5], where the $\ell$ items satisfying the query form $O(\mathrm{polylog}(n))$ intervals of consecutive entries each. For example, string searching in suffix trees and tries goes along a path leading to a node $v$, whose descending leaves represent the $\ell$ occurrences to report, say, from leaf $v_1$ to leaf $v_2$ in symmetrical order. In one-dimensional range searching, two paths leading to two leaves $v_1$ and $v_2$ identify the $\ell$ items lying in the range. In both cases, we locate an interval of consecutive entries in the symmetrical order, from $v_1$ to $v_2$. In two-dimensional range trees, we locate $O(\lg n)$ such (disjoint) intervals. For higher dimensions, we have $O(\mathrm{polylog}(n))$ disjoint intervals.

As previously said, for this class of output-sensitive data structures, we obtain the retrieved items as the union of $O(\mathrm{polylog}(n))$ disjoint intervals. We provide a framework for transforming such intervals into rank-sensitive data structures from which we choose the top $k$ best-ranking items satisfying a query. We aim at a cost dependency on the parameter $k$ specified by the user rather than on the query-dependent term $\ell$. Let *rank* denote a ranking function such that $rank(v_1) < rank(v_2)$ signifies that item $v_1$ should be preferred to item $v_2$. We do not enter into a discussion of the quality of the ranking adopted; for us, it just induces a total order on the importance of the items to store. Let $s(n)$ be the number of items (including their copies) stored in any such data structure $D$. Let $O(t(n) + \ell)$ be its query time and let $|D|$ be the number of memory words of space it occupies, each word composed of $O(\lg n)$ bits. We obtain a rank-sensitive data structure $D'$, with $O(t(n) + k)$ query time, increasing the space to $|D'| = |D| + O(s(n) \lg^\epsilon n)$ memory words, for any positive constant $\epsilon < 1$.

We allow for changing *rank* on the fly during the lifetime of the data structure $D'$, with ranking values in the range from 0 to $O(n)$. In this case, query time becomes $O(t(n) + k)$ plus $O(\lg n/\lg\lg n)$ per interval and each change in the ranking takes $O(\lg n)$ time per item copy. Our solution operates in *real time* as we discuss later. When $D$ allows for insert and delete operations on the set of items, we obtain an additive cost of $O(\lg n/\lg\lg n)$ time per query operation and $O(\lg n)$ time per update operation in $D'$. The space occupancy is $|D'| = |D| + O(s(n) \lg n/\lg\lg n)$ memory words. The preprocessing cost of $D'$ is dominated by sorting the items according to *rank*, plus the preprocessing cost of $D$.

**Attacking the Problem.** While ranking itself has been the subject of intense theoretical investigation in the context of search engines [17,18,24], we could not find any explicit study pertaining to ranking in the context of data structures. The only published data structure of this kind is the inverted lists [31] in which the documents are sorted according to their rank order. McCreight's paper on priority search trees [19] refers to enumeration in increasing order along the y-axis but it does not indeed discuss how to report the items in sorted order along the y-axis. An indirect form of ranking can be found in the (dynamic) rectangular intersection with priorities [15] and in the document listing problem [21].

For our class of output-sensitive data structures, we can formulate the ranking problem as a geometric problem. We are given a (dynamic) list $L$ of $n$ entries, where each entry $e \in L$ has an associated value $rank(e) \in 0\ldots O(n)$. The list induces a linear order on its entries, such that $e_i < e_j$ if and only if $e_i$ precedes $e_j$ in $L$. Let us indicate by $pos(e)$ the (dynamic) position of $e$ in $L$ (but we do not maintain $pos$ explicitly). Hence, $e_i < e_j$ if and only if $pos(e_i) < pos(e_j)$. We associate point $(pos(e), rank(e))$ with each entry $e \in L$. Then, given $e_i$ and $e_j$, let $e'$ be the $k$th entry in rank order such that $pos(e_i) \le pos(e') \le pos(e_j)$. We perform a three-sided or $1\frac{1}{2}$-dimensional query on $pos(e_i)\ldots pos(e_j)$ along the x-axis, and $0\ldots rank(e')$ along the y-axis.

Priority search trees [19] and Cartesian trees [29] are among the prominent data structures supporting these queries, but do not provide items in sorted order (they can end up with half of the items unsorted during their traversal). Since we can identify the aforementioned $e'$ by a variation of [10], in $O(k)$ time, we can retrieve the top $k$ best-ranking items in $O(k + \lg n)$ time in *unsorted order*. Improvements to get $O(k)$ time can be made using scaling [12] or persistent data structures [6,8,16]. Subsequent sorting reports the items in $O(k \lg k)$ time using $O(n)$ words of memory.

What if we adopt the above solution in a *real-time* setting? Think of a server that provides items in rank order on the fly, or any other similar real-time application in which guaranteed response time is mandatory. Given a query, the above solution and its variants can only start listing the first items after $O(t(n)+k\lg k)$ time! In contrast, our solution works in real-time by using more space. After $O(t(n))$ time, it provides each subsequent item in $O(1)$ worst-case time according to the rank order (i.e. the $q$th item in rank order is listed after $c_1 t(n) + c_2 q$ steps, for $1 \le q \le k$ and constants $c_1, c_2 > 0$).

Persistent data structures can attain real-time performance, in an efficient way, only in a static setting. Let us denote $L$'s entries by $e_0, e_1, \ldots, e_{n-1}$, and build persistent sublists of $2^r$ consecutive entries, for $r = 0, 1, \ldots, \lg n$. Namely, for fixed $r$, we start from the sublist containing $e_0, e_1, \ldots, e_{2^r-1}$ in *sorted rank order*. Then, for $i = 2^r, \ldots, n-1$, we remove entry $e_{i-2^r}$ and add $e_i$ using persistence to create a new version of the sorted sublist. Now, given our query with $e_i$ and $e_j$, we compute the largest $r$ such that $2^r \le j - i$. Among the versions of the sublists for $2^r$ entries, we take the one starting at $e_i$ and the one ending in $e_j$. Merging these two lists on the fly for $k$ steps solves the ranking problem. This solution has two drawbacks. First, it uses more space than our solution. Second, it is hard to *dynamize* since a single entry changing in $L$ can affect $\Theta(n)$ versions in the worst case. (Also the previous solutions based on persistence, priority search trees and Cartesian trees suffer similar problems in the dynamic setting.) We extend the notion of Q-heap [11] to implement our solution, introducing *multi-Q-heaps* described in Section 3.

## 2   The Static Case and Its Dynamization

Our starting point is a well-known scheme adopted for two-dimensional range trees [5]. Following the global rebuilding technique described in [23], we can restrict our attention to values of $n$ in the range $0 \ldots O(N)$ where $n = \Theta(N)$. Consequently, we use lookup tables tailored for $N$, so that when the value of $N$ must double or halve, we also rebuild these tables in $o(N)$ time. Our word size is $O(\lg N)$. As can be seen from [23], time bounds can be made worst-case.

We recall that the interval is taken from the list of items $L = e_0, e_1, \ldots, e_{n-1}$, indicating with $pos(e_i)$ the dynamic position of $e_i$ in $L$ (but we do not keep $pos$ explicitly) and with $rank(e_i)$ its rank value in $0 \ldots O(N)$. We use a special rank value $+\infty$ that is larger than the other rank values; multiple copies of $+\infty$ are each different from the other (and take $O(\lg N)$ bits each).

### 2.1   Static Case on a Single Interval

We employ a weight-balanced B-tree $W$ [2] as the skeleton structure. At the moment, suppose that $W$ has degree exactly two in the internal nodes and that the $n$ items in $L$ are stored in the leaves of $W$, assuming that each leaf stores a single item. For each node $u \in W$, let $R(u)$ denote the explicit sorted list of the items in the leaves descending from $u$, according to *rank* order. If $u_0$ and $u_1$ are the two children of $u$, we have that $R(u)$ is the merge of $R(u_0)$ and $R(u_1)$. Therefore, we can use 0s and 1s to mark the entries in $R(u)$ that originate, respectively, from $R(u_0)$ and $R(u_1)$. We obtain $B(u)$, a bitstring of $|R(u)|$ bits, totalizing $O(n \lg n)$ bits, hence $O(n)$ words of memory, for the entire $W$ (see [4]).

Rank query works as expected [5]. Given entries $e_i$ and $e_j$ in $L$, we locate their leaves in $W$, say $v_i$ and $v_j$. We find their least common ancestor $w$ in $W$ (the case $v_i = v_j$ is trivial). On the path from $w$ to $v_i$, we traverse $O(\lg n)$

internal nodes. If during this traversal, we go from node $u$ to its left child $u_0$, we consider the list $R(u_1)$, where $u_1$ is the right child of $u$. Analogously, on the path from $w$ to $v_j$, if we go from node $u$ to its right child $u_1$, we consider list $R(u_0)$ for its left child. In all other cases, we skip the nodes (including $w$ and its two children). Clearly, we include $v_i$ and $v_j$ if needed.

At this point, we reduce the rank-sensitive query for $v_i$ and $v_j$ to the problem of selecting the top $k$ best-ranking items from $O(\lg n)$ rank-sorted lists $R()$, containing integers in $0 \ldots O(N)$. Following Chazelle's approach, we do not explicitly store the lists $R()$, but keep only the bitstrings $B()$ and the additional machinery for translating bits in $B()$ into entries in $R()$, which occupies $O(n \lg^\epsilon n)$ words of memory, for any positive constant $\epsilon < 1$. (See Lemma 2 in Section 4 of [4].) As a result, we can retrieve the sorted items of lists $R()$ using Chazelle's approach.

## 2.2 Polylog Intervals in the Dynamic Case

In the general case, we are left with the problem of selecting the top $k$ best-ranking items from $O(\mathrm{polylog}(n))$ rank-sorted *dynamic* lists $R()$, containing integers in $0 \ldots O(N)$. We cannot use Chazelle's machinery in the dynamic setting. We maintain the degree $b$ of the nodes in the weight-balanced B-tree $W$, such that $(\beta/4) \lg n / \lg \lg n \le b \le (4\beta) \lg n / \lg \lg n$, for a suitable constant in $0 < \beta < 1$. As a result from [2], the height of the tree is $O(\lg n / \lg b) = O(\lg n / \lg \lg n)$. We also explicitly store the lists $R()$, totalizing $O(n)$ words per level of $W$, and thus yielding $O(n \lg n / \lg \lg n)$ words of memory. Note that the cost of splitting/merging a node $u \in W$ along with $R(u)$ can be deamortized [2].

To enable the efficient updating of all the lists $R()$, we use a variation of dynamic fractional cascading described in [25], which performs efficiently on graphs of a non-constant degree. Fractional cascading does not increase the overall space complexity. At the same time, for a given element $e$ of list $R(u)$, it allows locating the predecessor (in rank order) of $e$ in $R(u')$ when $u'$ is a child or parent of $u$. This locating is performed in time $O(\lg b + \lg \lg n)$ which amounts to $O(\lg \lg n)$ under our assumption concerning $b$, the degree of the tree.

Let us consider a single interval identified by a rank query. It is described by two leaves $v_i$ and $v_j$, along with their least common ancestor $w \in W$. However, we encounter $O(\lg n / \lg \lg n)$ lists $R()$ in each node $u$ along the path from $w$ to either $v_i$ or $v_j$. For any such node $u$, we must consider the lists for $u$'s siblings either to its left or its right. So we have to merge $O((\lg n / \lg \lg n)^2)$ lists on the fly. But we can only afford $O(\lg n / \lg \lg n)$ time.

We solve this multi-way merging problem by introducing *multi-Q-heaps* in Section 3, extending Q-heaps [11]. A multi-Q-heap stores $O(\lg n / \lg \lg n)$ items from a bounded universe $0 \ldots O(N)$, and performs constant-time search, insertion, deletion, and find-min operations. In particular, searching and finding can be *restricted* to any *subset* of its entries, still in $O(1)$ time. Each instance of a multi-Q-heap requires just $O(1)$ words of memory. These instances share common lookup tables occupying $o(N)$ memory words. We refer the reader to Theorem 2 in Section 3 for more details.

We employ our multi-Q-heap for the rank values in each node $u \in W$. This does not change the overall space occupancy, since it adds $O(n)$ words, but it allows us to handle rank queries in each node $u$ in $O(1)$ time per item as follows. Let $d = \alpha \lg N / \lg \lg N$ be the maximum number of items that can be stored in a multi-Q-heap (see Theorem 2). We divide the lists $R()$ associated with $u$'s children into $d$ clusters of $d$ lists each. For each cluster, we repeat the task recursively, with a constant number of levels and $O(\text{polylog}(n)/d)$ multi-Q-heaps. We organize these multi-Q-heaps in a hierarchical pipeline of constant depth. For the sake of discussion, let's assume that we have just depth 2. We employ a (first-level) multi-Q-heap, initially storing $d$ items, which are the minimum entry for each list in the cluster. We employ further $d$ (second-level) multi-Q-heap of $d$ entries each, in which we store a copy of the minimum element of each cluster. To select the top $k$ best-ranking leaves, we extract the $k$ smallest entries from the lists by using the above multi-Q-heaps: We first find the minimum entry, $x$, in one of the second-level multi-Q-heaps, and identify the corresponding first-level multi-Q-heap. From this, we identify the list containing $x$. We take the entry, $y$, following $x$ in its list. We extract $x$ from the first-level multi-Q-heap and insert $y$. Let $z$ be the new minimum thus resulting in the first level. We extract $x$ from the suitable second-level multi-Q-heap and insert $z$. By repeating this task $k$ times, we return the $k$ leaves in rank-sensitive fashion.

This does not yet solve our problem. Consider the path from, say, $v_i$ to its ancestor $w$. We have $O(\lg n / \lg \lg n)$ lists for each node along the path. Fortunately, our multi-Q-heaps allow us to handle any subset of these lists, in constant time. The net result is that we need to use just $O(\lg n / \lg \lg n)$ multi-Q-heaps for the entire path. For each node $u$ in the path, the find-min operation is limited to the lists corresponding to a subset of $u$'s sibling at its right. They form a contiguous range, which we can easily manage with multi-Q-heaps. Hence, we can apply the above 2-level organization, in which we have $O(\lg n / \lg \lg n)$ multi-Q-heaps in the path from $v_i$ to $w$ in the second level. (An analogous approach is for the path from $v_j$ to $w$.) In this way, we can perform a multi-way merging on the fly for finding the least $k$ keys in sorted rank order, in $O(k + \lg n / \lg \lg n)$ time. Note that the bound is real-time as claimed. In the case of polylog intervals, we use an additional multi-Q-heap hierarchical organization (of constant depth) to merge the items resulting from processing each interval separately.

We now describe how to handle rank changes of entries in $L$, as well as insertions and deletions in $L$. Changing the rank of entry $e_i$, say in leaf $v_i \in W$ is performed in a top-down fashion. It affects the nodes on the path from the root of $W$ to $v_i$. The list $R(u)$ for each node $u$ along this path contains a copy of $e_i$ but whose rank no longer complies with the ordering of the list. This element is extracted from the list and inserted into the correct place on this list. Both the element itself and the new correct place can be located in the list associated with the root in $O(\lg n)$ time. Next, using the fractional cascading structure, we can relocate the copy of $e_i$ in the list for the next node in the downward path to $v_i$, having already done it in the current node. This takes $O(\lg \lg n)$ time per node, thus yielding $O(\lg n)$ total time to relocate the copy of $e_i$ in all the lists of the path. As for the insertions in $L$ (and also in $W$), they follow the approach in [2]; moreover, the input item $e$ has its $rank(e)$ value, in the range $0 \ldots O(N)$, inserted into the lists $R()$ of the ancestor nodes as described above. Deletions are simply implemented as weak, changing the rank value of deleted items to $+\infty$. When their number is sufficiently large, we apply rebuilding as in [23]. If the original data structure contains multiple copies of the same item (as in the case of a range tree) then the update in the rank-sensitive structure is applied separately to the individual copies.

We obtain the following result. Let $D$ be an output-sensitive data structure for $n$ items, where the $\ell$ items satisfying a query on $D$ form $O(\text{polylog}(n))$ intervals of consecutive entries. Let $O(t(n) + \ell)$ be its query time and $s(n)$ be the number of items (including their copies) stored in $D$.

**Theorem 1.** *We can transform $D$ into a static rank-sensitive data structure $D'$, where query time is $O(t(n) + k)$ for any given $k$, thus reporting the top $k$ best-ranking items among the $\ell$ listed by $D$. We increase the space by an additional term of $O(s(n) \lg^\epsilon n)$ memory words of space, each of $O(\lg n)$ bits, for any positive constant $\epsilon < 1$. For the dynamic version of $D$ and $D'$, we allow for changing the ranking of the items, with ranking values in $0 \ldots O(n)$. In this case, query time becomes $O(t(n) + k)$ plus $O(\lg n / \lg \lg n)$ per interval. Each change in the ranking and each insertion/deletion of an item take $O(\lg n)$ time for each item copy stored in the original data structure. The additional term in space occupancy increases to $O(s(n) \lg n / \lg \lg n)$.*

## 3   Multi-Q-Heaps

The multi-Q-heap is a relative of the Q-heap [11]. Q-heaps provide a way to represent a sub-logarithmic set of elements in the universe $[N] = 0 \ldots O(N)$, so that such operations as inserting, deleting or finding the smallest element can be executed in $O(1)$ time in the worst case. The price to pay for the speed is the need to set up and store lookup tables in $o(N)$ time and space. These tables, however, need only to be computed once as a bootstrap cost and can be shared among any number of Q-heap instances. Our multi-Q-heap is functionally more powerful than Q-heap, as it allows performing operations on any *subset* of $d$ common elements, where $d \le \alpha \lg N / \lg \lg N$ for a suitable positive constant $\alpha < 1$. Naturally, this could be emulated by maintaining Q-heaps for all the different subsets of the elements, but that solution would be exponential in $d$ (for each instance!), while our multi-Q-heap representation requires two or three memory words and still supports constant-time operations. Our implementation based on lookup tables is quite simple and does not make use of multiplications or special instructions (see [9,26] for a thorough discussion of this topic). We first describe a simpler version (to be later extended) supporting the following:

- Create a heap for a given list of elements.
- Find the minimum element within a given range.
- Find an element within a given range of items.
- Update the element at a given position.

In the rest of the section, we prove the following result.

**Theorem 2.** *There exists a constant $\alpha < 1$ such that $d$ distinct integers in $0 \ldots O(N)$ (where $d \leq \alpha \lg N / \lg \lg N$) can be maintained in a multi-Q-heap supporting search, insert, delete, and find-min operations in constant time per operation in the worst case, with $O(d)$ words of space. The multi-Q-heap requires a set of pre-computed lookup tables taking $o(N)$ construction time and space.*

## 3.1   High-Level Implementation

The $d$ elements are integers from $[N]$. We can refer to their binary representations of $w = \lceil \lg[N] \rceil$ bits each. These strings can be used to build a compacted trie on binary strings of length $w$. However, instead of labeling the leaves of the compact trie with the strings (elements) they correspond to, we keep just the trie shape and the skip values contained in its internal nodes, like in [1,7]. We store the $d$ elements and their satellite data in a separate table. To provide a connection between the trie and the values, we store a permutation which describes the relation between the order of elements in the trie and the order in which they are stored in the table.

When searching for an element, we first perform a blind search on the trie [1,7]. Next we access the table corresponding to the found element and we compare it with the sought one. Note that this way we only access the table of values in one place, while the rest of the search is performed on the trie. With an assumption about the maximum number of elements stored in the multi-Q-heap, we can encode both the trie and the permutation as two single memory words. The operations are then performed on these encodings and only the relevant entries in the value table are accessed, which guarantees constant time. The operations on the encoded structures are realized using lookup tables and bit operations.

To support multi-Q-heap operations, we store a single structure containing all the elements. We implement all the extended operations so as to consider only the given subset of the elements while maintaining constant time. We assume a word size of $w = O(\lg N)$ bits. We use $d$ to refer to the number of items stored in the multi-Q-heap. We assume $d \leq \alpha \lg N / \lg \lg N$ for some suitable constant $\alpha < 1$. We use $x_0, x_1, \ldots, x_{d-1}$ to refer to the list of items stored in the multi-Q-heap. For our case, the order defined by the indices is relevant (when using multi-Q-heaps in the nodes of the weight-balanced B-tree of Section 2).

## 3.2   Multi-Q-Heap: Representation

The multi-Q-heap can be represented as a triplet $(S, \tau, \sigma)$, where $S$ is the array of elements stored in the structure, $\tau$ is the encoding of the compact trie and $\sigma$ is an encoding of the permutation. The array $S$ stores the elements $x_0, x_1, \ldots, x_{d-1}$ in that order and their satellite data. Each element occupies a word of space.

The encoding of the trie, $\tau$, can be defined in the following fashion. First, let us encode the shape of the binary tree of which it consists. This tree is binary, with no unary nodes and edges implicitly labeled with either 0 or 1. We can encode it by traversing the tree in inorder (visiting first 0 edges and then 1 edges) and outputting the labels of the edges traversed. This encoding can be decoded unambiguously and requires $4d - 4$ bits, since each edge is traversed twice and there are $2d - 2$ edges in the trie. Next, we encode the skip values. The internal nodes (in which the skip values are stored) are ordered according to their inorder which leads to an ordered list of skip values. Each skip value is stored in $\lceil \lg w \rceil$ bits, so the encoding of the list takes $(d - 1)\lceil \lg w \rceil$ bits. For a suitable value of $\alpha$ the complete encoding of the trie does not exceed $1/4 \lg N$ bits and hence can be stored in one word of memory.

The permutation $\sigma$ reflects the array order $x_0, x_1, \ldots, x_{d-1}$ with respect to the order of these elements sorted by their values (which is the same as the inorder of the corresponding leaves in the trie). There are $d!$ possible permutations, so we choose $\alpha$ so that $\lg d! < 1/4 \lg N$ and the encoding on the permutation fits in one word of memory. We use the encoding described in [22], which takes linear time to rank and unrank a permutation, hence to encode and decode it.

## 3.3   Multi-Q-Heap: Supported Operations

The *Init* operation sets up all the lookup tables required for implementing the multi-Q-heap. It needs to be performed only once. See section 3.4 for details concerning the lookup tables. These lookup tables are used in the implementations of the operations described below. If invoked multiple times, only the first is effective.

The *Create* operation takes the array $S$ of values $x_0, x_1, \ldots, x_{d-1}$ and sets up the structures $\tau$ and $\sigma$. It takes the time required to construct the compact trie for $d$ elements, hence $O(d)$.

The function *Findmin* returns the smallest element among the elements $x_i, \ldots, x_j$ stored in the multi-Q-heap. We implement it using the lookup table *Subheap* and *Index*. We use $Subheap[\tau, \sigma, i, j]$ to obtain $\tau'$ and $\sigma'$, the structure for elements $x_i, \ldots, x_j$. We then use $Index[\sigma', 1]$ to obtain the array index of the smallest element in the range.

The function *Search* searches the subset of elements $x_i, \ldots, x_j$ stored in the multi-Q-heap and returns the index of the element in the multi-Q-heap which is smallest among those not smaller than $y$, where $y \in [N]$ can be any value. As previously, we use $Subheap[\tau, \sigma, i, j]$ to obtain $\tau'$ and $\sigma'$, the subheap for elements $x_i, \ldots, x_j$. We then search the reduced trie for $x'$, the first half (bitwise) of $x$, by looking up $u = Top[\tau', x']$. Next, using $LDescendant[\tau', u]$, we identify one of the strings descending from $u$ and compare this string with $x'$ to compute their longest common prefix length $lcp$. This computation can be done in constant time with another lookup table, which is standard and is not described. If $lcp < 1/2 \lg N$, then $LDescendant[\tau', u]$ identifies the sought element. If $lcp = 1/2 \lg N$, we continue the search in the bottom part of the trie by setting $u = Top[\tau', x', u]$. Also here $LDescendant[\tau', u]$ provides the answer.

The *Update* operation replaces the element $x_r$ in the array $S$ with $y$, where $y \in [N]$ can be any value. It updates $\tau$ and $\sigma$ accordingly. We first simulate the search for $y$ in $\tau$, as described in the previous paragraph to find the rank $i$ of $y$

among $x_0, \ldots x_d$ and use this together with the table $UpdatePermutation[\sigma, r, i]$ to produce the updated permutation. We then use values obtained during the simulated blind search for $y$ in $\tau$ to obtain values needed to access the $UpdateTrie$ table. During the search we find the node $u$ at which the search for $y$ ends (in the second half of the trie in the case the search gets that far) and the $lcp$ obtained by comparing its leftmost descendant with $y$. We use $Ancestor[\tau, u, lcp]$ for identifying the node whose parent edge is to be split for inserting. The $lcp$ is the skip value the parameter $c$ depends on the bit at position $lcp + 1$ of $y$. With this information, we access $UpdateTrie$.

### 3.4   Multi-Q-Heap: Lookup Tables

This section describes the lookup tables required to perform the operations described in the previous section. The number of tables can be reduced, but at the expense of the clarity of the implementation description.

The *Index* table provides a way for obtaining the array index of an element given the inorder position of its corresponding leaf in the trie (let us call this the trie position). It contains the appropriate array index entry for every possible permutation and trie position. The space occupancy is $2^{1/4 \lg N} \times d \times \lg d = N^{1/4} \times d \times \lg d = o(N)$.

The $Index^{-1}$ table is the inverse of *Index* in the sense that it provides a way of obtaining a trie position from an index, by containing a position entry for every possible permutation and index. The space occupancy is the same as for *Index*.

The *Subheap* table provides a means of obtaining a new subheap structure, $(S, \tau', \sigma')$, from a given one $(S, \tau, \sigma)$. The new subheap structure uses the same array $S$, but takes into account only the subset $x_i, \ldots, x_j$ of its items. Note that only $\tau$, $\sigma$, $i$, and $j$ are needed to determine $\tau'$ and $\sigma'$ and not the values stored in $S$. The new trie $\tau'$ is obtained from the old trie $\tau$ by removing leaves not corresponding to $x_i, \ldots, x_j$ (these can be identified using $\sigma$). The new permutation $\sigma'$ is obtained from the old one $\sigma$ by extracting all the elements with values $i, \ldots, j$ and moving them to the beginning of the permutation (without changing their relative order) so that they now correspond to the appropriate $j - i + 1$ leaves of the reduced trie. The space occupancy of *Subheap* is $2^{1/4 \lg N} \times 2^{1/4 \lg N} \times d \times d \times 1/4 \lg N \times 1/4 \lg N = N^{1/2} \times d^2 \times (1/4 \lg N)^2 = o(N)$.

The *Top* and *Bottom* tables allow searching for a value in the trie. The searching for a value must be divided into two stages, because a table which in one dimension is indexed with a full value, one of $O(N)$ possible, would occupy too much space. We therefore set up two tables: *Top* for searching for the first $1/2 \lg N$ bits of the value and *Bottom* for the remaining. The table *Top* contains entries for every possible trie $\tau$ and $x'$, the first $1/2 \lg N$ bits of some sought value $x$. The value in the table specifies the node of $\tau$ (with nodes specified by their inorder position) at which the blind search [1,7] for $x'$ (starting from the root of the trie) ends. The table *Top* contains entries for every possible trie $\tau$, $x''$ (the second $1/2 \lg N$ bits of some sought value $x$) and an internal node of the trie $v$. The value in the table specifies the node of $\tau$ at which the blind search [1,7] for $x''$ ends, but in this case the blind search starts from $v$ instead of from the root of the trie. The space occupancy of *Top* is $2^{1/4 \lg N} \times 2^{1/2 \lg N} \times \lg d = N^{3/4} \times \lg d = o(N)$ and the space occupancy of *Bottom* is $2^{1/4 \lg N} \times 2^{1/2 \lg N} \times d \times \lg d = N^{3/4} \times d \times \lg d = o(N)$.

The *UpdateTrie* table specifies a new multi-Q-heap and permutation which is created from a given one by removing the leaf number $i$ from $\tau$ and inserting instead a new leaf. The new leaf is the $c$ child of a node inserted on the edge leading to $u$. This new node has skip value $s$. The space occupancy is $2^{1/4 \lg N} \times d \times 2 \times 2^{\lg \lg N} \times d \times 1/4 \lg N \times 1/4 \lg N = N^{1/4} \times d^2 \times 1/8 \lg^3 N = o(N)$.

The *UpdatePermutation* table specifies the permutation obtained from $\sigma$ if the element with index $r$ is removed and an element ranking $i$ among the original elements of the multi-Q-heap is inserted in its place. The space occupancy is $2^{1/4 \lg N} \times d \times d \times 1/4 \lg N = N^{1/4} \times d^2 \times 1/4 \lg N = o(N)$.

The *LDescendant* table specifies the leftmost descending leaf of node $u$ in $\tau$. Its space occupancy is $2^{1/4 \lg N} \times d \times \lg d = N^{1/4} \times d \times \lg d = o(N)$.

The *Ancestor* table specifies the shallowest ancestor of $u$ having a skip value equal to or greater than $s$. The space occupancy is $2^{1/4 \lg N} \times d \times 2^{\lg \lg N} \times \lg d = N^{1/4} \times d \times \lg N \times \lg d = o(N)$.

We will describe the general case of multi-Q-heap and discuss an example in the full version. Here we only say that we need also to encode a permutation $\pi$ in a single word since $x_0, \ldots, x_{d-1}$ can be further permuted due to the insertions and deletions. An arbitrary subset is represented by a bitmask that replaces the two small integers $i$ and $j$ delimiting a range. The sizes of the lookup tables in Section 3.4 increase but still remain $o(N)$.

# References

1. M. Ajtai, M. Fredman, and J. Komlòs. Hash functions for priority queues. *Information and Control*, 63(3):217–225, December 1984.
2. Lars Arge and Jeffrey S. Vitter. Optimal external memory interval management. *SIAM Journal on Computing*, 32:1488–1508, 2003.
3. Ricardo A. Baeza-Yates and Berthier Ribeiro-Neto. *Modern Information Retrieval*. Addison-Wesley Longman Publishing Co., Inc., 1999.
4. Bernard Chazelle. A functional approach to data structures and its use in multidimensional searching. *SIAM Journal on Computing*, 17(3):427–462, June 1988.
5. Mark de Berg, Marc van Kreveld, Mark Overmars, and Otfried Schwartzkopf. *Computational Geometry: Algorithms and Applications*. Springer, 1997.
6. James R. Driscoll, Neil Sarnak, Daniel D. Sleator, and Robert E. Tarjan. Making data structures persistent. *J. Computer and System Sciences*, 38(1):86–124, 1989.
7. P. Ferragina and R. Grossi. The string B-tree: A new data structure for string search in external memory and its applications. *J. ACM*, 46:236–280, 1999.
8. Amos Fiat and Haim Kaplan. Making data structures confluently persistent. *J. Algorithms*, 48(1):16–58, 2003.
9. Faith E. Fich. Class notes CSC 2429F: Dynamic data structures, 2003. Department of Computer Science, University of Toronto, Canada.
10. Greg N. Frederickson. An optimal algorithm for selection in a min-heap. *Inf. Comput.*, 104(2):197–214, 1993, June.

11. Michael L. Fredman and Dan E. Willard. Trans-dichotomous algorithms for minimum spanning trees and shortest paths. *JCSS*, 48(3):533–551, 1994.
12. Harold N. Gabow, Jon Louis Bentley, and Robert E. Tarjan. Scaling and related techniques for geometry problems. In *STOC '84*, 135–143, Washington, D.C., 1984.
13. Dan Gusfield. *Algorithms on strings, trees, and sequences: computer science and computational biology*. Cambridge University Press, 1997.
14. D. Hearn and M. Baker. *Computer Graphics with OpenGL*. Prentice-Hall, 2003.
15. Haim Kaplan, Eyal Molad, and Robert E. Tarjan. Dynamic rectangular intersection with priorities. In *STOC '03*, 639–648. ACM Press, 2003.
16. Kitsios, Makris, Sioutas, Tsakalidis, Tsaknakis, and Vassiliadis. 2-D spatial indexing scheme in optimal time. In *ADBIS: East European Symposium on Advances in Databases and Information Systems*. LNCS, 2000.
17. Jon M. Kleinberg. Authoritative sources in a hyperlinked environment. *Journal of the ACM*, 46(5):604–632, September 1999.
18. R. Lempel and S. Moran. SALSA: the stochastic approach for link-structure analysis. *ACM Transactions on Information Systems*, 19(2):131–160, 2001.
19. Edward M. McCreight. Priority search trees. *SIAM Journal on Computing*, 14(2):257–276, 1985.
20. C. W. Mortensen. Fully-dynamic two dimensional orthogonal range and line segment intersection reporting in logarithmic time. In *SODA '03*, 618–627, 2003.
21. S. Muthukrishnan. Efficient algorithms for document retrieval problems. In *SODA '02: Proceedings of the thirteenth annual ACM-SIAM symposium on Discrete algorithms*, pages 657–666. Society for Industrial and Applied Mathematics, 2002.
22. Wendy Myrvold and Frank Ruskey. Ranking and unranking permutations in linear time. *Information Processing Letters*, 79(6):281–284, September 2001.
23. Mark H. Overmars. *The Design of Dynamic Data Structures*, volume 156 of *Lecture Notes in Computer Science*. Springer-Verlag, 1983.
24. L. Page, S. Brin, R. Motwani, and T. Winograd. The pagerank citation ranking: Bringing order to the web. *Tech. rep, Stanford University, Stanford, CA*, 1998.
25. Rajeev Raman. *Eliminating amortization: on data structures with guaranteed response time*. PhD thesis, Rochester, NY, USA, 1993.
26. Mikkel Thorup. On $AC^0$ implementations of fusion trees and atomic heaps. In *Proceedings of the fourteenth Annual ACM-SIAM Symposium on Discrete Algorithms (SODA-03)*, pages 699–707, New York, January 12–14 2003. ACM Press.
27. E. Ukkonen. On-line construction of suffix trees. *Algorithmica*, 14:249–260, 1995.
28. Jeffrey Scott Vitter. External memory algorithms and data structures: dealing with massive data. *ACM Computing Surveys*, 33(2):209–271, June 2001.
29. Jean Vuillemin. A unifying look at data structures. *Communications of the ACM*, 23(4):229–239, April 1980.
30. P. Weiner. Linear pattern matching algorithms. In *Conference Record, IEEE 14th Annual Symposium on Switching and Automata Theory*, pages 1–11, 1973.
31. Ian H. Witten, Alistair Moffat, and Timothy C. Bell. *Managing gigabytes: Compressing and indexing documents and images*. Morgan Kaufmann Pubs. Inc., 1999.

# Cache-Conscious Collision Resolution
# in String Hash Tables

Nikolas Askitis and Justin Zobel

School of Computer Science and Information Technology,
RMIT University, Melbourne, Australia 3000
{naskitis, jz}@cs.rmit.edu.au

**Abstract.** In-memory hash tables provide fast access to large numbers of strings, with less space overhead than sorted structures such as tries and binary trees. If chains are used for collision resolution, hash tables scale well, particularly if the pattern of access to the stored strings is skew. However, typical implementations of string hash tables, with lists of nodes, are not cache-efficient. In this paper we explore two alternatives to the standard representation: the simple expedient of including the string in its node, and the more drastic step of replacing each list of nodes by a contiguous array of characters. Our experiments show that, for large sets of strings, the improvement is dramatic. In all cases, the new structures give substantial savings in space at no cost in time. In the best case, the overhead space required for pointers is reduced by a factor of around 50, to less than two bits per string (with total space required, including 5.68 megabytes of strings, falling from 20.42 megabytes to 5.81 megabytes), while access times are also reduced.

## 1   Introduction

In-memory hash tables are a basic building block of programming, used to manage temporary data in scales ranging from a few items to gigabytes. For storage of strings, a standard representation for such a hash table is a *standard chain*, consisting of a fixed-size array of pointers (or *slots*), each the start of a linked list, where each node in the list contains a pointer to a string and a pointer to the next node.

For strings with a skew distribution, such as occurrences of words in text, it was found in earlier work [10] that a standard-chain hash table is faster and more compact than sorted data structures such as tries and binary trees. Using move-to-front in the individual chains [27], the load average can reach dozens of strings per slot without significant impact on access speed, as the likelihood of having to inspect more than the first string in each slot is low.

Thus a standard-chain hash table has clear advantages over open-addressing alternatives, whose performance rapidly degrades as the load average approaches 1 and which cannot be easily re-sized. However, the standard-chain hash table is not cache-conscious, as it does not make efficient use of CPU cache. There is little spatial locality, as the nodes in each chain are scattered across memory. While there is some temporal locality for skew distributions, due to the pattern of frequent re-access to the common-est strings, the benefits are limited by the overhead of requiring two pointers per string and by the fact that there is no opportunity for hardware prefetch. Yet the cost of cache

M. Consens and G. Navarro (Eds.): SPIRE 2005, LNCS 3772, pp. 91–102, 2005.

inefficiency is serious: on typical current machines each cache miss incurs a delay of hundreds of clock cycles while data is fetched from memory.

A straightforward way of improving spatial locality is to store each string directly in the node instead of using a pointer, that is, to use *compact chaining*. This approach both saves space and eliminates a potential cache miss at each node access, at little cost. In experiments with large sets of strings drawn from real-world data, we show that, in comparison to standard chaining, compact-chain hash tables can yield both space savings and reductions in per-string access times.

We also propose a more drastic measure: to eliminate the chain altogether, and store the sequence of strings in a contiguous *array* that is dynamically re-sized as strings are inserted. With this arrangement, multiple strings are fetched simultaneously — caches are typically designed to use blocks of 32, 64, or 128 bytes — and subsequent fetches have high spatial locality. Compared to compact chaining, array hash tables can yield substantial further benefits. In the best case (a set of strings with a skew distribution) the space overhead can be reduced to less than two bits per string while access speed is consistently faster than under standard chaining. Contiguous storage has long been used for disk-based structures such as inverted lists in retrieval systems, due to the high cost of random accesses. Our results show that similar factors make it desirable to use contiguous, pointer-free storage in memory.

These results are an illustration of the importance of considering cache in algorithm design. Standard chaining was considered to be the most efficient structure for managing strings, but we have greatly reduced total space consumption while simultaneously reducing access time. These are dramatic results.

## 2   Background

To store a string in a hash table, a hash function is used to generate a slot number. The string is then placed in the slot; if multiple strings are hashed to the same location, some form of *collision* resolution is needed. (It is theoretically impossible to find find a hash function that can uniquely distinguish keys that are not known in advance [12].) In principle the cost of search of a hash table depends only on load average — though, as we show in our experiments, factors such as cache efficiency can be more important — and thus hashing is asymptotically expected to be more efficient than a tree.

Therefore, to implement a hash table, there are two decisions a programmer must make: choice of hash function and choice of collision-resolution method. A hash function should be from a universal class [22], so that the keys are distributed as well as possible, and should be efficient. The fastest hash function for strings that is thought to be universal is the bitwise method of Ramakrishna and Zobel [18]; we use this function in our experiments.

Since the origin of hashing — proposed by H.P. Luhn in 1953 [12] — many methods have been proposed for resolving collisions. The best known are separate or *standard* chaining and open addressing. In chaining, which was also proposed by Luhn, linear linked lists are used to resolve collisions, with one list per slot. Linked lists can grow indefinitely, so there is no limit on the *load average*, that is, the ratio of items to slots.

Open addressing, proposed by Peterson [17], is a class of methods where items are stored directly in the table and collisions are resolved by searching for another vacant slot. However, as the load average approaches 1, the performance of open addressing drastically declines. These open addressing schemes are surveyed and analyzed by Munro and Celis [16], and have recently been investigated in the context of cache [9]. Alternatives include coalesced chaining, which allows lists to coalesce to reduce memory wasted by unused slots [25], and to combine chaining and open-addressing [7]. It is not clear that the benefits of these approaches are justified by the difficulties they present.

In contrast, standard chaining is fast and easy to implement. Moreover, in principle there is no reason why a chained hash table could not be managed with methods designed for disk, such as linear hashing [13] and extensible hashing [19], which allow an on-disk hash table to grow and shrink gracefully. Zobel et al. [27] compared the performance of several data structures for in-memory accumulation of the vocabulary of a large text collection, and found that the standard-chain hash table, coupled with the bitwise hash function and a self-organizing list structure [12], move-to-front on access, is the fastest previous data structure available for maintaining fast access to variable-length strings. However, a standard-chain hash table is not particularly cache-efficient. With the cost of a memory access in a current computer being some hundreds of CPU cycles, each cache miss potentially imposes a significant performance penalty.

A cache-conscious algorithm has high locality of memory accesses, thereby exploiting system cache and making its behavior more predictable. There are two ways in which a program can be made more cache-conscious: by improving its *temporal* locality, where the program fetches the same pieces of memory multiple times; and by improving its *spatial* locality, where the memory accesses are to nearby addresses [14]. Chains, although simple to implement, are known for their inefficient use of cache. As nodes are stored in random locations in memory, and the input sequence of hash table accesses is unpredictable, neither temporal nor spatial locality are high. Similar problems apply to all linked structures and randomly-accessed structures, including binary trees, skiplists, and large arrays accessed by binary search.

Prefetch is ineffective in this context. Hardware prefetchers [6,2] work well for programs that exhibit regular access patterns. Array-based applications are prime candidates, as they exhibit stride access patterns that can be easily detected from an address history that is accumulated at run time. Hardware based prefetches however, are not effective with pointer-intensive applications, as they are hindered by the serial nature of pointer dereferences.

For such situations, techniques such as software prefetches [11,3] and different kinds of pointer cache [5,21,26] have been proposed. To our knowledge, there has been no practical examination of the impact of these techniques on the standard-chain hash table, nor is there support for these techniques on current platforms.

Moreover, such techniques concern the manifestation of cache misses, as opposed to the cause, being poor access locality. Chilimbi et al. [4] demonstrates that careful data layout can increase the access locality of some pointer-intensive programs. However, use of their techniques requires work to determine the suitability for tree-like data structures, which could prove limiting. Chilimbi et al. [4] note the applicability of their methods to chained hashing, but not with move-to-front on access, which is likely to be a limiting factor, as move-to-front is itself an effective cost-adaptive reordering

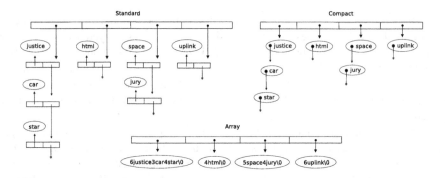

**Fig. 1.** The standard-chain (left), compact-chain (right) and array (below) hash tables

scheme. Nor is it clear that such methods will be of benefit in the kinds of environments we are concerned with, where the volume of data being managed may be many hundred times large than cache. However, significant gains should be available through cache-conscious algorithms, such as those we propose in this paper.

## 3    Cache-Conscious Hash Tables

Every node access in a standard-chain hash table incurs two pointer traversals, one to reach the node and one to reach the string. As these are likely to be randomly located in memory, each access is likely to incur a cache miss. In this section we explain our proposals for eliminating these accesses. We assume that strings are sequences of 8-bit bytes, that a character such as null is available as a terminator, and a 32-bit CPU and memory address architecture.

A straightforward step is to store each string in its node, rather than storing it in separate space. This halves the number of random accesses and saves 4 bytes per string, but requires that the nodes themselves be of variable length. Each node consists of 4 initial bytes, containing a pointer to the next node, followed by the string itself. We call this variant of chaining *compact*. The cache advantages of a compact-chain hash table are obvious, especially in the context of a skew distribution and move-to-front: each hash table lookup will involve only a single memory access, and the reduction in total size will improve the likelihood that the next node required is already cached.

We propose a novel alternative — to eliminate the chain altogether, and store the strings in a contiguous array. Prefetching schemes are highly effective with array-based structures, so this *array* hash table (shown, with the alternatives, in Figure 1) should maximize spatial access locality, providing a cache-conscious alternative to standard and compact chaining. Each array can be seen as a resizable *bucket*. The cost of access is a single pointer traversal, to fetch a bucket, which is then processed linearly.

Use of copying to eliminate string pointers for string sorting was proposed by Sinha et al. [23], who demonstrate that doing so can halve the cost of sorting a large set of strings [24]. It is plausible that similar techniques can lead to substantial gains for hash tables.

A potential disadvantage is that these arrays must be of variable size; whenever a new string is inserted in a slot, it must be resized to accommodate the additional bytes. (Note that this method does not change the size of the hash table; we are not proposing extendible arrays [20].) Another potential disadvantage is that move-to-front— which in the context of chaining requires only a few pointer assignments — involves copying of large parts of the array. On the other hand, the space overheads of an array hash table are reduced to the table itself and any memory fragmentation due to the presence of variable-length resizable objects. We explore the impact of these factors in our experiments.

A further potential disadvantage of array hash tables is that such contiguous storage appears to eliminate a key advantage of nodes — namely, that they can contain multiple additional fields. However, sequences of fixed numbers of bytes can easily be interleaved with the strings, and these sequences can be used to store the fields. The impact of these fields is likely to be much the same on all kinds of hash table. In our experiments, no fields were present.

We have explored two variants of array hash tables, *exact-fit* and *paging*. In exact-fit, when a string is inserted the bucket is resized by only as many bytes as required. This conserves memory but means that copying may be frequent. In paging, bucket sizes are multiples of 64 bytes, thus ensuring alignment with cache lines. As a special case, buckets are first created with 32 bytes, then grown to 64 bytes when they overflow, to reduce space wastage when the load average is low. (Note that empty slots have no bucket.) This approach should reduce the copying, but uses more memory. The value of 64 was chosen to match the cache architecture of our test machine, a Pentium IV.

The simplest way to traverse a bucket is to inspect it a character at a time, from beginning to end, until a match is found. Each string in a bucket must be null terminated and a null character must follow the last string in a bucket, to serve as the end-of-bucket flag. However, this approach can cause unnecessary cache misses when long strings are encountered; note that, in the great majority of cases, the string comparison in the matching process will fail on the first character.

Instead, we have used a skipping approach that allows the search process to jump ahead to the start of the next string. With skipping, each string is preceded by its length; that is, they are length-encoded [1]. The length of each string is stored in either one or two bytes, with the lead bit used to indicate whether a 7-bit or 15-bit value is present. It is not sensible to store strings of more than 32,768 characters in a hash table, as the cost of hashing will utterly dominate search costs.

## 4   Experimental Design

To evaluate the efficiency of the array, compact-chain, and standard-chain hash tables, we measured the elapsed time required for construction and search over a variety of string sets, as well as the memory requirements and the number of cache misses. The datasets used for our experiments are shown in Table 1. They consist of null-terminated variable length strings, acquired from real-world data repositories. The strings appear in order of first occurrence in the data; they are, therefore, unsorted. The trec datasets trec1 and trec2 are a subset of the complete set of word occurrences, with duplicates, in the first two TREC CDs [8]. These datasets are highly skew, containing a relatively

**Table 1.** Characteristics of the datasets used in the experiments

| Dataset | Distinct strings | String occs | Average length | Volume (MB) of distinct | Volume (MB) total |
|---|---|---|---|---|---|
| trec1 | 612,219 | 177,999,203 | 5.06 | 5.68 | 1,079.46 |
| trec2 | 411,077 | 155,989,276 | 5.00 | 3.60 | 937.27 |
| urls | 1,289,854 | 9,987,316 | 30.92 | 46.64 | 318.89 |
| distinct | 20,000,000 | 20,000,000 | 9.26 | 205.38 | 205.38 |

small set of distinct strings. Dataset `distinct` contains twenty million distinct words (that is, without duplicates) extracted from documents acquired in a web crawl and distributed as the "large web track" data in TREC. The `url` dataset, extracted from the TREC web data, is composed of non-distinct complete URLs. Some of our experiments with these sets are omitted, for space reasons, but those omitted showed much the same characteristics as those that are included.

To measure the impact of load factor, we vary the number of slots made available, using the sequence 10,000, 31,622, 100,000, 316,227 and so forth up until a minimal execution time is observed. Both the compact-chain and standard-chain hash tables are most efficient when coupled with move-to-front on access, as suggested by Zobel et al. [27]. We therefore enabled move-to-front for the chaining methods but disable it for the array, a decision that is justified in the next section.

We used a Pentium IV with 512 KB of L2 cache with 64-byte lines, 2 GB of RAM, and a Linux operating system under light load using kernel 2.6.8. We are confident — after extensive profiling — that our implementation is of high quality. We found that the hash function was a near-insignificant component of total costs.

## 5   Results

*Skewed data.* A typical use for a hash table of strings is to accumulate the vocabulary of a collection of documents. In this process, in the great majority of attempted insertions the string is already present, and there is a strong skew: some strings are much more common than others. To evaluate the performance of our hash tables under skewed access, we first construct then search using `trec1`. When the same dataset is used for both construction and search, the search process is called a self-search.

Figure 2 shows the relationship between time and memory for the hash tables during construction for `trec1`; the times in some of these results are also shown in Table 3 and the space in Table 5. Array hashing was the most efficient in both memory and speed, requiring in the fastest case under 24 seconds and around six megabytes of memory — an overhead of 0.41 MB or about 5 bits per string. This efficiency is achieved despite a load average of 20. Remarkably, increasing the number of slots (reducing the load average) has no impact on speed. Having a high number of strings per slot is efficient so long as the number of cache misses is low; indeed, having more slots can reduce speed, as the cache efficiency is reduced because each slot is accessed less often. Thus the usual assumption, that load average is a primary determinant of speed, does not always hold.

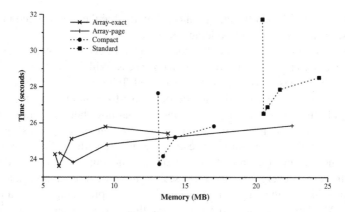

**Fig. 2.** Time versus memory when `trec1` is used for construction. Each point represents a different number of slots: 10,000, 31,622, 100,000, 316,228, and 1,000,000 respectively.

**Table 2.** The number of L2 cache misses during the construction and self-search with `trec1`

| Slots | Array | Compact | Standard |
|---|---|---|---|
| 10,000 | 68,506,659 | 146,375,946 | 205,795,945 |
| 100,000 | 89,568,430 | 80,612,383 | 105,583,144 |
| 1,000,000 | 102,474,094 | 94,079,042 | 114,275,513 |

The exact-fit and paging methods are compared in Figure 2. The exact-fit model is much more space-efficient. The relationship with speed is more complex, with paging faster in some cases, but degrading relative to exact-fit due to a more rapid increase in space consumption. We found that there is little relationship between speed and number of instructions executed.

The chaining hash tables use much more space than the array hash tables, and are no faster. The best case for chaining was with the compact chains, which in the best case required over 23.5 seconds and 13 MB of memory — an overhead of about 12 bytes per string. The standard-chain hash table was markedly inferior to both the array and compact approaches for both time and space. Given that the standard hash table was the fastest-known data structure for this task, we have strong evidence that our new structures are a significant advance.

The extent to which the speed is due to cache misses is shown in Table 2, where the `valgrind` tool (available online) is used to measure the number of L2 cache misses. (L1 misses have only a small performance penalty.) There is a reasonable correlation between the number of misses and the evaluation time. As can be seen, increasing the number of slots to 100,000 can reduce the number of misses for the chained methods, but as the table grows — and exceed cache size — performance again falls. When the load average is below 1, chaining becomes slightly more cache-efficient than the array, due to the fact that the array search function is slightly more complex.

In another experiment, the hash tables were constructed using dataset *trec1* and then searched using *trec2*. The array offered both the best time and space at 19.6 seconds with 31,662 slots. The compact chain was also able to offer a search time of about 20 seconds, but required 100,000 slots and more than double the memory required by the array. The standard chain was considerably slower.

Despite its high memory usage, the compact chain performed well under skewed access, partly due to the use of move-to-front on access. With buckets, move-to-front is computationally expensive as strings must be copied. Table 3 compares the construction and self-search costs of *trec1* with and without move-to-front on access, and includes the comparable figures for compact and standard chaining (both with move-to-front). The use of move-to-front results in slower construction and search times for the array hash; even though the vast majority of searches terminate with the first string (so there is no string movement), the cases that do require a movement are costly. Performing a move-to-front after every *k*th successful search might be more appropriate. Alternatively, the matching string can be interchanged with the preceding string, a technique proposed by McCabe [15]. However, we believe that move-to-front is unnecessary for the array, as the potential gains seem likely to be low.

*URLs.* Our next experiment was a repeat of the experiment above, using urls, a data set with some skew but in which the strings were much longer, of over thirty characters on average. As in the skewed search experiments discussed previously, our aim was to find the best balance between execution time and memory consumption. Construction results are shown in Figures 3; results for search were similar.

As for trec1, array hashing greatly outperformed the other methods. However, the optimum number of slots was much larger and the best load average was less than 1. Exact-fit achieved its fastest time of 4.54 seconds, using 3,162,228 slots while consuming 60 MB; paging was slightly faster, at 4.38 seconds, but used much more space. The best speed offered by the standard chain was 4.84 seconds with 3,162,228 slots, consuming over 90 MB. Compact chaining had similar speed with 74 MB. Again, array hashing is clearly the superior method.

**Table 3.** Elapsed time (seconds) when *trec1* is used for construction and search, showing the impact of move-to-front in the array method, and comparing to compact and standard chaining. Self-search times for paged array hash tables are omitted as they are indistinguishable from the times for exact array hash tables.

|  | Slots | Array | | Array-MTF | | Compact | Standard |
|---|---|---|---|---|---|---|---|
|  |  | page | exact | page | exact |  |  |
|  | 10,000 | 24.34 | 24.27 | 23.94 | 25.29 | 27.65 | 31.73 |
|  | 31,622 | 23.82 | 23.61 | 23.70 | 23.68 | 23.73 | 26.53 |
| Construction | 100,000 | 24.82 | 25.13 | 24.96 | 25.08 | 24.16 | 26.89 |
|  | 316,228 | 25.20 | 25.80 | 25.23 | 26.07 | 25.22 | 27.86 |
|  | 1,000,000 | 25.88 | 25.43 | 25.71 | 25.45 | 25.83 | 28.52 |
|  | 10,000 | — | 23.25 | — | 24.58 | 27.10 | 31.71 |
|  | 31,622 | — | 23.17 | — | 24.23 | 22.62 | 25.51 |
| Self-search | 100,000 | — | 24.93 | — | 25.94 | 22.68 | 25.60 |
|  | 316,228 | — | 25.26 | — | 25.98 | 23.75 | 26.56 |
|  | 1,000,000 | — | 25.27 | — | 26.07 | 24.26 | 27.02 |

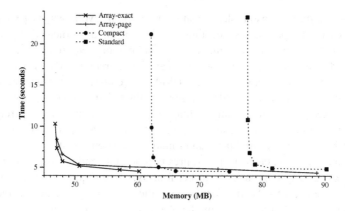

**Fig. 3.** Time versus memory when `urls` is used for construction. Each point represents a different number of slots: 10,000, 31,622, 100,000, 316,228, 1,000,000, and 3,162,287 respectively.

**Table 4.** Elapsed time (in seconds) when `distinct` is used for construction and self-search

|              | Slots       | Array page | Array exact | Compact | Standard |
|--------------|-------------|-----------|-------|---------|----------|
|              | 10,000      | 133.26 | 275.69 | 3524.32 | 3936.41 |
|              | 100,000     | 50.70  | 59.45  | 370.30  | 419.45  |
| Construction | 1,000,000   | 13.54  | 18.70  | 44.71   | 51.05   |
|              | 10,000,000  | 9.79   | 10.80  | 11.57   | 13.92   |
|              | 100,000,000 | 9.20   | 8.65   | 8.60    | 10.97   |
|              | 10,000      | —      | 109.11 | 3516.36 | 3915.26 |
|              | 100,000     | —      | 21.99  | 366.20  | 413.59  |
| Self-search  | 1,000,000   | —      | 11.14  | 42.47   | 47.08   |
|              | 10,000,000  | —      | 8.90   | 9.73    | 10.34   |
|              | 100,000,000 | —      | 6.96   | 6.67    | 6.94    |

*Distinct Data.* We then used the `distinct` dataset for construction and search. This dataset contains no repeated strings, and thus every insertion requires that the slot be fully traversed. Results for construction and self-search are shown in Table 4.

The difference in performance between the array and chaining methods is startling. This is an artificial case, but highlights the fact that random memory accesses are highly inefficient. With only 10,000 slots, the exact-fit array hash tables is constructed in about 275 seconds, whereas the compact and standard chains required about 3524 and 3936 seconds respectively. Paging requires only 133 seconds, a saving due to the lack of copying. This speed is despite the fact that the average load factor is 2000.

The results for self-search are similar to those for construction, with the array being up to 97% faster than the chaining methods. Once again, increasing the number of slots allows the chaining methods to be much faster, but the array is competitive at all table sizes. The chaining methods approach the efficiency of the array only when given surplus slots. For instance, with 100,000,000 slots, the compact chain is by a small margin the fastest method. However, the compact chain required 845 MB and the standard chain 1085 MB. The array used only 322 MB, a dramatic saving.

*Memory Consumption.* Hash tables consume memory in several ways: space allocated for the strings and for pointers; space allocated for slots; and overhead due to compiler-generated structures and space fragmentation. The memory required by the hash tables was measured by accumulating the total number of bytes requested with an estimated 8-byte overhead per memory allocation call. (With a special-purpose allocator, the 8-byte overhead for fixed-size nodes could be largely eliminated, thus saving space in standard chaining. On the other hand, in many architectures 8 rather than 4 bytes are required for a pointer.) We compared our measure with the total memory reported by the operating system under the `/proc/stat/` table and found it to be consistent.

For a standard chain, each string requires two pointers and (in a typical implementation) two `malloc` system calls. A further four bytes are required per slot. The space overhead is therefore $4n + 24s$ bytes, where $n$ is the number of slots and $s$ is the number of strings inserted. In a compact chain, the overhead is $4n + 12s$ bytes.

The memory consumed by the array hash table is slightly more complicated to model. First consider exact-fit. Apart from the head pointers leading from slots, there are no further pointers allocated by the array. The space overhead is then notionally $4n$ bytes plus 8 bytes per allocated array — that is, up to $12n$ bytes in total — but the use of copying means that there is unknown amount of space fragmentation; fortunately, inspection of the actual process size shows that this overhead is small. The array uses length-encoding, so once a string exceeds 128 characters in length an additional byte is required. For likely data the impact of these additional bytes is insignificant. Even in `urls` only 5,214 strings required this extra byte.

With paging, assume that the block size is $B$ bytes. When the load average is high, on average each slot has one block that is half empty, and the remainder are fully used; thus the overhead is $12n + B/2$ bytes. When the load average is low — that is, $s < n$ — most slots are either empty, at a cost of $4n$ bytes, or contain a single block, at a cost of $B - l + 8$, where $l$ is the average string length. For short arrays, we allow creation of blocks of length $B/2$. Thus the wastage is around $4n + s(B - l + 8)$ bytes.

The memory consumed is shown in Table 5. As can be seen by comparison with Table 1, the overhead for the exact-fit is in the best case less than two bits per string. This is a dramatic result — the space overhead is around one-hundredth of that required for even compact chaining, with, on `trec1`, only minimal impact on access times.

Exact-fit is considerably more space-efficient than paging, in particular when the table is sparse; for large tables, compact chaining is preferable. Again, there are no cases where standard chaining is competitive. These results are a conclusive demonstration that our methods are a consistent, substantial improvement.

**Table 5.** Comparison of the memory (in megabytes) consumed by the hash tables

| Slots | trec1 | | urls | | distinct | | |
|---|---|---|---|---|---|---|---|
| | 10,000 | 1,000,000 | 10,000 | 1,000,000 | 10,000 | 1,000,000 | 10,000,000 |
| array-exact | 5.81 | 13.78 | 46.77 | 57.16 | 205.52 | 218.39 | 322.64 |
| array-page | 6.13 | 22.51 | 47.04 | 72.94 | 205.83 | 249.66 | 463.29 |
| compact | 13.07 | 17.03 | 62.16 | 66.12 | 445.43 | 449.39 | 485.39 |
| standard | 20.42 | 24.38 | 77.64 | 81.60 | 685.43 | 689.39 | 725.39 |

## 6   Conclusions

We have proposed new representations for managing collisions in in-memory hash tables used to store strings. Such hash tables, which are a basic data structure used in programs managing small and large volumes of data, have previously been shown to be faster and more compact than sorted structures such as trees and tries. Yet in-memory hash tables for strings have attracted virtually no attention in the research literature.

Our results show that the standard representation, a linked list of fixed-size nodes consisting of a string pointer and a node pointer, is not cache-efficent or space-efficient. In every case, the simple alternative of replacing the fixed-length string pointer with the variable-length string, yielding a compact chain, proved faster and smaller.

We proposed the novel alternative of replacing the linked list altogether by storing the strings in a contiguous array. Despite what appears to be an obvious disadvantage — whenever a string is inserted, the array must be dynamically resized — the resulting cache efficiency means that the array method can be dramatically faster. In most cases, the difference in speed compared to the compact chain is small, but the space savings are large; in the best case the total space overhead was less than two bits per string, a reduction from around 100 bits for compact chaining, while speed was similar. We explored cache-aligned storage, but even in the best case the further gains were small. Our results also show that, in an architecture with cache, and in the presence of a skew data distribution, the load average can be very high with no impact on speed of access.

Our new cache-conscious representations dramatically improve the performance of this fundamental data structure, reducing both time and space and making a hash table by far the most efficient way of managing a large volume of strings.

## References

1. A. V. Aho, J. E. Hopcroft, and J. D. Ullman. *The Design and Analysis of Computer Algorithms*. Addison-Wesley, January 1974.
2. J. Baer and T. Chen. Effective hardware-based data prefetching for high-performance processors. *IEEE Transactions on Computers*, 44(5):609–623, 1995.
3. D. Callahan, K. Kennedy, and A. Porterfield. Software prefetching. In *Proc. ASPLOS Int. Conf. on Architectural Support for Programming Languages and Operating Systems*, pages 40–52. ACM Press, New York, 1991.
4. T. M. Chilimbi, M. D. Hill, and J. R. Larus. Cache-conscious structure layout. In *Proc. ACM SIGPLAN conf. on Programming Language Design and Implementation*, pages 1–12. ACM Press, New York, 1999.
5. J. Collins, S. Sair, B. Calder, and D. M. Tullsen. Pointer cache assisted prefetching. In *Proc. Annual ACM/IEEE MICRO Int. Symp. on Microarchitecture*, pages 62–73. IEEE Computer Society Press, 2002.
6. J. W. C. Fu, J. H. Patel, and B. L. Janssens. Stride directed prefetching in scalar processors. *SIGMICRO Newsletter*, 23(1-2):102–110, 1992.
7. C. Halatsis and G. Philokyprou. Pseudochaining in hash tables. *Communications of the ACM*, 21(7):554–557, 1978.
8. D. Harman. Overview of the second text retrieval conference (TREC-2). In *Information Processing & Management*, pages 271–289. Pergamon Press, Inc., 1995.

9. G. L. Heileman and W. Luo. How caching affects hashing. In *Proc. ALENEX Workshop on Algorithm Engineering and Experiments*, January 2005.

10. S. Heinz, J. Zobel, and H. E. Williams. Self-adjusting trees in practice for large text collections. *Software—Practice and Experience*, 31(10):925–939, 2001.

11. M. Karlsson, F. Dahlgren, and P. Stenstrom. A prefetching technique for irregular accesses to linked data structures. In *Proc. Symp. on High-Performance Computer Architecture*, pages 206–217, January 2000.

12. D. E. Knuth. *The Art of Computer Programming: Sorting and Searching*, volume 3. Addison-Wesley Longman, second edition, 1998.

13. P. Larson. Performance analysis of linear hashing with partial expansions. *ACM Transactions on Database Systems*, 7(4):566–587, 1982.

14. A. R. Lebeck. Cache conscious programming in undergraduate computer science. In *Proc. SIGCSE Technical Symp. on Computer Science Education*, pages 247–251. ACM Press, New York, 1999.

15. J. McCabe. On serial files with relocatable records. *Operations Research*, 13:609–618, 1965.

16. J. I. Munro and P. Celis. Techniques for collision resolution in hash tables with open addressing. In *Proc. ACM Fall Joint Computer Conf.*, pages 601–610. IEEE Computer Society Press, 1986.

17. W. W. Peterson. Open addressing. *IBM J. Research & Development*, 1:130–146, 1957.

18. M. V. Ramakrishna and J. Zobel. Performance in practice of string hashing functions. In *Proc. DASFAA Symp. on Databases Systems for Advanced Applications*, volume 6, pages 215–224. World Scientific, April 1997.

19. A. Rathi, H. Lu, and G. E. Hedrick. Performance comparison of extendible hashing and linear hashing techniques. In *Proc. ACM SIGSMALL/PC Symp. on Small Systems*, pages 178–185. ACM Press, New York, 1990.

20. A. L. Rosenberg and L. J. Stockmeyer. Hashing schemes for extendible arrays. *Jour. of the ACM*, 24(2):199–221, 1977.

21. A. Roth and G. S. Sohi. Effective jump-pointer prefetching for linked data structures. In *Proc. Int. Symp. on Computer Architecture*, pages 111–121. IEEE Computer Society Press, 1999.

22. D. V. Sarwate. A note on universal classes of hash functions. *Information Processing Letters*, 10(1):41–45, 1980.

23. R. Sinha, D. Ring, and J. Zobel. Cache-efficient string sorting using copying. *In submission*.

24. R. Sinha and J. Zobel. Cache-conscious sorting of large sets of strings with dynamic tries. *ACM Jour. of Exp. Algorithmics*, 9, 2005.

25. J. S. Vitter. Analysis of the search performance of coalesced hashing. *Jour. of the ACM*, 30(2):231–258, 1983.

26. C. Yang, A. R. Lebeck, H. Tseng, and C. Lee. Tolerating memory latency through push prefetching for pointer-intensive applications. *ACM Trans. Architecture Code Optimisation*, 1(4):445–475, 2004.

27. J. Zobel, H. E. Williams, and S. Heinz. In-memory hash tables for accumulating text vocabularies. *Information Processing Letters*, 80(6):271–277, December 2001.

# Measuring the Difficulty
# of Distance-Based Indexing

Matthew Skala

University of Waterloo, Waterloo, Ontario, N2L 3G1, Canada
mskala@cs.uwaterloo.ca

**Abstract.** Data structures for similarity search are commonly evaluated on data in vector spaces, but distance-based data structures are also applicable to non-vector spaces with no natural concept of dimensionality. The intrinsic dimensionality statistic of Chávez and Navarro provides a way to compare the performance of similarity indexing and search algorithms across different spaces, and predict the performance of index data structures on non-vector spaces by relating them to equivalent vector spaces. We characterise its asymptotic behaviour, and give experimental results to calibrate these comparisons.

## 1   Introduction

Suppose we wish to index a database for similarity search. For instance, we might have a database of text documents which we query with an example document to find others close to the example. Speaking of closeness implies we must have a distance function applicable to the objects in the database. Maybe our objects are actually vectors of real numbers with a Minkowski $L_p$ metric. Many effective data structures are known for that case, including $R$-trees and variants [3,11,17], $SR$-trees [13], and pyramid-trees [4].

But maybe the objects are not vectors; and maybe the distance function is not an $L_p$ metric. Edit distance on strings, for instance, forms a metric space that is not a vector space. Structures for indexing general metric spaces include $VP$-trees [20], $MVP$-trees [5], $GH$-trees [19], and $FQ$-trees [2]. Such structures are called "distance-based" because they rely exclusively on the distances between the query point and other points in the space.

The problem of distance-based indexing seems to become harder in spaces with more dimensions, but we cannot easily count dimensions in a non-vector space. Even when we represent our documents as long vectors, indexing algorithms behave much differently on real document databases from the prediction for similar-length randomly generated vectors. In this work we consider how to predict indexing performance on practical spaces by comparison with random vector spaces of similar difficulty.

### 1.1   Intrinsic Dimensionality

Suppose we have a general space, from which we can choose objects according to a fixed probability distribution, and measure the distance between any two

M. Consens and G. Navarro (Eds.): SPIRE 2005, LNCS 3772, pp. 103–114, 2005.

objects, but the objects are opaque: all we know about an object is its distance from other objects. We might wish to assume that we have a metric space, with the triangle inequality, but even that might only hold in an approximate way—for instance, only to within a constant factor, as with the "almost metrics" defined by Sahinalp and others [16]. Some functions we might like to use do not naturally obey the triangle inequality—such as relative entropy measured by compression, proposed in bioinformatics applications [10,14].

Given such a space, the only way we can describe the space or distinguish it from other general spaces is by choosing random points and considering the probability distribution of distances between them. Chávez and Navarro introduce a statistic called "intrinsic dimensionality" for describing spaces in terms of the distribution of the distance between two randomly chosen points. Where $\mu$ and $\sigma^2$ are the mean and variance of that distance, the intrinsic dimensionality $\rho$ is defined as $\mu^2/(2\sigma^2)$ [6]. Squaring the mean puts it in the same units as the variance; and as we prove, the constant 2 makes $\rho$ equal the number of vector dimensions for uniform random vectors with $L_1$ and approach it for normal random vectors with $L_2$.

Chávez and Navarro prove bounds on the performance of several kinds of distance-based index structures for metric spaces in terms of $\rho$. Spaces that are easy to index have small $\rho$, and the statistic increases as the spaces become harder to index. They also give an argument (using a proof of Yianilos) for why intrinsic dimensionality ought to be proportional to the number of vector components when applied to points chosen uniformly at random from vector spaces [6,21]. To calibrate the dimensionality measurement, they show experimental results for low-dimensional spaces to estimate the asymptotic constant of proportionality for $\rho$ in terms of $n$, with $n$-component vectors having each component chosen from a uniform distribution and using $L_p$ metrics [6].

We analyse the behaviour of $\rho$ for vectors chosen with independent identically distributed real components, and distance measured by an $L_p$ metric; the result is exact for $L_1$. We find that $\rho(n)$ is $\Theta(n)$ for $L_p$ with finite $p$, but not necessarily for $L_\infty$. We show $\rho$ to be $\Theta(\log^2 n)$ in the case of normally-distributed random vectors with the $L_\infty$ metric. We also present experimental results corroborating our theory. The slopes of the lines are found to be significantly greater than predicted by previous experiments, because the true asymptotic behaviour only shows itself at large $n$. The behaviour of the asymptotic lines as $p$ varies is seen to be counter-intuitive, with the $L_\infty$ metric on uniform vectors much different from the $L_p$ metric for large but finite $p$.

## 1.2   Notation

Following the notation used by Arnold, Balakrishnan, and Nagaraja, [1] we write $X \stackrel{d}{=} Y$ if $X$ and $Y$ are identically distributed, $X \stackrel{d}{\to} Y$ if the distribution of $X(n)$ converges to the distribution of $Y$ as $n$ goes to positive infinity, and $X \stackrel{d}{\leftrightarrow} Y$ if the distributions of both $X$ and $Y$ depend on $n$ and converge to each other. We also write $f(n) \to x$ if $x$ is the limit of $f(n)$ as $n$ goes to positive infinity, $E[X]$ and $V[X]$ for the expectation and variance of $X$ respectively, $\log x$ for the natural logarithm of $x$, and $\Gamma(x)$ for the standard gamma function (generalised factorial). Random variables that are **independent and identically distributed** are called iid, a random variable's **probability density function** is called its pdf, and its **cumulative distribution function** is called its cdf.

Let $X = Y$ be a real random variate realised as random variables $X_i$ and $Y_i$. Let $\boldsymbol{x}_n = \langle X_1, X_2, \ldots, X_n \rangle$ and $\boldsymbol{y}_n = \langle Y_1, Y_2, \ldots, Y_n \rangle$ be vector random variables with $n$ iid components each, each component drawn from the variate. Let $D_{p,n}$ be the distance between $\boldsymbol{x}$ and $\boldsymbol{y}$ under the $L_p$ metric $\|\boldsymbol{x} - \boldsymbol{y}\|_p$, defined as $\left(\sum_{i=1}^{n} |X_i - Y_i|^p\right)^{1/p}$ for real $p > 0$ or $\max_{i=1}^{n} |X_i - Y_i|$ where $p = \infty$. We are concerned with the distribution of the random variable $D_{p,n}$, and in particular the asymptotic behaviour for large $n$ of the intrinsic dimensionality statistic $\rho_p(n) = E[D_{p,n}]^2 / 2V[D_{p,n}]$ [6].

When discussing $L_\infty$, which is defined in terms of the maximum function, it is convenient to define for any real random variate $Z$ random variates $\max^{(k)}\{Z\}$ and $\min^{(k)}\{Z\}$ realised as random variables $\max_i^{(k)}\{Z\}$. and $\min_i^{(k)}\{Z\}$ respectively. Each $\max_i^{(k)}\{Z\}$ is the maximum, and each $\min_i^{(k)}\{Z\}$ the minimum, of $k$ random variables from $Z$.

### 1.3   Extreme Order Statistics

Extreme order statistics of collections of random variables (the maximum, the minimum, and generalisations of them) have been thoroughly studied [1,9]. If $F(x)$ is the cdf of $Z$, then $F^n(x)$ is the cdf of $\max^{(n)}\{Z\}$. We say that the random variable $W$ with non-degenerate cdf $G(x)$ is the limiting distribution of the maximum of $Z$ if there exist sequences $\{a_n\}$ and $\{b_n > 0\}$ such that $F^n(a_n + b_n x) \to G(x)$. There are only a few possible distributions for $W$, if it exists at all.

**Theorem 1 (Fisher and Tippett, 1928).** *If* $(\max^{(n)}\{Z\} - a_n)/b_n \xrightarrow{d} W$, *then the cdf $G(x)$ of $W$ must be of one of the following types, where $\alpha$ is a constant greater than zero [1, Theorem 8.3.1] [8]:*

$$G_1(x; \alpha) = \exp(-x^{-\alpha}) \text{ for } x > 0 \text{ and } 0 \text{ otherwise;} \tag{1}$$

$$G_2(x; \alpha) = \exp(-(-x)^{\alpha}) \text{ for } x < 0 \text{ and } 1 \text{ otherwise; or} \tag{2}$$

$$G_3(x) = \exp(-e^{-x}) \ . \tag{3}$$

## 2   Intrinsic Dimensionality of Random Vectors

Even though intrinsic dimensionality is most important for non-vector spaces, like strings with edit distance, we wish to know the behaviour of the intrinsic dimensionality statistic on familiar vector spaces so we can do meaningful comparisons. Let $\boldsymbol{x}$ and $\boldsymbol{y}$ be random $n$-component vectors as described above, using the $L_p$ metric. We will compute the asymptotic behaviour of the intrinsic dimensionality $\rho_p(n)$ as $n$ goes to infinity, based on the distribution of $|X - Y|$. Let $\mu'_k$ represent the $k$-th raw moment of $|X - Y|$; that is, the expected value of $|X - Y|^k$.

## 2.1   The $L_p$ Metric for Finite $p$

We would like the intrinsic dimensionality statistic to be proportional to the length of the vectors when applied to random vectors with iid components and distance measured by $L_p$ metrics. For finite $p$ as the number of components goes to infinity, it does indeed behave that way.

**Theorem 2.** *With the $L_p$ metric for fixed finite $p$, when the $|X_i - Y_i|$ are iid with raw moments $\mu'_k$, then $\rho_p(n) \to [p^2(\mu'_p)^2/(2(\mu'_{2p} - (\mu'_p)^2))]n$.*

*Proof.* The $L_p$ metric for finite $p$ is computed by taking the sum of random variables $|X_i - Y_i|^p$; call the result $S$. Then the metric is $S^{1/p}$. We have $V[|X_i - Y_i|^p] = \mu'_{2p} - (\mu'_p)^2$, $E[S] = n\mu'_p$, and $V[S] = n(\mu'_{2p} - (\mu'_p)^2)$.

Since the mean and variance both increase linearly with $n$, the standard deviation will eventually become small in relation to the mean. For large $n$ we can approximate the function $x^{1/p}$ with a tangent line:

$$E[S^{1/p}] \to E[S]^{1/p} = n^{1/p}(\mu'_p)^{1/p} \tag{4}$$

$$V[S^{1/p}] \to V[S]\left(\frac{d}{dS}S^{1/p}\right)^2\Bigg|_{S=E[S]} = \frac{\mu'_{2p} - (\mu'_p)^2}{np^2(\mu'_p)^2}n^{2/p}(\mu'_p)^{2/p} \tag{5}$$

$$\rho_p(n) = \frac{E[S^{1/p}]^2}{2V[S^{1/p}]} \to n\frac{p^2(\mu'_p)^2}{2(\mu'_{2p} - (\mu'_p)^2)} \tag{6}$$

$\square$

If we are using the $L_1$ metric, the analysis is even better:

**Corollary 1.** *When $p = 1$, the approximation given by Theorem 2 becomes exact: $\rho_1(n) = [(\mu'_1)^2/(2(\mu'_2 - (\mu'_1)^2))]n$.*

*Proof.* When $p = 1$, then $E[S^{1/p}] = E[S] = E[S]^{1/p}$ and $V[S^{1/p}] = V[S] = V[S]^{1/p}$, regardless of $n$, and the limits for large $n$ in the proof of Theorem 2 become equalities. $\square$

## 2.2   Binary Strings with Hamming Distance

Binary strings under Hamming distance are an easy case for the theory, and are of interest in applications like the Nilsimsa spam filter [7]. We can find the intrinsic dimensionality of the space of $n$-bit binary strings under Hamming distance by treating the strings as vectors with each component a Bernoulli random variable, equal to one with probability $q$ and zero otherwise. Then the Hamming distance (number of bits with differing values) is the same as the $L_1$ distance (sum of absolute component-wise differences), and by Corollary 1, $\rho_1(n) = nq(1 - q)/(1 - 2q + 2q^2)$.

Note that $q = 1/2$ produces the maximum value of $\rho_1(n)$, namely $n/2$. Substituting into the lower bound of Chávez and Navarro, we find that a pivot-based algorithm using random pivots on a database of $m$ strings each $n$ bits long, with the Hamming metric, must use at least $\frac{1}{2}(\sqrt{n} - 1/\sqrt{f})^2 \ln m$ distance evaluations on average per query, to satisfy random queries returning at most a fraction $f$ of the database [6].

## 2.3   The $L_\infty$ Metric

The distance $D_{\infty,n}$ is the maximum of $n$ variables drawn from $|X - Y|$. We can eliminate the absolute value function with the following lemma.

**Lemma 1.** *If $Z$ is a real variate with distribution symmetric about zero, and $W, a_n$, and $b_n$ exist with $(\max^{(n)}\{Z\} - a_n)/b_n \overset{d}{\to} W$, then $\max^{(n)}\{|Z|\} \overset{d}{\leftrightarrow} \max^{(2n)}\{Z\}$.*

*Proof.* Instead of taking the maximum absolute value of a set of $n$ variables from $Z$, we could find the maximum and the negative of the minimum and then take the maximum of those two. But as described by Arnold, Balakrishnan, and Nagaraja, the maximum and minimum of a collection of random variables are asymptotically independent [1, Theorem 8.4.3]. Thus $\max^{(n)}\{|Z|\} \overset{d}{\leftrightarrow} \max\{\max^{(n)}\{Z\}, -\min^{(n)}\{Z\}\}$; and by symmetry of $Z$,

$$\max\{\max^{(n)}\{Z\}, -\min^{(n)}\{Z\}\} \overset{d}{=} \max^{(2n)}\{Z\} \tag{7}$$

□

Given the distribution of $X - Y$ or $|X - Y|$, we can obtain the limiting distribution for $D_{\infty,n} = \max^{(n)}\{|X-Y|\}$; and if it exists, it will be in one of the three forms stated in Theorem 1. We can then integrate to find the expectation and variance, and standard results give acceptable choices for the norming constants $a_n$ and $b_n$, giving the following theorem.

**Theorem 3.** *For random vectors with the $L_\infty$ metric, when Theorem 1 applies to $\max^{(n)}\{|X - Y|\}$, we have:*

$$\rho_\infty(n) \to \frac{(a_n + b_n\Gamma(1 - 1/\alpha))^2}{2b_n^2(\Gamma(1 - 2/\alpha) - \Gamma^2(1 - 1/\alpha))} \text{ for } G_1(x;\alpha), \alpha > 2; \tag{8}$$

$$\rho_\infty(n) \to \frac{(a_n + b_n\Gamma(1 + 1/\alpha))^2}{2b_n^2(\Gamma(1 + 2/\alpha) - \Gamma^2(1 + 1/\alpha))} \text{ for } G_2(x;\alpha); \text{ and} \tag{9}$$

$$\rho_\infty(n) \to \frac{3(a_n + b_n\gamma)^2}{b_n^2\pi^2} \text{ for } G_3(x); \tag{10}$$

*where $\gamma = 0.57721\,56649\,015\ldots$, the Euler-Mascheroni constant.* □

Unlike in the finite-$p$ case, $\rho_\infty(n)$ does not necessarily approach a line.

## 2.4   Uniform Vectors

Let $X$ and $Y$ be uniform real random variates with the range $[0, 1)$, as used by Chávez and Navarro in their experiment [6]. The pdf of $|X - Y|$ is $2 - 2x$ for $0 \le x < 1$. Simple integration gives the raw moments $\mu'_p = 2/(p+1)(p+2)$ and $\mu'_{2p} = 1/(2p+1)(p+1)$, and then by Theorem 2, $\rho_p(n) \to [(4p+2)/(p+5)]n$.

For the $L_\infty$ metric, we note that the cdf of $|X - Y|$ is $F(x) = 2x - x^2$. Then standard results on extreme order statistics [1, Theorems 8.3.2(ii), 8.3.4(ii)] give us that $(\max^{(n)}\{|X - Y|\} - a_n)/n_b \overset{d}{\to} W$ where $a_n = 1, b_n = 1/\sqrt{n}$, and the cdf of $W$ is $G_2(x; \alpha)$ with $\alpha = 2$. By (9), $\rho_\infty(n) \to n/(2 - (\pi/2))$. So as $n$ increases, $\rho_\infty(n)$ approaches a line with slope $1/(2 - (\pi/2)) = 2.32989\,61831\,6\ldots$; the same line approached by $\rho_{\tilde{p}}(n)$ where $\tilde{p} = (1 + \pi)/(7 - 2\pi) = 5.77777\,31051\,9\ldots$.

We repeated the experiment described by Chávez and Navarro [6, Fig. 3], of randomly choosing one million pairs of points, finding their distances, and computing the intrinsic dimensionality. The results are shown in Fig. 1. Examination reveals an apparent linear trend for each metric, but the points seem to be on much shallower lines than the theory predicts. The points for $L_{256}$ match those for $L_\infty$, supporting the intuition that $L_p$ for large $p$ should have the same asymptotic behaviour as $L_\infty$.

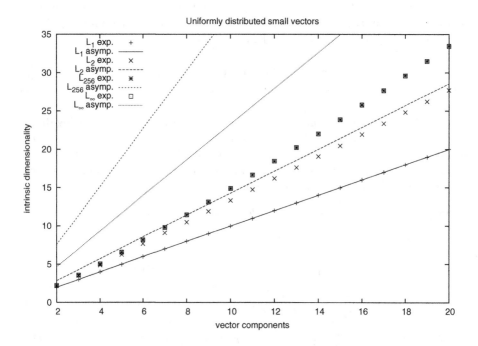

**Fig. 1.** Experimental results for short vectors with uniform random components

Intuition turns out to be wrong. Repeating the experiment with vectors of up to one million components (Fig. 2), we see that the line for $L_p$ does approach a slope of four as $p$ increases, but with the $L_\infty$ metric, the line drops to coincide with the line for $L_{\tilde{p}}$, $\tilde{p} \approx 5.778$, just as predicted by the theory. This phenomenon is actually not quite so strange as it may seem: this is simply a situation where we are taking two limits and it matters which order we take them.

## 2.5   Normal Vectors

Consider a similar case but let $X$ and $Y$ be standard normal random variates. Since $X$ and $Y$ are standard normal, their difference $X - Y$ is normal with mean

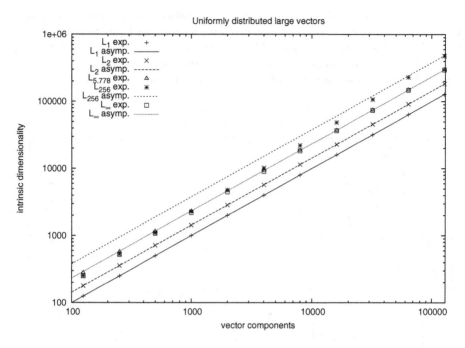

**Fig. 2.** Experimental results for long vectors with uniform random components

zero and variance two. Then each $|X - Y|$ has a "half-normal" distribution with pdf $(\sqrt{2}/\pi)e^{-x^2/2\pi}$. As before, we compute the raw moments and substitute into the intrinsic dimensionality formula, finding that $\mu_p' = \pi^{(p-1)/2}2^{p/2}\Gamma((p+1)/2)$, $\mu_{2p}' = \pi^{p-1/2}2^p\Gamma(p+1/2)$, and so $\rho_p(n) \to n[p^2\Gamma^2((p+1)/2)/2(\sqrt{\pi}\Gamma(p+1/2) - \Gamma^2((p+1)/2))]$. As in the uniform case, $\rho_p(n) = \Theta(n)$, but the slope is quite different. The maximum slope is one, with the $L_2$ metric; $L_1$ and $L_3$ give slopes of approximately 0.9; and for larger $p$ the slope rapidly approaches zero.

Now, $D_{\infty,n} = \max^{(n)}\{|X - Y|\}$. By Lemma 1 we can instead consider $\max^{(2n)}\{X - Y\}$. Each $X - Y$ is normal with mean zero and variance two. Standard results on the maximum of normal random variables give us that $(D_{\infty,n} - a_{2n})/b_{2n} \xrightarrow{d} W$ where the cdf of $W$ is $G_3(x) = \exp(-e^{-x})$ and the norming constants are $a_{2n} = 2\sqrt{\log 2n} - (\log(4\pi \log 2n))/2\sqrt{\log 2n}$ and $b_{2n} = 1/\sqrt{\log 2n}$ [1,9,12]. Then we can substitute into (10) to find the asymptotic intrinsic dimensionality $\rho_\infty(n) \to (3/4\pi^2) \cdot [4 \log n - \log \log 2n + \log(4/\pi) + 2\gamma]^2$, which is $\Theta(\log^2 n)$.

As with uniform vector components, the intrinsic dimensionality shows markedly different asymptotic behaviour with the $L_\infty$ metric from its behaviour with $L_p$ metrics for finite $p$; but here, instead of being linear with a surprising slope, it is not linear at all. The argument for linear behaviour from Yianilos [21, Proposition 2] only applies to finite $p$.

To verify these results, we generated one million pairs of randomly-chosen vectors for a number of combinations of vector length and $L_p$ metric, and calculated the intrinsic dimensionality. The results are shown in Figs. 3 and 4 along with the theoretical asymptotes. As with uniform components, the true asymptotic behaviour for some metrics is only shown at the largest vector sizes.

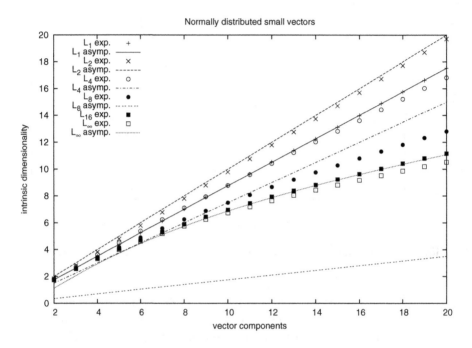

**Fig. 3.** Experimental results for short vectors with normal random components

Normal distributions in high-dimensional vector spaces have smaller intrinsic dimensionality than uniform distributions with vectors of the same length, when considered with $L_\infty$ and $L_p$ for large $p$. Does that mean normal distributions are easier to index, or only that intrinsic dimensionality is a poor measure of indexing difficulty? We argue that normal distributions really are easier to index.

A random vector $x$ from a high-dimensional normal distribution will typically have many small components and one, or a few, of much greater magnitude. Comparing $x$ to another random point $y$, the greatest components of $x$ will usually correspond to small components of $y$ and vice versa, so the $L_\infty$ distance between the two will usually be approximately equal to the one largest component of either vector. At high enough dimensions we could closely approximate the distances between points in almost all cases by only examining the index and magnitude of the single greatest component of each vector. We could achieve good indexing by just putting the points into bins according to index of greatest component, and using cheap one-dimensional data structures within bins. That is how pyramid-trees work [4], and they work well in this case.

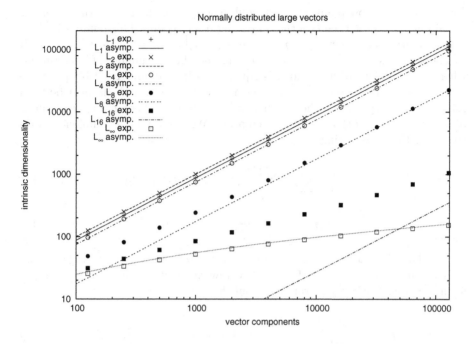

**Fig. 4.** Experimental results for long vectors with normal random components

However, when the vectors are selected from a uniform distribution, then componentwise differences have a triangular distribution, with heavier tails. More components of the difference vector are likely to be large and have a chance of determining the distance, so the indexing structure must represent more information per vector.

## 3   Other Spaces

Random vector spaces are of interest for calibrating the intrinsic dimensionality statistic, but practical spaces may be more difficult to analyse. Here we show the application of the statistic to some other spaces of interest.

### 3.1   Balls in Hamming Spaces

Consider a ball of radius $r$ in the space of $n$-bit binary strings; that is, a fixed $n$-bit string $c$ and all the strings with Hamming distance from $c$ equal to or less than $r$. If we consider this set as a metric space itself, using the Hamming distance and choosing points uniformly at random from the set, what is its intrinsic dimensionality?

**Theorem 4.** *For a ball in the space of $n$-bit strings of constant radius $r$ using the Hamming metric and choosing strings uniformly at random, $\rho \to [r/(2r+1)]n$.*

*Proof.* Consider how many ways we could choose $i$ of the $n$ bits, then $j$ of the remaining $n - i$ bits, then $k$ of the remaining $n - i - j$ bits. This number is given by the multichoose function $(i, j, k, n - i - j - k)! = n!/i!j!k!(n - i - j - k)!$. If we choose two strings $x$ and $y$ from the ball, let $i$ be the number of bit positions where $x$ is different from $c$ and $y$ is equal, let $j$ be the number of bit positions where $y$ is different from $c$ and $x$ is equal, and then let $k$ (which must be from zero to $r - \max\{i, j\}$) be the number of bit positions where $x$ and $y$ are both different from $c$ and thus equal to each other. We can count the number of ways to choose these two strings as

$$N = \sum_{i=0}^{r} \sum_{j=0}^{r} \sum_{k=0}^{r-\max\{i,j\}} (i, j, k, n - i - j - k)! \tag{11}$$

$$= \frac{1}{r!^2} n^{2r} - \frac{r - 3}{r!(r - 1)!} n^{2r-1} + o(n^{2r-1}) . \tag{12}$$

Similarly, by finding the leading terms of the sums and applying long division, we can find expressions for the first two raw moments of the distance for two strings chosen uniformly at random from the ball:

$$\mu'_1 = \frac{1}{N} \sum_{i=0}^{r} \sum_{j=0}^{r} \sum_{k=0}^{r-\max\{i,j\}} (i + j)(i, j, k, n - i - j - k)! \tag{13}$$

$$= 2r - 2r(r + 1)n^{-1} + o(n^{-1}) \tag{14}$$

$$\mu'_2 = \frac{1}{N} \sum_{i=0}^{r} \sum_{j=0}^{r} \sum_{k=0}^{r-\max\{i,j\}} (i + j)^2 (i, j, k, n - i - j - k)! \tag{15}$$

$$= 4r^2 - 2r(4r^2 + 2r + 1)n^{-1} + o(n^{-1}) . \tag{16}$$

Then by substitution into the intrinsic dimensionality formula, we obtain $\rho \to [r/(2r + 1)]n$.  □

## 3.2   An Image Database

We constructed an image database by selecting frames at random from a selection of commercial DVD motion pictures, choosing each frame with $1/200$ probability to create a database of 3239 images, which were converted and scaled to give 259200-element vectors representing the RGB colour values for $360 \times 240$ pixels. Sampling $10^5$ pairs of these vectors using each of the $L_2$ and $L_\infty$ metrics produced $\rho$ values of 2.759 for $L_2$ and 38.159 for $L_\infty$. These results suggest that the $L_2$ metric reveals much stronger clumping structure on this database than the $L_\infty$ metric does; and with $L_2$, this database is approximately as hard to index as a three-dimensional normal distribution in $L_2$ space ($\rho = 2.813$, from the experiment shown in Fig. 3). If we have a choice about which metric to use, the $L_2$ metric will produce a much more efficient index than the $L_\infty$ metric.

### 3.3   A Text Database

We obtained a sample of 28999 spam email messages from SpamArchive.org [18], and added 2885, or approximately 10 percent, non-spam messages from locally collected outgoing email, to simulate the database a practical spam-filtering application might process. We sampled $10^5$ pairs of messages, computed their distances using the Perl Digest::Nilsimsa [15] 256-bit robust hash, and Hamming distance, and computed the intrinsic dimensionality $\rho = 10.338$. For the spam messages alone, and for the non-spam messages alone, we obtained $\rho$ values of 10.292 and 11.262 respectively, again with sampling of $10^5$ pairs for each database. An index of the email database based on Hamming distance of the Nilsimsa hashes would perform better than a similar index on uniform random 256-bit strings, but answering queries would still be quite difficult, a little more difficult than for random data normally distributed in 10-dimensional $L_2$ space.

## 4   Conclusions and Future Work

Intrinsic dimensionality answers questions about spaces: which spaces have comparable indexing difficulty, which metrics will allow good indexing, and lower bounds on query complexity. We have characterised the asymptotic behaviour of the intrinsic dimensionality statistic for randomly chosen vectors with the components having uniform or normal distributions, and the $L_p$ metrics for both finite and infinite $p$. As our theoretical results show, uniform and normal components produce vastly different results, especially for $L_p$ with large $p$ and $L_\infty$. In those metrics, high-dimensional normal distributions are easier to index than uniform distributions of the same dimension. We have also given results for more complicated spaces: balls in Hamming space, and practical databases of images and email messages, demonstrating the flexibility of the technique.

The ultimate question for indexing difficulty measurement is how much making a distance measurement reduces our uncertainty about the query point's distance to points in the index. Intrinsic dimensionality attempts to answer the question based on the mean and variance of the distribution of a single distance; but we might obtain a more useful statistic by considering the joint distribution of distances among more than two randomly chosen points. Such a statistic could allow the proof of highly general bounds on indexing performance.

## References

1. Arnold, B.C., Balakrishnan, N., Nagaraja, H.N.: A First Course in Order Statistics. Wiley series in probability and mathematical statistics. John Wiley & Sons, Inc., New York (1992)
2. Baeza-Yates, R.A., Cunto, W., Manber, U., Wu, S.: Proximity matching using fixed-queries trees. In: CPM (Combinatorial Pattern Matching). Volume 807 of Lecture Notes in Computer Science., Springer (1994) 198–212
3. Beckmann, N., Kriegel, H.P., Schneider, R., Seeger, B.: The R*-tree: An efficient and robust access method for points and rectangles. In: SIGMOD (International Conference on Management of Data). (1990) 322–331

4. Berchtold, S., Böhm, C., Kriegel, H.P.: The pyramid-tree: Breaking the curse of dimensionality. In: SIGMOD (International Conference on Management of Data). (1998) 142–153

5. Bozkaya, T., Ozsoyoglu, M.: Indexing large metric spaces for similarity search queries. ACM Transactions on Database Systems **24** (1999) 361–404

6. Chávez, E., Navarro, G.: Measuring the dimensionality of general metric spaces. Technical Report TR/DCC-00-1, Department of Computer Science, University of Chile (2000) Submitted. Online `ftp://ftp.dcc.uchile.cl/pub/users/gnavarro/metricmodel.ps.gz`.

7. Damiani, E., De Capitani di Vimercati, S., Paraboschi, S., Samarai, P.: An open digest-based technique for spam detection. In: 2004 International Workshop on Security in Parallel and Distributed Systems, San Francisco, CA, USA (2004)

8. Fisher, R.A., Tippett, L.H.C.: Limiting forms of the frequency distribution of the largest or smallest member of a sample. Proceedings of the Cambridge Philosophical Society **24** (1928) 180–190

9. Galambos, J.: The Asymptotic Theory of Extreme Order Statistics. Second edn. Robert E. Krieger Publishing Company, Malabar, Florida, U.S.A. (1987)

10. Grumbach, S., Tahi, F.: A new challenge for compression algorithms: Genetic sequences. Journal of Information Processing and Management **30** (1994) 875–886

11. Guttman, A.: R-trees: a dynamic index structure for spatial searching. SIGMOD Record (ACM Special Interest Group on Management of Data) **14** (1984) 47–57

12. Hall, P.: On the rate of convergence of normal extremes. Journal of Applied Probability **16** (1979) 433–439

13. Katayama, N., Satoh, S.: The SR-tree: an index structure for high-dimensional nearest neighbor queries. In: SIGMOD (International Conference on Management of Data). (1997) 369–380

14. Li, M., Badger, J.H., Xin, C., Kwong, S., Kearney, P., Zhang, H.: An information based sequence distance and its application to whole mitochondrial genome phylogeny. Bioinformatics **17** (2001) 149–154

15. Norwood, C., cmeclax: Digest::Nilsimsa 0.06. Computer software (2002) Online `http://search.cpan.org/~vipul/Digest-Nilsimsa-0.06/`.

16. Sahjnalp, S.C., Tasan, M., Macker, J., Ozsoyoglu, Z.M.: Distance based indexing for string proximity search. In: ICDE (International Conference on Data Engineering), IEEE Computer Society (2003)

17. Sellis, T.K., Roussopoulos, N., Faloutsos, C.: The R+-tree: A dynamic index for multi-dimensional objects. In: VLDB'87 (International Conference on Very Large Data Bases), Morgan Kaufmann (1987) 507–518

18. SpamArchive.org: Donate your spam to science. Web site (2005) Online `http://www.spamarchive.org/`.

19. Uhlmann, J.K.: Satisfying general proximity/similarity queries with metric trees. Information Processing Letters **40** (1991) 175–179

20. Yianilos, P.N.: Data structures and algorithms for nearest neighbor search in general metric spaces. In: SODA (Symposium on Discrete Algorithms), SIAM (1993) 311–321

21. Yianilos, P.N.: Excluded middle vantage point forests for nearest neighbour search. In: ALENEX (Algorithm Engineering and Experimentation: International Workshop). (1999)

# N-Gram Similarity and Distance

Grzegorz Kondrak

Department of Computing Science, University of Alberta,
Edmonton, AB, T6G 2E8, Canada
kondrak@cs.ualberta.ca
http://www.cs.ualberta.ca/~kondrak

**Abstract.** In many applications, it is necessary to algorithmically quantify the similarity exhibited by two strings composed of symbols from a finite alphabet. Numerous string similarity measures have been proposed. Particularly well-known measures are based are edit distance and the length of the longest common subsequence. We develop a notion of $n$-gram similarity and distance. We show that edit distance and the length of the longest common subsequence are special cases of $n$-gram distance and similarity, respectively. We provide formal, recursive definitions of $n$-gram similarity and distance, together with efficient algorithms for computing them. We formulate a family of word similarity measures based on $n$-grams, and report the results of experiments that suggest that the new measures outperform their unigram equivalents.

## 1 Introduction

In many applications, it is necessary to algorithmically quantify the similarity exhibited by two strings composed of symbols from a finite alphabet. For example, for the task of automatic identification of confusable drug names, it is helpful to recognize that the similarity between *Toradol* and *Tegretol* is greater than the similarity between *Toradol* and *Inderal*. The problem of measuring string similarity occurs in a variety of fields, including bioinformatics, speech recognition, information retrieval, machine translation, lexicography, and dialectology [9]. A related issue of computing the similarity of texts as strings of words has also been studied.

Numerous string similarity measures have been proposed. A particularly widely-used method is *edit distance* (EDIT), also known as Levenshtein distance, which is defined as the minimum number of elementary edit operations needed to transform one string into another. Another, closely related approach relies on finding the length of the *longest common subsequence* (LCS) of the two strings. Other similarity measures are based on the number of shared $n$-grams, i.e., substrings of length $n$.

In this paper, we develop a notion of $n$-gram similarity and distance.[1] We show that edit distance and the length of the LCS are special cases of $n$-gram

---

[1] This is a different concept from the $q$-gram similarity/distance [12], which is simply the number of common/distinct $q$-grams ($n$-grams) between two strings.

M. Consens and G. Navarro (Eds.): SPIRE 2005, LNCS 3772, pp. 115–126, 2005.

distance and similarity, respectively. We provide formal, recursive definitions of $n$-gram similarity and distance, and efficient algorithms for computing them. We formulate a family of word similarity measures based on $n$-grams, which are intended to combine the advantages of the unigram and the $n$-gram measures. We evaluate the new measures on three different word-comparison tasks: the identification of genetic cognates, translational cognates, and confusable drug names. The results of our experiments suggest that the new $n$-gram measures outperform their unigram equivalents.

We begin with $n$-gram similarity because we consider it to be conceptually simpler than $n$-gram distance. The latter notion is then defined by modifying the formulation of the former.

## 2　Unigram Similarity

In this section, we discuss the notion of the length of the LCS, which we view as *unigram similarity*, in the context of its applicability as a string similarity measure. After defining the longest common subsequence problem in a standard way, we provide an alternative but equivalent formulation of the length of the LCS. The recursive definition not only elucidates the relationship between the LCS length and edit distance, but also generalizes naturally to $n$-gram similarity and distance.

### 2.1　Standard Definition

The standard formulation of the LCS problem is as follows [3]. Given a sequence $X = x_1 \ldots x_k$, another sequence $Z = z_1 \ldots z_m$ is a subsequence of X if there exist a strictly increasing sequence $i_1, \ldots, i_m$ of indices of X such that for all $j = 1, \ldots, m$, we have $x_{i_j} = z_j$. For example, *tar* is a subsequence of *contrary*. Given two sequences X and Y, we say that a sequence Z is a common subsequence of X and Y if Z is a subsequence of both X and Y. In the LCS problem, we are given two sequences and wish to find their maximum-length common subsequence. For example, the LCS of *natural* and *contrary* is *ntra*. The LCS problem can be solved efficiently using dynamic programming.

For the purpose of measuring string similarity, which is our focus here, only the *length* of the LCS is important; the actual longest common subsequence is irrelevant. The length of the LCS as a function of two strings is an interesting function in itself [2].

### 2.2　Recursive Definition

We propose the following formal, recursive definition of the function $\mathbf{s}(X, Y)$, which is equivalent to the length of the LCS. Let $X = x_1 \ldots x_k$ and $Y = y_1 \ldots y_l$ be strings of length $k$ and $l$, respectively, composed of symbols of a finite alphabet. In order to simplify the formulas, we introduce the following notational shorthand, borrowed from Smyth [10]. Let $\Gamma_{i,j} = (x_1 \ldots x_i, y_1 \ldots y_j)$ be a pair of prefixes of $X$ and $Y$, and $\Gamma_{i,j}^* = (x_{i+1} \ldots x_k, y_{j+1} \ldots y_l)$ a pair of suffixes of $X$ and $Y$.

For strings of length one or less, we define $s$ directly:

$$\mathbf{s}(x, \epsilon) = 0, \qquad \mathbf{s}(\epsilon, y) = 0, \qquad \mathbf{s}(x, y) = \begin{cases} 1 \text{ if } x = y \\ 0 \text{ otherwise} \end{cases}$$

where $\epsilon$ denotes an empty string, $x$ and $y$ denote single symbols.

For longer strings, we define $s$ recursively:

$$\mathbf{s}(X, Y) = \mathbf{s}(\Gamma_{k,l}) = \max_{i,j}(\mathbf{s}(\Gamma_{i,j}) + \mathbf{s}(\Gamma_{i,j}^*))$$

The values of $i$ and $j$ in the above formula are constrained by the requirement that both $\Gamma_{i,j}$ and $\Gamma_{i,j}^*$ are non-empty. More specifically, the admissible values of $i$ and $j$ are given by the following set of pairs:

$$D(k, l) = \{0, \ldots, k\} \times \{0, \ldots, l\} - \{(0, 0), (k, l)\}$$

For example, $D(2, 1) = \{(0, 1), (1, 0), (1, 1), (2, 0)\}$.

It is straightforward to show by induction that $\mathbf{s}(X, Y)$ is always equal to the length of the longest common subsequence of strings $X$ and $Y$.

## 2.3 Rationale

Our recursive definition exploits the semi-compositionality of the LCS. Clearly, LCS is not compositional in the usual sense, because the LCS of concatenated strings is not necessarily equal to the sum of their respective LCS. For example, $\|LCS(ab, a)\| = 1$ and $\|LCS(c, bc)\| = 1$, but $\|LCS(abc, abc)\| = 3$. What is certain is that the LCS of concatenated strings is always at least as long as the concatenation of their respective LCS:

$$\mathbf{s}(X_1, Y_1) + \mathbf{s}(X_2, Y_2) \leq \mathbf{s}(X_1 + X_2, Y_1 + Y_2)$$

Loosely speaking, $\mathbf{s}(X, Y)$ is superadditive, rather than compositional. It is always possible to compose the LCS of two strings by concatenating the LCS of their substrings, provided that the decomposition of the strings into substrings preserves all identity matches in the original LCS. Such a decomposition can always be found (cf. Figure 1, left pair).

## 2.4 A Reduced Set of Decompositions

Any decomposition of a pair of strings can be unambiguously defined by a pair of indices. The set $D$ contains all distinct decompositions of a pair of strings. The number of distinct decompositions of a pair of strings is $(k+1) * (l+1) - 2$.

The set of decomposition can be reduced without affecting the values of the function $s$. Let $D'$ be the following set of decompositions:

$$D'(k, l) = \{k - 1, k\} \times \{l - 1, l\} - \{(0, 0), (k, l)\}$$

**Fig. 1.** A decompositions of a unigram alignment that preserves all identity matches (left), and a decomposition of a bigram alignment with various levels of bigram similarity (right)

For example, $D'(2,1) = \{(1,0),(1,1),(2,0)\}$. $D'$ never contains more than three decompositions. By substituting $D$ by $D'$ in the recursive definition given in section 2.2, we obtain an alternative, equivalent formulation of $\mathbf{s}$:

$$\mathbf{s}(X,Y) = \mathbf{s}(\Gamma_{k,l}) = \max(\mathbf{s}(\Gamma_{k-1,l}), \mathbf{s}(\Gamma_{k,l-1}), \mathbf{s}(\Gamma_{k-1,l-1}) + \mathbf{s}(x_k, y_l))$$

The alternative formulation directly yields the well-known efficient dynamic-programming algorithm for computing the length of the LCS [14].

### 2.5   Beyond Unigram Similarity

The main weakness of the LCS length as a measure of string similarity is its insensitivity to context. The problem is illustrated in Figure 2. The two word pairs on the left demonstrate that neighbouring identity matches are a stronger indication of similarity than identity matches that are far apart. The two word pairs on the right show that parallel identity matches are a stronger indication of similarity than identity matches that are separated by unmatched symbols.

A family of similarity measures that do take context into account is based on *Dice coefficient* [1]. The measures are defined as the ratio of the number of $n$-grams that are shared by two strings and the total number of $n$-grams in both strings:

$$\frac{2 \times |n\text{-}grams(X) \cap n\text{-}grams(Y)|}{|n\text{-}grams(X)| + |n\text{-}grams(Y)|}$$

where *n-grams(X)* is a multi-set of letter $n$-grams in $X$. Dice coefficient with bigrams (DICE) is a particularly popular word similarity measure. For example, DICE(*Zantac, Contac*) = $(2 \cdot 3)/(5+5) = 0.6$ because three of the bigrams are shared.

Although more sensitive to context that LCS length, DICE has its own problems. First, because of its "low resolution", it often fails to detect any similarity

```
A M A R Y L      D I O V A N          C A R D U R A          O S M I T R O L
| |               \      |             \ \ |                       | | |
A M I K I N      A M I K I N          B E N A D R O L        B E N A D R O L
```

**Fig. 2.** Two pairs of words with different levels of similarity and the same length of the longest common subsequence

between strings that are very much alike; for example, the pair *Verelan/Virilon* have no $n$-grams in common. Second, DICE can return the maximum similarity value of 1 for strings that are non-identical; for example, both *Xanex* and *Nexan* are composed of the same set of bigrams: {an,ex,ne,xa}. Finally, the measure often associates $n$-grams that occur in radically different word positions, as in the pair *Voltaren/Tramadol*.

Brew and McKelvie [1] propose an extension of DICE, called XXDICE, in which the contribution of matching $n$-grams to the overall score depends on their absolute positions in the strings. XXDICE performed best among several tested measures, and it has subsequently been used by other researchers (e.g., [11]). Unfortunately, the definition of XXDICE is deficient: it does not specify which matching bigrams are to be selected for the calculation of the score when bigrams are not unique. There are a number of ways to amend the definition, but it then becomes implementation-dependent, which means that the results are no longer fully replicable. The case of XXDICE serves as an illustration that it is essential to define string similarity measures rigorously.

In the next section, we formulate the notion of $n$-gram similarity $\mathbf{s_n}$, which is intended to combine the advantages of the LCS length and Dice coefficient while eliminating their flaws.

## 3     $n$-Gram Similarity

The main idea behind $n$-gram similarity is generalizing the concept of the longest common subsequence to encompass $n$-grams, rather than just unigrams. We formulate $n$-gram similarity as a function $\mathbf{s_n}$, where $n$ is a fixed parameter. $\mathbf{s_1}$ is equivalent to the unigram similarity function $\mathbf{s}$ defined in Section 2.2.

### 3.1   Definition

For the purpose of providing a concise recursive definition of $n$-gram similarity, we slightly modify our convention regarding $\Gamma$. When dealing with $n$-grams for $n > 1$, we require $\Gamma_{i,j}$ and $\Gamma_{i,j}^*$ to contain at least one complete $n$-gram. This requirement is consistent with our previous convention for $n = 1$. If both strings are shorter than $n$, $\mathbf{s_n}$ is undefined.

In the simplest case, when there is only one complete $n$-gram in either of the strings, $n$-gram similarity is defined to be zero:

$$\mathbf{s_n}(\Gamma_{k,l}) = 0 \text{ if } (k = n \wedge l < n) \vee (k < n \wedge l = n)$$

Let $\Gamma_{i,j}^n = (x_{i+1} \ldots x_{i+n}, y_{j+1} \ldots y_{j+n})$ be a pair of $n$-grams in $X$ and $Y$. If both strings contain exactly one $n$-gram, our initial definition is strictly binary: 1 if the $n$-grams are identical, and 0 otherwise. (Later, we will consider modifying this part of the definition.)

$$\mathbf{s_n}(\Gamma_{n,n}) = \mathbf{s_n}(\Gamma_{0,0}^n) = \begin{cases} 1 & \text{if } \forall_{1 \leq u \leq n} \; x_u = y_u \\ 0 & \text{otherwise} \end{cases}$$

For longer strings, we define $n$-gram similarity recursively:

$$\mathbf{s}(X,Y) = \mathbf{s_n}(\Gamma_{k,l}) = \max_{i,j}(\mathbf{s_n}(\Gamma_{i+n-1,j+n-1}) + \mathbf{s_n}(\Gamma_{i,j}^*))$$

The values of $i$ and $j$ in the above formula are constrained by the requirement that both $\Gamma_{i,j}$ and $\Gamma_{i,j}^*$ contain at least one complete $n$-gram. More specifically, the admissible values of $i$ and $j$ are given by $D(k - n + 1, l - n + 1)$, where $D$ is the set defined in Section 2.2.

## 3.2    Computing $n$-Gram Similarity

As in the case of $\mathbf{s}$, a set of three decompositions is sufficient for computing $\mathbf{s_n}$.

$$\mathbf{s_n}(\Gamma_{k,l}) = \max(\mathbf{s_n}(\Gamma_{k-1,l}), \mathbf{s_n}(\Gamma_{k,l-1}), \mathbf{s_n}(\Gamma_{k-1,l-1}) + \mathbf{s_n}(\Gamma_{k-n,l-n}^n))$$

An efficient dynamic-programming algorithm for computing $n$-gram similarity can be derived directly from the alternative formulation. For $n = 1$, it reduces to the well-known algorithm for computing the length of the LCS. The algorithm is discussed in detail in Section 5.

## 3.3    Refined $n$-Gram Similarity

The binary $n$-gram similarity defined above is quite crude in the sense that it does not differentiate between slightly different $n$-grams and totally different $n$-grams. We consider here two possible refinements to the similarity scale. The first alternative, henceforth referred to as *comprehensive* $n$-gram similarity, is to compute the standard unigram similarity between $n$-grams:

$$\mathbf{s_n}(\Gamma_{i,j}^n) = \frac{1}{n}\mathbf{s_1}(\Gamma_{i,j}^n)$$

The second alternative, henceforth referred to as *positional* $n$-gram similarity, is to simply to count identical unigrams in corresponding positions within the $n$-grams:

$$\mathbf{s_n}(\Gamma_{i,j}^n) = \frac{1}{n}\sum_{u=1}^{n}\mathbf{s_1}(x_{i+u}, y_{j+u})$$

The advantage of the positional $n$-gram similarity is that it can be computed faster than the comprehensive $n$-gram similarity.

Figure 1 (right) shows a bigram decomposition of a pair of words with various levels of bigram similarity. The solid link denotes a complete match. The dashed links are partial matches according to both positional and comprehensive $n$-gram similarity. The dotted link indicates a partial match that is detected by the comprehensive $n$-gram similarity, but not by the positional $n$-gram similarity.

## 4   $n$-Gram Distance

Since the standard edit distance is almost a dual notion to the length of the LCS, the definition of $n$-gram distance differs from the definition of $n$-gram similarity only in details:

1. Recursive definition of edit distance:

$$\mathbf{d}(x, \epsilon) = 1, \qquad \mathbf{d}(\epsilon, y) = 1, \qquad \mathbf{d}(x, y) = \begin{cases} 0 \text{ if } x = y \\ 1 \text{ otherwise} \end{cases}$$

$$\mathbf{d}(X, Y) = \mathbf{d}(\Gamma_{k,l}) = \min_{i,j}(\mathbf{d}(\Gamma_{i,j}) + \mathbf{d}(\Gamma_{i,j}^*))$$

2. An alternative formulation of edit distance with a reduced set of decompositions:

$$\mathbf{d}(X, Y) = \mathbf{d}(\Gamma_{k,l}) = \min(\mathbf{d}(\Gamma_{k-1,l}) + 1, \mathbf{d}(\Gamma_{k,l-1}) + 1, \mathbf{d}(\Gamma_{k-1,l-1}) + \mathbf{d}(x_k, y_l))$$

3. Definition of $n$-gram edit distance:

$$\mathbf{d_n}(\Gamma_{k,l}) = 1 \text{ if } (k = n \wedge l < n) \vee (k < n \wedge l = n)$$

$$\mathbf{d_n}(\Gamma_{n,n}) = \mathbf{d_n}(\Gamma_{0,0}^n) = \begin{cases} 0 \text{ if } \forall_{1 \leq u \leq n} x_u = y_u \\ 1 \text{ otherwise} \end{cases}$$

$$\mathbf{d_n}(\Gamma_{k,l}) = \min_{i,j}(\mathbf{d_n}(\Gamma_{i+n-1,j+n-1}) + \mathbf{d_n}(\Gamma_{i,j}^*))$$

4. An alternative formulation of $n$-gram distance:

$$\mathbf{d_n}(\Gamma_{k,l}) = \min(\mathbf{d_n}(\Gamma_{k-1,l}) + 1, \mathbf{d_n}(\Gamma_{k,l-1}) + 1, \mathbf{d_n}(\Gamma_{k-1,l-1}) + \mathbf{d_n}(\Gamma_{k-n,l-n}^n))$$

5. Three variants of $n$-gram distance $\mathbf{d_n}(\Gamma_{i,j}^n)$:
   (a) The binary $n$-grams distance, as defined in 3.
   (b) The comprehensive $n$-grams distance: $\mathbf{d_n}(\Gamma_{i,j}^n) = \frac{1}{n} d_1(\Gamma_{i,j}^n)$.
   (c) The positional $n$-gram distance: $\mathbf{d_n}(\Gamma_{i,j}^n) = \frac{1}{n} \sum_{u=1}^{n} d_1(x_{i+u}, y_{j+u})$.

The positional $n$-gram distance is equivalent to the the comprehensive $n$-gram distance for $n = 2$. All three variants are equivalent for $n = 1$.

## 5   $n$-Gram Word Similarity Measures

In this section, we define a family of word similarity measures (Table 1), which include two widely-used measures, the longest common subsequence ratio (LCSR) and the normalized edit distance (NED), and a series of new measures based on $n$-grams, $n > 1$. First, however, we need to consider two measure-related issues: normalization and affixing.

Normalization is a method of discounting the length of words that are being compared. The length of the LCS of two randomly-generated strings grows with

**Table 1.** A classification of measures based on $n$-grams

|            | $n = 1$ | $n = 2$ | $n = 3$ | $\ldots$ | $n$ |
|------------|---------|---------|---------|----------|-----|
| Similarity | LCSR    | BI-SIM  | TRI-SIM | $\ldots$ | $n$-SIM |
| Distance   | NED$_1$ | BI-DIST | TRI-DIST| $\ldots$ | $n$-DIST |

the length of the strings [2]. In order to avoid the length bias, a normalized variant of the LCS is usually preferred. The longest common subsequence ratio (LCSR) is computed by dividing the length of the longest common subsequence by the length of the longer string [8]. Edit distance is often normalized in a similar way, i.e. by the length of the longer string (e.g., [5]). However, Marzal and Vidal [6] propose instead to normalize by the length of the editing path between strings, which requires a somewhat more complex algorithm. We refer to these two variants of Normalized Edit Distance as NED$_1$ and NED$_2$, respectively.

Affixing is a way of increasing sensitivity to the symbols at string boundaries. Without affixing, the boundary symbols participate in fewer $n$-grams than the internal symbols. For example, the word *abc* contains two bigrams: *ab* and *bc*; the initial symbol *a* occurs in only one bigram, while the internal symbol *b* occurs in two bigrams. In the context of measuring word similarity, this is a highly undesirable effect because the initial symbols play crucial role in human perception of words. In order to avoid the negative bias, extra symbols are sometimes added to the beginnings and/or endings of words.

The proposed $n$-gram similarity and distance measures N-SIM and N-DIST incorporate both normalization and affixing (Figure 3). Our affixing method is aimed at emphasizing the initial segments, which tend to be much more

**Algorithm** N-SIM $(X, Y)$

$K \leftarrow \text{length}(X)$
$L \leftarrow \text{length}(Y)$
**for** $u \leftarrow 1$ **to** $N - 1$ **do**
  $X \leftarrow x_1' + X$
  $Y \leftarrow y_1' + Y$
**for** $i \leftarrow 0$ **to** $K$ **do**
  $S[i, 0] \leftarrow 0$
**for** $j \leftarrow 1$ **to** $L$ **do**
  $S[0, j] \leftarrow 0$
**for** $i \leftarrow 1$ **to** $K$ **do**
  **for** $j \leftarrow 1$ **to** $L$ **do**
    $S[i, j] \leftarrow max($
      $S[i - 1, j],$
      $S[i, j - 1],$
      $S[i - 1, j - 1] + \mathbf{s_N}(\Gamma_{i-1,j-1}^N))$
**return** $S[K, L] / \max(K, L)$

**Algorithm** N-DIST $(X, Y)$

$K \leftarrow \text{length}(X)$
$L \leftarrow \text{length}(Y)$
**for** $u \leftarrow 1$ **to** $N - 1$ **do**
  $X \leftarrow x_1' + X$
  $Y \leftarrow y_1' + Y$
**for** $i \leftarrow 0$ **to** $K$ **do**
  $D[i, 0] \leftarrow i$
**for** $j \leftarrow 1$ **to** $L$ **do**
  $D[0, j] \leftarrow j$
**for** $i \leftarrow 1$ **to** $K$ **do**
  **for** $j \leftarrow 1$ **to** $L$ **do**
    $D[i, j] \leftarrow min($
      $D[i - 1, j] + 1,$
      $D[i, j - 1] + 1,$
      $D[i - 1, j - 1] + \mathbf{d_N}(\Gamma_{i-1,j-1}^N))$
**return** $D[K, L] / \max(K, L)$

**Fig. 3.** The algorithms for computing N-SIM and N-DIST of strings X and Y

important than final segments in determining word similarity. A unique special symbol is defined for each letter of the original alphabet. Each word is augmented with a prefix composed of $n - 1$ copies of the special symbol that corresponds to the initial letter of the word. For example, if $n = 3$, *amikin* is transformed into *ââamikin*. Assuming that the original words have lengths $K$ and $L$ respectively, the number of $n$-grams is thus increased from $K + L - 2(n-1)$ to $K + L$. The normalization is achieved by simply dividing the total similarity score by $\max(K, L)$, the original length of the longer word. This procedure guarantees that the new measures return 1 if and only if the words are identical, and 0 if and only if the words have no letters in common.

# 6    Experiments

In this section we describe three experiments aimed and comparing the effectiveness of the standard unigram similarity measures with the proposed $n$-gram measures. The three experiments correspond to applications in which the standard unigram measures have been used in the past.

## 6.1    Evaluation Methodology

Our evaluation methodology is the same in all three experiments. The underlying assumption is that pairs of words that are known to be related in some way (e. g., by sharing a common origin) exhibit on average much greater similarity than unrelated pairs. We evaluate the effectiveness of several similarity measures by calculating how well they are able to distinguish related word pairs from unrelated word pairs. In order for a measure to achieve 100% accuracy, any related pair would have to be assigned a higher similarity value than any unrelated pair. In practice, most of the related pairs should occur near the top of a list of pairs sorted by their similarity value.

The evaluation procedure is as follows:

1. Establish a gold standard set $G$ of word pairs that are known to be related.
2. Generate a much larger set $C$ of candidate word pairs, $C \supset G$.
3. Compute the similarity of all pairs in $C$ using a similarity measure.
4. Sort the pairs in $C$ according to the similarity value, breaking ties randomly.
5. Compute the 11-point interpolated average precision on the sorted list.

The 11-point interpolated average precision is an information-retrieval evaluation technique. Precision is computed for the recall levels of 0%, 10%, 20%, ..., 100%, and then averaged to yield a single number. We uniformly set the precision value at 0% recall to 1, and the precision value at 100% recall to 0.

## 6.2    Data

**Genetic Cognates.** Cognates are words of the same origin that belong to distinct languages. For example, English *father*, German *vater*, and Norwegian *far* constitute a set of cognates, since they all derive from a single Proto-Germanic word (reconstructed as *faδēr). The identification of cognates is one of the principal tasks of historical linguistics. Cognates are usually similar in their phonetic form, which makes string similarity an important clue for their identification.

In the first experiment, we extracted all nouns from two machine-readable word lists that had been used to produce an Algonquian etymological dictionary [4]. The two sets contain 1628 Cree nouns and 1023 Ojibwa nouns, respectively. The set $C$ of candidate pairs was created by generating all possible Cree-Ojibwa pairs (a Cartesian product). An electronic version of the dictionary, which contains over four thousand Algonquian cognate sets, served as the gold standard $G$. The task was to identify 409 cognate pairs among 1,650,780 candidate word pairs (approx. 0.025%).

**Translational Cognates.** Cognates are usually similar in form *and* meaning, which makes string similarity a useful clue for word alignment in statistical machine translation. Both LCSR and edit distance have been employed for cognate identification in bitext-related tasks (e.g., [8]).

In the second experiment, we used *Blinker*, a word-aligned French-English bitext containing translations of 250 sentences taken from the Bible [7]. For the evaluation, we manually identified all cognate pairs in the bitext, using word alignment links as clues. The candidate set of pairs was generated by taking a Cartesian product of words in corresponding sentences. This time, the task was to identify those 959 pairs among 36,879 candidate pairs (approx. 2.6%).

**Confusable Drug Names.** Many drug names either look or sound similar, which causes potentially dangerous errors. An example of a confusable drug name pair is *Zantac* and *Zyrtec*. Orthographic similarity measures have been applied in the past for detecting confusable drug names. For example, Lambert et al. [5] tested edit distance, normalized edit distance, and LCS, among other measures.

In the final experiment, we extracted 582 unique drug names form an online list of confusable drug names [13]. The candidate set of pairs was the Cartesian product of the names. The list itself served as the gold standard. The task was to identify 798 confusable pairs among 338,142 candidate pairs (approx. 0.23%).

## 6.3    Results and Discussion

Table 2 compares the average precision achieved by various measures in all three experiments. The similarity-based measures are given first, followed by the distance-based measures. PREFIX is a baseline-type measure that returns the length of the common prefix divided by the length of the longer string. Three values are given for the $N$-SIM and $N$-DIST measures corresponding to the binary, positional, and comprehensive variants, respectively.

Although the average precision values vary depending on the difficulty of a particular task, the relative performance of the measures is quite consistent across the three experiments. The positional and comprehensive variants of the $n$-gram measures outperform the standard unigram measures (the only exception is that NED slightly outperforms TRI-DIST on genetic cognates). The difference

**Table 2.** The average interpolated precision for various measures on three word-similarity tasks

|  | DICE | XXDICE | LCS | LCSR | BI-SIM | | | TRI-SIM | | |
|---|---|---|---|---|---|---|---|---|---|---|
|  |  |  |  |  | bin | pos | com | bin | pos | com |
| Drug names | .262 | .308 | .152 | .330 | .377 | .403 | .400 | .356 | .393 | .396 |
| Genetic cognates | .394 | .519 | .141 | .564 | .526 | .597 | .595 | .466 | .593 | .589 |
| Transl. cognates | .775 | .815 | .671 | .798 | .841 | .841 | .846 | .829 | .838 | .832 |

|  | PREFIX | EDIT | NED$_1$ | NED$_2$ | BI-DIST | | | TRI-DIST | | |
|---|---|---|---|---|---|---|---|---|---|---|
|  |  |  |  |  | bin | pos | com | bin | pos | com |
| Drug names | .256 | .275 | .364 | .369 | .389 | .399 | .399 | .352 | .391 | .391 |
| Genetic cognates | .276 | .513 | .592 | .592 | .545 | .602 | .602 | .468 | .589 | .589 |
| Transl. cognates | .721 | .681 | .821 | .823 | .840 | .838 | .838 | .828 | .829 | .830 |

is especially pronounced in the drug names experiment. The bigram methods are overall somewhat more effective than the trigram methods. The differences between the positional and the comprehensive *n*-gram variants, where they exist, are insignificant, but the binary variant is sometimes much worse. The normalized versions substantially outperform the un-normalized versions in all cases. NED consistently outperforms LCSR, but the differences between the similarity-based methods and the distance-based methods for $n > 1$ are minimal.

### 6.4 Similarity vs. Distance

Interestingly, there is a considerable variation in performance among the unigram measures, but not among the multigram measures. The reason may lie in LCSR's lack of context-sensitivity, which we mentioned in Section 2.5. Consider again the two pairs on the right side of Figure 2. The LCS lengths are identical in both cases (3), but edit distances differ (7 and 5, respectively). Notice the highly parallel arrangement of the identity links between the second pair, a phenomenon which usually positively correlates with perceptual similarity. Since by definition LCSR is concerned only with the number of identity matches, it cannot detect such a clue. The multigram measures, on the other hand, are able to recognize the difference, because *n*-grams provide an alternative mechanism for taking context into account.

## 7  Conclusion

We have formulated a new concept of *n*-gram similarity and distance, which generalizes the standard unigram string similarity and distance. On that basis, we have formally defined a family of new measures of word similarity, We have evaluated the new measures on three different word-comparison tasks. The experiments suggest that the new *n*-gram measures outperform the unigram measures. In general, normalization by word length is a must. With respect to the unigram measures, we have argued that the normalized edit distance may be more appropriate than LCSR. For $n \geq 2$, BI-SIM with positional $n$-gram similarity is recommended as it combines relative speed with high overall accuracy.

## Acknowledgments

Thanks to Bonnie Dorr for collaboration on the confusable drug names project. This research was supported by Natural Sciences and Engineering Research Council of Canada.

## References

1. Chris Brew and David McKelvie. 1996. Word-pair extraction for lexicography. In *Proc. of the 2nd Intl Conf. on New Methods in Language Processing*, pages 45–55.
2. Vaclàv Chvátal and David Sankoff. 1975. Longest common subsequences of two random sequences. *Journal of Applied Probability*, 12:306–315.
3. Thomas H. Cormen, Charles E. Leiserson, Ronald L. Rivest, and Clifford Stein. 2001. *Introduction to Algorithms*. The MIT Press, second edition.
4. John Hewson. 1993. *A computer-generated dictionary of proto-Algonquian*. Hull, Quebec: Canadian Museum of Civilization.
5. Bruce L. Lambert, Swu-Jane Lin, Ken-Yu Chang, and Sanjay K. Gandhi. 1999. Similarity As a Risk Factor in Drug-Name Confusion Errors: The Look-Alike (Orthographic) and Sound-Alike (Phonetic) Model. *Medical Care*, 37(12):1214–1225.
6. A. Marzal and E. Vidal. 1993. Computation of normalized edit distance and applications. *IEEE Trans. Pattern Analysis and Machine Intelligence*, 15(9):926–932.
7. I. Dan Melamed. 1998. Manual annotation of translational equivalence: The Blinker project. Technical Report IRCS #98-07, University of Pennsylvania.
8. I. Dan Melamed. 1999. Bitext maps and alignment via pattern recognition. *Computational Linguistics*, 25(1):107–130.
9. D. Sankoff and J. B. Kruskal, editors. 1983. *Time warps, string edits, and macromolecules: the theory and practice of sequence comparison*. Addison-Wesley.
10. Bill Smyth. 2003. *Computing Patterns in Strings*. Pearson.
11. Dan Tufis. 2002. A cheap and fast way to build useful translation lexicons. In *Proc. of the 19th Intl Conf. on Computational Linguistics*, pages 1030–1036.
12. Esko Ukkonen. 1992. Approximate string-matching with $q$-grams and maximal matches. *Theoretical Computer Science*, 92:191–211.
13. Use caution — avoid confusion. *United States Pharmacopeial Convention Quality Review*, No. 76, March 2001. Available from *http://www.bhhs.org/pdf/qr76.pdf*.
14. Robert A. Wagner and Michael J. Fischer. 1974. The string-to-string correction problem. *Journal of the Association for Computing Machinery*, 21(1):168–173.

# Using the $k$-Nearest Neighbor Graph
# for Proximity Searching in Metric Spaces*

Rodrigo Paredes[1] and Edgar Chávez[2]

[1] Center for Web Research, Dept. of Computer Science, University of Chile,
Blanco Encalada 2120, Santiago, Chile
`raparede@dcc.uchile.cl`
[2] Escuela de Ciencias Físico-Matemáticas, Univ. Michoacana, Morelia, Mich. México
`elchavez@fismat.umich.mx`

**Abstract.** Proximity searching consists in retrieving from a database, objects that are *close* to a query. For this type of searching problem, the most general model is the *metric space*, where proximity is defined in terms of a *distance* function. A solution for this problem consists in building an *offline* index to quickly satisfy *online* queries. The ultimate goal is to use as few distance computations as possible to satisfy queries, since the distance is considered expensive to compute. Proximity searching is central to several applications, ranging from multimedia indexing and querying to data compression and clustering.

In this paper we present a new approach to solve the proximity searching problem. Our solution is based on indexing the database with the $k$-nearest neighbor graph ($k$NNG), which is a directed graph connecting each element to its $k$ closest neighbors.

We present two search algorithms for both range and nearest neighbor queries which use navigational and metrical features of the $k$NNG graph. We show that our approach is competitive against current ones. For instance, in the document metric space our nearest neighbor search algorithms perform 30% more distance evaluations than AESA using only a 0.25% of its space requirement. In the same space, the pivot-based technique is completely useless.

## 1 Introduction

Proximity searching is the search for *close* or *similar* objects in a database. This concept is a natural extension of the classical problem of exact searching. It is motivated by data types that cannot be queried by exact matching, such as multimedia databases containing images, audio, video, documents, and so on. In this new framework the exact comparison is just a type of query, while close or similar objects can be queried as well. There exists a large number of computer applications where the concept of similarity retrieval is of interest. This applications include *machine learning and classification*, where a new element

---

* This work has been supported in part by the Millennium Nucleus Center for Web Research, Grant P04-067-F, Mideplan, Chile, and CYTED VII.19 RIBIDI Project.

M. Consens and G. Navarro (Eds.): SPIRE 2005, LNCS 3772, pp. 127–138, 2005.

must be classified according to its closest existing element; *image quantization and compression*, where only some samples can be represented and those that cannot must be coded as their closest representable one; *text retrieval*, where we look for words in a text database allowing a small number of errors, or we look for documents which are similar to a given query or document; *computational biology*, where we want to find a DNA or protein sequence in a database allowing some errors due to typical variations; and *function prediction*, where past behavior is extrapolated to predict future behavior, based on function similarity. See [6] for a comprehensive survey on proximity searching problems.

Proximity/similarity queries can be formalized using the metric space model, where a distance function $d(x, y)$ is defined for every object pair in $\mathbb{X}$. Objects in $\mathbb{X}$ do not necessarily have coordinates (for instance, strings and images).

The distance function $d$ satisfies the metric properties: $d(x, y) \geq 0$ (positiveness), $d(x, y) = d(y, x)$ (symmetry), $d(x, y) = 0$ iff $x = y$ (reflexivity), and $d(x, y) \leq d(x, z) + d(z, y)$ (triangle inequality). The distance is considered expensive to compute (for instance, when comparing two documents or fingerprints).

We have a finite *database* of interest $\mathbb{U}$ of size $n$, which is a subset of the universe of objects $\mathbb{X}$ and can be preprocessed to build a search index.

A proximity query consists in retrieving objects from $\mathbb{U}$ which are close to a new object $q \in \mathbb{X}$. There are two basic proximity queries:

**Range query** $(q, r)_d$**:** Retrieve all elements in $\mathbb{U}$ which are within distance $r$ to $q \in \mathbb{X}$. This is, $(q, r)_d = \{u \in \mathbb{U} \ / \ d(q, u) \leq r\}$.
**Nearest neighbor query** $NN_k(q)_d$**:** Retrieve the $k$ closest elements in $\mathbb{U}$ to $q \in \mathbb{X}$. This is, $|NN_k(q)_d| = k$, and $\forall \ u \in NN_k(q)_d, v \in \mathbb{U} - NN_k(q)_d$, $d(u, q) \leq d(v, q)$.

There are some considerations about $NN_k(q)_d$. In case of ties we choose any $k$-element set that satisfies the query. The query *covering radius* $cr_q$ is the distance from $q$ towards the farthest neighbor in $NN_k(q)_d$. Finally, a $NN_k(q)_d$ algorithm is called *range-optimal* if it uses the same number of distance evaluations than a range search with radius the distance to the $k$-th closest element [11].

An *index* is a data structure built *offline* over $\mathbb{U}$ to quickly solve proximity queries *online*. Since the distance is considered expensive to compute the goal of an index is to save distance computations. Given the query, we use the index to discard as many objects from the database as we can to produce a small set of candidate objects. Later, we check exhaustively the candidate set to obtain the query outcome.

There are three main components in the cost of computing a proximity query using an index, namely: the number of distance evaluations, the CPU cost of side computations (other than computing distances) and the number of I/O operations. However, in most applications the distance is the leader complexity measure, and it is customary to just count the number of computed distances to compare two algorithms. This measure applies to both index construction and object retrieval. For instance, computing the cosine distance [3] in the document metric space takes 1.4 msecs in our machine (Pentium IV, 2 GHz), which is really costly.

An important parameter of a metric space is its intrinsic dimensionality. In $\mathbb{R}^D$ with points distributed uniformly the intrinsic dimension is simply $D$. In metric spaces or in $\mathbb{R}^D$ where points are not chosen uniformly, the intrinsic dimensionality can be defined using the distance histogram [6]. In practice, the proximity query cost worsens quickly as the space dimensionality grows. In fact, an efficient method for proximity searching in low dimensions may become painfully slow in high dimensions. For large enough dimensions, no proximity search algorithm can avoid comparing the query against all the database.

## 1.1   A Note on $k$-Nearest Neighbor Graphs

The *k-nearest neighbors graph* ($k$NNG) is a directed graph connecting each element to its $k$ nearest neighbors. That is, given the element set $\mathbb{U}$ the $k$NNG is a graph $G(\mathbb{U}, E)$ such that $E = \{(u, v, d(u, v)), v \in NN_k(u)_d\}$, where each $NN_k(u)_d$ represent the outcome of the nearest neighbor query for each $u \in \mathbb{U}$.

The $k$NNG is interesting *per se* in applications like cluster and outlier detection [9,4], VLSI design, spin glass and other physical process simulations [5], pattern recognition [8], and query or document recommendation systems [1,2]. This contribution starts with the $k$NNG graph already built, we want to prove the searching capabilities of this graph. However, we show some specific $k$NNG construction algorithms for our present metric space application in [14].

Very briefly, the $k$NNG is a direct extension of the well known *all-nearest-neighbor* (ANN) problem. A naïve approach to build $k$NNG uses $\frac{n(n-1)}{2} = O(n^2)$ distance computations and $O(kn)$ memory. Although there are several alternatives to speed up the procedure, most of them are unsuitable for metric spaces, since they use coordinate information that is not necessarily available in general metric spaces. As far as we know, there are three alternatives in our context.

Clarkson generalized the ANN problem for general metric spaces solving the ANN by using randomization in $O(n(\log n)^2(\log \Gamma(\mathbb{U}))^2)$ expected time, where $\Gamma(\mathbb{U})$ is the distance ratio between the farthest and closest pair of points in $\mathbb{U}$ [7]. The technique described there is mainly of theoretical interest, because the implementation requires $o(n^2)$ space.

Later, Figueroa proposes build the $k$NNG by using a pivot-based index so as to solve $n$ range queries of decreasing radius [10]. As it is well known, the performance of pivot-based algorithms worsen quickly as the space dimensionality grows, thus limiting the applicability of this technique.

Recently, we propose two approaches for the problem which exploit several graph and metric space features [14]. The first is based on recursive partitions, and the second is an improvement over the Figueroa's technique. Our construction complexity for general metric spaces is around $O(n^{1.27}k^{0.5})$ for low and medium dimensionality spaces, and $O(n^{1.90}k^{0.1})$ for high dimensionality ones.

## 1.2   Related Work

We have already made another attempt about using graph based indices for metric space searching by exploring the idea of indexing the metric space with a

$t$-spanner [12,13]. In brief, a $t$-spanner is a graph with a bounded stretch factor $t$, hence the distance estimated through the graph (the length of the shortest path) is at most $t$ times the original distance. We show that the $t$-spanner based technique has better performance searching real-world metric spaces than the obtained with the classic pivot-based technique. However, the $t$-spanner can require much space. With the $k$NNG we aim at similar searching performance using less space.

In the experiments, we will compare the performance of our searching algorithms against the basic pivot-based algorithm and AESA [15]. It is known that we can trade space for time in proximity searching in the form of more pivots in the index. So, we will compare our $k$NNG approach to find out how much memory a pivot-based algorithm need to use to be as good as the $k$NNG approach. Note that all the pivot-based algorithms have similar behavior in terms of distance computations, being the main difference among them the CPU time of side computations. On the other hand, we use AESA just like a baseline, since its huge $O(n^2)$ memory requirement makes this algorithm suitable only when $n$ is small. See [6] for a comprehensive explanations of these algorithms.

### 1.3   Our Contribution

In this work we propose a new class of proximity searching algorithms using the $k$NNG as the data structure for searching $\mathbb{U}$. This is the first approach, up to the best of our knowledge, using the $k$NNG for metric searching purposes.

The core of our contribution is the use of the $k$NNG to estimate both an upper bound and a lower bound of the distance to the query from the database elements. Once we compute $d(q, u)$ for some $u$ we can upper bound the distance from $q$ to many database objects (if the graph is connected, we upper bound the distance to all the database objects). We can also lower bound the distance from the query to the neighbors of $u$. The upper bound allows the elimination of far-from-the-query elements whilst the lower bound can be used to test if an element can be in the query outcome.

As we explain later (Sections 2 and 3), this family of algorithms have a large number of design parameters affecting its efficiency (not the correctness). We tried to explore all the parameters experimentally in Section 4.

We selected two sets of heuristics rising two metric range query algorithms, and building on top of them we designed two nearest neighbor search algorithms.

The experiments confirm that our algorithms are efficient in distance evaluations. For instance, in the document metric space with cosine distance our nearest neighbor query algorithms just perform 30% more distance evaluations than AESA, but only using a 0.25% of its space requirement. In the same space, the pivot-based technique is completely useless.

## 2   $k$NNG-Based Range Query Algorithms

Given an arbitrary subgraph of the distance matrix of $\mathbb{U}$, one can upper bound the distance between two objects by using the shortest path between them.

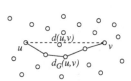 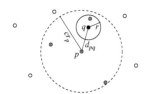 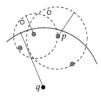

(a) We upper bound the distance $d(u, v)$ with the length of the shortest path $d_G(u, v)$

(b) If the ball $(p, cr_p)$ covers $(q, r)$, we make $C = (p, cr_p)$

(c) If we find an object $p \in (q, r)_d$, we check its neighborhood

**Fig. 1.** Using the $k$NNG features. In 1(a), approximating the distance in the graph. In 1(b), using the container. In 1(c), checking the neighborhood.

Formally, $d(u, v) \leq d_G(u, v)$ where $d_G(u, v)$ is the distance in the graph, that is, the length of the shortest path between the objects. Figure 1(a) shows this.

A generic graph-based approach for solving range queries consists in starting with a set of candidate nodes $C$ of the provable smallest set containing $(q, r)_d$. A fair choice for an initial $C$ is the whole database $\mathbb{U}$. Later, we iteratively extract an object $u$ from $C$ and if $d(u, q) \leq r$ we report $u$ as part of $(q, r)_d$. Otherwise, we delete all the objects $v$ such that $d_G(u, v) < d(u, q) - r$. Figure 2(a) illustrates this, we discard all the gray nodes because their distance estimations are small enough. We repeat the above procedure as long as $C$ have candidate objects. In this paper we improve this generic approach using the $k$NNG properties.

*Using Covering Radius.* Notice that each node in $k$NNG has a covering radius $cr_u$ (the distance towards its $k$-th neighbor). If the query ball $(q, r)_d$ is contained in $(u, cr_u)_d$ we can make $C = (u, cr_u)_d$, and proceed iteratively from there. Furthermore, we can keep track of the best fitted object, considering both its distance to the query and its covering radius using the equation $cr_u - d(u, q)$. The best fitted object will be the one having the largest difference, we call *container* this difference. Figure 1(b) illustrates this. So, once the container is larger than the searching radius we can make $C = (u, cr_u)_d$ as stated above. The probability of hitting a case to apply this property is low; but it is simply to check and the low success rate is compensated with the dramatic shrink of $C$ when applied.

*Propagating in the Neighborhood of the Nodes.* Since we are working over a graph built by an object closeness criterion, if an object $p$ is in $(q, r)_d$ it is likely that some of its neighbors are also in $(q, r)_d$. Moreover, since the out-degree of a $k$NNG is a small constant, spending some extra distance evaluations on neighbors of processed nodes do not add a large overhead to the whole process.

So, when we found an object belonging to $(q, r)_d$, it is worth to examine its neighbors, and, as with any other examination update the container. Note that every time we hit an answer we recursively check all of its neighbors. Special care must be taken to avoid multiple checks or cycles. Figure 1(c) illustrates this.

Note also that since we can lower bound the distance from the query to the neighbors of a processed object, we can discard some neighbors without directly computing the distance. Figure 2(b) illustrates this.

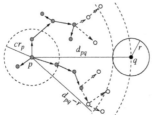
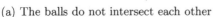
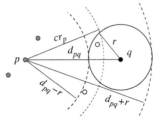

(a) The balls do not intersect each other     (b) The balls intersect each other

**Fig. 2.** In 2(a), we extract gray objects which have a distance estimation lower that $d_{pq} - r$ and count visits to the white ones which have estimations lower than $d_{pq}$. In 2(b), we use $p$ as a pivot discarding its gray neighbors when the distance from $p$ towards them is not in $[d_{pq} - r, d_{pq} + r]$, else, we count the visit to the white nodes.

*Working Evenly in All Graph Regions.* Since we use path expansions from some nodes it is important to choose them scattered in the graph to avoid concentrating efforts in the same graph region. Otherwise, we will compute a path several times. A good idea is to select elements far apart from $q$ and the previous selected nodes, because these nodes would have major potential of discarding non-relevant objects. Unfortunately, the selection of distant objects cannot be done by directly computing the distance to $q$. However, we can estimate "how visited" is some region. In fact, our two range query algorithms differ essentially in the way we select the next node to review.

### 2.1   First Heuristic for Metric Range Query ($k$NNGRQ1)

In this heuristic we prefer to start shortest path computations from nodes with few discarded neighbors, since these nodes have major discarding potential. Additionally, we also consider two criteria so as to break ties. The second criterion is to use nodes with small covering radius, and the third is that we do not want to restart elimination in an already visited node, and between two visited nodes we will choose the least traversed. So, to select a node, we consider the following:

1. How many neighbors already discarded has the node. Nodes with few discarded neighbors have major discarding potential, so they can reduce heavily the number of distance computations performed to solve the query.
2. The size of the covering radius. Objects having small covering radius, that is, very close neighbors, have major chance of discarding them (since if $cr_u < d(u, q) - r$, all its neighbors are discarded). Moreover, it is also likely that distance estimations computed from $u$ would have tighter upper bounds.
3. The number of times the node was checked in a path expansion (when computing the graph distance). We prefer a node that it had been checked few times in order to scatter the search effort on the whole graph.

The above heuristics are condensed in Eq. (1).

$$p = \operatorname{argmin}_{u \in C}\{|\mathbb{U}| \cdot (dn_u + f(u)) + \#visit\} \qquad (1)$$

With $f(u) \in [0,1]$, $f(u) = \frac{cr_u - cr_{min}}{cr_{max} - cr_{min}}$, and $cr_{min} = \min_{u \in U}\{cr_u\}$, $cr_{max} = \max_{u \in U}\{cr_u\}$, and $dn_u$ represents the number of discarded neighbors of $u$. Note that in Eq. (1) the leading term selects nodes with few discarded neighbors, the second term is the covering radius and the last term the number of visits.

The equation is computed iteratively for every node in the graph. For each node we save the value of Eq. (1) and every time we visit a node we update the heuristic value accordingly. Figure 2(a) illustrates this. Note that when we start a shortest path expansion we can discard some nodes (the gray ones), but for those that we cannot discard (the white nodes) we update their value of Eq. (1).

Please note that when we compute the graph distance (the shortest path between two nodes), we use a variation of Dijkstra's all shortest path algorithm which limits the propagation up to an estimation threshold, since a distance estimation grater that $d(u,q) - r$ cannot be used to discard nodes.

### 2.2   Second Heuristic for Metric Range Query ($k$NNGRQ2)

A different way to select a scattered element set is by using the graph distance. More precisely we assume that if two nodes are far apart according to the graph distance, they are also far apart using the original distance. The idea is to select the object with the largest sum of graph distances to all the previously selected objects. From other point of view, this heuristic tries to start shortest path computations from outliers.

## 3   $k$NNG-Based Nearest Neighbor Queries

Range query algorithms naturally induce nearest neighbor searching algorithms. To this end, we use the following ideas:

- We simulate the nearest neighbor query using a range query of decreasing radius, which initial radius $cr_q$ is $\infty$.
- We manage an auxiliary set of nearest neighbor candidates of $q$ known up to now, so the radius $cr_q$ is the distance from $q$ to its furthest nearest-neighbor candidate.
- Each non-discarded object reminds its own lower bound of the distance from itself to the query. For each node its initial lower bound is 0.

Note that, each time we find and object $u$ such that $d(u,q) < cr_q$, we replace the farthest nearest-neighbor candidate by $u$, so this can reduce $cr_q$. Note also that, if $d(u,q) < cr_q$ it is likely that some of the neighbors of $u$ can also be relevant to the query, so we check all the $u$ neighbors. However, since the initial radius is $\infty$ we change a bit the navigational schema. In this case, instead of propagating in the neighborhood, we start the navigation from the node $u$ towards the query $q$ by jumping from one node to another if the next node is closer to $q$ to than the previous one. Figure 3 illustrates this. In the figure, we start in $p$, and we navigate towards $q$ until we reach $p_c$.

On the other hand, unlike range queries, we split the discarding process in two stages. In the first, we compute the lower bound of all the non-discarded

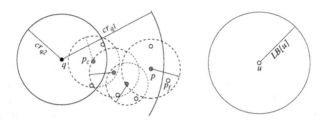

**Fig. 3.** If we find an object $p \in NN_l(q)_d$ we traverse through the graph towards $q$. Later, as $cr_q$ decreases, it is possible to discard the node $u$ when $LB[u] > cr_q$.

nodes. In the second, we extract the objects such that their lower bound are big enough, that is, we discard $u$ if $LB[u] > cr_q$. This is also illustrated in Figure 3. Note that, when we start in $p$ the covering radius is $cr_{q1}$. However, upon we reach $p_c$ the covering radius has been reduced to $cr_{q2} < LB[u]$, so we discard $u$.

$LB[u]$ is computed as the $\max_p\{d(p,q) - d_G(p,u)\}$, where $p$ is any of the previously selected nodes. Note that $LB[u]$ allows us to delay the discarding of $u$ until $cr_q$ is small enough, even if we only update $LB[u]$ once.

With these modifications we produce the algorithms $k$NNG$k$**NNQ1** which selects the next node according to Eq. (1), and $k$NNG$k$**NNQ2** which selects nodes far apart from each other.

## 4    Experimental Results

We have tested our algorithms on synthetic and real-world metric spaces. The synthetic set consists of 32,768 points distributed uniformly in the $D$-dimensional unitary cube $[0,1]^D$, under the Euclidean distance. This space allows us to measure the effect of the space dimension $D$ on our algorithms. Of course, we have not used the coordinates for discarding purposes, but just treated the points as abstract objects in an unknown metric space.

We also included two real-world examples. The first is a string metric space using the edit distance (the minimum number of character insertions, deletions and replacements needed to make two strings equal). The strings came from an English dictionary, where we index a random subset of 65,536 words. The second is a document metric space of 25,000 objects under the cosine distance. Both spaces are of interest in Information Retrieval applications.

Each point in the plots represents the average of 50 queries $q \in \mathbb{X} - \mathbb{U}$. For shortness we have called RQ the range query and NNQ the nearest neighbor query. We have compared our algorithms against AESA and a pivot-based algorithm (only in this case have we used range-optimal NNQs). For a fair comparison, we provided the same amount of memory for the pivot index and for our $k$NNG index (that is, we compare a $k$NNG index against a 1.5$k$ pivot set size).

With our experiments we tried to measure the behavior of our technique varying the vector space dimension, the query outcome size (by using different radii in RQs or different number of retrieved neighbors in NNQs), and the graph size (that is, number of neighbors per object) to try different index size.

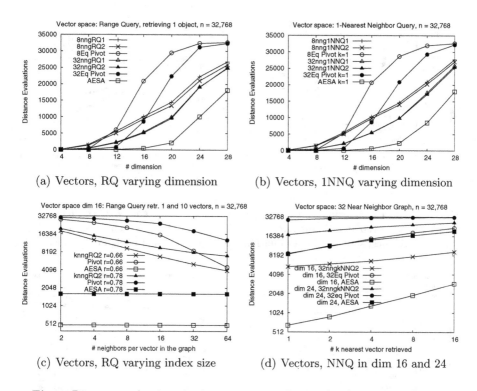

(a) Vectors, RQ varying dimension     (b) Vectors, 1NNQ varying dimension

(c) Vectors, RQ varying index size     (d) Vectors, NNQ in dim 16 and 24

**Fig. 4.** Distance evaluations in the vector space for RQ (left) and NNQ (right)

Figure 4 shows results in the vector space. Figure 4(a) shows RQs using radii that retrieve 1 object per query in average indexing the space with 8NNG and 32NNG graphs versus the dimension; and the Figure 4(b) shows the equivalent experiment for NNQs retrieving 1 neighbor. As can be seen from these plots, even though our NNQ algorithms are not range-optimal *per se*, they behave as if they were. Due to both RQ $k$NNG based variants behave very similar, we only show the better of them in the following plots in order to simplify the reading. We do the same in the NNQ plots.

Figure 4(c) shows RQs retrieving 1 and 10 vector in average per query versus the index size. Figure 4(d) shows NNQs over a 32NNG in dimension 16 and 24, versus the size of the query outcome. It is very remarkable that $k$NNG based algorithms are more resistant to both the dimension effect (Figures 4(a) and 4(b)) and the query outcome size (Figures 4(c) and 4(d)). As we can expect, the bigger the index size (that is, the more the neighbors in the $k$NNG), the better the searching performance (Figure 4(c)). Furthermore, all the plots in Figure 4 show that our algorithms have better performance than the classic pivot based approach for medium and high dimension metric spaces, that is $D > 8$.

Figure 5 shows results in the string space. Figure 5(a) shows RQs using radii $r = 1$, 2, and 3, and Figure 5(b) shows NNQs retrieving 2 and 16 nearest neighbors, both of them versus the index size. They confirm that $k$NNG based

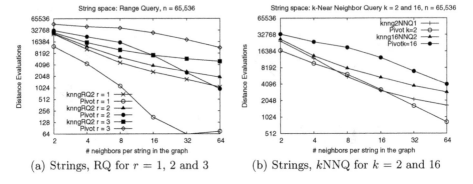

(a) Strings, RQ for $r = 1$, 2 and 3        (b) Strings, $k$NNQ for $k = 2$ and 16

**Fig. 5.** Distance evaluations in the string space for RQ (left) and NNQ (right). In RQs, AESA needs 25, 106 and 713 distance evaluations for radii $r = 1$, 2 and 3 respectively. In NNQs, AESA needs 42 and 147 evaluations to retrieve 2 and 16 neighbors respectively.

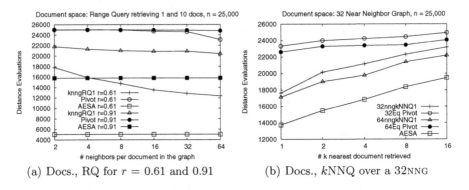

(a) Docs., RQ for $r = 0.61$ and 0.91        (b) Docs., $k$NNQ over a 32NNG

**Fig. 6.** Distance evaluations in the document space for RQ (left) and NNQ (right)

search algorithms are resistant against the query result size, as expected from the synthetic space experiments. With radii $r = 1$, 2 and 3, we retrieve approximately 2, 29 and 244 strings per query in average, however the performance of our algorithms do not degrade so strongly as the pivot-based one. With radius 1 the pivot based technique has better performance than our algorithms. However, with radius $r = 2$ and 3, our algorithms outperform the pivot-based algorithm. In this figures, we do not plot the AESA results because it uses too few distances evaluations, however recall that the AESA index uses $O(n^2)$ memory which is impractical in most of the scenarios. Note that the difference between pivot range queries of radius 1 and the 2-nearest neighbor queries appears because there are strings that have much more than 2 neighbors at distance 1, for example the query word "cams" retrieves "jams", "crams", "cam" and seventeen others, so these words distort the average for radius 1. We also verify that, the bigger the index size, the better the performance.

Figure 6 shows results in the document space. Figure 6(a) shows RQs using radii $r = 0.61$ and 0.91 versus the index size. Figure 6(b) shows NNQs over a 32NNG versus the query outcome size. This space is particularly difficult to manage, please observe that the pivot-based algorithms check almost all the database. Even in this difficult scenario, our algorithms handle to retrieve object checking a fraction of the database. It is remarkable that in NNQs, our algorithms perform 30% more distance evaluation than AESA using only a 0.25% of its space requirement.

Note that in the three spaces, the grater the *k*NNG index size, the better the behavior of our algorithms. However, the search performance improves strongly as we add more space to the graph only when we use small indices, that is, *k*NNG graphs with few neighbors. Fortunately, our algorithms behave better than the classic pivot technique in low memory scenarios with medium or high dimensionality. According to our experiments for $k \leq 32$ we obtain better results than the equivalent memory space pivot-based algorithm in $D$-dimensional vector spaces of $D > 8$ and the document space. In the string space we obtain better results in RQ using radii $r > 2$ or in NNQ retrieving more than 4 nearest strings.

## 5   Conclusions

We have presented four metric space searching algorithms that use the $k$-nearest neighbor graph *k*NNG as a metric index.

Our algorithms have practical applicability in low memory scenarios for metric spaces of medium or high dimensionality. For instance, in the document metric space with cosine distance our nearest neighbor algorithm uses just 30% more distance computations than AESA only using a 0.25% of its space requirement. In same space, the pivot-based technique is completely useless.

The future work involves the development of range-optimal nearest neighbor queries and the researching of *k*NNG optimizations tuned for our metric applications. For instance, we want to explore other local graphs, like the *all range r graph* where we assign to each node all the nodes within distance $r$. This way also allow us to control the size of the neighbor ball.

Since our data structure can efficiently search for nearest neighbor queries, it is natural to explore an incremental construction of the graph itself. To do this end we need to solve *reverse* nearest neighbor problem with this data structure. Incremental construction is very realistic in many real-world applications.

## References

1. R. Baeza-Yates, C. Hurtado, and M. Mendoza. Query clustering for boosting web page ranking. In *Proc. AWIC'04*, LNCS 3034, pages 164–175, 2004.
2. R. Baeza-Yates, C. Hurtado, and M. Mendoza. Query recommendation usign query logs in search engines. In *Proc. EDBT Workshops'04*, LNCS 3268, pages 588–596, 2004.
3. R. Baeza-Yates and B. Ribeiro-Neto. *Modern Information Retrieval*. Addison-Wesley, 1999.

4. M. Brito, E. Chávez, A. Quiroz, and J. Yukich. Connectivity of the mutual $k$-nearest neighbor graph in clustering and outlier detection. *Statistics & Probability Letters*, 35:33–42, 1996.
5. P. Callahan and R. Kosaraju. A decomposition of multidimensional point sets with applications to $k$ nearest neighbors and $n$ body potential fields. *JACM*, 42(1):67–90, 1995.
6. E. Chávez, G. Navarro, R. Baeza-Yates, and J.L. Marroquín. Searching in metric spaces. *ACM Computing Surveys*, 33(3):273–321, September 2001.
7. K. Clarkson. Nearest neighbor queries in metric spaces. *Discrete Computational Geometry*, 22(1):63–93, 1999.
8. R. O. Duda and P. Hart. *Pattern Classification and Scene Analysis*. Wiley, 1973.
9. D. Eppstein and J. Erickson. Iterated nearest neighbors and finding minimal polytopes. *Discrete & Computational Geometry*, 11:321–350, 1994.
10. K. Figueroa. An efficient algorithm to all $k$ nearest neighbor problem in metric spaces. Master's thesis, Universidad Michoacana, Mexico, 2000. In Spanish.
11. G. Hjaltason and H. Samet. Incremental similarity search in multimedia databases. Technical Report TR 4199, Dept. of Comp. Sci. Univ. of Maryland, Nov 2000.
12. G. Navarro and R. Paredes. Practical construction of metric $t$-spanners. In *Proc. ALENEX'03*, pages 69–81, 2003.
13. G. Navarro, R. Paredes, and E. Chávez. $t$-Spanners as a data structure for metric space searching. In *Proc. SPIRE'02*, LNCS 2476, pages 298–309, 2002.
14. R. Paredes and G. Navarro. Practical construction of $k$ nearest neighbor graphs in metric spaces. Technical Report TR/DCC-2005-6, Dept. of Comp. Sci. Univ. of Chile, May 2005. `ftp://ftp.dcc.uchile.cl/pub/users/gnavarro/knnconstr.ps.gz`.
15. E. Vidal. An algorithm for finding nearest neighbors in (approximately) constant average time. *Pattern Recognition Letters*, 4:145–157, 1986.

# Classifying Sentences Using Induced Structure[*]

Menno van Zaanen, Luiz Augusto Pizzato, and Diego Mollá

Division of Information and Communication Sciences (ICS),
Department of Computing, Macquarie University,
2109 North Ryde, NSW Australia
{menno, pizzato, diego}@ics.mq.edu.au

**Abstract.** In this article we will introduce a new approach (and several implementations) to the task of sentence classification, where pre-defined classes are assigned to sentences. This approach concentrates on structural information that is present in the sentences. This information is extracted using machine learning techniques and the patterns found are used to classify the sentences. The approach fits in between the existing machine learning and hand-crafting of regular expressions approaches, and it combines the best of both. The sequential information present in the sentences is used directly, classifiers can be generated automatically and the output and intermediate representations can be investigated and manually optimised if needed.

## 1  Introduction

Sentence classification is an important task in various natural language processing applications. The goal of this task is to assign pre-defined classes to sentences (or perhaps sequences in the more general case). For example, document summarisers that are based on the method of text extraction, identify key sentences in the original document. Some of these summarisers (e.g. [4,8]) classify the extracted sentences to enable a flexible summarisation process.

Sentence classification is also a central component of question-answering systems (e.g. the systems participating in the question answering track of TREC[1]). Different types of questions prompt different procedures to find the answer. Thus, during a question analysis stage the question is classified among a set of predefined expected answer types (EAT). Classes can be, for instance, "number" or "location", but they can also be more fine-grained, such as "number-distance", "number-weight", "location-city", or "location-mountain".

Here, we will introduce a novel approach to the problem of sentence classification, based on structural information that can be found in the sentences. Reoccurring structures, such as *How far ...* may help finding the correct sentence class (in this case a question with EAT "distance"). The approach described here

---

[*] This work is supported by the Australian Research Council, ARC Discovery grant no. DP0450750.

[1] See http://trec.nist.gov/ for more information.

M. Consens and G. Navarro (Eds.): SPIRE 2005, LNCS 3772, pp. 139–150, 2005.

automatically finds these structures during training and uses this information when classifying new sentences.

In the next section, the main approach will be described, as well as two different systems (with corresponding implementations) based on this idea. To get a feeling for whether the approach is feasible, we have applied implementations of the systems to real data in the context of question classification. The results published here indirectly provide insight in the approach. We will conclude with an overview of the advantages and disadvantages of these systems and finish with a discussion of future work.

## 2   Approach

Past approaches to sentence classification can be roughly divided into two groups: *machine learning* and *regular expression* approaches.

Grouping sentences in a fixed set of classes is typically a classification task for which "standard" machine learning techniques can be used. In fact, this has been done in the past [4,6,14]. However, an inconvenience of these methods is that the intermediate data produced by most machine learning techniques is difficult to interpret by a human reader. Therefore, these methods do not help to understand the classification task, and consequently the determination of the features to use by the classifier is usually reduced to a process of trial and error, where past trials do not help much to determine the optimal set of features.

Also, many features that are or have been used do not allow a description of the inherently sequential aspect of sentences. For example, bags of words, occurrence of part-of-speech (POS) tags, or semantically related words, all lose the sequential ordering of words. The order of words in a sentence contains information that may be useful for the classification.

The alternative approach, handcrafted regular expressions that indicate when sentences belong to a certain class do (in contrast to the machine learning approach) maintain word order information. As an example, one might consider the regular expression /^How far/ that can be used to classify questions as having an EAT of "distance". This method brings fairly acceptable results when there are specific structures in the sentence that clearly mark the sentence type, as is the case with question classification into a coarse-grained set of EAT [9,13]. The regular expression approach also has the advantage that regular expressions are human readable and easy to understand. However, creating them is not always trivial. Having many fine-grained classes makes it extremely hard to manually create regular expressions that distinguish between closely related classes.

Our method aims at using the advantages of both methods based on machine learning and methods based on regular expressions. Using machine learning, patterns are extracted from the training data. The resulting patterns are easy to read by humans, and we believe they give insight about the structure of the sentences and the ordering of words therein. These patterns serve as regular expressions during the classification task. An overview of the approach is given in Figure 1.

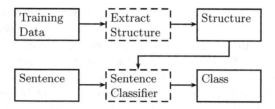

**Fig. 1.** Overview of the structure induction classification process

In the next two sections, we will describe two classifiers that fit into this approach. The alignment-based learning classifier uses a grammatical inference system to find structure in the sentences explicitly. This structure is converted into a set of regular expressions that are fed to the classifier. The trie classifier builds a trie structure that represents the sentences in the training corpus. It finds regularities in the trie and classifies new sentences using these regularities.

## 2.1   Alignment-Based Learning Classifier

The structure extraction phase of the first system is done by Alignment-Based Learning (ABL). ABL is a generic grammatical inference framework, that learns structure using plain text only. It has been applied to several corpora in different languages with good results [10] and it also compares well against other grammatical inference systems [12].

The underlying idea of ABL is that constituents in sentences can be interchanged. To give an example, if we substitute the noun phrase *the man* in the sentence *He sees the man* with another noun phrase *a woman*, we get another valid sentence: *He sees a woman*. This process can be reversed (by aligning sentences) and possible constituents, called hypotheses, are found and marked as such by grouping them. This is called the alignment learning phase. Table 1 shows the structure that would be learned from a sample of four sentences. Grouping is marked by parentheses.

The ABL framework consists of two more phases: clustering and selection learning. The clustering phase groups similar non-terminal labels that are assigned to the hypotheses during the alignment learning phase. The selection learning phase resolves a problem of the alignment learning phase. The alignment learning phase may introduce overlapping hypotheses. These are unwanted if the underlying grammar is considered context-free and the resulting structure is taken as a parse of the sentence using the underlying grammar. However, we are not interested in this now, we only need the structure that is found. Therefore, neither clustering nor selection learning are used here and we only concentrate on the results of the alignment learning phase.

Several methods of the alignment learning phase have been described in the literature [3,10]. Due to practical restrictions of time and space (memory) [11], we have used the suffixtree implementation [3]. Here, the alignments are found by building a suffixtree, which uncovers re-occurring sequences of words in the sentences. Using these sequences, the hypotheses are introduced.

Once the structure has been found, it is extracted and used in classification. Each training sentence is associated with its corresponding class, so the structure extracted from a certain sentence provides some evidence that this structure is related to the class. If a structure occurs with several classes (i.e. it occurs in multiple sentences that have different classes associated with them), the structure is stored with all classes and their respective frequencies.

The stored structures can be seen as regular expressions, so during classification, all structures can be matched to the new sentence. If a regular expression matches, its class and corresponding frequency information is remembered and when all structures have been tried, the class with the highest frequency is selected as output.

We will explain the different implementations by walking through an example. For this, we take a look at question classification with a set of four training questions and two EAT. Consider the questions combined with the structure found by ABL and their EAT depicted in Table 1.

The first implementation, which we will call *hypo*, takes the words in the hypotheses (i.e. the words between the brackets), turns them into regular expressions and stores them together with the corresponding EAT of the questions. The resulting regular expressions can be found in Table 2. For example, the first two questions both generate the regular expression /What/ combined with EAT "DESC", so it has frequency 2.

The second implementation, called *unhypo*, takes each hypothesis and removes it from the question when it is turned into a regular expression. The regular expressions extracted from the example questions can also be found in Table 2. For example, the first two questions would introduce /What/ combined with EAT "DESC" and frequency 2. The last question would introduce /What/ with class "LOC" and frequency 1.

**Table 1.** Example questions with ABL structure

"DESC" (What) (is (caffeine) )
"DESC" (What) (is (Teflon) )
"LOC"    (Where) is (Milan)
"LOC"    What (are the twin cities)

**Table 2.** Regular expressions found by ABL-based methods

| hypo | | unhypo | |
|---|---|---|---|
| caffeine | "DESC" 1 | What is | "DESC" 2 |
| is caffeine | "DESC" 1 | What | "DESC" 2 |
| What | "DESC" 2 | is caffeine | "DESC" 1 |
| Teflon | "DESC" 1 | is Teflon | "DESC" 1 |
| is Teflon | "DESC" 1 | Where is | "LOC" 1 |
| Milan | "LOC" 1 | is Milan | "LOC" 1 |
| Where | "LOC" 1 | What | "LOC" 1 |
| are the twin cities | "LOC" 1 | | |

Once the structures are stored with their class and frequency, the actual classification can start. We have built two methods that combine the evidence for classes of the sentence differently. Both start with trying to match each of the stored regular expressions against the sentence. If a regular expression matches, the first method (called *default*) will increase the counts of the classes of the question with the frequency that is stored with the classes of the regular expression.

The second method (called *prior*) will increase the counts of the classes of the sentence that are related to the regular expression with 1 (where the default method would increase it with the frequency count of each class).

When all regular expressions are tried, the default method selects the class with the highest frequency (if there are more with the same frequency, one is chosen at random), whereas the prior method also selects the one with the highest count, but if there are more than one, it selects one based on the class with the highest overall frequency.

## 2.2   Trie Classifier

The other system we describe here is based on finding structure using a trie. By searching this data structure, it can be used directly to classify new, unseen sentences. The work discussed here is an extension of the work in [7].

A trie $T(S)$ is a well-known data structure that can be used to store a set of sentences in an efficient way. It is defined by a recursive rule $T(S) = \{T(S/a_1), T(S/a_2), \ldots, T(S/a_r)\}$, where $S$ is a set of sequences (sentences, in our case). $S/a_n$ is the set of sequences that contains all sequences of $S$ that start with $a_n$, but stripped of that initial element [2].

In each node, local information extracted during training is stored. This includes the word, class and frequency information. Since each node represents part of a unique path in the trie, frequency information is the number of sentences that use that particular node in a path in the trie.

To get a feeling for what a trie looks like, Figure 2 illustrates the trie containing the questions of Table 1 (without the bracketed structure generated by ABL). Extra information is stored per node, so for instance, the node that is reached by following the path *What* contains "DESC": 2 and "LOC": 1 as frequency information, which is similar to the information found by the unhypo system.

The classification of new sentences is performed by extracting the class information stored in the trie nodes. The words of the new sentence are used to

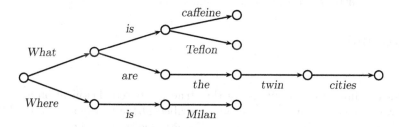

**Fig. 2.** The trie containing the example questions

find a path in the trie. The class information can then be extracted from the final node of the path through the trie. To be more concrete, consider a sentence $S = (s_1, s_2, \ldots, s_n)$ and a trie $T(S)$. The classification of the sentence is done by finding the path $T(S/s_1/s_2/\ldots/s_n)$. The class information stored in the final node of the path is returned.

Notice that only with the sentences present in the training data, we can find a complete path in the trie (ending in a final node). To find a path for sentences that have not been seen in the training data, we skip non-matching words wherever needed to complete the path. This is called the *look-ahead* process.

The look-ahead process works as follows. Let us say that the words of sentence $S$ match up to $s_k$, $k < n$, and the next word does not match $s_{k+1}$. That is, $T(S/s_1/s_2/\ldots/s_{k-1}/s_k) \neq \emptyset$, and $T(S/s_1/s_2/\ldots/s_{k-1}/s_k/s_{k+1}) = \emptyset$. The look-ahead process then builds a set of sub-tries $T(S/s_1/s_2/\ldots/s_{k-1}/s_k/\beta/s_{k+2})$, with $\beta \in \{x | T(S/s_1/s_2/\ldots/s_{k-1}/s_k/x) \neq \emptyset\}$.

One of these sub-tries is selected based on the frequency of the prefix defined by the sub-trie in the training data. In other words, for each sub-trie in the set, the prefix is extracted $(s_1, s_2, \ldots, s_{k-1}, s_k, \beta, s_{k+2})$ and the frequency of this prefix in the training data is looked up. The sub-trie that has the prefix with the highest frequency associated to it is selected. This defines $\beta$ and the process continues with the next word until all words in the sentence are consumed. In a sense we are replacing $s_k$ with $\beta$. This defines a final node for the particular sentence and the class information present in that node is used to find a class.

Here, we will describe two different implementations of the classification system using a trie structure. The first uses a method called *strict*, where a word can only be replaced (if needed) by another word if $s_{k+2}$ in the sentence matches the information in the trie. If there are several options, the one that has the highest frequency is selected. If no option is found, the search process stops at that node and the class information is retrieved from node $s_k$.

The second trie-based implementation, called *flex*, allows a word in the sentence to be replaced with one or more words in the trie. Initially, it works similar to the strict method. However, if no $\beta$ set of sub-tries can be found, because no match on $s_{k+2}$ can be found, it considers $\delta$ as in $T(S/s_1/s_2/\ldots/s_{k-1}/s_k/\beta/\delta)$, where $\beta$ can be anything and $\delta$ should match $s_{k+2}$. If no such $\delta$ can be found, it tries to match $\delta$ with the next sentence token, i.e. $s_{k+3}$. This continues until a match is found, or the end of the sentence is reached (and the class information present in $T(S/s_1/s_2/\ldots/s_{k-1}/s_k)$ is used). Again, if at some point multiple options are introduced, the most frequent is selected.

## 3   Results

### 3.1   Data

To get a feeling of the feasibility of the structure-based classification approach, we have applied the implementations described above to the annotated TREC questions [6]. This is a collection of 5,452 questions that can be used as training data and 500 questions testing data. The mean question length is 10.2 words in the training data and 7.5 words in the testing data. The corpus contains coarse-grained and fine-grained EAT information for each of the questions. In total, there are 6 coarse-grained classes and 50 fine-grained classes. Note that 8 fine-grained classes do not occur in the testing data. Some example questions are:

"NUM:dist"     *How tall is the Sears Building ?*
"NUM:period"  *How old was Elvis Presley when he died ?*
"LOC:city"     *What city had a world fair in 1900 ?*
"LOC:other"    *What hemisphere is the Philippines in ?*

There are some minor errors in the data that will affect the result of the structural, sequential systems described in this article.[2] In addition to an odd incorrect character, a few questions in the training data have POS tags instead of words:

*Who wrote NN DT NNP NNP " ?*
*What did 8 , CD NNS VBP TO VB NNP POS NN .*
*Why do USA fax machines not work in UK , NNP ?*

In [6], some additional problems with the data are described. Firstly, the training and testing data are gathered from different sources. In initial experiments, we found that this has quite an effect on the performance. The systems described in this article perform much better during development on the training data (using ten fold cross-validation) than on the testing data.[3] However, to make the most of the limited amount of data that is available, we have decided to stick with this division.

Secondly, in [6] it is mentioned that some questions are ambiguous in their EAT. For example, *What do bats eat?* can be classified in EAT "food", "plant", or "animal". They (partially) solve this by assigning multiple answers to a question. An adjusted precision metric is then needed to incorporate the multiple answers. We have decided not to do this (even though it is possible to do so), because we investigate the feasibility of the approach, which is not only dependent on the results of the systems applied to a particular problem. The approach encompasses many different systems of which only a few are tested here. Instead of focusing on the actual results, we want to show the validity of the structure induction approach in general.

**Part-of-Speech.** In addition to applying the systems on the plain text, we also apply them to tokens consisting of the word combined with their POS tag. This illustrates that the algorithms in the classification framework are not just limited to plain text. Extra information can be added to the data.

We create the POS information using the Brill tagger [1]. The POS information is simply combined with the plain words. However, adjacent words that have the same POS are combined into one token. Thus, the question *Who is Federico*

---

[2] We have decided to use the original data, without correcting the errors.
[3] The difference in performance of the systems displayed here is similar to that during development.

*Fellini?* is, after POS tagging, divided into three tokens: (*Who*, WP), (*is*, VBZ) and (*Federico Fellini*, NNP). *Federico* and *Fellini* are combined in one unique token, because *Federico* and *Fellini* are adjacent words that have the same POS.

The ABL approach simply uses these complex tokens as elementary units, so for tokens to be aligned, both the word and its POS tag have to match. The trie approach is slightly modified. When searching for a matching $\beta$ (i.e. when the word and POS tag do not match in the trie), the additional information of that position in the sub-tries should match. This means that in this case, the POS tag of the $\beta$ nodes in the trie should match the POS tag of the word in that position in the sentence.

## 3.2 Numerical Results

The different systems are first allowed to learn using the training data, and after learning, they are applied to the testing data. The output of the systems is compared against the correct class (as present in the testing data) and the precision is computed

$$\text{precision} = \frac{\#\text{ correctly classified questions}}{\text{total }\#\text{ of questions}}$$

The results of all implementations can be found in Table 3. To be able to compare the results, we have also computed a baseline. This simple baseline always selects the most frequent class according to the training data. This is the "HUM:ind" class for the fine-grained data and "ENTY" for the coarse-grained data. This baseline is, of course, the same for the plain words and POS tagged data. All implementations perform well above the baseline.

Looking at the coarse-grained results of the implementations using ABL, it shows that adding POS information to the words helps improving performance. We suspect that the POS information guides the alignment learning phase in that the POS restricts certain alignments. For example, words that can be a noun or a verb can now only be matched when their POS is the same. However, slightly fewer regular expressions are found because of this.

The trie-based implementations do not benefit from the POS information. We think that this is because the POS information is too coarse-grained for $\beta$ matching. Instead of trying to find a correct position to continue walking through the trie by finding the right word, only a rough approximation is found, namely a word that has the same POS.

Overall, it shows that the trie-based approach outperforms the ABL implementations. We expect that it has to do with how the trie-based implementations keep track of the sequence of the words in a question, whereas the ABL-based implementations merely test if word sequences occur in the question.

The results on the fine-grained data show similar trends. Adding POS generally improves the ABL-based system, however, this is not the case for the unhypo implementation. Again, the performance of the trie-based system decreases when POS information is added.

**Table 3.** Results of the ABL and Trie systems on question classification

| | | | coarse | | fine | |
|---|---|---|---|---|---|---|
| | | | words | POS | words | POS |
| Baseline | | | 0.188 | 0.188 | 0.110 | 0.110 |
| ABL | hypo | default | 0.516 | 0.682 | 0.336 | 0.628 |
| | | prior | 0.554 | 0.624 | 0.238 | 0.472 |
| | unhypo | default | 0.652 | 0.638 | 0.572 | 0.558 |
| | | prior | 0.580 | 0.594 | 0.520 | 0.432 |
| Trie | strict | | 0.844 | 0.812 | 0.738 | 0.710 |
| | flex | | 0.850 | 0.794 | 0.742 | 0.692 |

To get a further idea of how the systems react to the data, we give some additional information. Each ABL-based classifier extracts roughly 115,000 regular expressions when applied to the training data and the trie that contains the set of training questions consists of nearly 32,000 nodes.

When investigating how the trie-based system works, we noticed that around 90% of the questions that performed $\beta$ replacement (15% of the total number of questions) in the strict method (on the POS tagged data) provided correct results. This indicates that $\beta$ replacement in this implementations is often performed correctly. However, in the flex method on the same data, the same test shows a much lower success rate (around 65%). We think that this is due to the fewer constraints for the $\beta$ substitution in the latter method, causing it to occur in more than a half of the questions.

Unfortunately, it is hard to compare these results to other published results, because the data and corresponding classes (EAT) are often different. Given that [6] uses more information during classification (among others, chunks and named entities), the comparison is not reliable. We decided to compare our results to those in [14], who used the same question data to train various machine learning methods on bag-of-words features and bag-of-$n$-grams features. Their results ranged from 75.6% (Nearest Neighbours on words) to 87.4% (SVM on $n$-grams) in the coarse-grained data and from 58.4% (Naïve Bayes on words) to 80.2% (SVM on words) in the fine-grained data. Our results fit near the top on course-grained and near the middle on fine-grained data.

## 4 Future Work

The results on the question classification task indicate that classification based on induced structure is an interesting new approach to the more general task of sentence classification. However, there are many more systems that fall into the structure-based classification framework. These systems can be completely different from the ones described here, but modifications or extensions of these systems are also possible.

For example, some preliminary experiments show that in the case of the ABL-based system, it may be useful to retain only the regular expressions that are uniquely associated with a class. Taking only these regular expressions makes

**Table 4.** Similar tokens found by the trie-based approach

| Galileo | triglycerides | calculator | Milan |
|---|---|---|---|
| Monet | amphibians | radio | Trinidad |
| Lacan | Bellworts | toothbrush | Logan Airport |
| Darius | dingoes | paper clip | Guam |
| Jean Nicolet | values | game bowling | Rider College |
| Jane Goodall | boxcars | stethoscope | Ocho Rios |
| Thucydides | tannins | lawnmower | Santa Lucia |
| Damocles | geckos | fax machine | Amsterdam |
| Quetzalcoatl | chloroplasts | fountain | Natick |
| Confucius | invertebrates | horoscope | Kings Canyon |

hand-tuning of the expressions easier and the amount of regular expressions is reduced, which makes them more manageable.

More complex regular expressions can be learned from the data, incorporating more information from the structure found by the system. For instance, there can be further restrictions on the words that are skipped by the systems or even systems that perform robust partial parsing using the extracted structure can be envisioned.

Perhaps other grammatical inference techniques are better suited for the task described here. Alternative trie-based implementations include those using finite-state grammatical inference techniques such as EDSM or blue-fringe [5].

An interesting characteristic we found in the trie-based system is the discovering of semantic relations between the replaced tokens and their $\beta$ substitutes. In Table 4 we give a few subsets of related tokens that have been found. Similar groups of syntactic/semantic related words or word groups can be found using the ABL framework as well, as shown in [10–pp. 76–77]. In fact, these are found using their context only. Words or word groups that tend to occur in the same contexts also tend to be used similarly, which is found in both systems. We plan to use this knowledge to place extra restrictions in the regular expressions.

Finally, we suspect that the structure (in the form of a trie or regular expressions) may help in tasks related to sentence classification, such as finding the focus of the question, which is another part of the question analysis phase of question answering systems. The learned structure may give an indication on where in the question this information can be found. For example, if the regular expression /How far is .* from .*/ is found, the system can understand that it should find the distance between the values of the variables (.*). This is one area that we intend to explore further.

## 5   Conclusion

In this article, we have introduced a new approach to sentence classification. From the (annotated) training data, structure is extracted and this is used to classify new, unseen sentences. Within the approach many different systems are possible and in this article we have illustrated and evaluated two different systems. One system uses a grammatical inference system, Alignment-Based Learning, the other system makes use of a trie structure.

The approach falls in between the two existing approaches: "standard" machine learning techniques and manually created regular expressions. Using machine learning techniques, regular expressions are created automatically, which can be matched against new sentences. The regular expressions are human-readable and can be adjusted manually, if needed.

The results on the annotated questions of the TREC10 data, which stand in line with the current state-of-the-art results, show that the approach is feasible and both systems generate acceptable results. We expect that future systems that fall in the structure induced sentence classification framework will yield even better performance.

# References

1. Eric Brill. A simple rule-based part-of-speech tagger. In *Proceedings of ANLP-92, third Conference on Applied Natural Language Processing*, pages 152–155, Trento, Italy, 1992.
2. J. Clément, P. Flajolet, and B. Vallée. The analysis of hybrid trie structures. In *Proceedings of the Ninth Annual ACM-SIAM Symposium on Discrete Algorithms*, pages 531–539, Philadelphia:PA, USA, 1998. SIAM Press.
3. Jeroen Geertzen and Menno van Zaanen. Grammatical inference using suffix trees. In Georgios Paliouras and Yasubumi Sakakibara, editors, *Grammatical Inference: Algorithms and Applications: Seventh International Colloquium, (ICGI); Athens, Greece*, volume 3264 of *Lecture Notes in AI*, pages 163–174, Berlin Heidelberg, Germany, October 11–13 2004. Springer-Verlag.
4. Ben Hachey and Claire Grover. Sentence classification experiments for legal text summarisation. In *Proceedings of the 17th Annual Conference on Legal Knowledge and Information Systems (Jurix 2004)*, 2004.
5. Kevin J. Lang, Barak A. Pearlmutter, and Rodney A. Price. Results of the Abbadingo One DFA learning competition and a new evidence-driven state merging algorithm. In V. Honavar and G. Slutzki, editors, *Proceedings of the Fourth International Conference on Grammar Inference*, volume 1433 of *Lecture Notes in AI*, pages 1–12, Berlin Heidelberg, Germany, 1998. Springer-Verlag.
6. Xin Li and Dan Roth. Learning question classifiers. In *Proceedings of the 19th International Conference on Computational Linguistics (COLING); Taipei, Taiwan*, pages 556–562. Association for Computational Linguistics (ACL), August 24–September 1 2002.
7. Luiz Pizzato. Using a trie-based structure for question analysis. In Ash Asudeh, Cécile Paris, and Stephen Wan, editors, *Proceedings of the Australasian Language Technology Workshop; Sydney, Australia*, pages 25–31, Macquarie University, Sydney, Australia, December 2004. ASSTA.
8. Simone Teufel and Marc Moens. Argumentative classification of extracted sentences as a first step towards flexible abstracting. In Inderjeet Mani and Mark Maybury, editors, *Advances in automatic text summarization*. MIT Press, 1999.
9. *Proceedings of the Twelfth Text Retrieval Conference (TREC 2003); Gaithersburg:MD, USA*, number 500-255 in NIST Special Publication. Department of Commerce, National Institute of Standards and Technology, November 18–21 2003.

10. Menno van Zaanen. *Bootstrapping Structure into Language: Alignment-Based Learning*. PhD thesis, University of Leeds, Leeds, UK, January 2002.
11. Menno van Zaanen. Theoretical and practical experiences with Alignment-Based Learning. In *Proceedings of the Australasian Language Technology Workshop; Melbourne, Australia*, pages 25–32, December 2003.
12. Menno van Zaanen and Pieter Adriaans. Alignment-Based Learning versus EMILE: A comparison. In *Proceedings of the Belgian-Dutch Conference on Artificial Intelligence (BNAIC); Amsterdam, the Netherlands*, pages 315–322, October 2001.
13. E.M. Voorhees and Lori P. Buckland, editors. *Proceedings of the Eleventh Text REtrieval Conference (TREC 2002); Gaithersburg:MD, USA*, number 500-251 in NIST Special Publication. Department of Commerce, National Institute of Standards and Technology, November 19–22 2002.
14. Dell Zhang and Wee Sun Lee. Question classification using support vector machines. In Charles Clarke, Gordon Cormack, Jamie Callan, David Hawking, and Alan Smeaton, editors, *Proceedings of the 26th annual international ACM SIGIR conference on Research and development in information retrieval*, pages 26–32, New York:NY, USA, 2003. ACM Press.

# Counting Lumps in Word Space:
# Density as a Measure of Corpus Homogeneity

Magnus Sahlgren and Jussi Karlgren

SICS, Swedish Institute of Computer Science,
Box 1263, SE-164 29 Kista, Sweden
{mange, jussi}@sics.se

**Abstract.** This paper introduces a measure of corpus homogeneity that indicates the amount of topical dispersion in a corpus. The measure is based on the density of neighborhoods in semantic word spaces. We evaluate the measure by comparing the results for five different corpora. Our initial results indicate that the proposed density measure can indeed identify differences in topical dispersion.

## 1   Introduction

Word space models use co-occurrence statistics to construct high-dimensional semantic vector spaces in which words are represented as *context vectors* that are used to compute semantic similarity between the words. These models are now on the verge of moving from laboratories to practical usage, but while the framework and its algorithms are becoming part of the basic arsenal of language technology, we have yet to gain a deeper understanding of the properties of the high-dimensional spaces.

This study is ment to cast some light on the properties of high-dimensional word spaces; we find that computing a measure for the density of neighborhoods in a word space provides a measure of topical *homogeneity* — i.e. of how topically dispersed the data is. This is a fortunate discovery, since there are no established measures for corpus homogeneity. The hitherto most influential proposal boils down to defining a measure of homogeneity based on the similarity between randomly allocated parts of a corpus: the more similar the parts, the more homogeneous the corpus [3].

As an experimental evaluation of our density measure, we apply it to five different types of text corpus, each of varying degrees of topical homogeneity. The results show that the measure can indeed identify differences in topical dispersion and thus help provide some amount of understanding of what a word space is in relation to the language and the collection of text it models.

## 2   The Density Measure

The intuition our measure is based upon is the idea that words in a topically homogeneous data are used in more uniform ways than words in a topically

M. Consens and G. Navarro (Eds.): SPIRE 2005, LNCS 3772, pp. 151–154, 2005.

dispersed data. This would imply that the words in a topically homogeneous data have sparser semantic neighborhoods (i.e. fewer semantically related words) than would their topically more promiscuous counterparts. As an example, consider the difference between the semantic neighborhoods of a word with many possible meanings, such as "bark", which has nine meanings in WordNet 2.0, and a word with very few possible meanings, such as "toxin", which has only one meaning in WordNet. Obviously, the semantic neighborhood of "bark" is more populated (in the WordNet space) than the semantic neighborhood of "toxin".

In analogy with such WordNet neighborhoods, we suggest a measure of the number of words that occur within some specified radius around a given word in the word space. A large resulting number means that the word has a dense neighborhood, which indicates that the word occurs in a large number of contexts in the data, while a small resulting number means that it has a sparse neighborhood resulting from occurences in a small number of contexts. We define the density of the neighborhood of a word as the number of unique words that occur within the ten nearest neighbors of its ten nearest neighbors.

## 3   The Word Space Model

We use the Random Indexing [1,2] word space methodology, which is an alternative to algorithms such as Latent Semantic Analysis [4] that use factor analytic dimensionality reduction techniques. Rather than first assembling a huge co-occurrence matrix and then transforming it using factor analysis, Random Indexing *incrementally* accumulates context vectors in a two-step operation:

1. First, each word in the text is assigned a unique and randomly generated representation called an *index vector*. These random index vectors have a fixed dimensionality $k$, and consist of a small number $\epsilon$ of randomly distributed +1s and −1s.
2. Next, context vectors are produced by scanning through the text, and each time a word occurs, the index vectors of the $n$ surrounding words are added to its context vector.

This methodology has a number of advantages compared to other word space algorithms. First, it is an *incremental* method, which means that the context vectors can be used for similarity computations even after just a few examples have been encountered. Most other algorithms require the entire data to be sampled and represented in a very-high-dimensional space before similarity computations can be performed. Second, it uses fixed dimensionality, which means that new data do not increase the dimensionality of the vectors. Increasing dimensionality can lead to significant scalability problems in other algorithms. Third, it uses implicit dimensionality reduction, since the fixed dimensionality is much lower than the number of contexts in the data. This leads to a significant gain in processing time and memory consumption as compared to algorithms that employ computationally expensive dimensionality reduction techniques. Fourth, it is comparably robust with regards to the choice of parameters. Other algorithms tend to be very sensitive to the choice of dimensionality for the reduced space.

## 4  Experiment

In order to experimentally validate the proposed measure of corpus homogeneity, we first build a 1,000-dimensional word space for each corpus using Random Indexing, with parameters $n = 4$, $k = 1,000$, and $\epsilon = 10$.[1] Then, for each corpus, we randomly select 1,000 words, find their ten nearest neighbors, and then those neighbors' ten nearest neighbors. For each of the 1,000 randomly selected words, we count the number of unique words thus extracted. The maximum number of extracted neighbors for a word is 100, and the minimum number is 10. In order to derive a single measure of the neighborhood sizes of a particular corpus, we average the neighborhood sizes over the 1,000 randomly selected words. The largest possible score for a corpus under these conditions is 100, indicating that it is severely topically dispersed, while the smallest possible score is 10, indicating that the terms in the corpus are extremely homogeneous.

We apply our measure to five different corpora, each with a different degree of topical homogeneity. The most topically homogeneous data in these experiments consist of abstracts of scientific papers about nanotechnology (NanoTech). Also fairly homogeneous are samples of the proceedings from the European parliament (EuroParl), and newswire texts (ReutersVol1). Topically much more dispersed data are two examples of general balanced corpora, the TASA and the BNC corpora. Since the NanoTech data is very small in comparison with the other corpora (only 384,199 words, whereas the other corpora contain several millions of words), we used samples of comparable sizes from the other data sets. This was done in order to avoid differences resulting from mere sample size. The sampling was done by simply taking the first $\approx$ 380,000 words from each data set. We did not use random sampling, since that would affect the topical composition of the corpora.

We report results as averages over three runs using different random index vectors. The results are summarized in Table 1.

**Table 1.** The proposed density measure, as compared with the number of word tokens, the number of word types, and the type-token ratio, for five different English corpora

| Corpus | Word tokens | Word types | Type token ratio | Average density measure | Standard deviation |
|---|---|---|---|---|---|
| **NanoTech** | 384,199 | 13,678 | 28.09 | **49.149** | 0.35 |
| EuroParl (sample) | 375,220 | 9,148 | 41.47 | 50.9736 | 0.38 |
| ReutersVol1 (sample) | 368,933 | 14,249 | 25.89 | 51.524 | 0.18 |
| TASA (sample) | 387,487 | 12,153 | 31.88 | 52.645 | 0.45 |
| BNC (sample) | 373,621 | 18,378 | 20.33 | 54.488 | 0.74 |

---

[1] These parameters were chosen for efficiency reasons, and the size of the context window $n$ was influenced by [2].

# 5   Provisional Conclusions

The NanoTech data, which is by far the most homogeneous data set used in these experiments, receives the lowest density count, followed by the also fairly homogeneous EuroParl and ReutersVol1 data. The two topically more dispersed corpora receive much higher density counts, with the BNC as the most topically dispersed. This indicates that the density measure does in fact reflect topical dispersion: in more wide-ranging textual collections, words gather more contexts and exhibit more promiscuous usage, thus raising their density score.

Note that the density measure does not correlate with simple type-token ratio. Type-token ratio differentiates between text which tends to recurring terminological usage and text with numerous introduced terms. This can be seen to indicate that terminological variation — in spite of topical homogeneity — is large in the EuroParl data, which might be taken as reasonable in view that individual variation between speakers addressing the same topic can be expected; *text style* and *expression* have less effect on the density measure than the topical homogeneity itself. The ranking according to the density measure:

$$\textbf{Nano} > \text{EuroParl} > \text{ReutersVol1} > \text{TASA} > \text{BNC}$$

and the ranking according to type-token ratio:

$$\text{EuroParl} > \text{TASA} > \textbf{Nano} > \text{ReutersVol1} > \text{BNC}$$

only show a 0.5 rank sum correlation by Spearman's Rho.

This is obviously only a first step in the investigation of the characteristics of the well established word space model. The present experiment has clearly demonstrated that there is more to the word space model than meets the eye: even such a simple measure as the proposed density measure does reveal something about the topical nature of the data. We believe that a stochastic model of the type employed here will give a snapshot of topical dispersal of the text collection at hand. This hypothesis is borne out by the first experimental sample shown above: text of very differing types shows clear differences in the score defined by us. We expect that other measures of more global character will serve well to complement this proposed measure which generalizes from the character of single terms to the character of the entire corpus and the entire word space.

# References

1. Kanerva, P., Kristofersson, J., Holst, A.: Random Indexing of text samples for Latent Semantic Analysis. CogSci'00 (2000) 1036.
2. Karlgren, J., Sahlgren, M.: From Words to Understanding. In Uesaka, Y., Kanerva, P., Asoh, H. (eds.): *Foundations of Real-World Intelligence*. CSLI Publications (2001) 294–308.
3. Kilgariff, A.: Comparing Corpora. *Int. Journal of Corpus Linguistics* **6** (2001) 1–37.
4. Landauer, T., Dumais, S.: A solution to Plato's problem: The Latent Semantic Analysis theory of acquisition, induction and representation of knowledge. *Psychological Review* **104** (1997) 211–240.

# Multi-label Text Categorization Using K-Nearest Neighbor Approach with M-Similarity[*]

Yi Feng, Zhaohui Wu, and Zhongmei Zhou

College of Computer Science, Zhejiang University, Hangzhou 310027, P.R. China
{fengyi, wzh, zzm}@zju.edu.cn

**Abstract.** Due to the ubiquity of textual information nowadays and the multi-topic nature of text, it is of great necessity to explore multi-label text categorization problem. Traditional methods based on vector-space-model text representation suffer the losing of word order information. In this paper, texts are considered as symbol sequences. A multi-label lazy learning approach named kNN-M is proposed, which is derived from traditional k-nearest neighbor (kNN) method. The flexible order-semisensitive measure, M-Similarity, which enables the usage of sequence information in text by swap-allowed dynamic block matching, is applied to evaluate the closeness of texts on finding k-nearest neighbors in kNN-M. Experiments on real-world OHSUMED datasets illustrate that our approach outperforms existing ones considerably, showing the power of considering both term co-occurrence and order on text categorization tasks.

## 1 Introduction

With the rapid growth of online information, the majority of which is textual, effective retrieval is difficult without good indexing and summarization of document content. Text categorization (TC), also known as text classification and topic spotting, is a solution to this problem. In definition, text categorization solves the problem of assigning text content to predefined categories. Most researches in TC addressed the single-label TC problem, where a document can belong to only one category. In real-word situation, however, a document is always of multiple topics. Thus, it is of great necessity to explore the multi-label TC problem, where a piece of text is classified into more than one category. This is the motivation behind our research.

Existing methods to multi-label TC include Naïve Bayes (NB), Rocchio, Support Vector Machine (SVM), k-nearest neighbor (kNN), Boosting, etc. Most of the above TC approaches are based on the VSM, which treats each document as a bag of words. Although this model behaves well in many applications, it suffers from two main problems: 1) Its ignorance of term order. 2) Its assumption of term independence. As an inherent part of information for text, term order should be taken into account on text categorization. Especially, when the text is of short to medium length, the sequence information should be given more consideration. That is to say, the term order

---

[*] This research is supported by National Basic Research Priorities Programme of China Ministry of Science and Technology (2004DKA20250) and China 973 project (2003CB317006).

M. Consens and G. Navarro (Eds.): SPIRE 2005, LNCS 3772, pp. 155 – 160, 2005.

should be considered on the calculation of text distance. This problem is related to the research area of string matching. However, most of the edit distance-based methods are unsuitable for TC due to the concentration on the character level and non-support for swap-based matching. Pattern matching with swap [1] partly solved this problem by allowing local swap. Tichy [6] firstly applied block-moving approach to handle string-to-string correction problem, which extended to allow blocks to swap. The usage of swap-allowed block matching methods might help TC tasks. However, few work has been done to combine string matching techniques into TC until now. In 2002, Lodhi et al.[3] used string kernels to classify texts and achieved good perform-ance, showing the power of combining string matching method into TC. As another attempt, we incorporate M-Similarity proposed in [2] into traditional kNN method, which is named kNN-M approach, to handle multi-label TC problem in this paper.

The major differences between kNN-M and traditional one lie in the representation of text and the calculation of text distance on finding k-nearest neighbors. Different from VSM representation, documents are simply considered as symbol sequences in kNN-M. To evaluate the closeness of documents in this case, we use the flexible hybrid measure M-similarity proposed by Feng et al.[2]. M-Similarity is capable to partly use sequence information by giving extra weight to continuous matching. Be-sides, the size of each matching block in M-Similarity is dynamically determined, which makes it applicable to many applications where phrase segmentation is not a trivial problem. The text comparison using M-Similarity is actually a process of swap-allowed dynamic block matching. Experimental results on OHSUMED datasets dem-onstrate the effectiveness of kNN-M combining the power of kNN and M-Similarity.

The rest of the paper is organized as follows. A description of M-Similarity is given in Section 2. In Section 3 this similarity is incorporated into kNN-M algorithm to handle TC problem. Section 4 provides experimental results of kNN-M on real-world multi-label OHSUMED datasets. Finally, we conclude in Section 5.

## 2  M-Similarity

In this section a brief introduction to M-Similarity is presented. According to the order perspective of text similarity proposed in [2], M-Similarity belongs to the cate-gory of order-semisensitive similarities. Text comparison using this kind of similarity is a process of swap-allowed dynamic block matching.

In definition, given two texts $S$ (with shorter length $N$) and $L$ (with longer length $M$), and current position $curpos$ within $S$, the sum of weights for maximum sequence matching and potential matching can be defined below:

$$\lambda_{lm}(S,L,curpos) = \begin{cases} 0 & (if\ \theta(S,L,curpos) = 0) \\ \theta(S,L,curpos) \times W_s + [\theta(S,L,curpos) - 1] \times W_{ec} & (otherwise) \end{cases} \quad (1)$$

$$\lambda_{pm}(S,L) = [(M-N) \times W_s + (M-N-1) \times W_{ec}] \times P_{pm} \quad (2)$$

where $W_s$ stands for weight of single item matching, and $W_{ec}$ represents extra weight of continuous matching. $P_{pm}$ is potential matching parameter, which stands for the possibility of two texts of different length would be totally same when the shorter one is expanded to the length of longer one. It can also be seen as a normalizing parameter for text length. Given current position *curpos* within S, the system scans L from left to right for longest possible matching with substring in S starting from S[*curpos*] and returns the matching length as $\theta(S, L, curpos)$. Thus, M-Similarity can be calculated as below:

$$\text{M-Similarity } (S, L) = \frac{\sum\limits_{curpos} \lambda_{lm}(S, L, curpos) + \lambda_{pm}(S, L)}{M \times W_s + (M - 1) \times W_{ec}} \tag{3}$$

To calculate M-Similarity, *curpos* is increased from 1 to *N+1*. Note that, not all of the increase value of *curpos* is 1 at every step, and it is controlled by $\theta(S, L, curpos)$. Starting from initial position 1, *curpos* will move to the right adjacent position whenever $\theta(S, L, curpos)$ = zero. However, *curpos* will advance $\theta(S, L, curpos)$ positions whenever $\theta(S, L, curpos) > 0$. This process continues until *curpos* equals *N+1*, and the M-Similarity value are cumulated through this process as in formula (3).

## 3 kNN-M

In this section, a novel *k* nearest neighbor based method for multi-label text categorization named kNN-M is presented. Given a testing document, the kNN classifier scans for the *k* nearest neighbors among the training documents under a pre-defined similarity metric. After that, the categories of *k* neighbors are used to weight the category candidates. Suppose *d* is a test document, $d_i$ is the *i*th training document, and $C_j$ is the *j*th category ($1 \le j \le |C|$, $|C|$ is the total number of distinct training-set categories), the score of *d* belonging to $C_j$ is calculated as in (4). When *score ($C_j$| d)* for *d* is calculated for all possible *j*, a ranked list for the test document can be obtained by sorting the scores of candidate categories. Afterwards, by thresholding on these scores or ranks, binary document-category assignment can be made.

$$score(C_j | d) = \sum_{d_i \in kNN \ \& \ d_i \in C_j} sim(d, d_i) \tag{4}$$

To extending the above process of kNN to kNN-M, there are three questions that should be answered. The first question is how to calculate *sim(d, $d_i$)*. Different from traditional cosine metric, we use M-Similarity in kNN-M, that is: *sim(d, $d_i$)= M-Similarity(d, $d_i$)*. The second question is what kind of thresholding strategy to use. There are several candidate strategies available, including RCut, PCut, SCut, etc. The appropriate thresholding strategy for kNN-M should be selected according to characteristics of application. The third question is how to set the parameter *k*. In practice, this parameter is usually tuned empirically.

## 4   Experiment Results and Evaluation

In this section we describe the experiments that compare the performance of kNN-M with other methods to handle multi-label text categorization problem. We start with a description of the datasets and settings we used in our experiments.

### 4.1   Datasets and Settings

The corpus used in this paper is the widely used multi-label collection OHSUMED, which is a subset of MEDLINE database. Since the number of subheadings in MeSH field of OHSUMED rarely changes with time, we use subheadings as categories on TC in this paper. To evaluate how kNN-M behaves on small and large datasets, we pick two subsets of OHSUMED. The small subset, named OH-2000, is formed by 2000 records in 1991. The first / last half 1000 records of OH-2000 are used as training / testing set, respectively. The large subset, named OH-8788, consists of all records in 1987 and 1988 with non-null subheadings. All together, there are 119,565 records in OH-8788, in which 52,242 records in 1987 constitute training set, while 67,323 records in 1988 form testing set. The process of subheading extraction from MeSH field is carried out on both OH-2000 and OH-8788 to form a new field named category. On average 3 categories are assigned to each document in both datasets.

The next issue is the content to be categorized. In this paper, we use the title of each record as the text to be classified. One reason is that only a part of records have abstracts. However, the main motivation behind this choice is that classification on short text like title is worthy noting. In this age of information explosion, a tremendous volume of new textual information is produced everyday, which is hard to organize and use. However, mostly it contains a title given by authors. Thus, it is ideal if we could automatically categorize these materials only based on their titles.

In preprocessing steps, removal of punctuation marks and stemming are carried out. The stop-word removal in kNN-M is not recommended, especially for short text. Actually, for text with short length like title, almost each word is informative. Thus, we skip stop-word removal and feature selection in kNN-M, as well as in AdaBoost.MH, NB and Rocchio. (The influence of eliminating stop-words to short text categorization is found to be almost negligible by experiments). Considering the trade-off between precision / recall, we use the widely accepted *micro-F₁* as the evaluation measure.

### 4.2   Experimental Results

In this section, kNN-M, as well as NB, Rocchio and AdaBoost.MH, are carried out on both OH-2000 and OH-8788. The algorithms NB and Rocchio are available as part of the publicly Bow text categorization system [4]. The boosting algorithm AdaBoost.MH comes from BoosTexter text categorization system [5].

The thresholding strategy and other parameters in kNN-M can greatly influence the performance of TC. Since the OHSUMED dataset have some rare categories, we use Pcut to compute the thresholding of each category in kNN-M. For the parameter k, the value of 10, 20, 30, 40, 50 are tested. However, we find the resulting differences in the *micro-F₁* of kNN-M are almost negligible for different *k*. Thus, only the best case is listed. Besides, $W_s$, $W_{ec}$, $P_{pm}$ are tuned empirically to 0.8, 0.2, 0.1. The triplet $P_m$ with form (value1, value2, value3) is used to denote $W_s$, $W_{ec}$, $P_{pm}$. By setting $P_m$ = (0.8, 0, 0), kNN-M simulates the behavior of conventional kNN (without TFIDF-weighting). As for AdaBoost.MH, the number of boosting setps is set to 1000 and we regard labels with positive weights as category assignment decisions.

The experiments here include both closed evaluation and open evaluation. In closed evaluation, the training set is also used as testing set. In open evaluation, a classifier $h$ is learned from training set, and this $h$ is used to categorize texts in testing set. The results of closed/open evaluation on OH-2000/OH-8788 are presented in Table 1.

**Table 1.** *Micro-F$_1$* of different multi-label text classifiers on OH-2000 / OH-8788

| Data-set | Text Classifiers | Micro-F$_1$ | |
|---|---|---|---|
| | | Closed evaluation | Open evaluation |
| OH-2000 | NB | 0.2793 | 0.1163 |
| | TFIDF/Rocchio | 0.2699 | 0.0915 |
| | AdaBoost.MH | **0.4525** | 0.1293 |
| | kNN-M (k=30, $P_m$=(0.8, 0.2, 0.1)) | 0.2212 | **0.2185** |
| | kNN-M (k=30, $P_m$=(0, 0.2, 0)) | 0.1696 | 0.1676 |
| | kNN (k=30) | 0.1842 | 0.1803 |
| OH-8788 | NB | 0.2514 | 0.1987 |
| | TFIDF/Rocchio | 0.1807 | 0.1571 |
| | AdaBoost.MH | 0.2954 | 0.2656 |
| | kNN-M (k=40, $P_m$=(0.8, 0.2, 0.1)) | **0.3803** | **0.3749** |
| | kNN (k=40) | 0.3186 | 0.3143 |

## 4.3 Analysis

From the figures above, we can draw some conclusions as follows:

- **Multi-label Text Categorization on short document is a difficult problem.** As Table 1 shows, the best *micro-F$_1$* of closed / open evaluation on OH-8788 we can achieve is less than 0.4. These numbers on OH-2000 are 0.4525 and 0.2185. The reason behind this low *micro-F$_1$* lies in that the information inherent in short text used to construct the classifiers is very limited.
- **kNN-M outperforms other algorithms on multi-label text categorization.** As we can see from Table 1, the best *micro-F$_1$* of open evaluation on both OH-2000 and OH-8788 are achieved by kNN-M (0.3803, 0.3749). Besides, the best *micro-F$_1$* of closed evaluation on OH-2000 is also obtained from kNN-M (0.2185), while this number on OH-8788 is generated from AdaBoost.MH (0.4525). Considering most of TC tasks in real-world applications are open evaluation, we can conclude that kNN-M outperforms other algorithms on multi-label TC. Especially, kNN-M outperforms NB, TFIDF/Rocchio and AdaBoost.MH considerably in open evaluation.

- **Incorporating M-Similarity into kNN-M can improve the performance.** From Table 1, we can see kNN-M outperforms kNN (without TFIDF-weighting) by 3% on OH-2000 and 6% on OH-8788. These figures reflect the influence of using M-Similarity in kNN-M. It is not good to weigh too much on position information in text (Setting $P_m$=(0, 0.2, 0) lowers *micro-F$_1$* by more than 5%), but it is not good either to neglect term order. Thus, we can conclude that taking term order into account can improve short text categorization performance (TC on long texts need further experiments) and a balance should be achieved between considering term co-occurrence and order. ($P_m$ is tuned to (0.8, 0.2, 0.1) empirically here).

## 5   Conclusion

In this paper, we have proposed a kNN-based method, kNN-M, to multi-label text categorization. The order-semisensitive measure, M-Similarity, is incorporated into kNN-M to evaluate the closeness of texts on finding k-nearest neighbors. Due to the usage of swap-allowed dynamic block matching, kNN-M is capable to use the information of both term co-occurrence and order. Experiments on real-world OHSUMED datasets show the effectiveness of kNN-M and the power of incorporating string matching techniques into conventional TC approaches. Since there are three changeable parameters in the calculation of M-Similarity in kNN-M, we can adjust these parameters according to the characteristics of application, which makes kNN-M very flexible. Future work includes evaluating kNN-M on long documents and testing combining kNN with other string matching approaches to achieve higher performance.

## References

1. Amir, A., Aumann, Y., Landau, G.M., Lewenstein, M., Lewenstein, N.: Pattern Matching with Swaps, Journal of Algorithms, 37. (2000) 247–266
2. Feng, Y., Wu, Z., Zhou, Z.: Combining an Order-semisensitive Text Similarity and Closest Fit Approach to Textual Missing Values in Knowledge Discovery, In: KES 2005, (2005)
3. Lodhi, H., Saunders, C., et. al.: Text Classification using String Kernels. Journal of Machine Learning Research 2. (2002) 419–444
4. McCallum, A.K.: Bow: A toolkit for statistical language modeling, text retrieval, classification and clustering. http://www-2.cs.cmu.edu/~mccallum/bow. (1996)
5. Schapire, R.E., Singer, Y.: BoosTexter: A boosting-based system for text categorization. Machine Learning, 39(2/3). (2000) 135–168
6. Tichy, W.F.: The string to string correction problem with block moves. ACM Trans. Comp. Sys. 2(4). (1984) 309–321

# Lydia: A System for Large-Scale News Analysis*
## (Extended Abstract)

Levon Lloyd, Dimitrios Kechagias, and Steven Skiena

Department of Computer Science,
State University of New York at Stony Brook,
Stony Brook, NY 11794-4400
{lloyd, dkechag, skiena}@cs.sunysb.edu

## 1  Introduction

Periodical publications represent a rich and recurrent source of knowledge on both current and historical events. The *Lydia* project seeks to build a relational model of people, places, and things through natural language processing of news sources and the statistical analysis of entity frequencies and co-locations. *Lydia* is still at a relatively early stage of development, but it is already producing interesting analysis of significant volumes of text. Indeed, we encourage the reader to visit our website (*http://www.textmap.com*) to see our analysis of recent news obtained from over 500 daily online news sources.

Perhaps the most familiar news analysis system is *Google News* [1], which continually monitors 4,500 news sources. Applying state-of-the-art techniques in topic detection and tracking, they cluster articles by event, group these clusters into groups of articles about related events, and categorize each event into pre-determined top-level categories, finally selecting a single representative article for each cluster. A notable academic project along these lines is Columbia University's *Newsblaster* [2,4,8], which goes further in providing computer-generated summaries of the day's news from the articles in a given cluster.

Our analysis is quite different from this. We track the temporal and spatial distribution of the entities in the news: who is being talked about, by whom, when, and where? Section 2 more clearly describes the nature of the news analysis we provide, and presents some global analysis of articles by source and type to demonstrate the power of *Lydia*.

*Lydia* is designed for high-speed analysis of online text. We seek to analyze thousands of curated text feeds daily. *Lydia* is capable of retrieving a daily newspaper like *The New York Times* and then analyzing the resulting stream of text in under one minute of computer time. We are capable of processing the entire 12 million abstracts of Medline/Pubmed in roughly two weeks on a single computer, covering virtually every paper of biological or medical interest published since the 1960's.

A block diagram of the *Lydia* processing pipeline appears in Figure 1. The major phases of our analysis are:

* This research was partially supported by NSF grants EIA-0325123 and DBI-0444815.

M. Consens and G. Navarro (Eds.): SPIRE 2005, LNCS 3772, pp. 161–166, 2005.

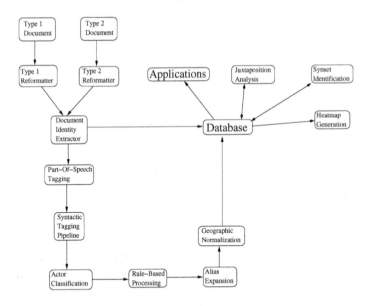

**Fig. 1.** Block diagram of the *Lydia* Pipeline

- *Spidering and Article Classification* – We obtain our newspaper text via spidering and parsing programs which require surprisingly little customization for different news sources.
- *Named Entity Recognition* – Identifying where *entities* (people, places, companies, etc.) are mentioned in newspaper articles is a critical first step in extracting information from them.
- *Juxtaposition Analysis* – For each entity, we wish to identify what other entities occur near it in an overrepresented way.
- *Synonym Set Identification* – A single news entity is often mentioned using multiple variations on their name. For example, *George Bush* is commonly referred to as *Bush, President Bush* and *George W. Bush.*
- *Temporal and Spatial Analysis* – We can establish local biases in the news by analyzing the relative frequency given entities are mentioned in different news sources. To compute the sphere of influence for a given newspaper, we look at its circulation, location, and the population of surrounding cities. We expand the radius of the sphere of influence until the population in it exceeds the circulation of the newspaper.

## 2   News Analysis with Lydia

In this section, we demonstrate the juxtapositional, spatial, and temporal entity analysis made possible by *Lydia*. We again encourage the reader to visit (*http://www.textmap.com*) to get a better feel of the power of this analysis on contemporary news topics.

**Table 1.** Top 10 Juxtapositions for Three Particular Entities

| Martin Luther King | | Israel | | North Carolina | |
|---|---|---|---|---|---|
| Entity | Score | Entity | Score | Entity | Score |
| Jesse Jackson | 545.97 | Mahmoud Abbas | 9,635.51 | Duke | 2,747.85 |
| Coretta Scott King | 454.51 | Palestinians | 9,041.70 | ACC | 1,666.92 |
| "I Have A Dream" | 370.37 | West Bank | 6,423.93 | Wake Forest | 1,554.92 |
| Atlanta, GA | 286.73 | Gaza | 4,391.05 | Virginia | 1,283.61 |
| Ebenezer Baptist Church | 260.84 | Ariel Sharon | 3,620.84 | Tar Heels | 1,237.39 |
| Saxby Chambliss | 227.47 | Hamas | 2,196.72 | Maryland | 1,029.20 |
| Douglass Theatre | 215.79 | Jerusalem, ISR | 2,125.96 | Raymond Felton | 929.48 |
| SCLC | 208.47 | Israelis | 1,786.67 | Rashad McCants | 871.44 |
| Greenville, SC | 199.27 | Yasser Arafat | 1,769.58 | Roy Williams | 745.19 |
| Harry Belafonte | 190.07 | Egypt | 1,526.77 | Georgia Tech | 684.07 |

Except where noted, all of the experiments in this paper were run on a set of 3,853 newspaper-days, partitioned among 66 distinct publications that were spidered between January 4, 2005 and March 15, 2005.

**Juxtaposition Analysis.** Our mental model of where an entity fits into the world depends largely upon how it relates to other entities. For each entity, we compute a significance score for every other entity that co-occurs with it, and rank its juxtapositions by this score. Table 1 shows the top 10 scoring juxtapositions (with significance score) for three popular entities. Some things to note from the table are:

- Many of the other entities in *Martin Luther King*'s list arise from festivities that surrounded his birthday.
- The position of *Mahmoud Abbas* at the top of *Israel*'s list reflects his ascent to the presidency of the *Palestinian National Authority*.
- The prominence of other universities and basketball terms in the *North Carolina* list reflects the quality and significance of the UNC basketball team.

There has been much work [5,6] on the similar problem of *recommender systems* for e-commerce. These systems seek to find what products a consumer is likely to purchase, given the products they have recently purchased. Our problem is also similar to the word collocation problem[7] from natural language processing. The goal there is to find which sets of two or more words occur close to each other more than they should by chance.

Developing a meaningful juxtapositionness function proved more difficult than anticipated. First, we discovered that if you simply use raw article counts, then the most popular entities will overly dominate the juxtapositions. Care must be taken, however, when punishing the popular entities against spurious juxtapositions dominated by the infrequently occurring entities. Our experience found that the popular scoring functions appearing in the literature [3] did not adequately correct for this problem.

To determine the significance of a juxtaposition, we bound the probability that two entities co-occur in the number of articles that they co-occur in if occurrences where generated by a random process. To estimate this probability we use a Chernoff Bound:

$$P(X > (1 + \delta)E[X]) \leq (\frac{e^\delta}{(1 + \delta)^{(1+\delta)}})^{E[X]}$$

where $\delta$ measures how far above the expected value the random variable is. If we set $(1+\delta)E[X] = F$ = number of co-occurrences, and consider $X$ as the number of randomized juxtapositions, we can bound the probability that we observe at least $F$ juxtapositions by calculating

$$P(X > F) \leq (\frac{e^{\frac{F}{E[X]} - 1}}{(\frac{F}{E[X]})^{(\frac{F}{E[X]})}})^{E[X]}$$

where $E[X] = \frac{n_a n_b}{N}$, $N$ = number of sentences in the corpus, $n_a$ = number of occurrences of entity a, and $n_b$ = number of occurrences of entity b, as the juxtaposition score for a pair of entities. We display $-$ log of this probability for numerical stability and ranking.

**Spatial Analysis.** It is interesting to see where in the country people are talking about particular entities. Each newspaper has a location and a circulation and each city has a population. These facts allow us to approximate a *sphere of influence* for each newspaper. The *heat* an entity is generating in a city is now a function of its frequency of reference in each of the newspapers that have influence over that city.

Figure 2 show the heatmap for *Washington DC* and *Phoenix*, in the news from over 500 United States news sources from April 11–May 30, 2005. The most intense heat for both city-entities focuses around their location, as should be expected. *Washington DC* generates a high level of interest throughout the United States. There is an additional minor concentration in the Pacific Northwest, which reflects the ambiguity between *Washington* the city and the state.

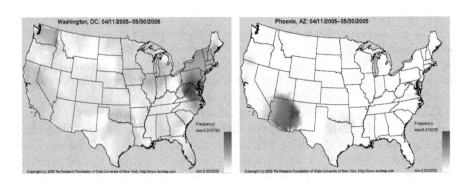

**Fig. 2.** Geographic News Distribution of two Spatially-Sensitive Entities

**Table 2.** Most Overrepresented Entities in Three Important U.S. Newspapers

| San Francisco Chronicle | | Chicago Tribune | | Miami Herald | |
|---|---|---|---|---|---|
| Entity | Score | Entity | Score | Entity | Score |
| Gavin Newsom | 10.84 | Chicago, IL | 8.57 | Miami, FL | 10.26 |
| San Francisco, CA | 10.56 | Richard Daley | 7.06 | South Florida | 9.53 |
| Bay Area | 8.44 | Joan Humphrey Lefkow | 5.20 | Fort Lauderdale, FL | 8.76 |
| Pedro Feliz | 5.36 | Aon Corp. | 4.69 | Cuba | 8.09 |
| BALCO | 5.29 | Salvador Dali | 4.54 | Caracas | 7.02 |
| Kimberly Bell | 5.02 | Wrigley Field | 4.42 | Florida Marlins | 6.91 |

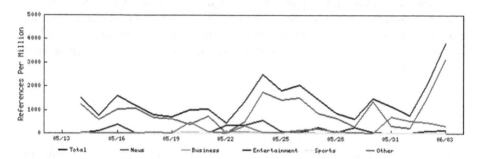

**Fig. 3.** Reference Frequency Time-Series for *Michael Jackson*, partitioned by article type

An alternate way to study relative geographic interest is to compare the reference frequency of entities in a given news source. Table 2 presents the most overrepresented entities in each of three major American newspapers, as scored by the number of standard deviations above expectation. These over-represented entities (primarily local politicians and sports teams) are all of stronger local interest than national interest.

**Temporal Analysis.** Our ability to track all references to entities broken down by article type gives us the ability to monitor trends. Figure 3 tracks the ebbs and flows in the interest in *Michael Jackson* as his trial progressed in May 2005. Note that the vast majority of his references are classified as news instead of entertainment, reflecting current media obsessions.

## 3   Conclusions and Future Work

We have presented the basic design and architecture of the *Lydia* text analysis system, along with experimental results illustrating its capabilities and performance. We are continuing to improve the entity recognition algorithms, particularly in synset construction, entity classification, and geographic normalization.

Future directions include dramatically increasing the scale of our analysis, as we anticipate moving from a single workstation to a 50-node cluster computer in the near future. With such resources, we should be able to do long-term historical news analysis and perhaps even larger-scale web studies. We are also exploring the use of the *Lydia* knowledge base as the foundation for a question answering system, and extracting semantic labels for explaining juxtaposition relationships.

## Acknowledgments

We thank Alex Kim for his help in developing the pipeline, Manjunath Srinivasaiah for his work on making the pipeline more efficient, Prachi Kaulgud for her work on markup and web interface design, Andrew Mehler for his Bayesian classifier and rules processor, Izzet Zorlu for his web interface design, Namrata Godbole for her work on text markup and spidering, Yue Wang, Yunfan Bao, and Xin Li for their work on Heatmaps, and Michael Papile for his geographic normalization routine.

## References

1. Google news. http://news.google.com.
2. R. Barzilay, N. Elhadad, and K. McKeown. Inferring strategies for sentence ordering in multidocument news summarization. *Journal of Artifical Intelligence Research (JAIR)*, 17:35–55, 2002.
3. W. Frakes and R. Baeza-Yates. *Information Retrieval: Data Structures and Algorithms.* Prentice-Hall, 1992.
4. V. Hatzivassiloglou, L. Gravano, and A. Maganti. An investigation of linguistic features and clustering algorithms for topical document clustering. In *Proceedings of the 23rd ACM SIGIR Conference on Research and Development in Information Retrieval*, pages 224–231, Athens, Greece, 2000.
5. W. Hill, L. Stead, M. Rosenstein, and G. Furnas. Recommending and evaluating choices in a virtual community of use. In *Proceedings of ACM Conference on Human Factors in Computing Systems(CHI'95)*, 1995.
6. T. Malone, K. Grant, F. Turbak, S. Brobst, and M. Cohen. Intelligent information-sharing systems. *Communications of the ACM*, 30:390–402, 1987.
7. C.D. Manning and H. Schutze. *Foundations of Statistical Natural Language Processing.* MIT Press. Cambridge, MA, 2003.
8. K. McKeown, R. Barzilay, D. Evans, V. Hatzivassiloglou, J. Klavans, A. Nenkova, C. Sable, B. Schiffman, and S. Sigelman. Tracking and summarizing news on a daily basis with columbia's newsblaster. In *Proceedings of HLT 2002 Human Language Technology Conference*, San Diego, California, USA, 2002.

# Composite Pattern Discovery
# for PCR Application

Stanislav Angelov[1,*] and Shunsuke Inenaga[2]

[1] Department of Computer and Information Science,
School of Engineering and Applied Sciences,
University of Pennsylvania, Philadelphia, PA 19104, USA
angelov@cis.upenn.edu
[2] Department of Informatics, Kyushu University, Fukuoka 812-8581, Japan
shunsuke.inenaga@i.kyushu-u.ac.jp

**Abstract.** We consider the problem of finding pairs of short patterns such that, in a given input sequence of length $n$, the distance between each pair's patterns is at least $\alpha$. The problem was introduced in [1] and is motivated by the optimization of multiplexed nested PCR.

We study algorithms for the following two cases; the special case when the two patterns in the pair are required to have the same length, and the more general case when the patterns can have different lengths. For the first case we present an $O(\alpha n \log \log n)$ time and $O(n)$ space algorithm, and for the general case we give an $O(\alpha n \log n)$ time and $O(n)$ space algorithm. The algorithms work for any alphabet size and use asymptotically less space than the algorithms presented in [1]. For alphabets of constant size we also give an $O(n\sqrt{n}\log^2 n)$ time algorithm for the general case. We demonstrate that the algorithms perform well in practice and present our findings for the human genome.

In addition, we study an extended version of the problem where patterns in the pair occur at certain positions at a distance at most $\alpha$, but do not occur $\alpha$-close anywhere else, in the input sequence.

## 1 Introduction

### 1.1 Composite Pattern Discovery

*Pattern discovery* is a fundamental problem in Computational Biology and Bioinformatics [2,3]. A large amount of effort was paid to devising efficient algorithms to extract interesting, useful, and surprising substring patterns from massive biological sequences [4,5]. Then this research has been extended to more complicated but very expressive pattern classes such as subsequence patterns [6,7], episode patterns [8,9], VLDC patterns [10], and their variations [11].

The demand for *composite pattern discovery* has recently arisen rather than simply finding single patterns. It is motivated by, for instance, the fact that many

---

* Supported in part by NSF Career Award CCR-0093117, NSF Award ITR 0205456 and NIGMS Award 1-P20-GM-6912-1.

M. Consens and G. Navarro (Eds.): SPIRE 2005, LNCS 3772, pp. 167–178, 2005.

of the actual regulatory signals are composite patterns that are groups of monad patterns occurring near each other [12]. The concept of composite patterns was introduced by Marsan and Sagot [13] as *structured motifs* which are two or more patterns separated by a certain distance. They introduced suffix tree [14] based algorithms for finding structured motifs, and Carvalho et al. [15] presented a new algorithm with improved running time and space.

In a similar concept, Arimura et al. [16,17] introduced *proximity patterns* and proposed algorithms to find these patterns efficiently. MITRA [12] is another method that looks for composite patterns. BioProspector [18] applies the Gibbs sampling strategy to discover gapped motifs. Bannai et al. [19] and Inenaga et al. [20] considered *Boolean combinations* of patterns, in order to find regulatory elements that cooperate, complement, or compete with each other in enhancing and/or silencing certain genomic functions.

Another application of composite pattern discovery is to find good adapters for primers used in *polymerase chain reaction (PCR)*. PCR is a standard technique for producing many copies of a DNA region [21]. It is routinely used for example in medicine to detect infections and in forensic science to identify individuals even from tiny samples. In PCR a pair of short fragments of DNA called primers is specifically designed for the amplified region so that each of them is complementary to the 3' end of one of two strands of the region (see also Fig. 1).

In order to achieve ultrasensitive detection, repeated PCR with nested primers, so-called *nested PCR*, is used. Also, detection tests are preferred to be carried out in a multiplexed fashion. Let $S$ denote any sequence taken from a sample of genome, and $S'$ denote the reverse complement of $S$. To obtain a good primer pair for multiplexed nested PCR, we are required to find a pair of patterns $(A, B)$ such that any occurrences of $A$ and $B$ are separated further than $\alpha$ in both sequences $S$ and $S'$, where $\alpha$ is a given threshold value. Then, the pair $(A, B)$ is called a *missing pattern pair* (or shortly a *missing pair*). For the application of multiplexed nested PCR, the patterns in a missing pair have to also satisfy that $|A| = |B| = k$ and $k$ is as short as possible. Namely, a missing pair with patterns of the same, and shortest length, is demanded. More details of the relationship between missing patterns and PCR can be found in [1].

## 1.2 Finding Missing Patterns

The problem of finding missing pattern pairs was firstly considered in [1]. The paper presented an algorithm which finds missing pattern pairs in $O(\alpha n \log \log n)$

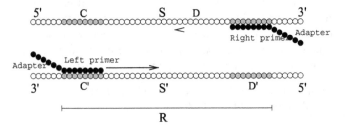

**Fig. 1.** Illustration for polymerase chain reaction (PCR)

time with $O(\alpha n)$ space, where $n$ denotes the length of the input sequence. For a more general case where the two patterns in a missing pair can have different lengths, the paper showed that the problem is solvable in $O(n^2)$ time and $O(n)$ space, or in $O(\alpha n \log n)$ time and $O(n \log n)$ space, both on a constant-size alphabet. We remark that the patterns considered in [1] were substring patterns, that is, exact match without errors was considered.

In this paper, we give simpler and more efficient algorithms that solve the stated problems for an arbitrary alphabet size $\sigma$. We give an $O(\alpha n \log \log_\sigma n)$ time algorithm for the case when the patterns in the missing pairs are of the same length, and $O(\min\{\alpha n \log_\sigma n, n\sqrt{n}(\sigma + \log n)\log_\sigma n\})$ time algorithm for the case when the two patterns can have different lengths. In both cases the space requirement is only $O(n)$.

See Tables 1 and 2 for a more detailed comparison between our algorithms and those in [1]. For patterns of the same length and constant-size alphabets, Algorithm 1 saves computational space by a factor of $\alpha$. It also improves the time complexity for arbitrary alphabet sizes. For pairs of patterns of different lengths, Algorithm 1 is superior to Suffix Tree Algorithm B on both constant and arbitrary alphabets. It is also noteworthy that although both Suffix Tree Algorithms heavily depend on manipulations on suffix trees [14], neither Algorithm 1 or 2 in this paper needs advanced data structures which can be rather expensive in practice.

Furthermore, since primers need to be present around the region to be amplified, we also study a natural extension of the problem where patterns in the pair occur at certain positions at a distance at most $\alpha$, but do not occur $\alpha$-close anywhere else, in the input sequence. We show how Algorithm 1 can be modified for this extended problem. Since the restriction can make "short" pattern pairs impossible, we also discuss a variant that allows for arbitrary pattern lengths. We note that for the case of primers, which typically have lengths in the range 17..25, the obtained algorithm runs in $O(\alpha n)$ time and $O(n)$ space.

**Table 1.** Summary of results for finding missing pairs of patterns of same length

| Algorithm | Time | Space |
|-----------|------|-------|
| Algorithm 1 | $O(\alpha n \log \log_\sigma n)$ | $O(n)$ |
| Bit Table Algorithm [1] | $O(\alpha n(\sigma + \log \log_\sigma n))$ | $O(\alpha n)$ |

**Table 2.** Summary of results for finding missing pairs of patterns of different length

| Algorithm | Time | Space |
|-----------|------|-------|
| Algorithm 1 | $O(\alpha n \log_\sigma n)$ | $O(n)$ |
| Algorithm 2 | $O(n\sqrt{n}(\sigma + \log n)\log_\sigma n)$ | $O(n)$ |
| Suffix Tree Alg. A [1] | $O(n^2)$ | $O(n)$ |
| Suffix Tree Alg. B [1]* | $O(\alpha n \log n)$ | $O(n \log n)$ |
| Suffix Tree Alg. B [1] | $O(\log \sigma \alpha n \log_\sigma n)$ | $O(\log \sigma \alpha n \log_\sigma n)$ |

* Constant size alphabet.

## 1.3  Organization

In Section 2 we formally introduce our model and state the considered problem. In Section 3 we present the main algorithm, and subsequent results. Next, in Section 4, we discuss natural extensions to the main algorithm and their implications. In Section 5 we discuss our findings on the human genome. Finally, Section 6 concludes this paper with possible directions for future work.

## 2  Preliminaries

A *string* $T = t_1 t_2 \cdots t_n$ is a sequence of *characters* from an ordered *alphabet* $\Sigma$ of size $\sigma$. We assume *w.l.o.g.* $\Sigma = \{0, 1, \ldots, \sigma - 1\}$ and that all characters occur in $T$. A *substring* of $T$ is any string $T_{i \ldots j} = t_i t_{i+1} \cdots t_j$, where $1 \leq i \leq j \leq n$. A *pattern* is a short string over the alphabet $\Sigma$. We say that pattern $P = p_1 p_2 \cdots p_k$ *occurs* at position $j$ of string $T$ iff $p_1 = t_j, p_2 = t_{j+1}, \ldots, p_k = t_{j+k-1}$. Such positions $j$ are called the occurrence positions of $P$ in $T$.

A *missing pattern* $P$ (with respect to sequence $T$) is such that $P$ is not a substring of $T$, i.e., $P$ does not occur at any position $j$ of $T$. Let $\alpha > 0$ be a threshold parameter. A *missing pattern pair* $(A, B)$ is such that if $A$ (resp. $B$) occurs at position $j$ of sequence $T$, then $B$ $(A)$ does not occur at any position $j'$ of $T$, such that $j - \alpha \leq j' \leq j + \alpha$. If $(A, B)$ is a missing pair, we say that $A$ and $B$ do not occur $\alpha$-*close* in $T$. These notions are illustrated in Fig. 2.

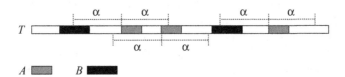

**Fig. 2.** Missing pattern pair $(A, B)$. No occurrences of $A$ and $B$ are at a distance closer than $\alpha$.

We study the following problem:

*Problem 1 (Missing Pattern Pair Problem).* Given a sequence $T$ and a threshold $\alpha$, find patterns $A$ and $B$ of minimum total length, such that $(A, B)$ is a missing pattern pair with respect to $T$, i.e., $A$ and $B$ do not occur $\alpha$-close in $T$.

## 3  Finding Missing Pattern Pairs

Missing pattern pairs can be formed by two processes. When a pattern does not occur in the input sequence $T$, it can be combined with any pattern to form a missing pair. Alternatively, both patterns in the pair may occur in the sequence, but always at least $\alpha$ positions away from each other. The first case, when a single pattern is missing, provides an insight to the upper bound on the missing pair length and is an interesting property on its own.

It is not hard to see that there is a missing pattern of length $\lceil \log_\sigma n \rceil$ from sequence $T$ with size $n$, where $\sigma$ is the input alphabet size. This is because there are at most $n - k + 1$ distinct patterns of length $k$ in $T$. In [1], a linear time algorithm based on suffix trees is proposed that finds the shortest missing pattern when $\sigma$ is a constant. The algorithm can be readily extended for the case of arbitrary alphabet sizes by a loss of $\log \sigma$ factor. Instead, we can compute a bit table of all patterns of length $\lfloor \log_\sigma n \rfloor$ that occur in $T$ using the natural bijective mapping of the patterns to the integers $0, 1, \ldots, \sigma^{\lfloor \log_\sigma n \rfloor} - 1$. This can be done in linear time by scanning the input sequence from left to right using the established technique of computing the entry of pattern $Yb$ knowing the entry of pattern $aY$ (see for example [22]). By examining consecutive runs of missing patterns of this length, one can compute the shortest string (the longest missing pattern prefix) that is missing from the input sequence. If all patterns of length $\lfloor \log_\sigma n \rfloor$ are present in $T$, then the shortest missing pattern is of length $\lceil \log_\sigma n \rceil$. In this case we can find a representative by computing the first $n$ entries of the corresponding bit table.

**Proposition 1.** *The shortest* single *missing pattern problem can be solved in linear time and space.*

In what follows, we let $m$ be the length of the shortest missing pattern.

### 3.1   Finding Missing Pairs of Fixed Lengths

We now present an $O(\alpha n)$ time and $O(n)$ space algorithm that finds a missing pattern pair $(A, B)$, where the lengths of $A$ and $B$ are given as input parameters. The algorithm serves as a basis for the missing pattern pairs algorithms that follow. Let $|A| = a$ and $|B| = b$ and assume *w.l.o.g.* $a \geq b$. We will consider the case when $a < m$, or else there is a pattern of length $m$ that is missing and by Proposition 1 it can be found in linear time. Let $N_1 = \sigma^a$ and $N_2 = \sigma^b$ be the number of distinct patterns of length $a$ and $b$ respectively. (Clearly, $n > N_1 \geq N_2$). The proposed algorithm heavily uses the bijective mapping of patterns to integers described in the previous subsection.

**Algorithm 1.** We now describe the steps of the algorithm.

1. Let $L$ be an array of length $N_1$, where $L[h]$ is the list of occurrence positions in $T$ of the pattern of length $a$ mapped to the integer $h$.
2. Compute an array $H$ s.t. $H[j]$ is the mapped value of the pattern of length $b$ at position $j$ of $T$.
3. For $h = 0 \ldots N_1 - 1$, count the number of distinct patterns $B$ of length $b$ that occur at distance at most $\alpha$ from the pattern $A$ of length $a$ mapped to $h$. We do this by maintaining a bit table of the distinct patterns $B$ that are $\alpha$-close to $A$.

At each iteration we perform the following sub-steps. Let $A$ be the pattern mapped to $h$.

(i) For each occurrence in $L[h]$ of pattern $A$, we mark in a table $M$ of size $N_2$ all patterns of length $b$ that occur at distance at most $\alpha$ by scanning the corresponding positions of the array $H$.

(ii) When a pattern of length $b$ is seen for the first time we increment a counter. The counter is set to 0 at the beginning of each iteration.

(iii) The iteration ends when the maintained counter becomes equal to $N_2$ to indicate that all patterns of length $b$ are $\alpha$-close to $A$, or when all of $L[h]$ is processed. At the end of an iteration, if the counter is less than $N_2$ we scan $M$ to output a missing pattern pair and the algorithm terminates.

**Analysis.** Step 1 of the algorithm can be performed in $O(n)$ time by scanning $T$ from left to right. Compute the value $h$ of the pattern at position $i$ from that of position $i - 1$ and append $i$ to the list $L[h]$. The total size of all lists is $n - a + 1$. The array $H$ in Step 2 can be computed in a similar fashion and takes $n - b + 1$ space. An iteration of Step 3 takes $O(\alpha|L[h]|)$ time for a total of $O(\alpha n)$ time and an additional $O(N_2) = O(n)$ space. We conclude the algorithm will output a missing pair $(A, B)$ with the desired pattern lengths, if such a pair exists, in $O(\alpha n)$ time and $O(n)$ space.

## 3.2    Finding Missing Pairs of the Same Length

We combine the algorithm from the previous subsection and the following property to obtain an efficient algorithm for the problem of finding missing pairs when the patterns are of the same length.

*Property 1 (Monotonicity Property).* If a pattern pair $(A, B)$ is missing, the pair $(C, D)$, where $A$ is a substring of $C$ and $B$ is a substring of $D$, is also missing.

We are now ready to state the following theorem.

**Theorem 1.** *The missing patterns problem on a sequence of length $n$ for patterns of the same length can be solved in $O(\alpha n \log \log_\sigma n)$ time and $O(n)$ space, where $\sigma$ is the alphabet size.*

*Proof.* Recall that there exists a missing pattern pair $(A, B)$, where $a = b = m \leq \lceil \log_\sigma n \rceil$. Therefore, such missing pair can be found in linear time by Proposition 1. In order to find the shortest pair, we can do binary search on the pattern length $1 \ldots m - 1$ and apply Algorithm 1 for each length. From the Monotonicity Property we are guaranteed to output the shortest missing pattern pair of the same length in $O(\alpha n \log \log_\sigma n)$ time and $O(n)$ space. $\square$

## 3.3    Finding Missing Pairs

We now consider the problem when the two patterns in a missing pair do not necessarily have the same length. From Proposition 1, there exists a missing pattern pair $(A, B)$, where $a = m$ and $b = 1$ for a combined pair length of $m + 1$. Recall that $m \leq \lceil \log_\sigma n \rceil$ is the length of the shortest missing pattern of the input sequence $T$ and can be found in linear time and space. Such missing pattern can be combined with any non-empty pattern to form a missing pattern pair. For $\alpha \geq m$, it is easy to see that for any missing pattern pair $(A, B)$ of length $\leq m$, the concatenation of $A$ and $B$ should also be missing, otherwise $A$ and $B$ occur at a distance $\leq \alpha$. Therefore, for any missing pattern pair $(A, B)$, $a + b \geq m$.

**Theorem 2.** *The missing patterns problem on sequence of length $n$ can be solved in $O(\alpha n \log_\sigma n)$ time and $O(n)$ space, where $\sigma$ is the alphabet size.*

*Proof.* We showed that the shortest missing pattern pair is of length at least $m$ and at most $m + 1$. To find if a pattern pair $(A, B)$ of length $m$ is missing it is enough to verify all possible combinations of $a + b = m$. This can be done by applying $m = O(\log_\sigma n)$ times Algorithm 1. Therefore, we obtain the desired running time and space.     □

The above analysis assumes $\alpha \geq m$. In the case when $\alpha < m$, there is also a solution to take in consideration of total length $2\alpha + 1$. Let $G, T \in \Sigma$ be two arbitrary letters of the input alphabet. Consider the pattern pair

$$(\underbrace{G \ldots G}_{\alpha+1}, \underbrace{T \ldots T}_{\alpha}) \ .$$

Trivially, it is a missing pair since the two patterns cannot occur $\alpha$-close.

*Remark 1.* Let the alphabet size $\sigma$ be a constant. Since there are $\sigma^m = O(n)$ pattern pairs of combined length $m = O(\log_\sigma n)$, one can adapt the bit-table algorithm from [1] to match the above running time and space requirements.

We now present an algorithm with running time independent of the threshold parameter $\alpha$. The algorithm finds for each pair of patterns $(A, B)$ of given length, the smallest $\alpha_{AB}$ s.t. the two patterns occur $\alpha_{AB}$-close. Therefore, a pattern pair $(A, B)$ is missing iff $\alpha_{AB} > \alpha$. The algorithm also finds the smallest $\alpha_{\min}$ s.t. all pattern pairs are $\alpha_{\min}$-close.

**Algorithm 2.** The algorithm takes advantage from the fact that there are not too many pattern pairs of total length $m$. More precisely, there are at most $\sigma^{\lceil \log_\sigma n \rceil} < \sigma n$ such pairs. Again, we present the algorithm for fixed lengths of the pattern pairs $(A, B)$ and adapt similar notation to Algorithm 1. We further assume $a + b = m$. The steps of the algorithm are as follow:

1. Let $L$ be an array of length $N_1$, where $L[h]$ is the list of occurrence positions in $T$ of the pattern of length $a$ mapped to the integer $h$. Let $R$ be an array of length $N_2$, where $R[h']$ is the list of occurrence positions in $T$ of the pattern of length $b$ mapped to the integer $h'$.
2. For each pattern pair $(A, B)$, merge efficiently the corresponding lists of occurrence positions (which are sorted by construction) to find the closest occurrence of $A$ and $B$ and therefore $\alpha_{AB}$.

**Analysis.** The algorithm clearly requires $O(n)$ space, and we claim it takes $O(n\sqrt{n}(\sigma + \log n))$ time. Step 1 of the algorithm can be performed in $O(n)$ time by scanning $T$ from left to right. We now analyze Step 2. For a given pattern, we will call its list of occurrence positions *long* if it has length at least $\sqrt{n}$. We note that there are at most $\sqrt{n}$ long lists in $L$ since the total length of all lists is at most $n$. Similarly, there are at most $\sqrt{n}$ long lists in $R$. All pairs of lists that are not long can be merged in $O(\sigma n\sqrt{n})$ time using merge sort since there are $O(\sigma n)$ such pairs. Let $I$ be the set of indices of long lists in $L$, i.e. for all $h \in I$, $|L[h]| \geq \sqrt{n}$. Fix $h \in I$. The list $L[h]$ can be merged using binary search with all lists in $R$ in time proportional to $\sum_{h'} |R[h']| \log |L[h]| = O(n \log |L[h]|)$. Summing over $h \in I$ we obtain $n \sum_{h \in I} \log |L[h]| = O(n\sqrt{n} \log n)$ since $|I| \leq \sqrt{n}$ and each list is of length at most $n$. Applying the same argument for the long lists in $R$ we obtain the desired running time.

The next theorem follows by an argument analogous to the proof of Theorem 2 but applying Algorithm 2.

**Theorem 3.** *The missing pattern problem on sequence of length $n$ can be solved in $O(n\sqrt{n}(\sigma + \log n) \log_\sigma n)$ time and $O(n)$ space, where $\sigma$ is the alphabet size.*

## 4    Extensions to the Missing Pattern Pair Problem

We discuss the following two extensions to the problem of finding missing pattern pairs of fixed lengths. First, we show how to find missing pairs when the patterns are restricted to occur at certain regions of the input sequence $T$. Next, in addition, we allow the patterns to be of length greater than $m$. We describe the required changes to Algorithm 1, and then state how it generalizes to the problem of finding the shortest pattern pairs of the same or different lengths.

**Localized Patterns.** Let $P_L$ ($P_R$) be the set of positions where pattern $A$ ($B$) of pair $(A, B)$ need to be present. The sets can be specified as interval lists or bit-tables. For simplicity we assume the latter representation, which can be obtained from the interval lists in $O(n)$ time and space[1]. We are interested in finding pattern pairs that occur at the restricted positions at a distance at most $\alpha$, but do not occur $\alpha$-close anywhere else.

We modify Algorithm 1 as follows. We restrict occurrence positions for $A$ patterns in lists in $L$ only to those in $P_L$ in a straightforward manner. In the same fashion, in Step 3, we count for each pattern $A$, the distinct patterns $B$ that occur at distance at most $\alpha$. If there is a pattern missing, we do an additional pass to look for $\alpha$-close unmarked pattern that start in $P_R$. It is not hard to see that with the described modifications the time and space requirements of the algorithm do not change.

**Long Patterns.** Since patterns are restricted to occur in the input sequence $T$, there are at most $n$ candidate patterns for each $A$ and $B$ irrespective to their given length. For patterns of length greater than $\lfloor \log_\sigma n \rfloor$, we can maintain the same framework of Algorithm 1 given a suitable (hash) function mapping valid $A$ and $B$ patterns to integers 0 to $O(n)$ in the corresponding lists $L$ (Step 1)

---

[1] The conversion from lists to a bit table can be done even when the intervals are overlapping by storing for each position the number of intervals starting and ending at that position, and then scanning the resulting array from left to right. We assume there are $O(n)$ lists.

and $H$ (Step 2). We obtain the desired properties by computing the suffix tree of $T$ and using the node indices corresponding to the patterns of length $a$ and $b$ in a standard way (see for details [14]). Computing the suffix tree only requires an additional $O(n \log \sigma)$ time and $O(n)$ space [14].

We are now ready to state the following theorem.

**Theorem 4.** *The generalized missing patterns problem on a sequence of length $n$ can be solved in*

- $O(\alpha n \ell)$ *time when the patterns are of the same length;*
- $O(\alpha n \ell^2)$ *time when the patterns are allowed to have different lengths,*

*where $\sigma$ is the alphabet size, and $\ell$ is the total length of the output pair. In both cases the space requirement is $O(n)$.*

*Proof.* (Sketch) Note that because of the condition that patterns must occur $\alpha$-close at specific positions, the Monotonicity Property does not hold. We therefore need to run the extended Algorithm 1 for all possible combinations of pattern lengths up to $\ell$. Furthermore, note that $\lfloor \log \sigma \rfloor \le \alpha$ since $\sigma \le n$, otherwise in the case when pattern length is greater than $\alpha$ there are trivial missing pattern pairs. Therefore, the term $O(\alpha n)$ in Algorithm 1 dominates the construction of the suffix tree of $O(n \log \sigma)$ time.                                     □

## 5   Experiments

We have performed preliminary tests with the human genome[2] to complement the results reported in [1] for the baker's yeast (*Saccharomyces cerevisiae*) genome. We set the threshold parameter $\alpha$ to a realistic value 5000 and searched for the shortest missing pattern pairs where the patterns are of the same length $k$. We found 238 pattern pairs for $k = 8$. Interestingly, the shortest pattern pairs found for the baker's yeast genome, which is about 250 times smaller, were also of length 8 [1]. From the 238 pattern pairs, 20 pairs are missing from both the human and the baker's yeast genome. Table 3 summarizes these missing pairs and the shortest distance between the patterns (or their reverse complements) of each pair in the corresponding genomes. For reference, the shortest (single) missing patterns from the human genome are of length 11 and are listed in Table 4. This is also surprising since the human genome length is roughly equal to $4^{16}$.

An implementation in Java of the used software is available at the author's homepage[3]. The program needed about 3 hours to process the baker's yeast genome on a 1GHz machine, and about 30 hours for the human genome. The stop condition of step 3 of Algorithm 1, namely when all pattern pairs are discovered for the current pattern, provides a significant optimization in practice which allows the software to run only 10 times slower (rather than 250 times) for the human genome compared to the yeast genome.

---

[2] Available at ftp://ftp.ensembl.org/pub/current_human/
[3] http://www.cis.upenn.edu/~angelov

**Table 3.** Unordered missing pattern pairs in both the human and baker's yeast genomes for $k = 8$. The reverse complements of the shown pattern pairs are also missing.

| Missing Pairs | Yeast $\alpha_{AB}$ | Human $\alpha_{AB}$ |
|---|---|---|
| (AATCGACG, CGATCGGT) | 5008 | 6458 |
| (CCGATCGG, CCGTACGG) | 5658 | 6839 |
| (CGACCGTA, TACGGTCG) | 13933 | 7585 |
| (CGACCGTA, TCGCGTAC) | 5494 | 5345 |
| (CGAGTACG, GTCGATCG) | 5903 | 8090 |
| (CGATCGGA, GCGCGATA) | 6432 | 6619 |

**Table 4.** (Single) missing patterns from the human genome of length 11. The reverse complements of the shown patters are also missing.

| Missing Patterns | | |
|---|---|---|
| ATTTCGTCGCG | CGGCCGTACGA | CGCGAACGTTA |
| CCGAATACGCG | CGTCGCTCGAA | CGTTACGACGA |
| CCGACGATCGA | CGACGCGATAG | GCGTCGAACGA |
| CGCGTCGATAG | CGATTCGGCGA | TATCGCGTCGA |

# 6    Conclusions and Further Work

The *missing pattern discovery problem* was first introduced in [1] for optimal selection of adapter primers for nested PCR. In this paper, we presented more simple and efficient algorithms to solve the problem. The presented algorithms only require linear space and thus are efficiently implementable, whereas most algorithms in [1] take super-linear space. Our algorithms also have advantages for running time compared to those in [1], especially when the alphabet size $\sigma$ is not constant. We implemented our algorithms and made experiments for the human genome and the baker's yeast genome, and we succeeded in finding shortest missing pairs of length 8 for both human and yeast genomes. In addition, we studied an extended version of the problem where patterns in the pair occur at certain positions at a distance at most $\alpha$, but do not occur $\alpha$-close anywhere else, in the input sequence.

As a generalization of the missing pattern discovery problem, the following problem that allows mismatches is worth to consider: Given sequence $T$, distance $\alpha$, and an error parameter $e$, find pattern pair $(A, B)$ such that any occurrence of $A$ and $B$ within $e$ mismatches in $T$ is not $\alpha$-close. In [13], they presented some algorithms to discover structured motifs with errors in the Hamming distance metric. Since the algorithms of [1] and [13] are both based on suffix trees, it might be possible to solve the above general missing pattern discovery problem by combining these algorithms.

# Acknowledgement

The authors would like to thank Teemu Kivioja of VTT Biotechnology and Veli Mäkinen of Bielefeld University, for their contributions to the pioneering work on finding missing pattern pairs in [1] and their fruitful comments to this work.

# References

1. Inenaga, S., Kivioja, T., Mäkinen, V.: Finding missing patterns. In: Proc. 4th Workshop on Algorithms in Bioinformatics (WABI'04). Volume 3240 of LNCS., Springer-Verlag (2004) 463–474
2. Apostolico, A.: Pattern discovery and the algorithmics of surprise. In: Artificial Intelligence and Heuristic Methods for Bioinformatics. (2003) 111–127
3. Shinohara, A., Takeda, M., Arikawa, S., Hirao, M., Hoshino, H., Inenaga, S.: Finding best patterns practically. In: Progress in Discovery Science. Volume 2281 of LNAI., Springer-Verlag (2002) 307–317
4. Shimozono, S., Shinohara, A., Shinohara, T., Miyano, S., Kuhara, S., Arikawa, S.: Knowledge acquisition from amino acid sequences by machine learning system BONSAI. Transactions of Information Processing Society of Japan 35 (1994) 2009–2018
5. Bannai, H., Inenaga, S., Shinohara, A., Takeda, M., Miyano, S.: Efficiently finding regulatory elements using correlation with gene expression. Journal of Bioinformatics and Computational Biology 2 (2004) 273–288
6. Baeza-Yates, R.A.: Searching subsequences (note). Theoretical Computer Science 78 (1991) 363–376
7. Hirao, M., Hoshino, H., Shinohara, A., Takeda, M., Arikawa, S.: A practical algorithm to find the best subsequence patterns. In: Proc. 3rd International Conference on Discovery Science (DS'00). Volume 1967 of LNAI., Springer-Verlag (2000) 141–154
8. Mannila, H., Toivonen, H., Verkamo, A.I.: Discovering frequent episode in sequences. In: Proc. 1st International Conference on Knowledge Discovery and Data Mining, AAAI Press (1995) 210–215
9. Hirao, M., Inenaga, S., Shinohara, A., Takeda, M., Arikawa, S.: A practical algorithm to find the best episode patterns. In: Proc. 4th International Conference on Discovery Science (DS'01). Volume 2226 of LNAI., Springer-Verlag (2001) 435–440
10. Inenaga, S., Bannai, H., Shinohara, A., Takeda, M., Arikawa, S.: Discovering best variable-length-don't-care patterns. In: Proc. 5th International Conference on Discovery Science (DS'02). Volume 2534 of LNCS., Springer-Verlag (2002) 86–97
11. Takeda, M., Inenaga, S., Bannai, H., Shinohara, A., Arikawa, S.: Discovering most classificatory patterns for very expressive pattern classes. In: Proc. 6th International Conference on Discovery Science (DS'03). Volume 2843 of LNCS., Springer-Verlag (2003) 486–493
12. Eskin, E., Pevzner, P.A.: Finding composite regulatory patterns in DNA sequences. Bioinformatics 18 (2002) S354–S363
13. Marsan, L., Sagot, M.F.: Algorithms for extracting structured motifs using a suffix tree with an application to promoter and regulatory site consensus identification. J. Comput. Biol. 7 (2000) 345–360
14. Gusfield, D.: Algorithms on Strings, Trees, and Sequences. Cambridge University Press (1997)

15. Carvalho, A.M., Freitas, A.T., Oliveira, A.L., Sagot, M.F.: A highly scalable algorithm for the extraction of cis-regulatory regions. In: Proc. 3rd Asia Pacific Bioinformatics Conference (APBC'05), Imperial College Press (2005) 273–282
16. Arimura, H., Arikawa, S., Shimozono, S.: Efficient discovery of optimal word-association patterns in large text databases. New Generation Computing **18** (2000) 49–60
17. Arimura, H., Asaka, H., Sakamoto, H., Arikawa, S.: Efficient discovery of proximity patterns with suffix arrays (extended abstract). In: Proc. the 12th Annual Symposium on Combinatorial Pattern Matching (CPM'01). Volume 2089 of LNCS., Springer-Verlag (2001) 152–156
18. Liu, X., Brutlag, D., Liu, J.: BioProspector: discovering conserved DNA motifs in upstream regulatory regions of co-expressed genes. In: Pac. Symp. Biocomput. (2001) 127–138
19. Bannai, H., Hyyrö, H., Shinohara, A., Takeda, M., Nakai, K., Miyano, S.: An $O(N^2)$ algorithm for discovering optimal Boolean pattern pairs. IEEE/ACM Transactions on Computational Biology and Bioinformatics **1** (2004) 159–170
20. Inenaga, S., Bannai, H., Hyyrö, H., Shinohara, A., Takeda, M., Nakai, K., Miyano, S.: Finding optimal pairs of cooperative and competing patterns with bounded distance. In: Proc. 7th International Conference on Discovery Science (DS'04). Volume 3245 of LNCS., Springer-Verlag (2004) 32–46
21. Alberts, B., Johnson, A., Lewis, J., Raff, M., Roberts, K., Walter, P.: Molecular Biology of the Cell, fourth edition. Garland Science (2002)
22. Karp, R., Rabin, M.: Efficient randomized pattern-matching algorithms. IBM Journal of Research and Development **31** (1987) 249–260

# Lossless Filter for Finding Long Multiple Approximate Repetitions Using a New Data Structure, the Bi-factor Array

Pierre Peterlongo[1], Nadia Pisanti[2,*],
Frederic Boyer[3], and Marie-France Sagot[3,4,**]

[1] Institut Gaspard-Monge, Universite de Marne-la-Vallée, France
[2] Dipartimento di Informatica, Università di Pisa, Italy
and LIPN Université Paris-Nord, France
[3] INRIA Rhône-Alpes and LBBE, Univ. Claude Bernard, Lyon, France
[4] King's College, London, UK

**Abstract.** Similarity search in texts, notably biological sequences, has received substantial attention in the last few years. Numerous filtration and indexing techniques have been created in order to speed up the resolution of the problem. However, previous filters were made for speeding up pattern matching, or for finding repetitions between two sequences or occurring twice in the same sequence. In this paper, we present an algorithm called NIMBUS for filtering sequences prior to finding repetitions occurring more than twice in a sequence or in more than two sequences. NIMBUS uses gapped seeds that are indexed with a new data structure, called a bi-factor array, that is also presented in this paper. Experimental results show that the filter can be very efficient: preprocessing with NIMBUS a data set where one wants to find functional elements using a multiple local alignment tool such as *GLAM* ([7]), the overall execution time can be reduced from 10 hours to 6 minutes while obtaining exactly the same results.

## 1   Introduction

Finding approximate repetitions (motifs) in sequences is one of the most challenging tasks in text mining. Its relevance grew recently because of its application to biological sequences. Although several algorithms have been designed to address this task, and have been extensively used, the problem still deserves investigation for certain types of repetitions. Indeed, when the latter are quite long and the number of differences they contain among them grows proportionally to their length, there is no exact tool that can manage to detect such repetitions efficiently. Widely used efficient algorithms for multiple alignment are heuristic, and offer no guarantee that false negatives are avoided. On the other hand, exhaustive inference methods cannot handle queries where the differences allowed

* Supported by the ACI IMPBio *Evolrep* project of the French Ministry of Research.
** Supported by the ACI Nouvelles Interfaces des Mathématiques $\pi$-*vert* project of the French Ministry of Research, and by the ARC *BIN* project from the INRIA.

M. Consens and G. Navarro (Eds.): SPIRE 2005, LNCS 3772, pp. 179–190, 2005.

among the occurrences of a motif represent as many as $5 - 10\%$ of the length of the motif, and the latter is as small as, say, 100 DNA bases. Indeed, exhaustive inference is done by extending or assembling in all possible ways shorter motifs that satisfy certain sufficient conditions. When the number of differences allowed is relatively high, this can therefore result in too many false positives that saturate the memory. In this paper, we introduce a preprocessing filter, called NIMBUS, where most of the data containing such false positives are discarded in order to perform a more efficient exhaustive inference. Our filter is designed for finding repetitions in $r \geq 2$ input sequences, or repetitions occurring more than twice in one sequence. To our knowledge, one finds in the literature filters for local alignment between two sequences [21,17,15], or for approximate pattern matching [19,3] only. Heuritic methods such as BLAST [1,2] and FASTA [16] filter input data and extend only *seeds* that are repeated short fragments satisfying some constraints. NIMBUS is based on similar ideas but uses different requirements concerning the seeds; among the requirements are frequency of occurrence of the seeds, concentration and relative position. Similarly to [17,15], we use also a concept related to gapped seeds that has been shown in [4] to be particularly efficient for pattern matching. The filter we designed is lossless: contrary to the filter in BLAST or FASTA, ours does not discard any repetition meeting the input parameters. It uses necessary conditions based on combinatorial properties of repetitions and an algorithm that checks such properties in an efficient way. The efficiency of the filter relies on an original data structure, the *bi-factor array*, that is also introduced in this paper, and on a labelling of the seeds similar to the one employed in [8]. This new data structure can be used to speed up other tasks such as the inference of structured motifs [18] or for improving other filters [14].

## 2   Necessary Conditions for Long Repetitions

A *string* is a sequence of zero or more symbols from an alphabet $\Sigma$. A string $s$ of length $n$ on $\Sigma$ is represented also by $s[0]s[1]\ldots s[n-1]$, where $s[i] \in \Sigma$ for $0 \leq i < n$. The length of $s$ is denoted by $|s|$. We denote by $s[i,j]$ the *substring*, or *factor*, $s[i]s[i+1]\ldots s[j]$ of $s$. In this case, we say that the string $s[i,j]$ occurs at position $i$ in $s$. We call $k$-factor a factor of length $k$. If $s = uv$ for $u, v \in \Sigma^*$, we say that $v$ is a *suffix* of $s$.

**Definition 1.** *Given $r$ input strings $s_1, \ldots, s_r$, a length $L$, and a distance $d$, we call a $(L, r, d)$-**repetition** a set $\{\delta_1, \ldots, \delta_r\}$ such that $0 \leq \delta_i \leq |s_i| - L$. For all $i \in [1, r]$ and for all $i, j \in [1, r]$ we have that*

$$d_H(s_i[\delta_i, \delta_{i+L-1}], s_j[\delta_j, \delta_{j+L-1}]) \leq d.$$

*where by $d_H$ we mean the Hamming distance between two sequences, that is, the minimum number of letter substitutions that transforms one into the other.*

Given $m$ input strings, the goal is to find the substrings of length $L$ that are repeated in at least $r \leq m$ strings with at most $d$ substitutions between each pair of the $r$ repetitions, with $L$ and $d$ given. In other words, we want to extract all the $(L, r, d)$-repetitions from a set of $r$ sequences among $m \geq r$ input sequences. The goal of the filter is therefore to eliminate from the input strings as many positions as possible that cannot contain $(L, r, d)$-repetitions. The value for parameter $d$ can be as big as 10% of $L$. The main idea of our filter is based on checking necessary conditions concerning the amount of *exact* $k$-factors that a $(L, r, d)$-repetition must share. A string $w$ of length $k$ is called a *shared $k$-factor* for $s_1, \ldots, s_r$ if $\forall i \in [1, r]$ $w$ occurs in $s_i$. Obviously, we are interested in shared $k$-factors that occur within substrings of length $L$ of the input strings. Let $p_r$ be the minimum number of non-overlapping shared $k$-factors that a $(L, r, d)$-repetition must have. It is intuitive to see that a $(L, 2, d)$-repetition contains at least $\frac{L}{k} - d$ shared $k$-factors, that is, $p_2 = \frac{L}{k} - d$. The value of $p_r$ for $r > 2$ is given in the following result whose proof is omitted due to space limitations. However, the intuition is that the positions where there are substitutions between each pair of sequences must appear clustered because if two sequences differ in a position, then a third sequence will, at this position, differ at least from one of the other two.

**Theorem 1.** *A $(L, r, d)$-repetition contains at least $p_r = \frac{L}{k} - d - (r - 2) \times \lfloor \frac{d}{2} \rfloor$ shared $k$-factors.*

The theorem above applies also to the case where one is interested in finding $(L, r, d)$-repetitions in a single string.

## 3   The Algorithm

NIMBUS takes as input the parameters $L$, $r$ and $d$, and $m$ (with $m \geq r$) input sequences. Given such parameters, it decides automatically the best $k$ to apply in order to filter for finding the $(L, r, d)$-repetitions either inside one sequence or inside a subset of $r$ sequences. In the following, we present the algorithm for finding $(L, r, d)$-repetitions in $r$ sequences. The algorithm can be adapted in a straightforward way to the case of finding $(L, r, d)$-repetitions occurring in a single sequence.

The goal of NIMBUS is to quickly and efficiency filter the sequences in order to remove regions which cannot contain a $(L, r, d)$-repetition applying the necessary conditions described in Section 2 and keeping only the regions which satisfy these conditions. We compute the minimum number $p_r$ of repeated $k$-factors each motif has to contain to possibly be part of a $(L, r, d)$-repetition. A set of $p_r$ $k$-factors contained in a region of length $L$ is called a $p_r$-set$_{\leq L}$. NIMBUS searches for the $p_r$-sets$_{\leq L}$ repeated in $r$ of the $m$ sequences. All the positions where a substring of length $L$ contains a $p_r$-set$_{\leq L}$ repeated at least once in $r$ sequences are kept by the filter, the others are rejected. To improve the search for the $p_r$-set$_{\leq L}$, we use what we call **bi-factors**, as defined below.

**Definition 2.** *A $(k, g)$-bi-factor is a concatenation of a factor of length $k$, a gap of length $g$ and another factor of length $k$. The factor $s[i, i + k - 1]s[i + k +$*

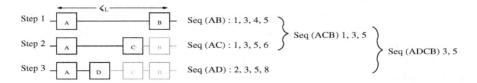

**Fig. 1.** Example of the construction of a 4-set$_{\leq L}$. At the first step, we find a bi-factor occurring at least once in at least $r = 2$ sequences among $m = 8$ sequences. During the second step, we add a bi-factor starting with the same $k$-factor (here called $A$), *included* inside the first one, and we merge the positions. We repeat this once and obtain a 4-set$_{\leq L}$ occurring in sequences 3 and 5. Actually not only the sequence numbers are checked during the merging but also the positions in the sequences, not represented in this figure for clarity.

$g, i + 2 \times k + g - 1]$ *is a bi-factor occurring at position* $i$ *in* $s$. *For simplicity's sake, we also use the term* bi-factor *omitting* $k$ *and* $g$.

For example, the $(2, 1)$-bi-factor occurring at position 1 in $AGGAGAG$ is $GGGA$. The bi-factors occurring in at least $r$ sequences are stored in a bi-factor array (presented in Section 4) that allows us to have access in constant time to the bi-factors starting with a specified $k$-factor. The main idea is to first find repeated bi-factors with the biggest allowed gap $g$ that may still contain $(p_r - 2)$ $k$-factors ($g \in [(p_r - 2)k, L - 2k]$). We call these *border bi-factors*. A border bi-factor is a 2-set$_{\leq L}$ that we then try to extend to a $p_r$-set$_{\leq L}$. To extend a $i$-set$_{\leq L}$ to a $(i + 1)$-set$_{\leq L}$, we find a repeated bi-factor (called an *extending bi-factor*) starting with the same $k$-factor as the border bi-factor of the $i$-set$_{\leq L}$ and having a gap length shorter than all the other gaps of the bi-factors already composing the $i$-set$_{\leq L}$. The occurring positions of the $(i + 1)$-set$_{\leq L}$ are the union of the extending bi-factor positions and of the positions of the $i$-set$_{\leq L}$. An example of this a construction is presented in Figure 1.

In order to extract all the possible $p_r$-set$_{\leq L}$, we iterate the idea described above on all the possible border bi-factors: all bi-factors with gap length in $[(p_r - 2)k, L - 2k]$ are considered as a possible border of a $p_r$-set$_{\leq L}$. Furthermore, while extending a $i$-set$_{\leq L}$ to a $(i + 1)$-set$_{\leq L}$, all the possible extending bi-factors have to be tested. The complete algorithm is presented in Figure 2.

Depending on the parameters, the $k$ value may be too small ($\leq 4$) leading to a long and inefficient filter. In this case, we start by running NIMBUS with $r = 2$, often allowing to increase the $k$ value, improving the sensitivity and the execution time. At the end, the remaining sequences are filtered using the initial parameters asked by the user. This actually results in an efficient strategy that we refer to later as the **double pass** strategy.

### 3.1 Complexity Analysis

Let us assume NIMBUS has to filter $m$ sequences each of length $\ell$. The total input size is then $n = \ell \times m$.

---

NIMBUS_**Initialise**()
1.    **for** $g$ in $[(p_r - 2)k, L - 2k]$
2.        **for** all $(k, g)$-bi-factors $bf$
3.            NIMBUS_Recursive($g - k$, positions($bf$), 2, firstKFactor($bf$))
NIMBUS_**Recursive** (gmax, positions, nbKFactors, firstKFactor)
1.    **if** $nbSequences(positions) < r$ **then** return //not in enough sequences
2.    **if** nbKFactors = p **then** save positions **and** return
3.    **for** $g$ in $[(p_r - (nbKFactors + 1)) \times k, gmax]$ // possible gaps length
4.        **for** $(k, g)$-bi-factors $bf$ starting with firstKFactor
5.            positions = merge(position, positions($bf$))
6.            NIMBUS_Recursive ($g - k$, positions, nbKFactors+1, firstKFactor)

---

**Fig. 2.** Extract the positions of all the $p_r$-sets$_{\leq L}$

For each possible gap length of the bi-factors considered by the algorithm, a bi-factor array is stored in memory (taking $O(n)$ as showed in section (4)). The bi-factor gap lengths are in $[0, L - 2k]$. The total memory used by NIMBUS is therefore in $O(n \times L)$. Let us assume that the time needed by the recursive extraction part of the NIMBUS algorithm depends only on a number of factors denoted by $nbKFactors$. We call this time $T(nbKFactors)$. With this notation NIMBUS takes $O(L \times \ell \times T(2))$. Furthermore $\forall\ nbKFactors < p$ :

$$T(nbKFactors) = \underbrace{L}_{\text{gap length}} \times \underbrace{\min(|\Sigma|^k, \ell)}_{\text{extending bi-factors}} \times (\underbrace{n}_{\text{merge}} + T(nbKFactors + 1))$$

$$\underbrace{\phantom{L \times \min(|\Sigma|^k, \ell)}}_{\text{replaced by } Z \text{ in the following}}$$

$$T(2) = Z \times (n + T(3)) = Z \times n + Z \times T(3)$$
$$= n \times Z + Z \times (Z \times n + Z \times T(4)) = n \times (Z + Z^2) + Z^2 \times T(4)$$

$$\vdots$$

$$= n \times \sum_{i=1}^{p-2} Z^i + Z^{p-2}T(p) = n \times \frac{Z^{p-1} - Z}{Z - 1} + Z^{p-2} \qquad (T(p) = O(1))$$
$$T(2) = O(n \times Z^{p-1})$$

The total time is therefore in $O(L \times \ell \times n \times Z^{p-1})$ with $Z = L \times \min(|\Sigma|^k, \ell)$. However, as we shall see later (Fig. 5), we have that this is just a rough upper bound of the worst-case. For instance, we do not take into consideration the fact that $T(i)$ decreases when $i$ increases because of the possible decrease in the gap lengths. Furthermore, a *balance* exists between lines 4 and 5 of the recursion algorithm. For instance, if the sequences are composed only by the letter $A$, lines 4 and 5 will do only one merge but for $n$ positions (in time $O(n)$). On the other hand, if the sequences are composed by $n$ different letters, lines 4 and 5 will do $n$ merges each in constant time, thus these two lines will be executed in time $O(n)$ as well. There can thus be a huge difference between the theoretical complexity and practical performance. The execution time strongly depends on the sequences composition. For sequences with few repetitions, the filter algorithm is very efficient. See Section 5 for more details.

Finally, we have that creating the bi-factor arrays takes $O(L \times n)$ time which is negligible w.r.t. to the extraction time.

## 4   The Bi-factor Array

Since we make heavy use of the inference of repeated bi-factors, we have designed a new data structure, called a **bi-factor array** (BFA), that directly indexes the bi-factors of a set of strings. The bi-factor array is a suffix array adapted for bi-factors (with $k$ and $g$ fixed) that stores them in lexicographic order (without considering the characters composing their gaps). This data structure allows to access the bi-factors starting with a specified $k$-factor in constant time. Notice that the same data structure can be used to index bi-factors where the two factors have different sizes (say, $(k_1, g, k_2)$-factors); we restrict ourselves here to the particular case of $k_1 = k_2$ because this is what we need for NIMBUS. For the sake of simplicity, we present the algorithm of construction of the bi-factor array for one sequence. The generalisation to multiple sequences is straightforward. We start by recalling the properties of a suffix array.

**The Suffix Array Data Structure.** Given a string $s$ of length $n$, let $s[i \ldots]$ denote the suffix starting at position $i$. Thus $s[i \ldots] = s[i, n-1]$. The **suffix array** of $s$ is the permutation $\pi$ of $\{0, 1, \ldots, n-1\}$ corresponding to the lexicographic order of the suffixes $s[i \ldots]$. If $\underset{l}{\leq}$ denotes the lexicographic order between two strings, then $s[\pi(0) \ldots] \underset{l}{\leq} s[\pi(1) \ldots] \underset{l}{\leq} \ldots \underset{l}{\leq} s[\pi(n-1) \ldots]$. In general, another information is stored in the suffix array: the length of the **longest common prefix** (lcp) between two consecutive suffixes ($s[\pi(i) \ldots]$ and $s[\pi(i+1) \ldots]$) in the array. The construction of the permutation $\pi$ of a text of length $n$ is done in linear time and space [12][9][11]. A linear time and space lcp row construction is presented in [10].

**BFA Construction.** We start by listing the ideas for computing the BFA using a suffix array and its lcp:

1. Give every $k$-factor a **label**. For instance, in a DNA sequence with $k = 2$, $AA$ has the label 0, $AC$ has label 1 and so on. A row is created containing, for every suffix, the label of its starting $k$-factor. In the remaining of this paper, we call a $(label_1, label_2)$-bi-factor a bi-factor of which the two $k$-factors are called $label_1$ and $label_2$.
2. For each suffix, the label of the $k$-factor occurring $k + g$ positions before the current position is known.
3. Construct the BFA as follows: let us focus for instance on the bi-factors starting with the $k$-factor called $label_1$. The predecessor label array is traversed from top to bottom, each time the predecessor label value is equal to

$label_1$, a new position is added for the part of the BFA where bi-factors start with the label $label_1$. Due to the suffix array properties, two consecutive bi-factors starting with the label $label_1$ are sorted w.r.t. the label of their second $k$-factor. The creation of the BFA is done such that for each $(label_1, label_2)$-bi-factor, a list of corresponding positions is stored.

We now explain in more detail how we perform the three steps above.

*Labelling the k-factors.* In order to give each distinct $k$-factor a different label, the $lcp$ array is read from top to bottom. The label of the $k$-factor corresponding to the $i^{th}$ suffix in the suffix array, called $label[i]$, is created as follows:

$$label[0] = 0$$

$$\forall\, i \in [1, n-1] : \; label[i] = \begin{cases} label[i-1] + 1 & \text{if } lcp[i] \le k \\ label[i-1] & \text{else} \end{cases}$$

*Giving each suffix a predecessor label.* For each suffix the label of the $k$-factor occurring $k+g$ positions before has to be known. Let $pred$ be the array containing the label of the predecessor for each position. It is filled as follows: $\forall\, i \in [0, |s| - 1]$, $pred[i] = label\left[\pi^{-1}[\pi[i] - k - g]\right]$ ($\pi^{-1}[p]$ is the index in the suffix array where the suffix $s[p\dots]$ occurs). Actually, the $pred$ array is not stored in memory. Instead, each cell is computed on line in constant time. An example of the $label$ and $pred$ arrays is given in Figure 3.

| $i$ | lcp | $\pi$ | associated suffix | $pred$ | label |
|---|---|---|---|---|---|
| 0 | 0 | 2 | $AACCAC$ | $\emptyset$ | 0 |
| 1 | 1 | 6 | $AC$ | 1 | 1 |
| 2 | 2 | 0 | $ACAACCAC$ | $\emptyset$ | 1 |
| 3 | 2 | 3 | $ACCAC$ | 1 | 1 |
| 4 | 0 | 7 | $C$ | 4 | 2 |
| 5 | 1 | 1 | $CAACCAC$ | $\emptyset$ | 3 |
| 6 | 2 | 5 | $CAC$ | 0 | 3 |
| 7 | 1 | 4 | $CCAC$ | 3 | 4 |

**Fig. 3.** Suffix array completed with the $label$ and the $pred$ arrays for $k = 2$ and $g = 1$ for the text $ACAACCAC$

*Creating the BFA.* The BFA contains in each cell a $(label_1, label_2)$-bi-factor. We store the $label_1$ and $label_2$ values and a list of positions of the occurrences of the $(label_1, label_2)$-bi-factor. This array is constructed on the observation that for all $i$, the complete suffix array contains the information that a $(pred[i], label[i])$-bi-factor occurs at position $\pi[i] - k - g$. Let us focus on one first $k$-factor, say $label_1$. Traversing the predecessor array from top to bottom each time $pred[i] = label_1$, we either create a new $(label_1 = pred[i], label[i])$-bi-factor at position $\pi[i] - k - g$, or add $\pi[i] - k - g$ as a new position in the list of positions of the previous bi-factor if the $label_2$ of the latter is equal to $label[i]$. Of course, this is done simultaneously for all the possible $label_1$. An example of a BFA is given in Figure 4.

| position(s) | $(label_1, label_2)$ | associated gapped-factor |
|:-----------:|:--------------------:|:------------------------:|
| 2 | (0,3) | $AACA$ |
| 0, 3 | (1, 1) | $ACAC$ |
| 1 | (3, 4) | $ACCC$ |

**Fig. 4.** BFA. Here $k = 2$ and $g = 1$. The text is $ACAACCAC$. One can notice that the $(2, 1)$-bi-factor $ACAC$ occurs at two different positions.

**Complexity for Creating a BFA.** The space complexity is in $O(n)$, as all the steps use linear arrays. Furthermore, one can notice that no more than four arrays are simultaneously needed, thus the effective memory used is $16 \times n$ bytes. The first two steps are in time $O(n)$ (simple traversals of the suffix array). The last step is an enumeration of the bi-factors found (no more than $n$). The last step is therefore in $O(n)$ as well. Hence the total time construction of the suffix array is in $O(n)$. With the following parameters: $L = 100$ and $k = 6$, NIMBUS has to construct BFAs for around 90 different $g$ values, which means 90 different BFAs. This operation takes for sequences of length 1Mb around 1.5 minutes on a 1.2 GHz Pentium 3 laptop.

## 5   Testing the Filter

We tested NIMBUS on a 1.2 GHz Pentium 3 laptop with 384 Mb of memory.

Figure 5 shows the time and memory usage in function of the input data length. We can observe that the memory usage is worse in the case of identical sequences. This is due to the fact that all the positions contain $L$ repeated bi-factors stored in memory. Furthermore, when the sequences are identical, all the positions are kept by the filter, representing the worst time complexity case. On the other hand, when all the sequences are distinct, the complexity is clearly linear.

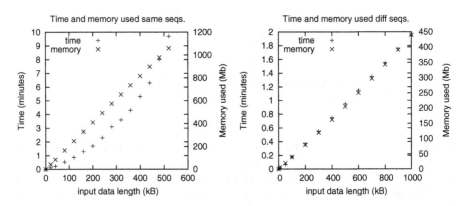

**Fig. 5.** Time and memory spent by NIMBUS $w.r.t.$ the input data length. The parameters are $L = 100$, $k = 6$, $d = 7$, $r = 3$ which implies $p_3 = 6$. The input file contains 10 DNA sequences of equal length. On the left part of the figure the sequences are the same, whereas on the right part they are randomised.

| Sequence filtered | | 2 Motifs | 5 Motifs | 100 Motifs | MC58 |
|---|---|---|---|---|---|
| Memory Used | | 675 | 675 | 681 | 943 |
| $r = 2$ | Time (Min.) | 4.8 | 4.8 | 5 | 53 |
| | Kept (Nb and Ratio) | 406: 0.04 % | 1078: 0.10 % | 22293: 2.2 % | 127782 : 12.7 % |
| | False Positive Ratio | 0.02 % | 0.08 % | 2.0 % | unknown |
| $r = 3$ | Time (Min.) | 4.8 + 0 | 4.8 + 0.1 | 5 + 0.5 | 53 + 0.9 |
| | Kept (Nb and Ratio) | 0: 0 % | 1078: 0.10 % | 21751: 2.2 % | 92069: 9.21 % |
| | False Positive Ratio | 0 % | 0.11 % | 2.0 % | unknown |
| $r = 4$ | Time (Min.) | 4.8 + 0 | 4.8 + 0.1 | 5 + 0.5 | 53 + 10 |
| | Kept (Nb and Ratio) | 0: 0 % | 1066: 0.11 % | 21915: 2.2 % | 106304: 10.63 % |
| | False Positive Ratio | 0.0 % | 0.09 % | 1.8 % | unknown |

**Fig. 6.** NIMBUS **behaviour** on four types of sequences while filtering in order to find $r = 2$, 3 and 4 repetitions

In Figure 6, we present the behaviour of the filter for four kinds of input DNA sequences. The first three sequences are randomised and contain respectively 2, 5 and 100 motifs of length 100 distant pairwise by 10 substitutions. For each of these three sequences we ran NIMBUS in order to filter searching for motifs of length $L = 100$ occurring at least $r = 2$, 3 and 4 times with less than $d = 10$ substitutions. The last DNA sequence is the genomic sequence of the *Neisseria meningitidis* strain MC58. *Neisseria* genomes are known for the abundance and diversity of their repetitive DNA in terms of size and structure [6]. The size of the repeated elements range from 10 bases to more than 2000 bases, and their number, depending on the type of the repeated element, may reach more than 200 copies. This fact explains why the *N. meningitidis MC58* genomic sequence has already been used as a test case for programs identifying repetitive elements like in [13]. We ran NIMBUS on this sequence in order to filter the search for motifs of length $L = 100$ occurring at least $r = 2$, 3 and 4 times with less than $d = 10$ substitutions.

For $r = 2$, we used $k = 6$ which gives a good result: less than 5 minutes execution time for all the randomised sequences. On can notice that for the MC58 sequence, the execution time is longer (53 to 63 minutes) due to its high rate of repetitions.

For $r = 3$ and 4, we apply the double pass strategy described earlier, and start the filtration with $r = 2$ and $k = 6$. The time results are therefore subdivided into two parts: the time needed for the first pass and the one needed for the second pass. The time needed for the second pass is negligible w.r.t. the time used for the first one. This is due to the fact that the first pass filters from 89 % to 99 % of the sequence, thus the second pass works on a sequence at least 10 times shorter than the original one. This also explains why no extra memory space is needed for the second pass. For $r = 3$, the second pass uses $k = 5$ while for $r = 4$, the second pass uses $k = 4$. With $k = 4$, the efficiency of the filter is lower than for superior values of $k$. That is why for MC58, more positions are kept while searching for motifs repeated 4 times, than for motifs repeated 3 times. Without using the double pass, for instance on MC58, with $r = 3$ the memory used is 1435 Mb (instead of 943 Mb) and the execution time is around 12 hours (instead of 54 minutes). The false positive ratio observed in practice (that is, the ratio, computed on random sequences with planted motifs, of non filtered data that are not part of a real motif) is very low (less than 1.2 %). In general, many of the false positives occur around a $(L, r, d)$-repetition motif and not anywhere in the sequences.

*Efficiency of the filter.* Although it depends on the parameters used and on the input sequences, the efficiency of the filter is globally stable. For instance, when asking for motifs of length $L \approx 100$ of which the occurrences are pairwise distant of $d \approx 10$, NIMBUS keeps also motifs of which the occurrences are pairwise distant up to $d + 7 \approx 17$. The smaller is $d$, the more efficient is NIMBUS. When $d = 1$ substitution, NIMBUS thus keeps motifs of which the occurrences are pairwise distant up to $d + 3 \approx 4$ only instead of $d + 7$ as for $d \approx 10$.

## 6   Using the Filter

In this section we show two preliminary but interesting applications of NIMBUS. The first concerns the inference of long biased repetitions, and the second multiple alignments.

### 6.1   Filtering for Finding Long Repetitions

When inferring long approximate motifs, the number of differences allowed among the occurrences of a motif is usually proportional to the length of the motif. For instance, for $L = 100$ and allowing for as many as $L/10$ substitutions, one would have $d = 10$ which is high. This makes the task of identifying such motifs very hard and, to the best of our knowledge, no exact method for finding such motifs with $r > 2$ exists. Yet such high difference rates are common in molecular biology. The NIMBUS filter can efficiently be used in such cases as it heavily reduces the search space. We now show some tests that prove this claim. For testing the ability of NIMBUS concerning the inference of long approximate repetitions, we ran an algorithm for extracting structured motifs called RISO [5] on a set of 6 sequences of total length 21 kB for finding motifs of length 40 occurring in every sequence with at most 3 substitutions pairwise. Using RISO, this test took 230 seconds. By previously filtering the data with NIMBUS, the same test took 0.14 seconds. The filtering time was 1.1 seconds. The use of NIMBUS thus enabled to reduce the overall time for extracting motifs from 230 seconds to 1.24 seconds.

### 6.2   Filtering for Finding Multiple Local Alignments

Multiple local alignment of $r$ sequences of length $n$ can be done with dynamic programming using $O(n^m)$ time and memory. In practice, this complexity limits the application to a small number of short sequences. A few heuristics, such as MULAN [20], exist to solve this problem. One alternative exact solution could be to run NIMBUS on the input data so as to exclude the non relevant information (*i.e.* parts that are too distant from one another) and then to run a multiple local alignment program. The execution time is hugely reduced. For instance, on a file containing 5 randomised sequences of cumulated size 1 Mb each containing an approximate repetition[1], we ran NIMBUS in approximatively 5 minutes. On the remaining sequences, we ran a tool for finding functional elements using multiple local alignment called GLAM [7]. This operation took about 15 seconds. Running GLAM without the filtering, we obtained the same results[2] in more than 10 hours. Thus by using NIMBUS, we reduced the execution time of GLAM from many hours to less than 6 minutes.

## 7  Conclusions and Future Work

We presented a novel lossless filtration technique for finding long multiple approximate repetitions common to several sequences or inside one single sequence. The filter localises the parts of the input data that may indeed present repetitions by applying a necessary condition based on the number of repeated $k$-factors the sought repetitions have to contain. This localisation is done using a new type of seeds called bi-factors. The data structure that indexes them, called a bi-factor array, has also been presented in this paper. It is constructed in linear time. This data structure may be useful for various other text algorithms that search for approximate instead of exact matches. The practical results obtained show a huge improvement in the execution time of some existing multiple sequence local alignment and pattern inference tools, by a factor of up to 100. Such results are in partial contradiction with the theoretical complexity presented in this paper. Future work thus includes obtaining a better analysis of this complexity.

Other important tasks remain, such as filtering for repetitions that present an even higher rate of substitutions, or that present insertions and deletions besides substitutions. One idea for addressing the first problem would be to use bi-factors (and the corresponding index) containing one or two mismatches inside the $k$-factors. In the second case, working with edit instead of Hamming distance implies only a small modification on the necessary condition and on the algorithm but could sensibly increase the execution time observed in practice.

## References

1. S. F. Altschul, W. Gish, W. Miller, E. W. Myers, and D. J. Lipman. A basic local alignment search tool. *J. Mol. Biol.*, 215:403–410, 1990.
2. S. F. Altschul, T. L. Madden, A. A. Schaffer, J. Zhang, Z. Zhang, W. Miller, and D. J. Lipman. Gapped BLAST and PSI–BLAST: a new generation of protein database search programs. *Nucleic Acids Res.*, 25:3389–3402, 1997.

---

[1] Repetitions of length 100 containing 10 substitutions pairwise.

[2] Since GLAM handles edit distance and NIMBUS does not, in the tests we have used randomly generated data where we planted repetitions allowing for substitutions only, in order to ensure that the output would be the same and hence the time cost comparison meaningful.

3. S. Burkhardt, A. Crauser, P. Ferragina, H.-P. Lenhof, E. Rivals, and M. Vingron. q-gram based database searching using a suffix array (quasar). In *proceedings of 3rd RECOMB*, pages 77–83, 1999.

4. S. Burkhardt and J. Karkkainen. Better filtering with gapped q-grams. In *Proceedings of the 12th Annual Symposium on Combinatorial Pattern Matching*, number 2089, 2001.

5. A. M. Carvalho, A. T. Freitas, A. L. Oliveira, and M-F. Sagot. A highly scalable algorithm for the extraction of cis-regulatory regions. *Advances in Bioinformatics and Computational Biology*, 1:273–282, 2005.

6. H. Tettelin *et al.* Complete genome sequence of Neisseria meningitidis serogroup B strain MC58. *Science*, 287(5459):1809–1815, Mar 2000.

7. M. C. Frith, U. Hansen, J. L. Spouge, and Z. Weng. Finding functional sequence elements by multiple local alignment. *Nucleic Acids Res.*, 32, 2004.

8. C. S. Iliopoulos, J. McHugh, P. Peterlongo, N. Pisanti, W. Rytter, and M. Sagot. A first approach to finding common motifs with gaps. *International Journal of Foundations of Computer Science*, 2004.

9. J. Karkkainen, P. Sanders, and S. Burkhardt. Linear work suffix array construction. *J. Assoc. Comput. Mach.*, to appear.

10. T. Kasai, G. Lee, H. Arimura, S. Arikawa, and K. Park. Linear-time longest-common-prefix computation in suffix arrays and its applications. In *Proceedings of the 12th Annual Symposium on Combinatorial Pattern Matching*, pages 181–192. Springer-Verlag, 2001.

11. D.K. Kim, J.S. Sim, H. Park, and K. Park. Linear-time construction of suffix arrays. In *Proceedings of the 14th Annual Symposium on Combinatorial Pattern Matching*, june 2003.

12. P. Ko and S. Aluru. Space efficient linear time construction of suffix arrays. *Journal of Discrete Algorithms*, to appear.

13. R. Kolpakov, G. Bana, and G. Kucherov. mreps: Efficient and flexible detection of tandem repeats in DNA. *Nucleic Acids Res*, 31(13):3672–3678, Jul 2003.

14. G. Krucherov, L.No, and M.Roytberg. Multi-seed lossless filtration. In *Proceedings of the 15th Annual Symposium on Combinatorial Pattern Matching*, 2004.

15. M. Li, B. Ma, D. Kisman, and J. Tromp. Patternhunter ii: Highly sensitive and fast homology search. *J. of Comput. Biol.*, 2004.

16. D. J. Lipman and W. R. Pearson. Rapid and sensitive protein similarity searches. *Sci.*, 227:1435–1441, 1985.

17. B. Ma, J. Tromp, and M. Li. Patternhunter: faster and more sensitive homology search. *Bioinformatics*, 18(3):440–445, 2002.

18. L. Marsan and M.-F. Sagot. Algorithms for extracting structured motifs using a suffix tree with application to promoter and regulatory site consensus identification. *J. of Comput. Biol.*, (7):345–360, 2000.

19. G. Navarro, E. Sutinen, J. Tanninen, and J. Tarhio. Indexing text with approximate q-grams. In *Proceedings of the 11th Annual Symposium on Combinatorial Pattern Matching*, number 1848 in Lecture Notes in Computer Science, pages 350–363, 2000.

20. I. Ovcharenko, G.G. Loots, B.M. Giardine, M. Hou, J. Ma, R.C. Hardison, L. Stubbs, , and W. Miller. Mulan: Multiple-sequence local alignment and visualization for studying function and evolution. *Genome Research*, 15:184–194, 2005.

21. K. R. Rasmussen, J. Stoye, and E. W. Myers. Efficient q-gram filters for finding all ε-matches over a given length. In *Proceedings of the 16th Annual Symposium on Combinatorial Pattern Matching*, 2005.

# Linear Time Algorithm for the Generalised Longest Common Repeat Problem

Inbok Lee* and Yoan José Pinzón Ardila

King's College London, Dept. of Computer Science,
London WC2R 2LS, United Kingdom
inbok@dcs.kcl.ac.uk, Yoan.Pinzon@kcl.ac.uk

**Abstract.** Given a set of strings $\mathcal{U} = \{T_1, T_2, \ldots, T_\ell\}$, the longest common repeat problem is to find the longest common substring that appears at least twice in each string of $\mathcal{U}$, considering direct, inverted, mirror as well as everted repeats. In this paper we define the generalised longest common repeat problem, where we can set the number of times that a repeat should appear in each string. We present a linear time algorithm for this problem using the suffix array. We also show an application of our algorithm for finding a longest common substring which appears only in a subset $\mathcal{U}'$ of $\mathcal{U}$ but not in $\mathcal{U} - \mathcal{U}'$.

**Keywords:** suffix arrays, pattern discovery, inverted repeats, DNA repeats, DNA Satellites.

## 1 Introduction

Genomic sequences are far from random. One of the more intriguing features of DNA is the extent to which it consists of repeated substrings. Repeated DNA sequences account for large portions of eukaryotic genomes that have been studied to date. The origin of these repeats, as well as their biological function, is not fully understood. Nevertheless, they are believed to play a crucial role in genome organization and evolution [3].

There are basically four repeat sequences types, differing by their orientation and localization. They are: direct, inverted, mirror and everted repeats. The easiest to understand is the direct repeat, in which a substring is duplicated (see Fig. 1(a)). Some multiple direct repeats have been associated with human genetic diseases. For example, the triplet[1] CGG is tandemly repeated 6 to 54 times in a normal FMR-1[2] gene. In patients with the Fragile X syndrome, the pattern occurs more than 200 times. People with this mutation fail to produce a protein involved in making cellular connections in the brain producing mental impediment or retardation. An estimated 1 in 2000 boys (girls are also affected, but the

---

* This work was supported by the Korea Research Foundation Grant funded by Korean Government(MOEHRD, Basic Research Promotion Fund) M01-2004-000-20344-0.
[1] 3 DNA bp, a.k.a *microsatellites*.
[2] Fragile Mental Retardation 1.

M. Consens and G. Navarro (Eds.): SPIRE 2005, LNCS 3772, pp. 191–201, 2005.

incidence rate is lower) are mentally weakened because of Fragile X. Kennedy's disease, Parkinson's disease, Huntington's disease and Myotonic Dystrophy are examples of other genetic diseases that have been associated with direct repeats [4,19,21].

Inverted repeats are another important element in the Genomes. An inverted repeat, also called palindrome, is except that the second half of the repeat is also found nearby on the opposite strand (complementary strand) of the DNA helix, as shown in Fig. 1(b). This implies that the substring and its reverse complement are both to be found on the same strand, which can thus fold back to base-pair with itself[3] and form a stem-and-loop structure, as shown on the righthand side of Fig. 1(b). Because of their nature, inverted repeats engages in intra- and intermolecular base pairing. The ability to form hairpin, recursive (see Fig. 1(e)), cruciform (see Fig. 1(f)) and pseudo-knot (see Fig. 1(g)) secondary structures is associated with the initiation of DNA replication and frameshift mutations.

An application that makes repeats an interesting research topic is related to the *multiple alignment problem*, producing multiple alignments becomes very complicated when the sequences to be aligned contain multiple repeats, because matches may be present in numerous places. As a precursor to multiple alignment, it is helpful to recognize all multiple repeats within the set of strings that ought to be aligned [15]. Other applications of repetitive DNA analysis range from genetic markers, DNA fingerprinting, mapping genes, comparative genomics and evolution studies [3,8,7].

In this paper we consider the problem which combines two: finding common repetitive substrings in a set of strings. In addition, we want to specify the number of occurrences in each string. We focus on finding the longest substring since the substrings of the longest substring also appear in each string. We also consider reversed and reverse-complemented strings in finding repeats.

Before beginning, we will establish some notation. We will uniformly adopt the four letter alphabet $\Sigma_{\text{DNA}} = \{\mathsf{a}, \mathsf{c}, \mathsf{g}, \mathsf{t}\}$, each letter standing for the first letter of the chemical name of the *nucleotide* in the polymer's chain, and let a bar notation represent an operation corresponding to simple base complementarity, *i.e.* indicating bases that are able to physically base-pair between strands of double-helical DNA:

$$\overline{\mathsf{g}} = \mathsf{c}, \overline{\mathsf{c}} = \mathsf{g}, \overline{\mathsf{a}} = \mathsf{t}, \text{ and } \overline{\mathsf{t}} = \mathsf{a}$$

Let $A$ be a string over $\Sigma_{\text{DNA}}$. $A[i]$ denotes the $i$-th character of $A$ and $A[i..j]$ is the substring $A[i]A[i+1]\cdots A[j]$ of $A$. We denote the *reverse* of $A$ by $\overleftarrow{A}$ and the *reverse complemented* of $A$ by $\overleftarrow{\overline{A}}$. Sequence $\overleftarrow{A}$ is obtained by reversing $A$ and mapping each character to its complement. If $A = \mathsf{gtaac}$, for example, $\overleftarrow{A} = \mathsf{caatg}$ and $\overleftarrow{\overline{A}} = \overline{\mathsf{caatg}} = \mathsf{gttac}$.

A *repeat* of $A$ is a substring of $A$ which appears at least twice in $A$. There are four kinds of repeats.

---

[3] Base-pairings within the same strand are called *secondary structures*.

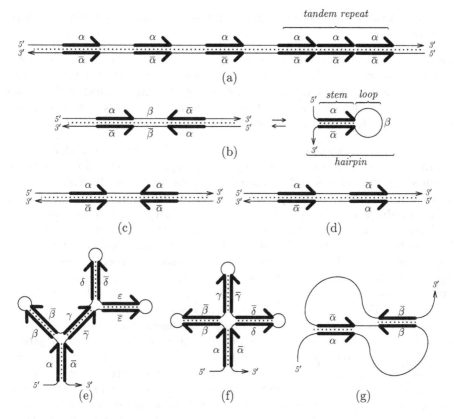

**Fig. 1.** Different types of repetitions present in genomic sequences. (a) Direct repeat (left: non-tandem direct repeat, right: tandem direct repeats — when two repeats immediately follow each other in the string —). (b) Inverted repeat (left: single-stranded hairpin structure containing an inverted repeat). (c) Mirror repeat. (d) Everted repeat. (e) Recursive secondary structure containing several inverted repeats. (f) Cruciform structure formed at inverted repeats by four hairpins and the four-arm junction. (g) Nested double pseudo-knot structure including two inverted repeats.

— DIRECT REPEAT: A string $p$ is called a *direct repeat* of $A$ if $p = A[i..i+|p|-1]$ and $p = A[i'..i'+|p|-1]$, for some $i \neq i'$.

— INVERTED REPEAT: A string $p$ is called an *inverted repeat* of $A$ if $p = A[i..i+|p|-1]$ and $\bar{p} = A[i'..i'+|p|-1]$, for some $i \neq i'$.

— MIRROR REPEAT: A string $p$ is called a *mirror repeat* (or *reverse repeat*) of $A$ if $p = A[i..i+|p|-1]$ and $\overleftarrow{p} = A[i'..i'+|p|-1]$, for some $i \neq i'$.

— EVERTED REPEAT: This type of repeat is equivalent to the direct repeat described above.

We will only consider direct, mirror and inverted repeats as standard prototypes. It is easy to see, that everted repeats are covered by direct repeats when considering multiple strings.

The generalised longest common repeat problem can subsequently be defined as follows.

*Problem 1.* Given a set of strings $\mathcal{U} = \{T_1, T_2, \ldots, T_\ell\}$, a set of positive integers $\mathcal{D} = \{d_1, d_2, \ldots, d_\ell\}$, and a positive integer $k$, the GENERALISED LONGEST COMMON REPEAT PROBLEM is to find the longest string $w$ which satisfies two conditions: (a) There is a subset $\mathcal{U}'$ of $\mathcal{U}$ such that $w$ appears at least $d_i$ times in every string $T_i$ in $\mathcal{U}'$, and (b) $|\mathcal{U}'| = k$.

This definition is an extension of that in [16]. The difference is that we can restrict the number of times that a repeat can appear in each string. Hence the frequency of the repeats will be bound according to our needs. Note that the above definition allows $w$ to appear just once in some strings.

Karp, Miller, and Rosenberg first proposed an $O(|T| \log |T|)$ time algorithm for finding the longest normal repeat in a text $T$. Also suffix trees [17,20,6] can be used to find it in $O(|T|)$ time. Landau and Schmidt gave an $O(k|T| \log k \log |T|)$ time algorithm for finding approximate squares where at most $k$ edit distance is allowed [14]. Schmidt also gave an $O(|T|^2 \log |T|)$ time algorithm for finding approximate tandem or non-tandem repeats [18]. Abouelhoda et al. [1] for finding various types of repeats using the enhanced suffix array.

In [16], a linear time algorithm for finding the longest common repeat problem using the generalised suffix tree was derived. Although it is the first approach for this problem, so far as we know, it has some drawbacks. First, it is not easy to implement. Second, it is based on the suffix tree data structure which is not memory-efficient. Besides, it requires a generalised suffix tree for all the strings and a suffix tree for each single string. Our new algorithm is easy to implement and requires only one suffix array[4]. And it can handle a generalised problem.

The outline of the paper is as follows: In Section 2 we explain some basics. In Section 3 we describe our algorithms for the generalised longest repeat problem. In Section 4 we explain an application of our algorithm, *i.e.* finding the longest repeat which appears only in a subset $\mathcal{U}'$ of $\mathcal{U}$ but not in $\mathcal{U} - \mathcal{U}'$. Conclusions are drawn in Section 5.

## 2    Preliminaries

The *suffix array* of a text $T$ is a sorted array $s[1..|T|]$ of all the suffixes of $T$. That is, $s[k] = i$ iff $T[i..|T|]$ is the $k$-th suffix of $T$. We also define the auxiliary *LCP array* as an array of the length of the longest common prefix between each substring in the suffix array and its predecessor, and define $lcp(a, b) = \min_{a \leq i \leq b} lcp[i]$ with the following properties.

**Fact 1.** $lcp(a, b) \leq lcp(a', b')$ if $a \leq a'$ and $b \geq b'$.

**Fact 2.** *The length of the longest common prefix of* $T[s[a]..|T|]$, $T[s[a + 1]..|T|], \ldots, T[s[b]..|T|]$ *is* $lcp(a + 1, b)$.

---

[4] A suffix array is more compact and amenable to store in secondary memory.

We can build the suffix array over a set of strings $\mathcal{U} = \{T_1, T_2, \ldots, T_\ell\}$ as in [5]. First we create a new string $T'' = T_1\%T_2\%\cdots T_\ell$ where $\%$ is a special symbol which is smaller than any other character in $\Sigma_{\mathrm{DNA}}$. Then we compute the $lcp$ array using the technique presented in [10] in $O(|T''|)$ time. By a simple trick we can use only one special character $\%$ instead of $\ell$ symbols, that is, $\%$ does not match itself. We also compute the $ids$ array such that $ids[k] = i$ if $T''[k]T''[k+1]\cdots\%$ is originally a suffix of $T_i$ (of course, $\%$ is the symbol that appears first after $T''[k]$ in $T''$). The $ids$ array can be calculated in $O(|T''|)$ time and space. Fig. 2 is an example of the suffix array over a set of strings. Note that in this example we discarded all the suffixes which begin with $\%$.

$T_1 = \text{acac}, T_2 = \text{aac}, T_3 = \text{caac}$

|   | 1 | 2 | 3 | 4 | 5 | 6 | 7 | 8 | 9 | 10 | 11 | 12 | 13 | 14 |
|---|---|---|---|---|---|---|---|---|---|----|----|----|----|----|
| $T'' =$ | a | c | a | c | % | a | a | c | % | c | a | a | c | % |

| $i$ | 1 | 2 | 3 | 4 | 5 | 6 | 7 | 8 | 9 | 10 | 11 |
|---|---|---|---|---|---|---|---|---|---|---|---|
| $ids$ | 3 | 2 | 3 | 1 | 2 | 1 | 3 | 1 | 2 | 3 | 1 |
| $s$ | 11 | 6 | 12 | 3 | 7 | 1 | 13 | 4 | 8 | 10 | 2 |
| $lcp$ | - | 3 | 1 | 2 | 2 | 2 | 0 | 1 | 1 | 1 | 2 |
| suffix | a a c % | a a c % | a c % | a c % | a c % | a c a c % | c % | c % | c % | c a a c % | c a c % |

**Fig. 2.** A suffix array over strings $T_1 = \text{acac}$, $T_2 = \text{aac}$, and $T_3 = \text{caac}$

Given a suffix array over a set of strings, a set of positive integers $\mathcal{D} = \{d_1, d_2, \ldots, d_\ell\}$, and a positive integer $k$, a *candidate range* is a range $(a, b)$ which contains $k$ distinct values in the set $\{ids[a], ids[a+1], \ldots, ids[b]\}$ and each value $i$ appears at least $d_i$ times in the set. A *critical range* is a candidate range that does not properly contain other candidate ranges. Based on these two definitions, it is straightforward to draw the following lemma.

**Lemma 1.** *The answer for the generalised longest common repeat problem is* $T''[s[a']]..T''[s[a'+lcp(a'+1,b')]]$ *such that* $(a',b')$ *is a critical range and* $lcp(a'+1,b')$ *is the greatest among all critical ranges.*

*Proof.* Fact 2 and the definition of the candidate range implies that if $(a, b)$ is a candidate range, then $T''[s[a]]..T''[s[b] + lcp(a+1,b)]$ is a common string of at least $k$ strings in $\mathcal{U}$. Fact 1 says that we should narrow the range which contains the $k$ different values in order to find the longest substring, which will correspond to the critical range. Among the critical ranges $(a', b')$, we select that one whose $lcp(a'+1,b')$ is the greatest as the answer to the problem.

## 3   Algorithm

The algorithm's framework for the generalised longest common repeat problem is as follows.

— **Step 1:** Create a new string $T_i'$ for each $1 \leq i \leq \ell$ to consider inverted, mirror and everted repeats, and another string $T''$ which is the concatenation of all $T_i'$'s.
— **Step 2:** Construct the suffix array of $T''$.
— **Step 3:** Find the critical ranges.
— **Step 4:** Find the longest common substring for each critical range.

Steps 1,2 and 4 are easy to compute. Step 3 is the key step and needs a bit more or care. Following, we show each step in more detail and show how to use Kim et al.'s algorithm [12] to find the answer for the generalised longest common repeat problem.

**Step 1:** We first modify each string in $\mathcal{U}$ to consider inverted, mirror and everted repeats. For each $i = 1, 2, \ldots, \ell$, we create a new string $T_i' = T_i \% \overleftarrow{T}_i \% \overleftarrow{T}_i$. And we create a string $T'' = T_1' \% T_2' \% \ldots \% T_\ell' \%$.

**Step 2:** We build the suffix array of $T''$. The construction of the suffix array takes $O(|T''|)$ time and space [9,11,13]. We also compute $lcp$ and $ids$ arrays in $O(|T''|)$ time and space. Note that the suffixes of $T_i$, $\overleftarrow{T}_i$ and $\overleftarrow{T}_i$ have the same value $i$ in the $ids$ array.

**Step 3:** We will present an uncomplicated explanation on how to find the critical ranges. For a more exhaustively detailed account, interested readers are directed to [12].

We maintain a range $(a, b)$ during this step. At first $a = 1$ and $b = 0$. We maintain $\ell$ counters $c_1, c_2, \ldots, c_\ell$ (initially $c_1 = c_2 = \cdots = c_\ell = 0$) and a counter $h$ which contains the number of $c_i$'s ($1 \leq i \leq \ell$) that are $\geq d_i$. Initially $h = 0$.

We now define the following two sub-steps: *expanding* and *shrinking*. In the expanding sub-step, we find a candidate range. We expand the range from $(a, b)$ to $(a, b + 1)$. We check $ids[b]$ and set $c_{ids[b]} = c_{ids[b]} + 1$. If $c_{ids[b]} = d_i$, then $h = h + 1$. We stop the expanding sub-step if $h = k$. Now $(a, b)$ is a candidate range and we move to the shrinking sub-step. To speed up, we use the following idea, if $lcp[b] = 0$, then we reset all counters $c_1, c_2, \cdots, c_\ell$ and $h$ to 0. We then start expanding sub-step again with $a = b$ and $b = b - 1$.

In the shrinking sub-step, we find a critical range from the candidate range $(a, b)$ found in previous sub-step. We start by shrinking the candidate range downwards. First, we set $c_a = c_a - 1$ and $a = a + 1$. If $c_a < d_a$, then $h = h - 1$. If $h < k$, then $(a - 1, b)$ is a critical range. We report it and go back to the expanding sub-step with $a$ and $b$. All these steps run in $O(|T''|)$ time and space.

**Step 4:** For each critical range $(a', b')$ found in Step 3, we use Bender and Farach-Colton's technique [2] to compute $lcp(a' + 1, b')$. After $O(|T''|)$ time and

$T_1$=acac, $T_2$=aac, $T_3$=caac     $\mathcal{D}$={2,1,1}

$T''$ =
| | 1 | 2 | 3 | 4 | 5 | 6 | 7 | 8 | 9 | 10 | 11 | 12 | 13 | 14 |
|---|---|---|---|---|---|---|---|---|---|---|---|---|---|---|
| | a | c | a | c | % | a | a | c | % | c | a | a | c | % |

| $i$ | 1 | 2 | 3 | 4 | 5 | 6 | 7 | 8 | 9 | 10 | 11 |
|---|---|---|---|---|---|---|---|---|---|---|---|
| $ids$ | 3 | 2 | 3 | 1 | 2 | 1 | 3 | 1 | 2 | 3 | 1 |
| $s$ | 11 | 6 | 12 | 3 | 7 | 1 | 13 | 4 | 8 | 10 | 2 |
| $lcp$ | - | 3 | 1 | 2 | 2 | 2 | 0 | 1 | 1 | 1 | 2 |
| suffix | aac% | aac% | ac% | ac% | ac% | acac% | c% | c% | c% | caac% | cac% |

Intervals: $I_1$; $I_2$ (ac); $I_3$; $I_4$; $I_5$ (c).

$T_1'$=acac%caca%gtgt, $T_2'$=aac%caa%gtt, $T_3'$=caac%caac%gttg

$T''$ = a c a c % c a c a % g t g t % a a c % c a a % g t t % c a a c % c a a c % g t t g %

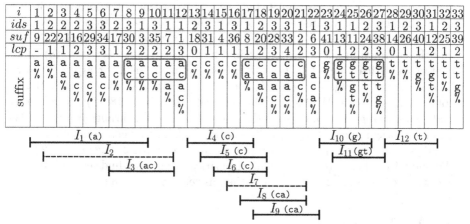

**Fig. 3.** An example of the generalised longest common repeat problem. The table at the top only considers direct repeats. The table at the bottom considers direct, inverted, mirror and everted repeats. The intervals depicts dashed lines for candidate ranges and bold lines for critical ranges.

space preprocessing, each query takes $O(1)$ time. Since it is easy to show that the number of critical ranges is $O(|T''|)$, the time complexity of Step 4 is $O(|T''|)$.

**Theorem 1.** *The generalised longest common repeat problem can be solved in* $O(\sum_{i=1}^{\ell} |T_i|)$ *time and space.*

*Proof.* We showed that all the steps run in $O(\sum_{i=1}^{\ell} |T_i'|)$ time and space. And $|T_i'| = O(|T_i|)$ for $1 \leq i \leq \ell$.

Fig. 3(top) is an example. We have three strings $T_1 = $ acac, $T_2 = $ aac, and $T_3 = $ caac, $k = 2$, and $\mathcal{D} = \{2, 1, 1\}$. For simplicity, we do not consider the inverted repeats here. We first build the suffix array, then we find the critical ranges. At first we find a candidate range $I_1 = (1, 6)$. Then we shrink this range to find a critical range $I_2 = (3, 6)$. Here $lcp(4, 6) = 2$, therefore this range shares a common prefix ac. Then we try to find the next candidate range from position 5, but we meet position 7 where $lcp[7] = 0$. So we discard $I_3 = (5, 7)$ and move to position 7 to find a new candidate range. From position 7, we find candidate range $I_4 = (7, 11)$. Then we shrink to find critical range $I_5 = (8, 11)$. $lcp(8, 11) = 2$, meaning that they share a common prefix c. Since ac is longer than c, the answer is ac.

Fig. 3(bottom) consider the previous example but taking into consideration all type of repeats. Once the suffix array is built, we proceed to find the critical ranges. At first we find a candidate range $I_1 = (1, 9)$. It is also a critical range and $lcp(2, 9) = 1$. Hence the suffixes in $I_1$ shares a common prefix a. Then we try to find the next candidate range from position 2. We find a candidate range $I_2 = (2, 11)$. Next, we shrink it to find a critical range $I_3 = (7, 11)$. $lcp(8, 11) = 2$ and it shares a common prefix ac, which is longer than a we found before. We begin again from position 9, but we meet position 13 where $lcp[13] = 0$ before finding a candidate range. So we begin again from position 13. We find a candidate range $I_4 = (13, 17)$ which is also a critical range. $lcp(14, 17) = 1$ and the common prefix is c. Next we find a candidate range $I_5 = (14, 18)$ which contains a critical range $I_6 = (15, 18)$, whose common prefix is c again. Then we find a candidate range $I_7 = (16, 21)$ containing a critical range $I_8 = (17, 21)$. Its common prefix is ca. We find another critical range $I_9 = (18, 22)$ and again we find ca. Since $lcp[23] = 0$, we start the search for the critical range from here. We find two critical ranges $I_{10} = (23, 26)$ and $I_{11} = (24, 27)$. Their common prefix is g and gt, respectively. From position 28, we have one critical range $I_{12} = (28, 31)$. We find t from here. After finding all the critical ranges, we find seven substrings a, c, g, t, ac, ca, and gt. In this case, the answer is ac, ca, and gt.

## 4    An Application

We extend our algorithm for the generalised longest common repeat problem, finding a longest *feature* from a set. A feature of a subset $\mathcal{U}' \subset \mathcal{U}$ is a common string which appears only in $\mathcal{U}'$ but not in $\mathcal{U} - \mathcal{U}'$. More formally, this problem can be defined as follows.

*Problem 2.* Given a set of strings $\mathcal{U} = \{T_1, T_2, \ldots, T_\ell\}$, and a set of non-negative integers $\mathcal{D} = \{d_1, d_2, \ldots, d_\ell\}$, THE GENERALISED LONGEST FEATURE PROBLEM is to find the longest string $w$ which satisfies two conditions: (a) There is a subset $\mathcal{U}'$ of $\mathcal{U}$ such that $w$ appears at least $d_i$ times in every string $T_i$ in $\mathcal{U}'$ if $d_i > 0$, (b) $|\mathcal{U}'| = k$, and (c) $w$ should not appear in $T_i$ if $d_i = 0$.

$T_1$=caca, $T_2$=aac, $T_3$=caac     $\mathcal{D}$={2,0,1}

|  | 1 | 2 | 3 | 4 | 5 | 6 | 7 | 8 | 9 | 10 | 11 | 12 | 13 | 14 |
|---|---|---|---|---|---|---|---|---|---|---|---|---|---|---|
| $T''=$ | c | a | c | a | % | a | a | c | % | c | a | a | c | % |

| $i$ | 1 | 2 | 3 | 4 | 5 | 6 | 7 | 8 | 9 | 10 | 11 |
|---|---|---|---|---|---|---|---|---|---|---|---|
| $ids$ | 1 | 3 | 2 | 3 | 2 | 1 | 3 | 2 | 1 | 3 | 1 |
| $s$ | 4 | 11 | 6 | 12 | 7 | 2 | 13 | 8 | 3 | 10 | 1 |
| $lcp$ | - | 1 | 3 | 1 | 2 | 2 | 0 | 1 | 1 | 2 | 2 |
| suffix | a % | a a c % | a a c % | a c % | a c a % | a c c a % | c % | c % | c a % | c a a c % | c a c a % |

$I_1$   $I_2$   $I_3$   $I_4$(ca)

**Fig. 4.** An example for the generalised longest feature problem. Bold lines where used to indicate critical ranges.

To find the longest feature, we modify our algorithm for the generalised longest repeat problem. First, we partition the suffix array into intervals so that each partition would contain only suffixes of strings in $\mathcal{U}'$. Then we find the candidates of the longest feature. We find critical ranges using the algorithm in Section 3. Finally, we check if the candidates are valid or not. A candidate is not valid if it appears in some string $T_i$ where $d_i = 0$. For an interval $[a, b]$ and a critical range $[a', b']$ where $a \le a'$ and $b \ge b'$, we compute $lcp(a, b')$ and $lcp(a' + 1, b + 1)$. If either of the two equals to $lcp(a' + 1, b')$, it means that it appears in some string $T_i$ where $d_i = 0$.

Fig. 4 is an example. We have three strings, $T_1 =$ caca, $T_2 =$ aac, and $T_3 =$ caac. And $\mathcal{D} = \{2, 0, 1\}$, meaning that a feature should appear at least twice in $T_1$, and at least once in $T_3$, but it should not appear in $T_2$. We partition the suffix array into four intervals, $I_1 = [1, 2]$, $I_2 = [4, 4]$, $I_3 = [6, 7]$, and $I_4 = [9, 11]$. There is no critical range in $I_1$, $I_2$, and $I_3$. $I_4$ contains a critical range $[9, 11]$ which represents a candidate ca. Since $lcp(9, 11) = 1$ ('c') and $lcp(10, 11) = 2$, we say that ca never appears in $T_2$.

**Theorem 2.** *The generalised longest common repeat problem can be solved in* $O(\sum_{i=1}^{\ell} |T_i|)$ *time and space.*

*Proof.* We showed that all the steps run in $O(\sum_{i=1}^{\ell} |T_i'|)$ time and space. And $|T_i'| = O(|T_i|)$ for $1 \le i \le \ell$.

## 5   Conclusions

We have defined the generalised longest common repeat problem and presented a linear time algorithm for the problem, allowing direct and inverted repeats.

A remaining work is to devise a more space-efficient algorithm for the problem. Another possibility is to incorporate the detection of degenerated repeats (which are called approximate repeats in the stringology literature).

# References

1. M. I. Abouelhoda, S. Kurtz, and E. Ohlebusch. The enhanced suffix array and its application to genome analysis. In *Proceedings of the 2nd Workshop on Algorithms in Bioinformatics (WABI 2002)*, pages 449–463, 2002.
2. M. A. Bender and M. Farach-Colton. The LCA problem revisited. In *Proceedings of the Fourth Latin American Symposium*, pages 88–94, 2000.
3. J. Beckman and M. Soller. Toward a unified approach to genetic mapping of eukaryotes based on sequence tagged microsatellite sites. *Biotechnology*. pages 8:930–932, 1990.
4. C. T. Caskey, *et al*. An unstable triplet repeat in a gene related to Myotonic Dystrophy, *Science*, pages 255:1256–1258, 1992.
5. S. Dori and G. M. Landau. Construction of aho-corasick automaton in linear time for integer alphabets. In *Proceedings of the 16th Annual Symposium on Combinatorial Pattern Matching (CPM 2005)*, page to appear, 2005.
6. M. Farach-Colton, P. Ferragina, and S. Muthukrishnan. On the sorting-complexity of suffix tree construction. *Journal of the ACM*, 47(6):987–1011, 2000.
7. K. Inman and N. Rudin. An introduction to forensic DNA analysis. CRC press, Boca Raton, Florida, 1997.
8. A. Jeffreys, D. Monckton, K. Tamaki, D. Neil, J. Armour, A. MacLeod, A. Collick, M. Allen, and M. Jobling. Minisatellite variant repeat mapping: application to DNA typing and mutation analysis. *In DNA Fingerprinting: State of the Science*. pages 125–139, Basel, 1993.
9. J. Kärkkäinen and P. Sanders. Simpler linear work suffix array construction. In *Proceedings of the 13th International Colloquim on Automata, Languages and Programming (ICALP 2003)*, pages 943–945, 2003.
10. T. Kasai, G. Lee, H. Arimura, S. Arikawa, and K. Park. Linear-time longest-common-prefix computation in suffix arrays and its applications. In *Proceedings of the 12th Annual Symposium on Combinatorial Pattern Matching (CPM 2001)*, pages 181–192, 2001.
11. D. K. Kim, J. S. Sim, H. Park, and K. Park. Linear-time construction of suffix arrays. In *Proceedings of the 14th Annual Symposium on Combinatorial Pattern Matching (CPM 2003)*, pages 186–199, 2003.
12. S.-R. Kim, I. Lee, and K. Park. A fast algorithm for the generalised $k$-keyword proximity problem given keyword offsets. *Information Processing Letters*, 91(3):115–120, 2004.
13. P. Ko and S. Aluru. Space-efficient linear time construction of suffix arrays. In *Proceedings of the 14th Annual Symposium on Combinatorial Pattern Matching (CPM 2003)*, pages 200–210, 2003.
14. G. M. Landau and J. P. Schmidt. An algorithm for approximate tandem repeats. In *Proceedings of the Fourth Combinatorial Pattern Matching*, pages 120–133, 1993.
15. G. M. Landau, J. P. Schmidt and D. Sokol. An algorithm for approximate tandem repeats. In *Journal of Computational Biology*, pages 8(1): 1–18 , 2001.
16. I. Lee, C. S. Iliopoulos, and K. Park. Linear time algorithm for the longest common repeat problem. In *Proceedings of the 11th String Processing and Information Retrieval (SPIRE 2004)*, pages 10–17, 2004.

17. E. M. McCreight. A space-economical suffix tree construction algorithm. *Journal of the ACM*, 23(2):262–272, April 1976.
18. J. P. Schmidt. All highest scoring paths in weighted grid graphs and its application to finding all approximate repeats in strings. *SIAM Journal on Computing*, 27(4):972–992, 1998.
19. R. H. Singer. Triplet-repeat transcripts: A role for RNA in disease. *Science*, 280(5364):696-697, 1998.
20. E. Ukkonen. On-line construction of suffix trees. *Algorithmica*, 14:249–260, 1995.
21. K. J. Woo, K. Sang-Ho, and C. Jae-Kwan. Association of the dopamine transporter gene with Parkinsons disease in Korean patients *Journal of Korean Medical Science*, 15(4), 2000.

# Application of Clustering Technique in Multiple Sequence Alignment

Patrícia Silva Peres and Edleno Silva de Moura

Universidade Federal do Amazonas,
Manaus, AM, Brasil
{psp, edleno}@dcc.ufam.edu.br

**Abstract.** This article presents a new approach using clustering technique for creating multiple sequence alignments. Currently, the most widely used strategy is the progressive alignment. However, each step of this strategy might generate an error which will be low for closely related sequences but will increase as sequences diverge. For that reason, determining the order in which the sequences will be aligned is very important. Following this idea, we propose the application of a clustering technique as an alternative way to determine this order. To assess the reliability of this new strategy, two methods were modified in order to apply a clustering technique. The accuracy of their new versions was tested using a reference alignment collection. Besides, the modified methods were also compared with their original versions, obtaining better alignments.

## 1 Introduction

The simultaneous alignment of many nucleotide or amino acid sequences is one of the commonest tasks in molecular biological analyses [5,6]. Multiple alignments are important in many applications, such as, predicting the secondary or tertiary structure of new sequences, demonstrating homology between new sequences and existing families, inferring the evolutionary history of a protein family, finding the characteristic motifs between biological sequences, etc.

The automatic generation of an accurate multiple sequence alignment is a tough task and many efforts have been made to achieve this goal. There is a well-known algorithm for pairwise alignment based on Dynamic Programming where the optimal alignment according to an objective-function is given. It is possible to adapt this method to multiple alignments, but it becomes impracticable, since it requires CPU time proportional to $n^k$, where $n$ is the average length of the sequences and $k$ is the number of sequences [4].

Some heuristics to align multiple sequences have been developed in the last years, with the purpose of accomplishing comparisons more quickly, even without the same precision of the Dynamic Programming algorithms. Nonetheless, even if these heuristic methods successfully provide the optimal alignments, there remains the problem of whether the optimal alignment really corresponds to the

M. Consens and G. Navarro (Eds.): SPIRE 2005, LNCS 3772, pp. 202–205, 2005.

biologically correct one, since there is not a standard to define the best multiple alignment [8].

Most of the heuristic methods are based on the progressive alignment strategy, especially the one proposed by Feng and Doolittle [3], in which a guide tree is built to specify the sequences alignment order. Examples of methods based on progressive alignment strategy are Muscle [2] and ClustalW [9].

The progressive alignment method is based on a greedy approach and the error introduced once during the alignment cannot be corrected subsequently. Each alignment step in the multiple alignment process might generate an error which will be low on average for closely related sequences but will increase as sequences diverge. Therefore, determining the order in which the sequences will be aligned is an important stage, where it is desired that the most similar sequences are aligned first, so the error would be low in the beginning of the process.

Based on that hypothesis, this work consists in a first effort to apply a clustering technique as an alternative way to determine the order in which the sequences will be aligned, with the purpose of achieving better results at the end of the process.

To develop this idea, two well-known methods were modified in order to apply a clustering technique to their algorithms. These two methods are Muscle[1] version 3.52 and ClustalW[2] version 1.83. The performance of these new modified methods was assessed using a set of reference alignments and they achieve better results than their original versions.

# 2    Applying the Clustering Strategy

The process of constructing the guide tree in the progressive alignment can also be considered as a clustering technique, since the multiple alignment is generated by clustering in each iteration only the closest pair of sequences or profiles[3] according to the branch order of the tree. However, this behavior may cause a loss of important information about the other sequences and their relationship, and such information can be very helpful in the whole alignment process.

Therefore, it is proposed the application of a clustering method, as a previous stage of the progressive alignment, whose main purpose is to determine the order in which the sequences will be aligned by performing a global analysis of the sequences similarities before clustering them.

It was used a simple global clustering algorithm fully explained in [7]. This global clustering algorithm is based on the entries of a distance matrix (part of the progressive alignment process) returning as a result a list of sequences clusters which will be aligned later.

---

[1] http://www.drive5.com/muscle/

[2] http://www.ebi.ac.uk/clustalw/

[3] A profile consists in a group of sequences already aligned. Such group, however, may contain just one sequence.

# 3 Experiments and Preliminary Results

To assess the reliability of our new strategy, Muscle version 3.52 and ClustalW version 1.83 were modified in order to apply the global clustering algorithm as a new stage of the whole multiple alignment process. Then, it was used the BAliBASE reference collection [1,10,11] to compare the results obtained by the new versions of Muscle and ClustalW and their original ones.

BAliBASE (Benchmark Alignment dataBASE) is a reference collection of manually refined protein sequence alignments categorized into eight different references, where each one characterizes a real problem. However, reference 6, 7 and 8 of BAliBASE are not used in this work because they characterize problems for which none of the tested algorithms is designed. A separate program is used to compute BAliBASE score, which is divided into *sum-of-pairs score* (SP score) and *total column score* (TC score) [11]. It was used version 2 of BAliBASE freely available in the Web[4].

The results are depicted in Table 1, in which there is a comparison between original Muscle and its modified version called C-Muscle; and a comparison between ClustalW and its modified version called C-ClustalW. The best results are underlined.

The results correspond to the SP and TC scores associated with five references of BAliBASE collection denoted here as Ref1 – Ref5. As it can be observed, the new version of Muscle achieved better results than its original one in references 1, 3 and 4. In reference 5, we have the same results and in reference 2 we have a decrease in the result.

**Table 1.** SP and TC scores on each reference of BAliBASE for each version of Muscle and ClustalW algorithms

| Method | Ref1 SP | Ref1 TC | Ref2 SP | Ref2 TC | Ref3 SP | Ref3 TC | Ref4 SP | Ref4 TC | Ref5 SP | Ref5 TC |
|---|---|---|---|---|---|---|---|---|---|---|
| C-Muscle | 78.10 | 67.83 | 87.58 | 46.30 | 70.54 | 41.75 | 68.33 | 38.00 | 82.78 | 65.17 |
| Muscle | 78.06 | 67.73 | 87.83 | 46.83 | 70.43 | 41.50 | 68.23 | 37.83 | 82.78 | 65.17 |
| C-ClustalW | 77.43 | 66.70 | 85.78 | 44.57 | 65.05 | 34.67 | 63.04 | 33.75 | 76.02 | 51.58 |
| ClustalW | 77.38 | 66.60 | 86.09 | 45.61 | 64.65 | 34.42 | 63.08 | 33.50 | 75.78 | 50.75 |

On the other hand, the new version of ClustalW achieved better results than its original one in references 1, 3, 4 and 5. In reference 2, we have a decrease in the result.

It is important to notice that there are some cases, especially when the sequences are very similar to each other, where the clustering technique does not interfere in the results. In these cases, the results are equal than those obtained by the original version of the methods.

---

[4] http://bips.u-strasbg.fr/en/Products/Databases/BAliBASE2/

# 4    Conclusions and Future Work

This work consists in a first effort to apply clustering techniques as a previous stage of the progressive multiple alignment to achieve better results. It was used a simple clustering method whose main purpose is to determine the order in which the sequences will be aligned by performing a global analysis of them, instead of the local one proposed by the guide tree [7].

In the previous section, it was depicted the preliminary results obtained by our proposed strategy compared with the original results of Muscle and ClustalW, using the BAliBASE reference collection. From the experiments, it is possible to conclude that applying clustering techniques in multiple sequence alignment methods can help to improve their results.

Therefore, in future work, we intend to study other clustering techniques and apply them to existing multiple sequence alignment methods, in order to assess their viability and reliability in achieving better alignment results and as a way to consolidate this first attempt presented in this work.

# References

1. A. Bahr, J. D. Thompson, J. C. Thierry, and O. Poch. Balibase (benchmark alignment database): enhancements for repeats, transmembrane sequences and circular permutations. *Nucleic Acids Res.*, 29(1):323–326, 2001.
2. R. C. Edgar. Muscle: multiple sequence alignment with high accuracy and high throughput. *Nucleic Acids Res.*, 32:1792–1797, 2004.
3. D. F. Feng and R. F. Doolittle. Progressive sequence alignment as a prerequisite to correct phylogenetic trees. *J. Mol. Evol.*, 25:351–360, 1987.
4. D. Gusfield. *Algorithms on strings, tress, and sequences: computer science and computational biology.* Cambridge University Press, 1997.
5. C. Korostensky and G. Gonnet. Near optimal multiple sequence alignments using a traveling salesman problem approach. *Proc. of the 6th International Symposium on String Processing and Information Retrieval, Cancun, Mexico*, pages 105–114, 1999.
6. J. Meidanis and J. C. Setúbal. Multiple alignment of biological sequences with gap flexibility. *Proc. of the 2nd South American Workshop on String Processing, Valparaiso, Chile*, pages 138–153, 1995.
7. P. S. Peres and E. S. Moura. Application of clustering technique in multiple sequence alignment. Unpublished manuscript.
8. J. C. Setúbal and J. Meidanis. *Introduction to coputational molecular biology.* PSW Publishing Company, 1997.
9. J. D. Thompson, D. G. Higgins, and T. J. Gibson. Clustal w: improving the sensitivity of progressive multiple sequence alignment through sequence weighting, position-specific gap penalties and weight matrix choice. *Nucleic Acids Res.*, 22:4673–4680, 1994.
10. J. D. Thompson, F. Plewniak, and O. Poch. Balibase: a benchmark alignment database for the evaluation of multiple sequence alignment programs. *Bioinformatics*, 15:87–88, 1999.
11. J. D. Thompson, F. Plewniak, and O. Poch. A comprehensive comparison of multiple sequence alignment programs. *Nucleic Acids Res.*, 27:2682–2690, 1999.

# Stemming Arabic Conjunctions and Prepositions

Abdusalam F.A. Nwesri, S.M.M. Tahaghoghi, and Falk Scholer

School of Computer Science and Information Technology,
RMIT University, GPO Box 2476V, Melbourne 3001, Australia
{nwesri, saied, fscholer}@cs.rmit.edu.au

**Abstract.** Arabic is the fourth most widely spoken language in the
world, and is characterised by a high rate of inflection. To cater for
this, most Arabic information retrieval systems incorporate a stemming
stage. Most existing Arabic stemmers are derived from English equivalents; however, unlike English, most affixes in Arabic are difficult to
discriminate from the core word. Removing incorrectly identified affixes
sometimes results in a valid but incorrect stem, and in most cases reduces
retrieval precision. Conjunctions and prepositions form an interesting
class of these affixes. In this work, we present novel approaches for dealing with these affixes. Unlike previous approaches, our approaches focus
on retaining valid Arabic core words, while maintaining high retrieval
performance.

## 1 Introduction

Arabic is a Semitic language, with a morphology based on building inflected
words from roots that have three, four, or sometimes five letters. For example, each verb can be written in sixty-two different forms [14]. Words are inflected and morphologically marked according to gender (masculine and femi-
nine); case (nominative, genitive, and accusative); number (singular, dual, and
plural); and determination (definite and indefinite).

Arabic has three types of affixation: prefixes, suffixes and infixes. In contrast
with English, some Arabic affixes are very difficult to remove without proper
identification. It is common to have multiple affixes on a word. A clear example
is the use of pronouns. Unlike English, where words and possessive pronouns are
written separately, Arabic possessive pronouns are attached at the end of the
word in most cases. For instance, the English sentence "they will teach it to you"
can be written in one Arabic word as سيعلمونكها (see Table 1 for the mapping
between this word and its English translation). A good stemmer identifies the
stem يعلم (teach) or علم (knew) for this word.

Arabic prefixes are widely used in Arabic text. Some of these prefixes are
used with verbs, some with nouns, and others with both. Affixes that are three
letters in length are easily identified; however, the shorter the affix is, the more
difficult it is to identify. A sub-class of prefixes is formed by prepositions and
conjunctions; this sub-class is of particular interest because, if they are identified and removed correctly, we will obtain valid Arabic core words. Some of

M. Consens and G. Navarro (Eds.): SPIRE 2005, LNCS 3772, pp. 206–217, 2005.
© Springer-Verlag Berlin Heidelberg 2005

**Table 1.** Mapping between the Arabic word سيعلمونكها and its English translation

| ها | ك | و ن | يعلم | س |
|---|---|---|---|---|
| It | you | they | teach | will |

these prefixes are letters attached to the beginning of the word. However, these same letters also frequently appear as part of affix-free words. For example, the letter و (waw) in the word وقار (respect) is part of the word, whereas, it is a conjunction in the word والطالب (and the student). Removing this letter from the first word results in the word قار (Asphalt); this should be avoided, because the meaning of the returned word is changed. However, in the second word, it is essential to remove the letter in order to obtain the stem طالب (student). Although it is important to remove such prefixes, current popular search engines do not do so. Moukdad [10] showed that searching for the words الجامعة (the university), والجامعة (and the university),and وبالجامعة (and by the university) by four well-known search engines, gives different results by each search engine for each particular word. All four search engines performed badly when searching for the second and the last words.

## 1.1 Arabic Conjunctions and Prepositions

Arabic has nine conjunctions. The majority of these are written separately, except the *inseparable conjunctions* و (waw) and ف (faa), which are usually attached to a noun or a verb.

In Arabic, prepositions are added before nouns. There are twenty prepositions. Five of these are usually attached to the beginning of a word. These *inseparable prepositions* [13] are: ل (lam), و (waw), ك (kaf), ب (baa), and ت (taa).

Prepositions and conjunctions occur frequently in Arabic text. To aid information retrieval, they should be removed, so that variant forms of the same word are conflated to a single form. Separable prepositions can be easily detected and removed as stopwords. However, inseparable prepositions are difficult to remove without inadvertently changing the meaning of other words in the text.

The set of inseparable prepositions and conjunctions consists of the six letters: *lam, waw, faa, taa, baa*, and *kaf*. These letters differ in terms of their function in Arabic text and can be further divided into three different groups:

- *waw* and *faa*, can be added to any Arabic word as they are conjunctions. While *waw* is also a valid preposition, the fact that it is a conjunction means that this letter can be added to any Arabic word.
- *kaf, taa*, and *baa* are prepositions that can only be used before a noun. *taa* is also used as a prefix for verbs in the present simple tense. It is rarely used as a preposition in modern standard Arabic.
- *lam*, in addition to its purpose as a preposition like *kaf* and *baa*, can also be used with verbs as the "lam of command". Here, it is usually prefixed to the third person to give it an imperative sense, for example لتقلها (say it). It is also used to indicate the purpose for which an action is performed [13].

In this paper, our main focus is on single-letter inseparable prepositions and conjunctions, and their effects on Arabic stemming. For the reminder of this paper, the term *particles* will be used to represent the class of inseparable prepositions and conjunctions together. The particle *taa* is not considered to be a member of this class, due to its rare usage as a preposition in modern standard Arabic.

The rest of this paper is structured as follows. We first present related background, examining previous approaches for dealing with particles. We then propose several new techniques for removing particles from Arabic text, with the aim of retaining correct core words. The effectiveness of these techniques is evaluated experimentally, both based on the characteristics of terms that are produced by the various algorithms, as well as based on the impact that they have on retrieval performance.

## 2   Arabic Stemming

There are two main classes of stemming algorithms: heavy stemming, and light stemming [2]. In both cases, the aim of stemming is to remove affixes from an input string, returning the stem of the word as an output.

Heavy — or root-based — stemming usually starts by removing well-known prefixes and suffixes. It aims to return the actual root of a word as the remaining stem, usually by applying patterns of fixed consonants [7]. The most well known pattern is فعل, which is often used to represent three-letters root words. For example: the word كتب (wrote) can be represented by the pattern فعل by mapping ك to ف, ت to ع, and ب to ل.

Many stems can be generated from this root using different standard patterns. For instance, فعال, فاعل, and يفعل are three different patterns to form the singular noun, nomina agentis, and present tense verb out of the pattern فعل respectively. By fixing the core letters and adding additional letters in each pattern, we can generate كتاب (book), كاتب (writer), يكتب (write) respectively. The new words can accept Arabic prefixes and suffixes.

Heavy stemmers usually reverse this process by first removing any prefixes and suffixes from the word. They then identify the pattern the remaining word corresponds to, and usually return the root by extracting letters that match the letters ف, ع and ل. For example, to find the root of the word والكاتب (and the writer), any heavy stemmer has to remove the prefixes وال to get the stem كاتب, then use the pattern فاعل which matches this (has the same length, and with the letter ا in the same position). The root كتب is then returned [1].

Heavy stemming has been shown to produce good results in the context of information retrieval. For example, Larkey et al. [8] show that mean average precision is improved by 75.77% using the Khoja heavy stemmer. We discuss retrieval metrics further in Sect. 4.

---

[1] The letter ع and ـع are two forms for the same letter.

Light stemming stops after removing prefixes and suffixes, and does not attempt to identify the actual root. It has been demonstrated that light stemming outperforms other techniques in Arabic information retrieval. Aljlayl et al. [2] demonstrate an increase in mean average precision of 87.7%, and Larkey et al. [8] report an increase in mean average precision of 100.52%.

The core of both approaches involves the removal of affixes. Generally, removing prefixes has been dealt with in the same manner as for many European languages, by matching the first character of the word to a pre-prepared list of prefixes, and truncating any letters that match without first checking whether or not it is a real prefix [3,4,7,8]. For Arabic text, this frequently results in incorrect root extraction in heavy stemming, and an incorrect stem in light stemming. It is therefore doubtful that these simple approaches are appropriate when dealing with Arabic text. For example, ولدين (two boys) returns the root لدن (soft) instead of root word ولد (gave birth) using the Khoja stemmer, and the stem لد (has no meaning) instead of the stem ولد (a boy) using the Larkey stemmer. In the last two cases, the incorrect root (stem) was due to removal of the first letter *waw* after incorrectly identifying it as a particle.

## 2.1   Current Approaches for Stemming Particles

Many stemmers have been developed for Arabic [1]. However, none deals with the removal of all particles. Some particles, such as *waw*, are removed by all existing stemmers; other particles, such as *kaf*, have never been considered on their own in existing stemming approaches. The way in which existing stemmers deal with particles can be grouped into three general categories:

- Matching the first letter with a pre-prepared list of particles. If a match is found, the first letter is removed as long as the remaining word consists of three or more letters. This approach is used by most of the current stemmers to deal with a small subset of particles [3,4,7,8]. We call this approach *Match and Truncate* (MT).
- Matching the first letter with a list of particles. If a particle is found, the remaining word is checked against the list of all words that occur in the document collection being stemmed. If the stemmed word occurs in the collection, the first letter is considered a particle and removed. This approach was used by Aito et. al. [3] in conjunction with the other two approaches. We call this approach *Remove and Check* (RC).
- Removing particles with other letters. For example, removing a combination of particles and the definite article ال (the), particularly, وال *wal*, فال *fal*, بال *bal*, كال *kal*. These combinations are removed whenever they occur at the beginning of any word, and this approach is used by most current stemmers. We call this approach *Remove With Other Letters* (RW).

Existing stemmers often use a combination of these approaches. They usually start by using the third approach, then continue by removing other particles, particularly *waw* and *lam*.

## 2.2   Evaluation of Current Approaches

To check the effectiveness of current approaches for particle removal in Arabic text, we extracted all correct words that start with a particle from a collection of Arabic documents used in the TREC 2001 Cross-Language Information Retrieval track, and the TREC 2002 Arabic/English CLIR track [6,11]. Further collection details are provided in Sect. 4.

The number of words start with a possible particle constitute 24.4% of this collection. To ensure that we extracted only correct words, we checked them using the Microsoft Office 2003 Arabic spellchecker [9]. Stopwords such as pronouns and separable particles were then removed. This procedure resulted in a list of 152,549 unique correct words that start with a possible inseparable particle.

We use three measures to evaluate the effectiveness of the above approaches:

- The number of incorrect words produced; Although correct words are not the main target of stemming, an incorrect stem can have a completely different meaning and correspond to a wrong index cluster. This is particularly true when a a core letter is removed from an Arabic word,
- The number of words that remain with an initial letter that could be a particle. This indicates how many possible particles remain after an approach is applied. In Arabic, the second character could possibly be a particle if the first character is a conjunction.
- The number of words actually changed; This shows the *strength* of each approach [5] by counting the stems that differ from the unstemmed words.

Using the assumption that a correct Arabic word with a particle should also be correct without that particle, we experimentally applied the MT, RC, and RW approaches to every word in our collection of unique correct words. The results are shown in Table 2.

**Table 2.** Removing particles using current approaches

| Approach | Incorrect words | Possible particles | Altered words |
|----------|-----------------|--------------------|---------------|
| MT | 5,164 | 21,945 | 151,040 |
| RC | 220 | 41,599 | 133,163 |
| RW | 724 | 122,878 | 33,847 |

It can be seen that the MT approach produces a large number of incorrect words (3.39% of all correct words). The results also show that when the MT approach truncates the first letter as a particle, there is a chance that the second letter is also a particle. The portion of words that still start with letters that could be particles constitutes 14.39% of the total number of correct words. Manual examination of the stemmed list showed that many words have another particle that should be removed, and that many words have their first letter removed despite this letter not being a particle.

The RC approach produces fewer incorrect words. This is because no prefix removal is carried out when the remaining word is not found in the collection. The incorrect words we obtain are due to the collection itself containing many incorrect words. Approximately twice as many words still start with possible particles as seen in the first approach. This implies that the RC approach leaves the first letter of many words unchanged. This might be desirable, since these might be valid words that do not actually start with a particle. Indeed, manual examination of the result list revealed that many words with particles have been recognised, and particles have been removed correctly. However, the result list also contained a large proportion of words that still start with particles as their first letter.

The RW approach produces a smaller number of incorrect words than the first approach, but generates a very large number of words still starting with possible particles (80.55% of the list of correct words). Moreover, many words are left entirely unchanged.

To conclude, the first approach is too aggressive. It affects Arabic words by removing their first letter, regardless of whether this letter is actually a particle. The second approach, while better at recognising particles in the text, leaves a considerable portion of words with real particles untouched. More importantly, in many cases a word is modified to one with completely different meaning. The third approach leaves a big portion of words without removing particles at all, and only deals with a small subset of particles in the text. It also affects words that start with the combination of particles and other letters especially proper nouns and foreign words such as فالوجة (the Iraqi city of Fallujah) and بالتيمور (the US city of Baltimore). It is also very hard to recognise such combinations if they are preceded by another particle (conjunction) such as وبالارض (and by the land).

## 3   New Approaches

Given the incomplete way in which particles have been dealt with in previous approaches, we have investigated techniques to identify and remove inseparable conjunctions and prepositions from core words in a principled manner. Our methods are based on removing particles using grammatical rules, aiming to decrease the number of incorrect words that are produced by the stemming process, and increasing the completeness of the process by reducing the number of words that still start with a particle after stemming.

A requirement for being able to recognise affixes in text is a good lexicon. We use the Microsoft Office 2003 Arabic lexicon; this contains more than 15,500,000 words covering mainly modern Arabic usage [9].

We introduce four rules, based on consideration of Arabic grammar, to identify particles in Arabic text. Let $L$ be an Arabic lexicon, $P$ be the set of prepositions {kaf, baa, lam}, $C$ be the set of two conjunctions {waw, faa}, $c$ be a letter in $C$, $p$ be a letter in $P$, and $w$ be any word in $L$. Then:

- Rule 1: Based on grammatical rules of the Arabic language, a correct Arabic word that is prefixed by a particle is also a correct word after that particle is removed. More formally:

$$\forall (p + w) \in L \Rightarrow w \in L$$

and

$$\forall (c + w) \in L \Rightarrow w \in L$$

- Rule 2: Any correct Arabic word should be correct if prefixed by either conjunction, *waw* or *faa*:

$$\forall w \in L \Rightarrow (c + w) \in L$$

- Rule 3: Based on the above two rules, any correct word with a particle prefix, should be correct if we replace that prefix with *waw* or *faa*:

$$\forall (p + w) \in L \Rightarrow (c + w) \in L$$

- Rule 4: Any correct Arabic word that is prefixed by a particle should not be correct if prefixed by the same particle twice, except the particle *lam* which could occur twice at the beginning of the word. Let $p_1$ and $p_2$ be two particles in $(P \cup C)$, and $p_1 = p_2 \neq lam$, then

$$\forall (p_1 + w) \in L \Rightarrow (p_2 + p_1 + w) \notin L$$

Based on these rules, we define three new algorithms: *Remove and Check in Lexicon* (RCL); *Replace and Remove* (RR); and *Replicate and Remove* (RPR).

Due to the peculiarities of the letter *lam*, we deal with this letter as a common first step before applying any of our algorithms. Removing the particle *lam* from words start with the combination لل results in some incorrect words. We therefore deal with this prefix before we deal with the particle *lam* by itself. The prefix لل is a result of adding the particle *lam* ل to one of the following:

- A noun that starts with the definite article. When the particle ل is added to a word whose first two letters are the definite article ال, the first letter ا is usually replaced with the letter *lam* ل. For example, الجامعة (the university) becomes للجامعة (for the university). However, if the letter following the definite article is also the letter *lam* ل, then next case applies.
- A noun that starts with the letter *lam*. For example, لقب (surname or championship) should be written as للقب when prefixed by the particle *lam* ل.
- A verb that starts with the letter *lam*. For example, لفّ (wrapped) should be written as للفّ when prefixed by the particle *lam* ل.

To stem this combination, we first check whether removing the prefix لل produces a correct word. If so, we remove the prefix. Otherwise, we try adding the letter ‏ا‎ before this word. If the new word is correct, we drop one *lam* from the original word.

To remove the particle *lam* from words that originally start with the definite article, we replace the first *lam* with the letter ‏ا‎ and check whether the word exists in the lexicon. If so, we can stem the prefix ‏لل‎ without needing to check whether the remaining part is correct. If not, we remove the first letter and check to see whether we can drop the the first *lam*. This algorithm is used before we start dealing with any other particles in the three following algorithms.

**Remove and Check in Lexicon (RCL).** In our first algorithm we start by checking the first letter of the word. If it is a possible particle — that is, it is a member of the set $P$ of $C$ — we remove it and check the remaining word in our dictionary. If the remainder is a valid word, the first letter is considered to be a particle, and is removed. Otherwise, the original word is returned unchanged. This approach differs from the RC approach in that we check the remaining word against a dictionary, rather than against all words occurring in the collection. We expect that this will allow us to better avoid invalid words.

**Replace and Remove (RR).** Our second algorithm is based on Rule 3. If the first letter of the word is a possible particle, we first test whether the remaining string appears in our dictionary. If it does, we replace the first letter of the original string with *waw* and *faa* in turn, and test whether the new string is also a valid word. If both of the new instances are correct, the evidence suggests that the original first letter was a particle, and it is removed, with the remainder of the string being returned. The string is returned unchanged if any of the new strings are incorrect.

Manual examination of the output list of the RR algorithm shows some interesting trends. The algorithm achieves highly accurate particle recognition (few false positives). However, it often fails to recognise that the first letter is an actual particle, because replacing the first letter with *faa* and *waw* will often produce valid new words. For example, consider the word ‏بارع‎ (clever). Applying the RR algorithm results in two valid words: ‏وارع‎ (and look after), and ‏فارع‎ (and look after). The first letter of the original word is therefore removed, giving the word ‏ارع‎ (look after), instead of the original word ‏بارع‎ (clever).

**Replicate and Remove (RPR).** Our third algorithm performs two independent tests on a candidate string. First, the initial letter is removed, and the remaining word is checked against the dictionary. Second, based on Rule 4 above, the initial letter is duplicated, and the result is tested for correctness against the dictionary. If either test succeeds, the unchanged original word is returned (no stemming takes place).

We have noticed that if the word is a verb starting with *baa* or *kaa*, the first letter is removed whether or not it is a particle, since these are particles that cannot precede verbs. Duplicating them in verbs produces incorrect words, and causes the first letter of the original word to be removed. We can use the letter *lam* to recognise verbs that start with those particles. Accordingly, we add a new step where we add the letter *lam* to the word and check it for correctness. If the word is incorrect with the letter *lam* and also incorrect with the first letter replicated, then we conclude that the word is not a verb, and we remove the first letter.

For words starting with the letter *lam*, we add both *baa* and *kaf* instead of replicating them, since replication will result in a correct word, and lead to the particle *lam* being preserved. If both new instances are incorrect, we remove the first *lam*.

The above algorithms may be applied repeatedly. In particular, if stemming a word starting with either *waw* or *faa* results in a new word of three or more characters that has either *waw*, *kaf*, *baa*, or *lam* as its first character, the particle removal operation is repeated.

## 3.1   Evaluation of Our Approaches

We have evaluated our new algorithms using the same data set described in Sect. 2.2. As seen from Table 3, all three algorithms result in a low number of incorrect words after stemming, with similar strength. However, they differ in the number of words with possible particles after stemming. The RPR approach leaves many words with possible particles (around 5,000 more than the RR approach and 3,000 more than RCL approach).

**Table 3.** Results of the new approaches, showing significantly lower incorrect words, lower possible particles, and a comparable strength over the baseline in Table2

| Approach | Incorrect words | Possible particles | Altered words |
|----------|-----------------|--------------------|--------------|
| RCL      | 82              | 17,037             | 146,032      |
| RR       | 82              | 15,907             | 146,779      |
| RPR      | 82              | 20,869             | 142,082      |

Compared to the previous approaches for handling particles, our algorithms result in 82 incorrect words, compared to 5,164 using MT, 724 using RW, and 220 using RC. The number of words that start with possible particles has also dropped dramatically using both RCL and RR.

Using the RPR approach we extracted all words that have not been stemmed (words still having a first letter as a possible particle). The list had 10,476 unique words. To check algorithm accuracy, we randomly selected and examined 250 of these. We found that only 12 words are left with particles that we believe should be stemmed; this indicates an accuracy of around 95%.

As stemming particles can result in correct but completely different words, we decided to pass the list we extracted using RPR approach to other approaches and check whether stemmed words would be correct. We extracted correctly stemmed words changed by each approach. Out of the 10,476 words, RR resulted in 4,864 new correct stems.

We noticed that about 90% of these are ambiguous, where the first character could be interpreted as a particle or a main character of the stemmed word; the meaning is different in the two cases. For example, the words فيلم (film) could also mean (and he collects) when considering the first letter as a particle. MT, and RCL resulted in 3,950 similar stems, while RC resulted in 2,706 stems. Examples are shown in Table 4.

**Table 4.** Words with different meaning when stemmed by RPR and RR

| Word | Stemmed using RPR | | Stemmed using RR | |
|---|---|---|---|---|
| | stem | Meaning | stem | meaning |
| وبطاقتي | بطاقتي | my ID card | طاقتي | my power |
| فاتتهم | فاتتهم | they missed it | اتتهم | it came to them |
| بوسادة | وسادة | pillow | سادة | masters |
| وليفي | وليفي | my mate | يفي | made his promise |
| وكفنها | كفنها | her coffin | فنها | her art |
| بوصفاتها | وصفاتها | her recipes | صفاتها | her characterstics |

RPR keeps any letter that is possibly a core part of the word, even though it might also be considered as a particle. In contrast, RR removes such letters. In most cases, keeping the letter appears to be the best choice.

# 4    Information Retrieval Evaluation

While the ability to stem particles into valid Arabic words is valuable for tasks such as machine translation application, document summarisation, and information extraction, stemming is usually applied with the intention of increasing the effectiveness of an information retrieval system. We therefore evaluate our approaches in the context of an ad-hoc retrieval experiment.

We use a collection of 383,872 Arabic documents, mainly newswire stories published by Agence France Press (AFP) between 1994 and 2000.

This collection was used for information retrieval experiments in the TREC 2001 and TREC 2002 Arabic tracks [6,11]. Standard TREC queries and ground truth have been generated for this collection: 25 queries defined as part of TREC 2001, and 50 additional queries as part of TREC 2002. Both sets of queries have corresponding relevance judgements, indicating which documents are correct answers for which queries.

As most stemmers in the literature start by using the RW approach and then proceed to stem ال prefixes, we decided to likewise not use this approach on its own, but instead use it in conjunction with other approaches.

To form our baseline collection, we preprocessed the TREC collection, by first removing all stopwords, using the Larkey light9 stopword list [2]. Then we

---

[2] http://www.lemurproject.org

**Table 5.** Performance of different approaches

|  | TREC 2001 | | | TREC 2002 | | |
|---|---|---|---|---|---|---|
|  | MAP | P10 | RP | MAP | P10 | RP |
| Baseline | 0.2400 | 0.5320 | 0.3015 | 0.2184 | 0.3200 | 0.2520 |
| MT | 0.2528 | 0.5400 | 0.3193 | 0.2405 | 0.3440 | 0.2683 |
| RC | 0.2382 | 0.5080 | 0.3037 | 0.2319 | 0.3360 | 0.2663 |
| LarkeyPR | 0.2368 | 0.4800 | 0.3102 | 0.2345 | 0.3280 | 0.2679 |
| AlstemPR | 0.2328 | 0.4800 | 0.2998 | 0.2194 | 0.3180 | 0.2582 |
| BerkeleyPR | 0.1953 | 0.4520 | 0.2460 | 0.2072 | 0.2680 | 0.2423 |
| RCL | 0.2387 | 0.5080 | 0.3041 | 0.2320 | 0.3360 | 0.2654 |
| RPR | 0.2586 | 0.5440 | 0.3246 | 0.2379 | 0.3420 | 0.2654 |
| RR | 0.2543 | 0.5320 | 0.3200 | 0.2394 | 0.3440 | 0.2681 |

removed all definite article combinations and ran each algorithm on this baseline collection.

For retrieval evaluation, we used the public domain Zettair search engine developed at RMIT University [3]. We evaluate retrieval performance based on three measures: mean average precision (MAP), precision at 10 documents (P10), and R-precision (RP) [6]. Table 5 shows the results recorded for each approach.

Both RC and RCL perform badly and result in lower precision than the baseline. In contrast, MT, RPR, and RR, showed an improvement over the baseline, for all measures. The improvement for MT, RPR and RR is statistically significant for the TREC 2001 and TREC 2002 queries when using the $t$-test; this test has been demonstrated to be particularly suited to evaluation of IR experiments [12].

The RR and RPR approaches produce results comparable to previous prefix removal approaches. By way of comparison, we also show the performance of the particle removal stages only of three well-known stemmers: Larkey stemmer, Alstem stemmer, and Berkeley stemmer. Our approaches performed better for both the TREC 2001 and TREC 2002 query sets.

## 5   Conclusion

In this work, we have presented three new approaches for the stemming of prepositions and conjunctions in Arabic text. Using a well-known collection of Arabic newswire documents, we have demonstrated that our algorithms for removing these affixes offer two significant advantages over previous approaches while achieving information retrieval results that are comparable to previous work. First, our algorithms identify particles more consistently than previous approaches; and second, they retain a higher ratio of correct words after removing particles. In particular, we believe that by producing correct words as an output, our approach will be of benefit for application to machine translation,

---

[3] http://www.seg.rmit.edu.au/zettair/

document summarisation, information extraction, and cross-language information retrieval applications. We plan to extend this work to handle suffixes in Arabic text.

## Acknowledgements

We thank Microsoft Corporation for providing us with a copy of Microsoft Office Proofing Tools 2003.

## References

1. I. A. Al-Sughaiyer and I. A. Al-Kharashi. Arabic morphological analysis techniques: A comprehensive survey. *Journal of the American Society for Information Science and Technology*, 55(3):189–213, 2004.
2. M. Aljlayl and O. Frieder. On Arabic search: improving the retrieval effectiveness via a light stemming approach. In *Proceedings of the International Conference on Information and Knowledge Management*, pages 340–347. ACM Press, 2002.
3. A. Chen and F. Gey. Building an Arabic stemmer for information retrieval. In *Proceedings of the Eleventh Text REtrieval Conference (TREC 2002)*. National Institute of Standards and Technology, November 2002.
4. K. Darwish and D. W. Oard. Term selection for searching printed Arabic. In *Proceedings of the ACM-SIGIR International Conference on Research and Development in Information Retrieval*, pages 261–268. ACM Press, 2002.
5. W. B. Frakes and C. J. Fox. Strength and similarity of affix removal stemming algorithms. *SIGIR Forum*, 37(1):26–30, 2003.
6. F. C. Gey and D. W. Oard. The TREC-2001 cross-language information retrieval track: Searching Arabic using English, French or Arabic queries. In *Proceedings of TREC10*, Gaithersburg: NIST, 2001.
7. S. Khoja and R. Garside. Stemming Arabic text. Technical report, Computing Department, Lancaster University, Lancaster, September 1999.
8. L. S. Larkey, L. Ballesteros, and M. E. Connell. Improving stemming for Arabic information retrieval: light stemming and co-occurrence analysis. In *Proceedings of the ACM-SIGIR International Conference on Research and Development in Information Retrieval*, pages 275–282. ACM Press, 2002.
9. Microsoft Corporation. Arabic proofing tools in Office 2003, 2002. URL: http://www.microsoft.com/middleeast/arabicdev/office/office2003/Proofing.asp.
10. H. Moukdad. Lost in cyberspace: How do search engine handle Arabic queries. In *Proceedings of CAIS/ACSI 2004 Access to information: Skills, and Socio-political Context*, June 2004.
11. D. W. Oard and F. C. Gey. The TREC-2002 Arabic/English CLIR track. In *TREC*, 2002.
12. M. A. Sanderson and J. Zobel. Information retrieval system evaluation: Effort, Sensitivity, and Reliability. In *Proceedings of the ACM-SIGIR International Conference on Research and Development in Information Retrieval*. ACM Press, 2005. to appear.
13. W. Wright. *A Grammar of the Arabic language*, volume 1. Librairie du Liban, 1874. third edition.
14. A. B. Yagoub. *Mausooat Annaho wa Assarf*. Dar Alilm Lilmalayn, 1988. third reprint.

# XML Multimedia Retrieval

Zhigang Kong and Mounia Lalmas

Department of Computer Science, Queen Mary, University of London
{cskzg, mounia}@dcs.qmul.ac.uk

**Abstract.** Multimedia XML documents can be viewed as a tree, whose nodes correspond to XML elements, and where multimedia objects are referenced in attributes as external entities. This paper investigates the use of textual XML elements for retrieving multimedia objects.

## 1 Introduction

The increasing use of eXtensible Mark-up Language (XML) [6] in document repositories has brought about an explosion in the development of XML retrieval systems. Whereas many of today's retrieval systems still treat documents as single large blocks, XML offers the opportunity to exploit the logical structure of documents in order to allow for more precise retrieval.

In this work, we are concerned with XML multimedia retrieval, where multimedia objects are referenced in XML documents. As a kind of hypermedia with controlled structure, XML multimedia documents are usually organized according to a hierarchical (tree) structure. We believe that exploiting this hierarchical structure can play an essential role in providing effective retrieval of XML multimedia documents, where indexing and retrieval is based on any textual content extracted from the XML documents.

An XML document can be viewed as a tree composed of nodes, i.e. XML elements. The root element corresponds to the document, and is composed of elements (i.e. its children elements), themselves composed of elements, etc., until we reach leaf elements (i.e. elements with no children elements). An XML multimedia element, which is an element that references in an attribute a multimedia object as an external entity, has a parent element, itself having a parent element, all of them constituting the ancestor elements of that multimedia element. It can also have its own (i.e. self) textual content, which is used to describe the referenced multimedia entity. Our aim is to investigate whether "hierarchically surrounding" textual XML elements of various sizes and granularities (e.g., self, parent, ancestor, etc.) in a document or any combination of them can be used for the effective retrieval of the multimedia objects of that document.

The exploitation of textual content to perform multimedia retrieval is not new. It has, for example, been used in multimedia web retrieval [1,4]. Our work follows the same principle, which is to use any available textual content to index and retrieve non-textual content; the difference here is that we are making use of the hierarchical tree structure of XML documents to delineate the textual content to consider.

M. Consens and G. Navarro (Eds.): SPIRE 2005, LNCS 3772, pp. 218–223, 2005.
© Springer-Verlag Berlin Heidelberg 2005

The paper is organized as follows. In Section 2, we describe our XML multimedia retrieval approach. In Section 3, we describe the test collection built to evaluate our approach. In Section 4, we present our experiments and results. Finally we conclude in Section 5.

## 2 Our Approach

A multimedia object is referenced as an external entity in the attribute of an XML element that is specifically designed for multimedia content. We call this element a multimedia element. Some textual content can appear within the element, describing (annotating) the multimedia object itself. The elements hierarchically surrounding the multimedia element can have textual content that provides additional description of the object. Therefore, the textual content within a multimedia element and the text of elements hierarchically surrounding it can be used to calculate a representation of the multimedia object that is capable of supplying direct retrieval of this multimedia data by textual (natural language) query.

We say that these hierarchically surrounding elements and the multimedia element itself form regions. The regions of a given multimedia object are formed upward following the hierarchical structure of the document containing that multimedia object: the self region, its sibling elements, its parent element, and so on; the largest region being the document element itself. We define the text content of the region used to represent the multimedia object as its region knowledge (RK).

As the elements of XML are organized in a hierarchical tree and nested within each other, the regions are defined as hierarchically disjoint. This is important to avoid repeatedly computing the text content. We therefore define N+1 disjoint RKs, where N is the maximum depth in the XML multimedia document collection:

- Self level RK: It is a sequence of one or more consecutive character information items in the element information item, which is a multimedia element in which the multimedia object is referenced as an external entity. This is the lowest level region knowledge of a given multimedia object.
- Sibling level RK: It is a sequence of one or more consecutive character information items in the element information items, which is at the same hierarchical level of the multimedia element and just before or after it.
- 1st ancestor level RK: It is a sequence of one or more consecutive character information items in the element information item, which is the parent element of the multimedia element, excluding those text nodes having been used for Self and Sibling RKs.
- ...
- Nth ancestor level RK: It is a sequence of one or more consecutive character information items in the element information item, which is the parent of the element of N-1th ancestor level RK, excluding those text nodes having been used for its lower level RKs.

The RKs are used as the basis for indexing and retrieving XML multimedia objects. At this stage of our work, we are only interested in investigating whether regions can indeed be used for effectively retrieving multimedia XML elements. Therefore, we use a simple indexing and retrieval method, where it is straightforward to perform experiments that will inform us on the suitability of our approach. For this purpose indexing is based on the standard tf-idf weighting and retrieval is based on the vector space model [3].

The weight of term t in the RK is given by the standard tf-idf, where idf, is computed across elements (and not across documents) as it was shown to lead to better effectiveness in our initial experiments. The weight of a term in the combination of RKs, which is then the representation of a given multimedia object, is calculated as the weighted sum of the tf-idf value of the term in the individual RKs. The weight is the importance associated with a given RK in contributing to the representation of the multimedia object.

## 3   The Test Collection

To evaluate the effectiveness of our proposed XML multimedia retrieval approach, we requires a test collection, with its set of XML documents containing non-textual elements, its set of topics, and relevance assessments stating which non-textual elements are relevant to which topics. We used the collection developed by INEX, the INitiative for the Evaluation of XML Retrieval [2], which consists of 12,107 articles, marked-up with XML, of the IEEE Computer Society's publications covering the period of 1995-2002, and totaling 494 mega-bytes in size. On average an article contains 1,532 XML elements, where the average depth of an element is 6.9 [2]. 80% of the articles contain at least one image, totaling to 81,544 images. There is an average of 6.73 images per articles. The average depth of an image is 3.62.

We selected six volumes of the INEX document set, in which the average number of images per XML document is higher than in others. Due to resource constraint, we restricted ourselves to those articles published in 2001. The resulting document collection is therefore composed of 7,864 images and 37 mega-bytes XML text contained in 522 articles. There is an average of 15.06 images per article. On average, the depth of an image ranges between 2 and 8, where the average depth is 3.92. In addition, we calculated the distribution of the images across various depths (levels). 62% of images are at depth 4, 23.5% of them have depth 3, 13.5% of them have depth 5. If an image has depth 4, its highest level RK is a 4th level RK, etc. In most cases, 4th level RK corresponds to the article excluding the body elements (it contains titles, authors and affiliation, i.e. heading information, classification keywords, abstracts); 3rd level RK corresponds to the body element excluding the section containing the image; 2nd level RK corresponds to the section excluding the sub-section containing the image; and 1st level RK corresponds to the sub-section excluding the caption and sibling RKs of the image.

The topics designed for this test collection are modified versions of 10 topics of the original INEX 2004 topics. These topics were chosen so that enough relevant XML elements (text elements) were contained in the 522 XML articles forming the created collection. These topics have a total of 745 relevant elements, with an average of 74.5 relevant elements per topic. Each topic was modified so that indeed images were searched for.

Our topics are based on actual INEX topics, for which relevance assessments are available. As such, for a given query, only articles that contained at least one relevant element were considered. This simplified greatly the relevance assessment process. The assessments were based on images and their captions, and performed by computer science students from our department following the standard TREC guidelines [5]. The relevance assessment identified a total of 199 relevant images (out of 7,864 images), and an average of 20 relevant images per topic.

# 4  Experiments, Results and Analysis

The purpose of our experiments is to investigate the retrieval effectiveness of the so-called RKs for retrieving multimedia objects referenced in XML elements, which in our test collection are image objects. Our experiments include self, sibling, and 1st ancestor level up to 6th ancestor level RKs, used independently or in combination to represent multimedia XML elements. We report average precision values for all experiments. The title component of the topics was used, stop-word removal and stemming were performed.

## 4.1  Individual RKs

Experiments were performed to investigate the types of RKs for retrieving image elements. The average precision values for self level, 1st level, ..., to 6th ancestor level RKs are, respectively: 0.1105, 0.1414, 0.1403, 0.2901, 0.2772, 0.2842, 0.0748, and 0.0009. Therefore using lower (self, sibling, 1st) level RK leads to low average performance. The reason could be two-fold: (1) the text content in self level RK are captions and titles that are small so the probability of matching caption terms to query terms is bound to be very low – the standard mismatch problem in information retrieval; and (2) captions tend to be very specific – they are there to describe the images, whereas INEX topics may tend to be more general, so the terms used in captions and the topics may not always be comparable in terms of vocabulary set.

The 2nd, 3rd, and 4th level RKs give the best performance. This is because they correspond to regions (1) not only in general larger, but also (2) higher in the XML structure. (1) seems to imply - obviously - that there is a higher probability to match query terms with these RKs, whereas (2) means that in term of vocabulary used, these RKs seems to be more suited to the topics. Furthermore, 2nd level RKs perform best, meaning usually the sections containing the images, and then 4th level RK performs second best, meaning the heading information, abstract and reference of the article containing the images.

Since a large number (i.e. 62%) of image objects in the INEX collection are within lower level elements (i.e. depth 4), the images within a document will have the same 4th level RK, i.e made of same abstract and heading elements. Our results seem to indicate that retrieving all the images of a document whose abstract and heading match the query is a better strategy than one based on exploiting text very near to the actual image.

Performance decreases when using higher levels RKs. This is because most images have a 4th level RK (since they have depth of 4), much fewer have 5th level RK

(13.5%), and less than 1% have a 6th level RKs. Thus nothing should be concluded from these results. We therefore do not discuss performance using these RKs.

We also looked at the amount of overlaps between the images retrieved using the various RKs (i.e. percentage of retrieved images that were also retrieved by another RK). Although not reported here, our investigation showed that a high number of images retrieved using the self RK are also retrieved using most of the other RKs. The reverse does not hold; many of the images retrieved using 2nd level RK are not retrieved using smaller RKs. This would indicate that higher level RKs have definitively an impact on recall. The "many but not all" is a strong argument for combining low level and high level RKs for retrieving multimedia objects.

## 4.2  Combination of RKs

This section investigates the combinations of various RKs for retrieving multimedia objects. The combinations are divided into three sets: (1) combinations from lower level up to higher level with the same weights for each participating level, i.e. self RK is combined with sibling RK, and together they are combined with 1st ancestor level RK, etc; (2) combinations of self, sibling, 1st up to 3rd level RKs, with the 4th ancestor level with the same weight for each participating level – level 4 RK was chosen as it led to good performance in Section 4.1 (although images within a document could be differentiated); and (3) combinations of all level RKs but with different weights to each level.

The average precision values for the first set are: 0.1832, 0.2329, 0.2748, 0.2897, and 0.3716. We can see that performance increases when a lower level RK is combined with an upper level RK. In addition, we can see that the combinations up to 4th level RK obtain much better performance than any single level (i.e. 2nd: 0.2901 and 4th: 0.2842). These results show clearly that combining RKs in a bottom-up fashion lead to better performance, as they indicate that the RKs seem to exhibit different (and eventually complementary) aspects, which should be combined for effective retrieval.

The average precisions for our second set of experiments are: 0.3116, 0.3061, 0.3336, 0.3512, and 0.3716, which are very comparable. We can see that by combining the self RK with the 4th ancestor level leads already to effectiveness higher than when using any single level RK. As discussed in Section 4.1, using the 4th level RK retrieves all the images in a document (as long as the RK matches the query terms), so our results show that using in addition lower level RKs - which is based on elements closer to the multimedia objects and thus will often be distinct for different images - should be used to differentiate among the images in the document.

To further justify our conclusion, that is to make sure that our results are not caused by the way our test collection was built, we looked at the relevant elements for the original topics in the INEX test collection: 3.7% of them are at the document level, whereas 81% have depth 3, 4 and 5. Therefore, this excludes the possibility that the document level RKs (4th level RK combined with all lower level RKs) lead to the best strategy for XML multimedia retrieval in our case, because they were the elements assessed relevant to the original topics. This further indicates that higher level RKs seem best to identify which documents to consider, and then using lower level RKs allows selecting which images to retrieve in those documents.

The last set of experiments aims to investigate if assigning different weights to different levels can lead to better performance. We did four combinations (including all RKs) and the average precisions for the four combinations are: 0.3904 (same weight to every level), 0.3796 (emphasize lower level RKs), 0.3952 (emphasize higher level RKs), and 0.3984 (emphasizes 2nd and 4th level RKs). The performances are better when weights are introduced - compared to previous experiments, although there is not a great difference with the various weights. However, this increase could also be due to the fact that the 5th and 6th ancestor level RKs are used, which corresponds for some (few) images to the abstract and heading elements, which were shown to lead to good performance.

## 5 Conclusions and Future Work

Our work investigates the use of textual elements to index and retrieve non-textual elements, i.e. multimedia objects. Our results, although based on a small data set, show that using elements higher in a document hierarchical structure works well in selecting the documents containing relevant multimedia objects, whereas elements lower in the structure are necessary to select the relevant images within a document. Our next step is to investigate these findings on larger and different data sets, as that being built by an XML multimedia track as INEX 2005.

## References

1. Harmandas, V., Sanderson, M., & Dunlop, M.D. (1997). Image retrieval by hypertext links. Proceedings of SIGIR-97, 20th ACM International Conference on Research and Development in Information Retrieval, Philadelphia, US, pp 296–303.
2. INEX. Initiative for the Evaluation of XML Retrieval http://inex.is.informatik.uni-duisburg.de/
3. Salton, G., & Buckley, C. (1988). Term-weighting approaches in automatic retrieval. Information Processing & Management, 24(5):513-523.
4. Swain, M. J., Frankel, C., and Athitsos, V. (1997). WebSeer: An image search engine for the World Wide Web. In Proceedings of the IEEE Conference on Computer Vision and Pattern Recognition (San Juan, Puerto Rico).
5. Text REtrieval Conference (TREC). http://trec.nist.gov/
6. XML (eXtensible Markup Language). http://www.w3.org/XML/

# Retrieval Status Values in Information Retrieval Evaluation

Amélie Imafouo and Xavier Tannier

Ecole Nationale Supérieure des Mines de Saint-Etienne,
158 Cours Fauriel - 42023 Saint-Etienne, Cedex 2, France
{imafouo, tannier}@emse.fr

**Abstract.** Retrieval systems rank documents according to their retrieval status values (RSV) if these are monotonously increasing with the probability of relevance of documents. In this work, we investigate the links between RSVs and IR system evaluation.

## 1 IR Evaluation and Relevance

Kagolovsk *et al* [1] realised a detailed survey of main IR works on evaluation. Relevance was always the main concept for IR Evaluation. Many works studied the relevance issue. Saracevic [2] proposed a framework for classifying the various notions of relevance. Some other works proposed some definitions and formalizations of relevance. All these works and many others suggest that there is no single relevance: relevance is a complex social and cognitive phenomenon [3].

Because of the collections growth nowadays, relevance judgements can not be complete and techniques like the pooling technique are used to collect a set of documents to be judged by human assessors. Some works investigated this technique, its limits and possible improvements [4].

To evaluate and classify IR systems, several measures have been proposed; most of them based on the ranking of documents retrieved by these systems, and ranking is based on monotonously decreasing RSVs. Precision and recall are the two most frequently used measures. But some others measures have been proposed (the Probability of Relevance, the Expected Precision, the E-measure and the Expected search length, etc). Korfhage [5] suggested a comparison between an IRS and a so-called ideal IRS. (the normalized recall and the normalized precision). Several user-oriented measures have been proposed (coverage ratio, novelty ratio, satisfaction, frustration).

## 2 IR Evaluation Measures and RSV

### 2.1 Previous Use of RSVs

Document ranking is based on the RSV given to each document by the IRS. Each IRS has a particular way to compute document RSV according to the IR

M. Consens and G. Navarro (Eds.): SPIRE 2005, LNCS 3772, pp. 224–227, 2005.

model on which it is based (0 or 1 for the Boolean model, $[0,1]$ for the fuzzy retrieval, $[0,1]$) or $\Re$ for the vector-space,etc). Little effort has been spent on analyzing the relationship between RSV and probability of relevance of documents. This relationship is described by *Nottelman et al.* [6] by a "normalization" function which maps the RSV onto the probability of relevance (linear and logistic mapping functions).

Lee [7] used a min-max normalization of RSVs and combined different runs using numerical mean of the set of RSVs of each run. *Kamps et al.* [8] and *Jijkoun et al.* [9] also used normalized RSVs to combine different kinds of runs.

## 2.2  Proposed Measures

We will use the following notation in the rest of this paper: $d_i$ is the document retrieved at rank $i$ by the system; $s_i(t)$ is, for a given topic $t$, the RSV that a system gives to the document $d_i$. Finally $n$ is the number of documents that are considered while evaluating the system.

We assume that all the scores are positive. Retrieved documents are ranked by their RSV and documents are given a binary relevance judgement (0 or 1).

RSVs are generally considered as meaningless system values. Yet we guess that they have stronger and more interesting semantics than the simple rank of the document. Indeed, two documents that have close RSVs are supposed to have close probabilities of relevance. In the same way, two distant scores suggest a strong difference in the probability of relevance, even if the documents have consecutive or close ranks. But the RSV scale depends on the IRS model and implementation. Different RSV scales should not act on the evaluation. Nevertheless, the relative distances between RSVs attributed by the same system are very significant; In order to free from the absolute differences between systems, we use a maximum normalization:

For a topic $t$, $\forall i \, s_i'(t) = \frac{s_i(t)}{s_1(t)}$. Thus, $\forall i \, s_i'(t)$, $s_i'(t) \in [0,1]$ and $s_i'(t) < s_{i+1}'(t)$.

$s_i'(t)$ gives an estimation by the system of the relative closeness of the document $d_i$ to the document considered as the most relevant by the system ($d_1$) for topic $t$. For $d_1$, $s_1' = 1$, we consider that $s_i = 0$ and $s_i' = 0$ for any non-retrieved document. We assume that a lower bound exists for the RSV and is equal to 0. If it is not the case we need to know (or to calculate) a lower bound and to perform a min-max normalization.

We propose a first pair of metrics, applicable to each topic; the figure $r$ determines a success rate while $e$ is a failure rate ($p_i$ is the binary assessed relevance of document $d_i$):

$$r_1(n) = \frac{\sum\limits_{i=1..n} s_i' \times p_i}{n} \text{ and } e_1(n) = \frac{\sum\limits_{i=1..n} s_i' \times (1 - p_i)}{n}$$

$r_1(n)$ (resp $e_1$) is the average normalized RSV (NRSV) considering only the relevant documents (resp non relevant documents). The second proposed pair of metrics is derived from $r_1$ and $e_1$:

$$\begin{cases} r_2(n) = \underbrace{\dfrac{\sum\limits_{i=1..n}(1-s_i')\times(1-p_i)}{n}}_{r_{2,1}} + \dfrac{\sum\limits_{i=1..n}s_i'\times p_i}{n} \\[3em] e_2(n) = \underbrace{\dfrac{\sum\limits_{i=1..n}(1-s_i')\times p_i}{n}}_{e_{2,1}} + \dfrac{\sum\limits_{i=1..n}s_i'\times(1-p_i)}{n} \end{cases}$$

$r_{2,1}(n)$ is a distance representing the estimation by the system of the "risk" of non relevance for the document. $e_{2,1}(n)$ is equivalent to $r_{2,1}(n)$ for relevant document. Documents with high NRSVs have a high influence on these metrics by increasing $r_i$ (if they are relevant) and by penalizing the system through $e_i$ (if they are not relevant).

A new problem arises at this step, if a document $d_i$ is assessed as relevant, it seems difficult to evaluate the system according to $s_i$. Indeed the assessor cannot say *how much* the document is relevant (in the case of binary judgment). One does not know if the confidence of the system was justified, whether this confidence was strong (high NRSV) or not (low NRSV). We can also notice that if a system retrieves $n$ relevant documents (out of $n$), the success rates $r_1$ and $r_2$ will be less than 1, which is unfair. Thus we propose a new measure

$$r_3(n) = \frac{\sum\limits_{i=1..n}p_i + \sum\limits_{i=1..n}(1-s_i)(1-p_i)}{n}$$

Any relevant document retrieved contributes to this measure for 1, and a non relevant document contributes by its distance to the top ranked document.

Measures $r_1$ and $r_2$ can be useful when comparing two IRSs, because they favor systems that give good RSVs to relevant documents. On the other hand, $r_3$ may allow a more objective evaluation of a single system performances.

## 3   Experiments

We experimented on TREC9 WebTrack results (105 IRSs). We used a correlation based on Kendall's $\tau$ in order to compare our measures with classical IR evaluation measures. IPR stands for Interpolated Precision at Recall level.

The ranking obtained with the measure $r_1$ which is based on the normalized RSV for relevant documents is highly correlated with precision on the first documents retrieved ($P@N$). This correlations decreases as $N$ increases.

Conversely, the ranking obtained with the $e_1$ which is based on the normalized RSVs for non relevant documents is inversely correlated with $P@N$ and with IPR at first recall levels (this was excepted, since $e_1$ represents a failure rate).

The measures $r_2$ (resp. $e_2$) that combines NRSVs for relevant documents (resp. for non relevant documents) with a value expressing the distance between

**Table 1.** Kendall tau between IRS ranking

| - | IPR at 0 | IPR at 0.1 | IPR at 0.2 | IPR at 1 | MAP | $P@5$ | $P@10$ | $P@100$ | $P@1000$ |
|---|---|---|---|---|---|---|---|---|---|
| $r_1$ | 0.92 | 0.83 | 0.80 | 0.87 | 0.81 | 0.90 | 0.83 | 0.64 | 0.53 |
| $e_1$ | −0.50 | −0.06 | 0.18 | 0.59 | 0.52 | −0.61 | −0.43 | −0.14 | −0.11 |
| $r_2$ | 0.31 | 0.29 | 0.31 | 0.46 | 0.55 | 0.71 | 0.50 | 0.15 | 0.20 |
| $e_2$ | −0.51 | −0.064 | 0.20 | 0.59 | 0.59 | −0.68 | −0.47 | −0.09 | −0.09 |
| $r_3$ | 0.31 | 0.31 | 0.33 | 0.46 | 0.54 | 0.64 | 0.46 | 0.18 | 0.21 |

non relevant documents (resp. relevant documents) and the first document are less (resp. less inversely) correlated with $P@N$ and with IPR at first recall levels.

The measure $r_3$ that combines contribution from relevant documents retrieved (1) and contribution from irrelevant documents retrieved (a value that expresses the way the IRS valuate the risk of mistaking when ranking this irrelevant documents at a given position) is even less correlated with $P@N$ and with IPR at first recall levels.

# 4  Conclusion

RSV is used to rank the retrieved documents. Despite this central place, it is still considered as a system value with no particular semantics. We proposed IR measures directly based on normalized RSVs. Experiments on the TREC9 results show a high correlation between these measures and some classical IR evaluation measures. These correlations indicate possible semantics besides documents RSVs. The proposed measures are probably less intuitive than precision and recall but they put forth the question of the real place of RSV in IR evaluation.

# References

[1] Kagolovsk, Y., Moehr, J.: Current status of the evaluation in information retrieval. Journal of medical systems **27** (2003) 409–424
[2] Saracevic, T.: Relevance: A review of and a framework for the thinking on the notion in information science. JASIS **26** (1975) 321–343
[3] Mizzaro, S.: How many relevances in information retrieval? Interacting with Computers **10** (1998) 303–320
[4] Zobel, J.: How reliable are the results of large scale information retrieval experiments. In: Proceedings of ACM SIGIR'98. (1998) 307–314
[5] Korfhage, R.: Information storage and retrieval. Wiley Computer publising (1997)
[6] Nottelman, H., Fuhr, N.: From retrieval status value to probabilities of relevance for advanced ir applications. Information retrieval **6** (2003) 363–388
[7] Lee, J.H.: Combining multiple evidence from different properties of weighting schemes. In: Proceedings of SIGIR '95. (1995) 180–188
[8] Kamps, J., Marx, M., de Rijke, M., Sigurbjrnsson, B.: The importance of morphological normalization for xml retrieval. In: Proceedings of INEX'03. (2003) 41–48
[9] Jijkoun, V., Mishne, G., Monz, C., de Rijke, M., Schlobach, S., Tsur, O.: The university of amsterdam at the trec 2003 question answering track (2003)

# A Generalization of the Method for Evaluation of Stemming Algorithms Based on Error Counting

Ricardo Sánchez de Madariaga, José Raúl Fernández del Castillo, and José Ramón Hilera

Dept. of Computer Science, University of Alcalá, 28805 Madrid, Spain

**Abstract.** Until the introduction of the *method for evaluation of stemming algorithms based on error counting*, the effectiveness of these algorithms was compared by determining their retrieval performance for various experimental test collections. With this method, the performance of a stemmer is computed by counting the number of identifiable errors during the stemming of words from various text samples, thus making the evaluation independent of Information Retrieval. In order to implement the method it is necessary to group *manually* the words in each sample into disjoint sets of words holding the same semantic concept. One single word can belong to only one concept. In order to do this grouping *automatically*, in the present work this constraint has been generalized, allowing one word to belong to several different concepts. Results with the generalized method confirm those obtained by the non-generalized method, but show considerable less differences between three affix removal stemmers. For first time evaluated four letter successor variety stemmers, these appear to be slightly inferior with respect to the other three in terms of general accuracy (ERRT, error rate relative to truncation), but they are weight adjustable and, most important, need no linguistic knowledge about the language they are applied to.

## 1 Introduction

Stemming algorithms are widely used to improve the efficiency of Information Retrieval (IR) Systems [1]. They are used to merge together words which share the same stem but differ in their endings [2]. The idea is to remove suffixes to produce the same stem, which is supposed to represent the basic concept held by a word. Thus, different words holding similar concepts are reduced to a common "stem" and can be treated as equivalent [3] by the IR system, resulting in improved retrieval performance.

No stemmer following one of the four automatic approaches [4] can be expected to work perfectly, since natural language is highly irregular. This results in the fact that semantically distinct words are merged to the same stem (this is called an *overstemming error*). Conversely there are many other cases in which words with the same meaning are reduced to different stems (an *understemming error*) [2]. Since different algorithms commit different errors the question raises as whether an algorithm can be said to be "better" than another one.

M. Consens and G. Navarro (Eds.): SPIRE 2005, LNCS 3772, pp. 228–233, 2005.

Until the introduction by Paice of the *method for evaluation of stemming algorithms based on error counting* [2], the effectiveness of these algorithms was evaluated mainly by their retrieval performance in IR systems. Moreover, in the vast majority of these studies there were difficulties to prove significant differences between the retrieval performance of different stemmers [5] [4]. Furthermore, IR performance measures provide no evidence about the detailed behaviour of a stemmer, preventing the design of better stemmers in the future. Finally, it must be said that stemming is useful not only in IR, but also in many other natural language applications like a semantic frame filler or an intelligent command interface [2].

In the method introduced by Paice a stemmer is evaluated by counting the actual understemming and overstemming errors which it makes. This approach makes stemmer evaluation independent of IR. On the other hand, it requires that the "correct" merging are made explicit, i.e. all the words of the word samples used to test the stemmer need to be grouped in sets of words holding the same or similar concepts. This means that the correct "merging are defined by human intellectual decision, which naturally raises questions about the consistency of the standard". Besides the "questions about the consistency" this means that, in order to evaluate a stemmer on a specific corpus, all the words in that corpus must be grouped manually into disjoint sets of words holding the same sense or concept. In order to avoid this manual grouping step and do it automatically we would need to use an automatic disambiguation procedure which would obtain the actual sense of every word in the given corpus. But in this case, why not use the automatic disambiguation procedure to obtain the stems i.e. to perform the stemming rather than evaluating it? Moreover, suppose that we have little knowledge about the language we are performing the stemming for. In this case we would have no evident possibility of implementing the automatic disambiguation procedure to perform the sophisticated stemmer. In this paper we are assuming that we have no knowledge about the language the stemmer is being applied to, or else, for some reason, we are just applying a simple, i.e. not sophisticated stemmer, but do not want to do the groupings manually for each distinct corpus.

Instead of obtaining the actual sense of every word in the corpus, manually or automatically, we let it have all its possible senses as they appear in one dictionary or thesaurus for the given language. So instead of assigning one concept group to each word in the corpus, these are allowed to belong to several concept groups as usual. This generalization imposes some changes when computing the *under- and overstemming indexes* as in the ERRT method, which are discussed in the following sections.

## 2 The Method Based on Error Counting

### 2.1 Understemming Errors

Given a concept group $g$ of $Ng$ words morphologically related (i.e. sharing the same stem) and holding the same semantic concept, suppose that the algorithm reduces all $Ng$ words to $fg$ distinct stems and each stem $i$ is obtained in $ngi$ cases. In this case we can compute the "unachieved merge total" $UMTg$ for such a group, which represents the total number of undertstemming errors, from the following formula:

$$UMT_g = \frac{1}{2} \sum_{i=1}^{f_g} n_{gi} \left( N_g - n_{gi} \right) \; . \tag{1}$$

Since the total number of possible committed errors in this group, called the "desired merge total" $DMTg$ can be computed by

$$DMT_g = \frac{1}{2} N_g \left( N_g - 1 \right) \tag{2}$$

we can compute the *understemming index UI* as the ratio $UI = GUMT/GDMT$ where $GUMT$ is the "global unachieved merge total" and $GDMT$ is the "global desired merge total" and are obtained by adding over all word groups $g$.

## 2.2 Overstemming Errors

Every distinct stem produced by the stemmer defines now a "stem group" whose members are derived from a number of different original words. Suppose that the stem group $s$ contains stems derived from $fs$ different concept groups. The total number of overstemming errors can be computed as a "wrongly-merged total" $WMTs$ for the stem following the formula

$$WMT_s = \frac{1}{2} \sum_{i=1}^{f_s} n_{si} \left( N_s - n_{si} \right) \tag{3}$$

where $Ns$ is the total number of stems in the stem group and $nsi$ is the number of stems derived from the $i$th concept group.

The overstemming index is normalized according to the actual number of word pairs which became identical after stemming. Following this, we can define an "actual merge total" $AMTs$ given by

$$AMT_s = \frac{1}{2} N_s \left( N_s - 1 \right) \tag{4}$$

## 2.3 ERRT (Error Rate Relative to Truncation)

The performance of a stemmer in terms of the number of errors may depend on its concrete application. For one application, a light stemmer (few overstemming errors) may be preferable than a heavy stemmer (few understemming errors), or viceversa. In order to measure performance taking both opposite numbers into account, a common baseline can be formed using the simple stemmers obtained by the *truncation* of words to a fixed maximum length. Values of $UI$ and $OI$ can be calculated for each of these stemmers and plotted into a two dimensional axis to form a *truncation line*. Given a stemmer for evaluation with its point $\mathbf{P}$ and point **X** as the intersection of line **OP** with the truncation line, its performance can be computed as

$$ERRT = (lengthOP) / (lengthOX) \ . \tag{5}$$

Moreover, it can be seen that the gradient of the **OP** line is a measure of the weight or strength of the stemmer, i.e. the relation between the number of over- and understemming errors: a strong or aggressive stemmer tends to produce more overstemming and less understemming errors than a light stemmer.

## 3   The Generalized Method

In the method outlined in the previous section the correct merging for the sample words must be made explicit in order to count the number of errors committed by the stemmer.

In order to calculate *UI* using Equations (1) and (2) one word can belong to only one concept group. In the generalized ERRT method, one word can belong to (many) different concept groups.

When calculating *OI* using Equations (3) and (4) every word belonging to each stem group also belongs to only one concept group. In the generalized method, one single word is allowed to belong to as many concept groups as necessary. Instead of doing this as a manual process, the classification is obtained directly from an English language thesaurus [6]. Clearly the manual reclassification approach would have made the method word sample adaptive, but its costs, in terms of effort for each evaluation, is also almost prohibitive.

In the English thesaurus used in the present work, every entry, which is a lemmatized word, is followed by a list of numbers that represent important concepts or "categories" it belongs to. A typical entry has a list of about 10 categories it belongs to. Of course a special file [7] was used to lemmatize the declined words in the corpora.

### 3.1   Understemming Errors

In order to calculate the understemming index *UI*, Equations (1) and (2) are still applicable. Now one single (declined) word form can belong to more than one concept group, which means that its corresponding stem $i$, obtained in $ngi$ cases, may appear also in more than one concept group (strictly speaking, this could happen in the non-generalized method too, but in much fewer cases).

### 3.2   Overstemming Errors

In this case Equation (3) needs to be rewritten, in order to reflect the fact that now each original (declined) word form corresponding to each entry in one stem group may belong to many distinct concept groups. The generalized formula is:

$$WMTG_s = \frac{1}{2} \sum_{i=1}^{g_s} (N_s - n_{gsi}) \tag{6}$$

where $Ns$ is the number of entries in each stem group $s$. Each entry in $s$ now has a set of concept groups for which it belongs to, instead of just one. If this set is called the "set of concept groups", then $gs$ is the total number of distinct sets of concept groups in $s$, and $ngsi$ is the number of entries in $s$ which belong to a set of concept groups which has at least one concept group in common with set of concept groups $i$.

## 4    Results

Table 1 shows results for three word sources and seven stemmers [8][9][10][11]. The Hafer stemmers are four variants of experiment number 11 reported by Hafer and Weiss in [11].

**Table 1.** ERRT values(%) and stemming weights for seven stemmers and corpora CISI Test Collection (A), MULTEXT JOC Corpus (B) and BLLIP WSJ Corpus (C)

|  | Word source A | | Word source B | | Word source C | |
|---|---|---|---|---|---|---|
|  | ERRT | SW | ERRT | SW | ERRT | SW |
| Porter | 52.7 | 0.293 | 48.1 | 0.290 | 50.6 | 0.205 |
| Lovins | 52.6 | 0.454 | 44.3 | 0.422 | 44.2 | 0.365 |
| Paice/Husk | 50.3 | 0.569 | 44.0 | 0.578 | 44.4 | 0.490 |
| Hafer_2_16 | 60.6 | 0.304 | 53.8 | 0.303 | 50.9 | 0.288 |
| Hafer_2_13 | 58.8 | 0.392 | 56.4 | 0.384 | 47.7 | 0.368 |
| Hafer_2_10 | 62.2 | 0.445 | 52.0 | 0.460 | 50.5 | 0.451 |
| Hafer_2_07 | 69.9 | 0.554 | 58.3 | 0.596 | 56.4 | 0.599 |

## 5    Conclusions

We have achieved an automatic IR-independent method for evaluation of stemming algorithms which does not need to perform the equivalence relation among every word of the corpus manually, allowing one word to have several different meanings as usual. The alternative to this is manual classification of every word from the corpus taking into account the actual sense of each word in the corpus. This is done in the non-generalized (ERRT) method reported in [2]. In order to avoid the manual step of the ERRT method, an automatic semantic disambiguation procedure could be used to find the actual sense of each word in the corpus [12]. In this case, why not use the automatic disambiguation to perform the stemming instead of evaluating it? This would really be a case of use of an evaluating method much more sophisticated than the evaluated one. Moreover, it could be used to rebuild the stemmer. And there are many cases in which there is no reasonable access to such a tool.

The results obtained by the generalized ERRT method confirm those obtained by the non-generalized ERRT method for the three affix removal stemmers in terms of stemming weight and in general accuracy (ERRT). However, in the generalized method the three affix removal stemmers show considerable less difference among them than in the non-generalized one – from 6% to 26%.

As in the non-generalized method, the Paice/Husk stemmer performs better than the other two, but the Lovins stemmer obtains slightly better results in the generalized method with respect to the non-generalized approach. This result must take into account that the weight of these stemmers is very different, and hence are not strictly comparable. The successor variety stemmers are evaluated with the (generalized) ERRT method for the first time, and seem to show slightly lower performance than the affix removal algorithms in terms of general accuracy. However, the former are easily weight-adjustable, cover a wider spectrum and, most important, do not need of linguistic knowledge about the language the corpus is written in.

# References

1. Kowalski, G., Maybury, M.T.: Information storage and retrieval. Theory and Implementation. Kluwer Academic Publishers (2000)
2. Paice, C.D.: Method for evaluation of stemming algorithms based on error counting. Journal of the American Society for Information Science **47** (8) (1996) 632–649
3. Salton, G., McGill, M.: Introduction to modern information retrieval. McGraw-Hill New York (1983)
4. Frakes, W., Baeza-Yates, R.: Information Retrieval: data structures and algorithms. Prentice-Hall Englewood Cliffs NJ (1992)
5. Lennon, M., Pierce, D. S., Tarry, B. D., Willet, P.: An evaluation of some conflation algorithms for information retrieval. Journal for Information Science **3** (1981) 177–183
6. Kirkpatrick, B. (ed.): Rogets thesaurus of English words and phrases. Penguin Books (2000)
7. MULTEXT project: MULTEXT lexicons. Centre National de la Recherche Scientifique (1996-1998)
8. Lovins, J. B.: Development of a stemming algorithm. Mechanical Translation and Computational Linguistics **11** (1968) 22–31
9. Porter, M.F.: An algorithm for suffix stripping. Program **14** (1980) 130–137
10. Paice, C.D.: Another stemmer. SIGIR Forum **24** (1990) 56–61
11. Hafer, M., Weiss, S.: Word segmentation by letter successor varieties. Information Storage and Retrieval **10** (1974) 371–385
12. Goldsmith, J.A., Higgins, D., Soglasnova, S.: Automatic language-specific stemming in information retrieval. LNCS **2069** Cross-Language Information Retrieval and Evaluation: Workshop CLEF 2000 C.Peters (Ed.) (2000) 273–284

# Necklace Swap Problem for Rhythmic Similarity Measures

Yoan José Pinzón Ardila[1], Raphaël Clifford[2,*], and Manal Mohamed[1]

[1] King's College London, Department of Computer Science,
London WC2R 2LS, UK
Yoan.Pinzon@kcl.ac.uk, manal@dcs.kcl.ac.uk
[2] University of Bristol, Department of Computer Science, UK
raphael@clifford.net

**Abstract.** Given two $n$-bit (cyclic) binary strings, $A$ and $B$, represented on a circle (necklace instances). Let each sequence have the same number $k$ of 1's. We are interested in computing the cyclic swap distance between $A$ and $B$, *i.e.*, the minimum number of swaps needed to convert $A$ to $B$, minimized over all rotations of $B$. We show that this distance may be computed in $O(k^2)$.

**Keywords:** repeated patterns, music retrieval, swap distance, cyclic strings, rhythmic/melodic similarity.

## 1  Introduction

Mathematics and music theory have a long history of collaboration dating back to at least Pythagoras[1] [9]. More recently the emphasis has been mainly on analysing string pattern matching problems that arise in music theory [2,3,4,5,6,7].

A fundamental problem of both theoretical and practical importance in music information retrieval is that of comparing arbitrary pieces of music. Here we restrict our attention to rhythm similarity, *i.e.* to what extent is rhythm $A$ similar to rhythm $B$? Long term goals of this research include content-based retrieval methods for large musical databases using such techniques as *query-by-humming* (QBH) [10,12] and finding music copyright infringements [8].

In geometry and other branches of mathematics, we often measures of the similarity of two objects that are in the same class but no identical. For example, the relative similarity of two real numbers can be computed as the difference, or the square of their differences. The similarity of two functions over some period might be computed as the unsigned integral between them over this period. We can say that two pieces of music are similar if their melody or rhythm are similar.

Six examples of 4/4 time clave and bell timelines are given in Fig. 1. Rhythms are usually notated for musicians using the standard western music notation (*see*

---

\* This work was carried out while the author was at King's College London.
[1] In the 5th century BC, Pythagoreas was quoted to have said,"*There is geometry in the humming of strings. There is music in the spacing of the spheres*".

M. Consens and G. Navarro (Eds.): SPIRE 2005, LNCS 3772, pp. 234–245, 2005.
© Springer-Verlag Berlin Heidelberg 2005

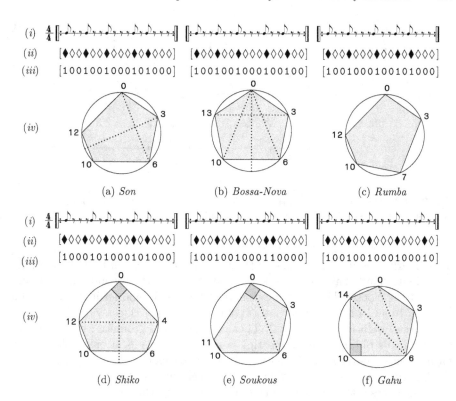

**Fig. 1.** Six fundamental 4/4 time *clave* rhythms. Each rhythm is depicted using (*i*) a standard western notation system, (*ii*) a box notation, (*iii*) a binary representation and (*iv*) a common geometric representation using convex polygons. The dotted lines indicate an axis of mirror symmetry (e.g. for the clave *son*, if the rhythm is started at location 3 then it sounds the same whether it is played forward or backwards, thus, clave *son* is a palindrome rhythm). It can be noticed that *son* rhythm is more like the others, thus offering an explanation for its world wide popularity.

Fig. 1(a–f)(*i*). A more popular way of representation is called the Box Notation Method intended for percussionists that do not read music. It was developed by Philip Harland and it is also known as TUBS (Time Unit Box System). The box notation method is convenient for simple-to-notate rhythms like bell and clave patterns as well as for experiments in the psychology of rhythm perception, where a common variant of this method is simply to use one symbol for the note (*e.g.* ♦) and another for the rest (*e.g.* ◊), For our purpose, A rhythm is represented as a cyclic binary sequence where a zero denotes a rest (silence) and a one represents a beat or note onset, for example, the clave *son* would be written as the 16-bit binary sequence[2]: [1001001000101000]. An even better representation for such cyclic rhythms is obtained by imagining a clock with 16

---

[2] This rhythm can also be thought as a point in a 16-dimensional space (the hypercube).

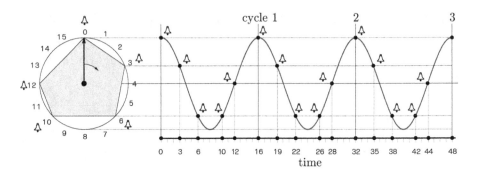

**Fig. 2.** 16-hours cycle clock representation of the clave *son* rhythm. The end of one cycle is the same spatial position as the beginning of the next.

hours marked on its face instead of the usual 12. Let us think that the hour and the minute hands have been broken off so that only the second-hand remains. Now set the clock ticking starting at noon (16 O'clock) and let it strike a bell at the 3, 6, 10 and 12 position for a total of five strikes per clock cycle. These times are marked with a bell in Fig. 2. Thereof, a common geometric representation of rhythms is obtained by connecting consecutive note locations with edges to form a convex polygon inscribed in our imaginary clock (*see* Fig. 1(a–f)(*iv*)).

A natural measure of the difference between two rhythms represented as binary sequences is the well known Hamming distance, which counts the number of positions in which the two rhythms disagree. Although the Hamming distance measures the existence of a mismatch, it does not measure how far the mismatch occurs, that is why, Toussaint [15] proposed a distance measure termed the swap distance. A swap is an interchange of a one and a zero (note duration and rest interval) that are adjacent in the sequence. The swap distance between two rhythms is the minimum number of swaps required to convert one rhythm to the other. The swap distance measure of dissimilarity was shown in [14] to be more appropriate than several other measures of rhythm similarity including the Hamming distance, the Euclidean interval-vector distance, the interval-difference distance measure of Coyle and Shmulevich, and the chronotonic distance measures of Gustafson and Hofmann-Engl.

In this paper we aim to find an efficient algorithm to measure the difference between two rhythms represented as binary sequences using the swap distance. More formally, given two $n$-bit (cyclic) binary strings, $A$ and $B$, represented on a circle (necklace instances). Let each sequence have the same number $k$ of 1's. We are interested in computing the cyclic swap distance between $A$ and $B$, *i.e.*, the minimum number of swaps needed to convert $A$ to $B$, minimized over all rotations of $B$. We show that this distance may be efficiently and elegantly computed in $O(k^2)$.

The outline of the paper is as follows: Some preliminaries are described in Section 2. An $O(k^3)$ solution is presented in Section 3 followed by a more efficient $O(k^2)$ solution in Section 4. Conclusion is drawn in Section 5.

## 2   Preliminaries

Let $X[0..n-1]$ be a necklace (circular string) of length $n$ over $\Sigma = \{0,1\}$. By $X[i]$ we denote the $(i+1)$-st bit in $X$, $0 \leq i < n$. We also denote by $k$, the number of 1's in $X$. Let $x = (x_0, x_1, \ldots, x_{k-1})$ be the *compressed* representation of $X$, such that $X[x_i] = 1$ for $0 \leq i < k$. For some integer $r$, let $x^{\langle r \rangle}$ be the *r-inverted-rotation* of $x$ such that $x_i^{\langle r \rangle} = x_{i \ominus r}$ for $0 \leq i < k$ and $i \ominus r = \mathrm{mod}(i - r, k)$. If $X = [10000100010001001]$, for example, $x = (0, 5, 9, 13, 16)$, $x^{\langle 1 \rangle} = (5, 9, 13, 16, 0)$, and $x^{\langle 3 \rangle} = (13, 16, 0, 5, 9)$.

We define a *mapping* $\pi{:}\{1, \ldots, k\} \to \{1, \ldots, k\}$ such that $\pi$ is a bijective (both onto and 1-1) function. We will show in Lemma 3 that only mappings that don't cross should be considered. We define the non-crossing mappings $\pi^0, \ldots, \pi^{k-1}$ as follows

$$\pi^h(i) = (i + h) \bmod k, \quad \text{for } 0 \leq i, h < k. \tag{1}$$

For $n = 5$, Fig. 3 graphically illustrates the $\pi^h$ mappings for $0 \leq h < 5$. Note that these mappings have the property that their arrows never cross.

We define the *median* of a sorted sequence $x = (x_0, \ldots, x_{k-1})$ as follows:

$$x_{\mathrm{med}} = \begin{cases} x_{(k-1)/2} , & k \text{ odd,} \\ x_{\lfloor (k-1)/2 \rfloor}, & k \text{ even.} \end{cases} \tag{2}$$

Note that, when $k$ is even, there are actually two medians, occurring at $\lfloor (k+1)/2 \rfloor$ *(lower median)* and $\lceil (k+1)/2 \rceil$ *(upper median)*. For simplicity, Eq. 2 considers the lower median as "the median" when $k$ is even.

Let $x = (x_0, x_1, ..., x_{k-1})$ and $y = (y_0, y_1, ..., y_{k-1})$ be two compressed representation of $X$ and $Y$, respectively. Then the *manhattan distance* $L_1(x, y)$ is defined as follows:

$$L_1(x, y) = \sum_{i=0}^{k-1} |x_i - y_i| \tag{3}$$

*Problem 1.* Given $X$ and $Y$, two necklaces both of length $n$ and same number of 1's, the MINIMUM NECKLACE SWAP PROBLEM is to find the cyclic swap distance between $X$ and $Y$, *i.e.*, the minimum number of swaps needed to convert $X$ to $Y$, minimized over all rotations of $Y$. A swap is an interchange of a one and a zero that are adjacent in the binary string.

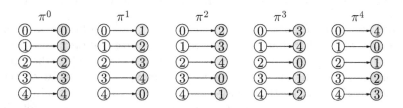

**Fig. 3.** $\pi^h$-mappings for $0 \leq h < 5$

## 3    An $O(k^3)$ Algorithm

The naive approach is to examine each mapping and calculate for each possible rotation the sum of number of swap operations between each pair of mapped 1's. This approach costs $O(nk^2)$ time. This is because there are $k$ possible mappings (*cf.* Lemma 3) and $n$ possible rotations per each mapping. The question is: Do we really need to examine all possible $n$ circular shifts for each mapping? Lemma 2 suggests that only $k$ circular shifts need to be checked for each mapping. This gives a total cost of $O(k^3)$ time. The algorithm works as follows:

Let $u$ and $v$ be the *rest-interval* sequences for $x$ and $y$, *resp.*, define as follows

$$u_i = \begin{cases} x_{i+1} - x_i & , \text{ if } 0 \le i < k-1 \\ n - x_{k-1} + x_0, & \text{if } i = k-1 \end{cases} \tag{4}$$

and

$$v_i = \begin{cases} y_{i+1} - y_i & , \text{ if } 0 \le i < k-1 \\ n - y_{k-1} + y_0, & \text{if } i = k-1 \end{cases} \tag{5}$$

Using this representation we can now compute the following sequences

$$x_i^{[h]} = \begin{cases} u_i^{\langle h \rangle} & , \text{ if } i = 0 \\ (x_{i-1}^{[h]} + u_i^{\langle h \rangle}) \bmod n, & \text{if } 0 < i < k \end{cases} \tag{6}$$

and

$$y_i^{[h]} = \begin{cases} v_i^{\langle h \rangle} & , \text{ if } i = 0 \\ (y_{i-1}^{[h]} + v_i^{\langle h \rangle}) \bmod n, & \text{if } 0 < i < k \end{cases} \tag{7}$$

with the characteristic that for $x^{[i]}$ and $y^{[j]}, 0 \le i, j < k$, the $(i+1)$st 1-bit in $X$ coincides with the $(j+1)$st 1-bit in $Y$.

*Example 1.* Let $x = (1, 6, 9, 12, 13)$ and $y = (0, 3, 4, 10, 16)$. To understand the meaning $u$ and $v$ we first show the representation of $x$ and $y$ as bit string $X$ and $Y$ in Table 1. Note that the 1's are numbered from 0 to $k$, so for example the 1st and 2nd bit in $X$ are located at positions 1 and 6. So, it'll be easier now to understand the *interval* sequences $u = (5, 3, 3, 1, 5)$ which corresponds to the intervals where there is a rest (0's) between two strokes (1's). In other words $u$ stores the gaps between the 1's in $X$. In the same way, we find $v = (3, 1, 6, 6, 1)$.

Now let's understand the meaning of, lets say, $x^{[3]}$. To find $x^{[3]}$, according with Eq. 6, we need to find $u^{\langle 3 \rangle}$, thus the *3rd-inverted-rotation* of $u$. So since $u = (5, 3, 3, 1, 5)$, then $u^{\langle 3 \rangle} = (1, 5, 5, 3, 3)$ (*i.e.* $u$ was rotated to the left by 3 positions). Notice that, by doing this we have moved the gap between the 4th and the 5th 1's in $X$, by three positions to the left. Using $u^{\langle 3 \rangle}$ and (6) we get $x^{[3]} = (1, 6, 11, 14, 0)$. In plain English words, $x^{[3]}$ corresponds to a rotation of $X$ such that its 4th 1-bit is set at location 0; this is equivalent to shifting $X$ to the left by 12 places. Table 1 also shows the representation of $Y^{[1]}$ which has

**Table 1.** Illustration of $x^{[3]}$ and $y^{[1]}$ computation for $x = (1, 6, 9, 12, 13)$ and $y = (0, 3, 4, 10, 16)$. $u = (5, 3, 3, 1, 5)$ and $v = (3, 1, 6, 6, 1)$

| $i$ | 0 | 1 | 2 | 3 | 4 | 5 | 6 | 7 | 8 | 9 | 10 | 11 | 12 | 13 | 14 | 15 | 16 |
|---|---|---|---|---|---|---|---|---|---|---|---|---|---|---|---|---|---|
| $X$ | 0 | $1^0$ | 0 | 0 | 0 | 0 | $1^1$ | 0 | 0 | $1^2$ | 0 | 0 | $1^3$ | $1^4$ | 0 | 0 | 0 |
| $Y$ | $1^0$ | 0 | 0 | $1^1$ | $1^2$ | 0 | 0 | 0 | 0 | 0 | $1^3$ | 0 | 0 | 0 | 0 | 0 | $1^4$ |
| $X^{[3]}$ | $1^3$ | $1^4$ | 0 | 0 | 0 | 0 | $1^0$ | 0 | 0 | 0 | 0 | $1^1$ | 0 | 0 | $1^2$ | 0 | 0 |
| $Y^{[1]}$ | $1^1$ | $1^2$ | 0 | 0 | 0 | 0 | 0 | $1^3$ | 0 | 0 | 0 | 0 | 0 | $1^4$ | $1^0$ | 0 | 0 |

**Table 2.** Computation of $L_1(x^{[i]}, y^{[j]})$ for $0 \leq i, j < 5$. $x = (1, 6, 9, 12, 13)$, $y = (0, 3, 4, 10, 16)$, $u = (5, 3, 3, 1, 5)$ and $v = (3, 1, 6, 6, 1)$

| $v^{(j)} \rightarrow y^{[j]}$  \  $u^{(i)} \rightarrow x^{[i]}$ | 5,3,3,1,5 → 5,8,11,12,0 | 3,3,1,5,5 → 3,6,7,12,0 | 3,1,5,5,3 → 3,4,9,14,0 | 1,5,5,3,3 → 1,6,11,14,0 | 5,5,3,3,1 → 5,10,13,16,0 |
|---|---|---|---|---|---|
| 3,1,6,6,1 → 3,4,10,16,0 | 11 | 9 | 3 | 7 | 11 |
| 1,6,6,1,3 → 1,7,13,14,0 | 9 | 11 | 9 | 3 | 9 |
| 6,6,1,3,1 → 6,12,13,16,0 | 11 | 19 | 17 | 15 | 3 |
| 6,1,3,1,6 → 6,7,10,11,0 | 4 | 8 | 10 | 10 | 12 |
| 1,3,1,6,6 → 1,4,5,11,0 | 15 | 7 | 9 | 11 | 23 |

the property of having the 2nd 1-bit in $Y$ moved at position 0. Note that if we compute $L_1(x^{[3]}, y^{[1]})$ we compute the swap distance for $\pi^3$ (*cf.* Fig. 3).

The minimum necklace swap problem is equivalent to calculating

$$s^* = \min_{0 \leq i, j < k} L_1(x^{[i]}, y^{[j]}).  \tag{8}$$

*Example 2.* For $x = (1, 6, 9, 12, 13)$ and $y = (0, 3, 4, 10, 16)$ Table 2 shows the computation of $L_1(x^{[i]}, y^{[j]})$ for $0 \leq i, j < 5$. 3 was the minimum swap distance given by, for example, (3,4,9,14,0) and (3,4,10,16,0). This is also the number of swaps needed to match [10011000010000100] and [10011000001000001].

Fig. 4 shows the main steps of the algorithm. Lines 5 can be compute in $O(k)$, hence Algorithm 1 runs in $O(k^3)$. By using the "high/low- frequency" technique [1], Indyk et. al. proposed an algorithm to calculate all values $L_1(x^{[i]}, y^{[j]}), 0 \leq i, j < k$ in $O(k^{(\omega+3)/2})$ time, where $O(k^\omega)$ is the running time required to multiply two matrices of size $k \times k$ [11]. In the following section, we will show that we don't need to calculate all values $L_1(x^{[i]}, y^{[j]})$ in order to find $s^*$.

## 4   An $O(k^2)$ Algorithm

In this section we present the main algorithm for solving the necklace problem. We show that in order to find the minimum swap distance we need only consider

---

**Algorithm 1**

---

**Input:** $x, y, k$
**Output:** $s^*$
1.     ▷ compute $u$ and $v$ using Eq. 4 and Eq. 5, *resp.*
2.     $s^* = \infty$
3.     **for** $i = 0$ **to** $k - 1$ **do**
4.         **for** $j = 0$ **to** $k - 1$ **do**
5.             ▷ compute $x^{[i]}$ and $y^{[j]}$ using Eq. 6 and Eq. 7, *resp.*
6.             **if** $s^* < L_1(x^{[i]}, y^{[j]})$ **then** $s^* = L_1(x^{[i]}, y^{[j]})$
7.     **return** $s^*$

---

**Fig. 4.** Algorithm 1

a column in Table 2 instead of the whole table, hence giving an overall time complexity of $O(k^2)$. Our strategy is to consider each different mapping $\pi^h$, for $0 \le h < k$, in turn and to find the rotation for each that minimises the swap distance. The algorithm works as follows:

If we define the *residual* sequence $c^{[h]}$ as

$$c_i^{[h]} = y_i^{[h]} - x_i^{[0]}, \text{ for } 0 \le i, h < k, \tag{9}$$

then the minimum necklace swap problem is equivalent to calculating

$$s^* = \min_{0 \le h < k} \sum_{i=0}^{k-1} |\delta_i^{[h]}|, \tag{10}$$

where

$$\delta_i^{[h]} = c_i^{[h]} - c_{\text{med}}^{[h]}, \text{ for } 0 \le i, h < k, \tag{11}$$

and $c_{\text{med}}^{[h]}$ is the median of $c^{[h]}$ as defined in Eq. 2.

Fig. 5 shows the main steps of the algorithm. Lines 1,2,5-6,8 can be computed in $O(k)$. Line 7 can be computed in $O(k)$ using [13], therefore Algorithm 2 runs in $O(k^2)$.

Before proving the correctness of Algorithm 2, we give an example.

*Example 3.* For $x = (1, 6, 9, 12, 13)$ and $y = (0, 3, 4, 10, 16)$, Table 3 shows the computation of $c^{[h]}, c_{\text{med}}^{[h]}, \delta^{[h]}$, and $\sum_{i=0}^{k-1} |\delta_i^{[h]}|$ for $0 \le h < k$. 3 was the minimum swap distance between $(5,8,11,12,0)$ and $(6,7,10,11,0)^{\langle -1 \rangle}$. This is also the number of swaps needed to match [10001001001100000] and [01000011001100000]. This example is also fully illustrated in Fig. 6.

**Lemma 1.** *For a given mapping $\pi$ and two cyclic bit-strings $X$ and $Y$, the rotation of $\theta$ that minimises the swap distance between $X$ and $Y$ can be found in $O(k)$ time using Algorithm 2.*

*Proof.* Let's consider mapping $\pi^0$, then we are trying to find the rotation $\theta$ that minimises the swap distance between $X$ and $Y$ under $\pi^0$. According with Eq. 1, $\pi^0$ pair's the 1's in $X$ with the 1's in $Y$ as follows:

**Algorithm 2**

**Input:** $x, y, k$
**Output:** $s^*$

1.     $\triangleright$ compute $u$ and $v$ using Eq. 4 and Eq. 5, *resp.*
2.     $\triangleright$ compute $x^{[0]}$ using Eq. 6.
3.     $s^* = \infty$
4.     **for** $h = 0$ **to** $k - 1$ **do**
5.         $\triangleright$ compute $y^{[h]}$ using Eq. 7
6.         $\triangleright$ compute $c^{[h]}$ using $y^{[h]}, x^{[0]}$ and Eq. 9.
7.         $c^{[h]}_{\text{med}} = \text{median}(c^{[h]})$
8.         $\triangleright$ compute $\delta^{[h]}$ using $c^{[h]}, c^{[h]}_{\text{med}}$ and Eq. 11.
9.         $d = 0$
10.        **for** $i = 0$ **to** $k - 1$ **do**
11.             $d = d + |\delta^{[h]}_i|$
12.             **if** $s^* < d$ **then** $s^* = d$
13.     **return** $s^*$

**Fig. 5.** Algorithm 2

**Table 3.** Computation of $c^{[h]}, c^{[h]}_{\text{med}}, \delta^{[h]}$, and $\sum_{i=0}^{k-1} |\delta^{[h]}_i|$ for $0 \leq h < k$. $x = (1, 6, 9, 12, 13)$, $y = (0, 3, 4, 10, 16)$, $u = (5, 3, 3, 1, 5)$, $v = (3, 1, 6, 6, 1)$.

| $h$ | $v^{\langle h \rangle} \to y^{[h]}$ | $x^{[0]}$ | $c^{[h]}$ | $c^{[h]}_{\text{med}}$ | $\delta^{[h]}$ | $\sum_{i=0}^{k-1} \|\delta^{[h]}_i\|$ |
|---|---|---|---|---|---|---|
| 0 | 3,1,6,6,1 → 3,4,10,16,0 | 5,8,11,12,0 | -2,-4,-1,4,0 | -1 | -1,-3,0,5,1 | 10 |
| 1 | 1,6,6,1,3 → 1,7,13,14,0 | 5,8,11,12,0 | -4,-1,2,2,0 | 0 | -4,-1,2,2,0 | 9 |
| 2 | 6,6,1,3,1 → 6,12,13,16,0 | 5,8,11,12,0 | 1,4,2,4,0 | 2 | -1,2,0,2,-2 | 7 |
| 3 | 6,1,3,1,6 → 6,7,10,11,0 | 5,8,11,12,0 | 1,-1,-1,-1,0 | -1 | 2,0,0,0,1 | 3 |
| 4 | 1,3,1,6,6 → 1,4,5,11,0 | 5,8,11,12,0 | -4,-4,-6,-1,0 | -4 | 0,0,-2,3,4 | 9 |

$$(0 \to 0), (1 \to 1), \dots, (k \to k).$$

Since the locations of the 1's in $X$ and $Y$ are stored in $x^{[0]}$ and $y^{[0]}$, *resp.*, vector $c^{[0]}$ stores the number of swaps needed to make each 1-bit in $Y$ coincides with its corresponding 1-bit in $X$ following $\pi^{[0]}$, but at the same time, since the signs are kept, it also "stores" the direction in which the swaps are to be done. Fig. 7 illustrates this fact.

Now we seek to find the value $\theta$ such that

$$\sum_{i=0}^{k-1} |c^{[0]}_i - \theta| \tag{12}$$

is minimum. To minimise (12) we need only set $\theta$ to be the median of $c^{[0]}$ which can be calculated in $O(k)$ time using a linear time selection algorithm [13]. Once

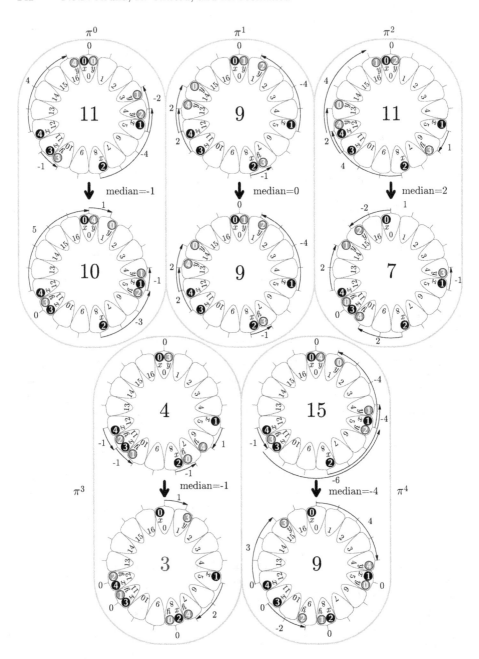

**Fig. 6.** Computation of the minimum cyclic swap distance for $X = $ [1000000100000101] and $Y = $ [0001000000101010]. $x = (0, 7, 13, 15)$, $y = (3, 10, 12, 14)$, $u = (5, 3, 3, 1, 5)$, $v = (3, 1, 6, 6, 1)$.

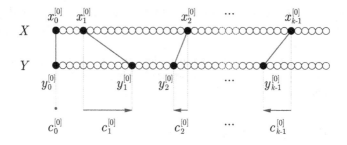

**Fig. 7.** Illustration of $c^{[0]}$ computation

$\theta$ has been found, (12) can be calculated in linear time. Hence, Eq. 10 produces the minimum swap distance and it is the optimal.                                        □

**Lemma 2.** *In order to find the global minimum, the only rotations $\theta$ that need to be consider are those where two 1's coincide.*

*Proof.* This follows from Lemma 1. For a given mapping $h$ and residual sequence $c^{[h]}$; $c^{[h]}_{\text{med}}$ is a value in $c^{[h]}$ therefore, there must be at least one $\delta^{[h]}_i = 0$, for $0 \le i < k$.                                        □

**Lemma 3.** *Consider cyclic string $x$ and a linearised string $y'$. A minimum mapping for any shift contains no crossings.*

*Proof.* Consider any linearisation of string $y$ and let $y'$ be the concatenation of $y$ to itself to make a string of length $2n$. Define $\pi':\{1,\ldots,n\} \to \{1,\ldots,2n\}$ to be a mapping from the position of the 1's in $x$ to the 1's in $y's$ such that $\pi'(1) \le \pi'(i)$ for all $i$ and $\max_{i,j} \pi'(j) - \pi'(i) \le n$. It is clear that there is an 1-1 correspondence between mappings of this type and those defined in Section 2. We further extend the definition so that $\pi'^h(1)$ is equal to the position of the $h$th 1 in $y'$.

We say that a mapping, $\pi'$, has a *crossing* if there exist $i$, $j$ such that $j > i$ and $\pi'(i) > \pi'(j)$. Given a mapping $\pi'^h$ (and a particular rotation of $x$), the swap distance with respect to $\pi'^h$ is simply $\Sigma|\pi'^h(i)-i|$. We define a *minimum mapping* for a particular value of $h$ to be any mapping that minimises the swap distance.

The proof is by contradiction. Consider a mapping $\pi'$ that both minimises the swap distance for the alignment and also contains a crossing. It follows that there exist $i$, $j$ such that $j > i$ and $\pi'(i) > \pi'(j)$. Now consider a new mapping $\pi'^*$, identical to $\pi'$ except that $\pi'^*(i) = \pi'(j)$ and $\pi'^*(j) = \pi'(i)$. Now $|\pi'^*(i) - i| < |\pi'(i) - i|$ and $|\pi'^*(j) - j| < |\pi'(j) - j|$ so the swap distance of $\pi'^*$ is less than the swap distance of $\pi'$. This gives us our contradiction.                                        □

We can now prove the main theorem.

**Theorem 1.** *Algorithm 2 solves problem 1 in $O(k^2)$ time.*

*Proof.* Every set of swap moves has a corresponding mapping from which its swap distance can be calculated. By Lemma 3, we need only consider crossing free mappings. By Lemma 1, Algorithm 2 finds the minimum swap distance for every mapping and as so by simply iterating over all crossing free mappings the necklace problem is solved.

The running time of each iteration is determined by the time taken by Algorithm 2 which Lemma 1 shows to be $O(k)$. There are $k$ iterations, giving an overall time complexity of $O(k^2)$. $\qquad\qquad\qquad\qquad\qquad\qquad\qquad\qquad\qquad\qquad\qquad$ $\square$

## 5    Conclusion

We have presented a new algorithm that solve the problem of cyclic swap distance between two $n$-bit (cyclic) binary strings in $O(k^2)$ where $k$ is the number of 1's (same) in both strings. We have also shown that the swap distance calculated by our algorithm is the optimal. Natural extensions to this problem could be (1) to consider unequal number of 1's among the input strings and (2) to allow scaling (increased/decrease the rest intervals of one of the input strings by a given constant).

## References

1. K. Abrahamson. Generalized string matching. *SIAM J. Comput.*, 16(6):1039-1051, 1987.
2. E. Cambouropoulos, M. Crochemore, C. S. Iliopoulos, L. Mouchard, and Y. J. Pinzon. Computing approximate repetitions in musical sequences. *International Journal of Computer Mathematics*, 79(11):1135-1148, 2002.
3. P. Clifford, R. Clifford and C. S. Iliopoulos. Faster algorithms for $\delta, \gamma$-matching and related problems. *In Proceedings of the 16th Annual Symposium on Combinatorial Pattern Matching (CPM'05)*, pages 68-78, 2005.
4. R. Clifford, T. Crawford, C. Iliopoulos, and D. Meredith. Problems in computational musicology. In C. S. Iliopoulos and Thierry Lecroq, editors, *String Algorithmics*, NATO Book series, King's College Publications, 2004.
5. R. Clifford and C. S. Iliopoulos. Approximate string matching for music analysis. *Soft Computing*, 8(9):597-603, 2004.
6. T. Crawford, C. S. Iliopoulos, R. Raman. String matching techniques for musical similarity and melodic recognition. *Computing in Musicology*, 11:71-100, 1998.
7. M. Crochemore, C. S. Iliopoulos, G. Navarro, and Y. Pinzon. A bit-parallel suffix automaton approach for $(\delta,\gamma)$-matching in music retrieval. In M. A. Nascimento, Edleno S. de Moura, and A. L. Oliveira, editors, *10th International Symposium on String Processing and Information Retrieval (SPIRE'03)*, ISBN 3-540-20177-7, Springer-Verlag, pages 211-223, 2003.
8. C. Cronin. Concepts of melodic similarity in music-copyright infringement suits. In W.B. Hewlett and E. Selfridge-Field, editors, *Melodic Similarity: Concepts, Procedures and Applications*, MIT Press, Cambridge, Massachusetts, 1998.
9. J. Godwin. The Harmony of the spheres: A sourcebook of the pythagorean tradition in music. Inner Traditions Intl. Ltd, 1993.
10. A. Ghias, J. Logan, D. Chamberlin and and B. C. Smith. Query by humming: Musical information retrieval in an audio database. *ACM Multimedia*, pages 231-236, 1995.

11. P. Indyk, M. Lewenstein, O. Lipsky, E. Porat. Closest pair problems in very high dimensions. *ICALP*, pages 782-792, 2004.
12. J-S. Mo, C. H. Han, and Y-S. Kim. A melody-based similarity computation algorithm for musical information. *Workshop on Knowledge and Data Engineering Exchange*, page 114, 1999.
13. A. Reiser. A Linear selection algorithm for sets of elements with weights. *Inf. Process. Lett.*, 7(3):159-162, 1978.
14. G. T. Toussaint. Computational geometric aspects of musical rhythm. *Abstracts of the 14th Annual Fall Workshop on Computational Geometry*, Massachussetts Institute of Technology, pages 47-48, 2004.
15. G. T. Toussaint. A comparison of rhythmic similarity measures. *Proceedings of the 5th International Conference on Music Information Retrieval (ISMIR'04)*, Universitat Pompeu Fabra, Barcelona, Spain, pages 242-245, 2004. A longer version also appeared in: *School of Computer Science*, McGill University, Technical Report SOCS-TR-2004.6, August 2004.

# Faster Generation of Super Condensed Neighbourhoods Using Finite Automata

Luís M.S. Russo* and Arlindo L. Oliveira

IST / INESC-ID, R. Alves Redol 9, 1000 LISBOA, Portugal
lsr@algos.inesc-id.pt, aml@inesc-id.pt

**Abstract.** We present a new algorithm for generating super condensed neighbourhoods. Super condensed neighbourhoods have recently been presented as the minimal set of words that represent a pattern neighbourhood. These sets play an important role in the generation phase of hybrid algorithms for indexed approximate string matching. An existing algorithm for this purpose is based on a dynamic programming approach, implemented using bit-parallelism. In this work we present a bit-parallel algorithm based on automata which is faster, conceptually much simpler and uses less memory than the existing method.

## 1   Introduction and Related Work

Approximate string matching is an important subject in computer science, with applications in text searching, pattern recognition, signal processing and computational biology.

The problem consists in locating all occurrences of a given pattern string in a larger text string, assuming that the pattern can be distorted by errors. If the text string is long it may be infeasible to search it on-line, and we must resort to an index structure. This approach has been extensively investigated in recent years [1,5,6,9,13,16,17].

The state of the art algorithms are hybrid, and divide their time into a *neighbourhood generation* phase and a *filtration* phase [12,9].

This paper is organised as follows: in section 2 we define the basic notation and the concept of strings and edit distance. In section 3 we present a high level description of hybrid algorithms for indexed approximate pattern matching. In section 4 we present previous work on the neighbourhood generation phase of hybrid algorithms. In section 5 we present our contribution, a new algorithm for generating Super Condensed Neighbourhoods. In section 6 we describe the bit-parallel implementation of our algorithm and present a complexity analysis. Section 7 presents the experimental results obtained with our implementation. Finally, section 8 presents the conclusions and possible future developments.

---

* Supported by the Portuguese Science and Technology Foundation through program POCTI POSI/EEI/10204/2001 and Project BIOGRID POSI/SRI/47778/2002.

M. Consens and G. Navarro (Eds.): SPIRE 2005, LNCS 3772, pp. 246–255, 2005.

# 2   Basic Concepts and Notation

## 2.1   Strings

**Definition 1.** *A string is a finite sequence of symbols taken from a finite alphabet $\Sigma$. The empty string is denoted by $\epsilon$. The size of a string $S$ is denoted by $|S|$.*

By $S[i]$ we denote the symbol at position $i$ of $S$ and by $S[i..j]$ the substring from position $i$ to position $j$ or $\epsilon$ if $i > j$. Additionally we denote by $S\langle i\rangle$ the point[1] in between letters $S[i-1]$ and $S[i]$. $S\langle 0\rangle$ represents the first point and $S\langle i-1..j\rangle$ denotes $S[i..j]$.

## 2.2   Computing Edit Distance

**Definition 2.** *The* edit *or* Levenshtein *distance between two strings ed($S$, $S'$) is the smallest number of edit operations that transform $S$ into $S'$. We consider as operations insertions (I), deletions (D) and substitutions (S).*

For example:

$ed(abcd, bedf) = 3$

```
D S I
abcd
bedf
```

The edit distance between strings $S$ and $S'$ is computed by filling up a dynamic programming table $D[i,j] = ed(S\langle 0..i\rangle, S'\langle 0..j\rangle)$, constructed as follows:

$$D[i,0] = i, \qquad D[0,j] = j$$
$$D[i+1,j+1] = D[i,j], \text{ if } S[i+1] = S'[j+1]$$
$$1 + \min\{D[i+1,j], D[i,j+1], D[i,j]\}, \text{otherwise}$$

The dynamic programming approach to the problem is the oldest approach to computing the edit distance. As such it has been heavily researched and many such algorithms have been presented, surveyed in [11].

One particularly important contribution was Myers proposal of an algorithm to compute the edit distance in a bit-parallel way [10]. The previous algorithm for computing Super Condensed Neighbourhoods [14] is based on this algorithm.

A different approach for the computation of the edit distance is to use a nondeterministic automaton(NFA). We can use a NFA to recognise all the words that are at *edit* distance $k$ from another string $P$, denoted $N_P^k$. Figure 1 shows an automaton that recognises words that are at distance at most one from *abbaa*. It should be clear that the word *ababaa* is recognised by the automaton since $ed(abbaa, ababaa) = 1$.

To find every match of a string $P$ in another string $T$ we can build an automaton for $P$ and restart it with every letter of $T$. This is equivalent to adding a loop labelled with all the character in $\Sigma$ to the initial state. We shall denote this new automaton by $N_P'^k$.

---

[1] The notion of point is superfluous but useful since it provides a natural way to introduce automata states.

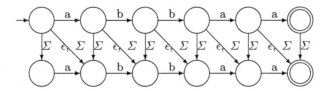

**Fig. 1.** Automaton for *abbaa* with at most one error

## 3    Indexed Approximate Pattern Matching

If we wish to find the occurrences of $P$ in $T$ in sub-linear time, with a $O(|T|^\alpha)$ complexity for $\alpha < 1$, we need to use an index structure for $T$. Suffix arrays [12] and q-grams have been proposed in the literature [6,9]. An important class of algorithms for this problem are hybrid in the sense that they find a trade-off between neighbourhood generation and filtration techniques.

### 3.1    Neighbourhood Generation

A first and simple-minded approach to the problem consists in generating all the words at distance $k$ from $P$ and looking them up in the index $T$. The set of generated words is the *k-neighbourhood* of $P$.

**Definition 3.** *The* k-neighbourhood *of $S$ is* $U_k(S) = \{S' \in \Sigma^* : ed(S, S') \le k\}$

Let us denote the language recognised by the automaton $N_P^k$ as $L(N_P^k)$. It should be clear that $U_k(P) = L(N_P^k)$. Hence computing $U_k(P)$ is achieved by computing $L(N_P^k)$, this can be done by performing a DFS search in $\Sigma^*$ that halts whenever all the states of $N_P^K$ became inactive.

### 3.2    Filtration Techniques

The classic idea of filtration is to eliminate text areas, by guaranteeing that there is no match at a given point, using techniques less expensive than dynamic programming. Since this approach has the obvious drawback that it cannot exclude all such areas, the remaining points have to be inspected with other methods.

In the indexed version of the problem, filtration can be used to reduce the size of neighbourhoods, hence speeding up the algorithm.

The most common filtration technique splits the pattern according to the following lemma:

**Lemma 1.** *If* ed$(S,S') \le k$ *and* $S = S_1 x_1 S_2 x_2 \ldots S_{l-1} x_{l-1} S_l$ *then $S_h$ appears in $S'$ with at most $\lfloor k/l \rfloor$ errors for some h.*

This lemma was presented by Navarro and Baeza-Yates [12]. Myers had also presented a similar proposition [9].

## 4   Neighbourhood Analysis

The $k$-*neighbourhood*, $U_k(S)$, turns out to be quite large. In fact $|U_k(S)| = O(|S|^k|\Sigma|^k)$ [15] and therefore we restrict our attention to the *condensed k-neighbourhood* [9,12].

**Definition 4.** *The* condensed k-neighbourhood *of $S$, $CU_k(S)$ is the largest subset of $U_k(S)$ whose elements $S'$ verify the following property: if $S''$ is a proper prefix of $S'$ then $ed(S, S'') > k$.*

The generation of $CU_k(P)$ can be done using automaton $N_P^k$ by testing the words of $\Sigma^*$ obtained from a DFS traversal of the lexicographic tree. The search backtracks whenever all states of $N_P^k$ became inactive or a final state becomes active.

The second criterion guarantees that no generated word is a prefix of another one.

Algorithm 1 generates $CU_k(P)$ by performing a controlled DFS that does not extend words of $L(N_P^k)$ found in the process [2]. [2]

---

**Algorithm 1.** Condensed Neighbourhood Generator Algorithm

---
```
1: procedure SEARCH(Search Point p, Current String v)
2:     if IS_MATCH_POINT(p) then
3:         REPORT(v)
4:     else if EXTENDS_TO_MATCH_POINT(p) then
5:         for z ∈ Σ do
6:             p' ← UPDATE(p, z)
7:             SEARCH(p', v.z)
8:         end for
9:     end if
10: end procedure
11: SEARCH(⟨0, 1, . . . , |P|⟩, ε)
```
---

The search point $p$ is a set of active states of $N_P^k$. The IS_MATCH_POINT predicate checks whether some state of $p$ is a final state. The EXTENDS_TO_MATCH_POINT predicate checks whether $p$ is non-empty. The UPDATE procedure updates the active states of $p$ by processing character $z$ with $N_P$.

It has been noted [14] that the *condensed neighbourhood* still contains some words that can be discarded without missing any matches.

**Definition 5.** *The* super condensed k-neighbourhood *of $S$, $SCU_k(S)$ is the largest subset of $U_k(S)$ whose elements $S'$ verify the following property:if $S''$ is a proper substring of $S'$ then $ed(S, S'') > k$.*

---

[2] We can shortcut the generate and search cycle by running algorithm 1 on the index structure. For example in the suffix tree this can be done by using a tree node instead of $v$.

In our example *ababaa* and *abaa* are in the *condensed neighbourhood* of *abbaa*, but only *abaa* is in the *super condensed neighbourhood*.

Figure 2 shows an example of the *1-neighbourhood*, the *1-condensed neighbourhood* and the *1-super condensed neighbourhood* of *abbaa*. Observe that $SCU_k(P) \subseteq CU_k(P) \subseteq U_k(P)$.

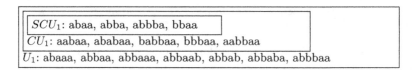

**Fig. 2.** Figure representing the one-neighbourhoods of *abbaa*

The *Super Condensed k-neighbourhood* is minimal in the sense that we can't have a set with a smaller number of words that can be used in the search without missing matches [14].

H. Hyyrö and G. Navarro [6] presented the notion of artificial *prefix-stripped length-q* neighbourhood, that is smaller than the condensed neighbourhood but it is not minimal.

## 5    Computing Super Condensed Neighbourhoods Using Finite Automata

We now present the main contribution of this paper, a new approach to compute *super condensed neighbourhoods*.

In order to compute the *super condensed neighbourhood* we define a new automaton. Consider the automaton $N_P''^k$ that results from $N_P^k$ by adding a new initial state with a loop labelled by all the characters of $\Sigma$ linked to the old initial state by a transition also labelled by all the characters of $\Sigma$. An example of $N_P''^k$ is shown in fig. 3. The language recognised by $N_P''^k$ consists of all the strings that have a **proper** suffix $S''$ such that $ed(P, S'') \leq k$.

The set $L(N_P^k) \backslash L(N_P''^k)$ is <u>not</u> a *super condensed neighbourhood* by the following two reasons:

**prefixes** Some words might still be prefixes of other words. For example both *abaa* and *abaaa* belong to $L(N_{abbaa}^k) \backslash L(N_{abbaa}''^k)$. This can be solved when performing the DFS traversal of the lexicographic tree, as before.
**substrings** The definition of $L(N_P^k) \backslash L(N_P''^k)$ will yield the subset of $L(N_P^k)$ such that no proper <u>suffix</u> is at distance at most $k$ from $P$. But this is not what we want, since we desire a subset of $L(N_P^k)$ that does not contain a string and a proper <u>substring</u> of that string. In order to enforce this requirement we must stop the DFS search whenever a final state of $N_P''^k$ is reached.

A point $p$ in the DFS search of the lexicographical tree now corresponds to two sets of states, one for $N_P^k$ and one for $N_P''^k$. The Is_MATCH_POINT predicate

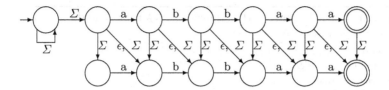

**Fig. 3.** Automaton $N_P''^k$ for *abbaa* that matches every proper suffix

checks that no active state of $N_P''^k$ is final and that there is one active state of $N_P^k$ that is final. The EXTENDS_TO_MATCH_POINT checks that no active state of $N_P''^k$ is final and that there is one active state of $N_P^k$ that is inactive in $N_P''^k$. The UPDATE procedure updates both automata using letter $z$.

Observe that, in this version of the algorithm, the string *ababaa* is no longer reported. In fact the DFS search backtracks after having reached *abab*. After reading *abab* the only active state of $N_P^k$ is the one corresponding to *abb* on the second column, since $ed(abab, abb) = 1$. This state is also active in $N_P''^k$ since $ed(ab, abb) = 1$ and *ab* is a proper suffix of *abab*. Interestingly, the dynamic programming DFS for this same example backtracked only after reaching *ababa*. Clearly, the dynamic programming algorithm could be improved to backtrack sooner but it is conceptually much simpler to use the automata approach.

It was shown by Myers [9] that $|CU_k(P)| = O(|P|^{pow(|P|/k)})$, where:

$$pow(\alpha) = \log_{|\Sigma|} \frac{(\alpha^{-1} + \sqrt{1 + \alpha^{-2}}) + 1}{(\alpha^{-1} + \sqrt{1 + \alpha^{-2}}) - 1} + \alpha \log_{|\Sigma|}(\alpha^{-1} + \sqrt{1 + \alpha^{-2}}) + \alpha$$

We establish no new worst case bound for the size of the *super condensed neighbourhood* so $|SCU_k(P)| = O(|P|^{pow(|P|/k)})$. However ours results do show a practical improvement in speed.

## 6    Bit Parallel Implementation and Complexity Analysis

We implemented $N_P^k$ and $N_P''^k$ by using bit-parallelism techniques that have been proposed for $N_P'^k$ [17,4].

Algorithm 2 describes the details of implementation of the necessary predicates.

The $F_i$ computer words store the $N_P^k$ for row $i$. The $S_i$ computer words store the $N_P''^k$ automata states for row $i$. The $B[z]$ computer words stores the bit mask of the positions of the letter $z$ in $P$.

Our implementation of the Wu and Manber algorithm stores the first column of the automata. Furthermore for automata $N_P''^k$ we don't need to store the artificial state that was inserted, since it is sufficient to initialise the $S_i$ state vectors to zero.

Since the UPDATE and EXTENDS_TO_MATCH_POINT procedures run in $O(k\lceil|P|/w\rceil)$ the final algorithm takes $O(k\lceil|P|/w\rceil |P| s)$ where $s = |SCU_k(P)| = O(|P|^{pow(|P|/k)})$ and $w$ is the size of the computer word. This is a conservative bound since it is easy to modify the algorithm so that it runs in $O((k\lceil|P|/w\rceil + |P|)s)$. This is achieved by using the KMP failure links and was

**Algorithm 2.** Bit-Parallel of the Algorithm. $N_P^k$ represented by $F_i$ and $N_P''^k$ by $S_i$. Bitwise operations in C-style.

```
1: procedure IS_MATCH_POINT(Search Point F₀, ..., Fₖ, S₀, ..., Sₖ)
2:     return Fₖ&&!Sₖ
3: end procedure
4: procedure EXTENDS_TO_MATCH_POINT(Search Point F₀, ..., Fₖ, S₀, ..., Sₖ)
5:     return ((F₀&~S₀)| ... |(Fₖ&~Sₖ))&&!Sₖ
6: end procedure
7: procedure UPDATE(Search Point F₀, ..., Fₖ, S₀, ..., Sₖ, letter z)
8:     F'₀ ← (F₀ << 1)&B[z]
9:     S'₀ ← ((S₀ << 1)|1)&B[z]
10:    for i ← 0, k do
11:        F'ᵢ₊₁ ← ((Fᵢ₊₁ << 1)&B[z])|Fᵢ|(Fᵢ << 1)|(F'ᵢ << 1)
12:        S'ᵢ₊₁ ← ((Sᵢ₊₁ << 1)&B[z])|Sᵢ|(Sᵢ << 1)|(S'ᵢ << 1)
13:    end for
14:    return F'₀, ..., F'ₖ, S'₀, ..., S'ₖ
15: end procedure
```

first presented by Myers [9]. Recently Heikki Hyyrö presented way of achieving the same result in a sequential way that is relevant for bit-parallel algorithms [8].

We also implemented a version based on Navarro and Baeza-Yates [3] variation of the NFA. The procedures are implemented in a similar way and the resulting algorithm runs in $O(\lceil k(|P| - k)/w \rceil \, |P| \, s)$. We improved this to $O((\lceil k(|P| - k)/w \rceil + |P|) \, s)$ using an approach similar to the one followed by Myers but found no time difference in practice. Usually $O(\lceil k(|P| - k)/w \rceil)$ is approximately constant for small patterns, which is the case for hybrid algorithms. We usually split the pattern into pieces of size $\Theta(\log_\sigma |T|)$.

In previous work [14], we reported a complexity of $O(|P| \lceil |P|/w \rceil s)$ which was too pessimistic, since it did not take into account the possible reduction in complexity that is possible to achieve by applying the method based on the KMP failure links of Myers [9].

Once again the Baeza-Yates and Navarro algorithm usually doesn't store the states below the first diagonal including the diagonal. We don't need to keep track of the states below the diagonal but we do need to keep track of the diagonal [3].

## 7   Experimental Results

We tested our approach by analysing its impact in the hybrid index [12]. Since we are only interested in the neighbourhood generation phase we set the $j$ option of the index to 1, preventing the pattern from getting split.

Tests were run in a 800MHz Power PC G3 processor with 512K level 2 cache 640MB SDRAM, Mac Os X 10.2.8 and gcc 3.3.

---

[3] Actually this could be reduced but the gains would be practically none.

Our implementation was based on the original implementation of Navarro and Baeza-Yates. The NFA is also based on the variation presented by Navarro and Baeza-Yates [3].

For each $(|P|, k)$ combination we tested 100 patterns randomly selected from the text and computed the average time to search for those patterns. The patterns were taken with sizes 10, 15 and 20. We used two source texts, an English text [18], that consists of cleaned up newsgroups text and a DNA file of 5.6 Mb, from the *S. cerevisiae* (baker's yeast) genome. Results are shown in figure 4.

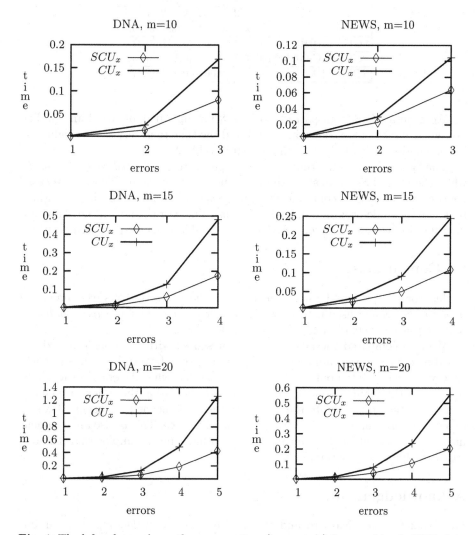

**Fig. 4.** The left column shows the average time (in seconds) for searching in DNA data and the right column shows the average time for searching in the Newsgroups data. The pattern size is indicated by $m$.

**Table 1.** Average size of $CU_k$ vs $SCU_k$

|         | $|\Sigma| = 2$ |         | $|\Sigma| = 4$ |          |
|---------|--------|--------|--------|--------|
|         | $k = 2$ | $k = 4$ | $k = 2$ | $k = 4$ |
| $CU_k$  | 67     | 42     | 810    | 21430  |
| $SCU_k$ | 22     | 14     | 320    | 591    |

**Table 2.** Bit-parallel and increased bit-parallel algorithms in milliseconds

|                      | $|\Sigma| = 2$ |         | $|\Sigma| = 4$ |         |
|----------------------|--------|--------|--------|--------|
|                      | $k = 2$ | $k = 4$ | $k = 2$ | $k = 4$ |
| $CU_k$               | 0.036  | 0.013  | 1.038  | 20.459 |
| $SCU_k$-CARRY        | 0.012  | 0.004  | 0.297  | 0.312  |
| $SCU_k$-INC-CARRY    | 0.009  | 0.003  | 0.125  | 0.142  |
| $SCU_k$-NFA          | 0.0043 | 0.0026 | 0.1048 | 0.171  |

We generated random patterns of size 8 for alphabets of size 2 and 4. The average size results are shown in table 1 and the time to generate the neighbourhoods blindly without the text are shown in table 2.

The first row shows the times needed to generate *Condensed Neighbourhoods*, while the next three rows show the times needed to generate *Super Condensed Neighbourhoods*. The second and third rows were obtained using the algorithms based on dynamic programming while the last line corresponds to the algorithm described in this article implemented using Wu and Manber bit-parallel algorithm.

## 8   Conclusions

In this work we proposed a new algorithm for the generation of *super condensed neighbourhoods* and used it to show the practical gains of using *super condensed neighbourhoods* instead of *condensed neighbourhoods*.

We also compared the algorithms we presented with the ones that existed based on dynamic programming. As expected, results favour this new approach. However it was pointed out by Heikki Hyyrö [7] that when generating the neighbourhoods the main time factor corresponds to accessing the index in memory and not in how we compute the *edit* distance. This means that using NFA's or dynamic programming makes little practical difference. This is also an argument in favour of this algorithm since it is conceptually much simpler than the one based on dynamic programming.

## Acknowledgements

We thank Gonzalo Navarro and Baeza-Yates for kindly lending us their implementation of the hybrid index. We also thank Gonzalo Navarro and Heikki Hyyrö for their suggestions and remarks. We thank Gene Myers for suggestions, corrections and remarks.

# References

1. Ricardo A. Baeza-Yates. Text-retrieval: Theory and practice. In Jan van Leeuwen, editor, *IFIP Congress (1)*, volume A-12 of *IFIP Transactions*, pages 465–476. North-Holland, 1992.
2. Ricardo A. Baeza-Yates and Gaston H. Gonnet. A new approach to text searching. *Commun. ACM*, 35(10):74–82, 1992.
3. Ricardo A. Baeza-Yates and Gonzalo Navarro. A faster algorithm for approximate string matching. In Daniel S. Hirschberg and Eugene W. Myers, editors, *CPM*, volume 1075 of *Lecture Notes in Computer Science*, pages 1–23. Springer, 1996.
4. Ricardo A. Baeza-Yates and Gonzalo Navarro. Faster approximate string matching. *Algorithmica*, 23(2):127–158, 1999.
5. Archie L. Cobbs. Fast approximate matching using suffix trees. In Zvi Galil and Esko Ukkonen, editors, *CPM*, volume 937 of *Lecture Notes in Computer Science*, pages 41–54. Springer, 1995.
6. H. Hyyrö and G. Navarro. A practical index for genome searching. In *Proceedings of the 10th International Symposium on String Processing and Information Retrieval (SPIRE 2003)*, LNCS 2857, pages 341–349. Springer, 2003.
7. Heikki Hyyrö. *Practical Methods for Approximate String Matching*. PhD thesis, Faculty of Information of the University of Tampere, 2003.
8. Heikki Hyyrö. An improvement and an extension on the hybrid index for approximate string matching. In *Proceedings of the 11th International Symposium on String Processing and Information Retrieval (SPIRE 2003)*, LNCS 3246, pages 208–209. Springer, 2004.
9. E. Myers. A sublinear algorithm for approximate keyword matching. *Algorithmica*, (12):345–374, 1994.
10. Gene Myers. A fast bit-vector algorithm for approximate string matching based on dynamic programming. In Martin Farach-Colton, editor, *CPM*, volume 1448 of *Lecture Notes in Computer Science*, pages 1–13. Springer, 1998.
11. G. Navarro. A guided tour to approximate string matching. *ACM Computing Surveys*, 33(1):31–88, 2001.
12. G. Navarro and R. Baeza-Yates. A hybrid indexing method for approximate string matching. *Journal of Discrete Algorithms*, 1(1):205–239, 2000.
13. G. Navarro, R. Baeza-Yates, E. Sutinen, and J. Tarhio. Indexing methods for approximate string matching. *IEEE Data Engineering Bulletin*, 24(4):19–27, 2001.
14. Luís M. S. Russo and Arlindo L. Oliveira. An efficient algorithm for generating super condensed neighborhoods. In Alberto Apostolico, Maxime Crochemore, and Kunsoo Park, editors, *CPM*, volume 3537 of *Lecture Notes in Computer Science*, pages 104–115. Springer, 2005.
15. E. Ukkonen. Finding approximate patterns in strings. *Journal of Algorithms*, pages 132–137, 1985.
16. Esko Ukkonen. Approximate string-matching over suffix trees. In Alberto Apostolico, Maxime Crochemore, Zvi Galil, and Udi Manber, editors, *CPM*, volume 684 of *Lecture Notes in Computer Science*, pages 228–242. Springer, 1993.
17. Sun Wu and Udi Manber. Fast text searching allowing errors. *Commun. ACM*, 35(10):83–91, 1992.
18. file 20ng-train-all-terms from http://www.gia.ist.utl.pt/~acardoso/datasets

# Restricted Transposition Invariant Approximate String Matching Under Edit Distance

Heikki Hyyrö

Department of Computer Sciences, University of Tampere, Finland
heikki.hyyro@gmail.com

**Abstract.** Let $A$ and $B$ be strings with lengths $m$ and $n$, respectively, over a finite integer alphabet. Two classic string mathing problems are computing the edit distance between $A$ and $B$, and searching for approximate occurrences of $A$ inside $B$. We consider the classic Levenshtein distance, but the discussion is applicable also to indel distance. A relatively new variant [8] of string matching, motivated initially by the nature of string matching in music, is to allow transposition invariance for $A$. This means allowing $A$ to be "shifted" by adding some fixed integer $t$ to the values of all its characters: the underlying string matching task must then consider all possible values of $t$. Mäkinen et al. [12,13] have recently proposed $O(mn \log \log m)$ and $O(dn \log \log m)$ algorithms for transposition invariant edit distance computation, where $d$ is the transposition invariant distance between $A$ and $B$, and an $O(mn \log \log m)$ algorithm for transposition invariant approximate string matching. In this paper we first propose a scheme to construct transposition invariant algorithms that depend on $d$ or $k$. Then we proceed to give an $O(n + d^3)$ algorithm for transposition invariant edit distance, and an $O(k^2 n)$ algorithm for transposition invariant approximate string matching.

## 1 Introduction

Let $\Sigma$ be a finite integer alphabet of size $\sigma$ so that each character in $\Sigma$ has a value in the range $0, \ldots, \sigma - 1$. We assume that strings are composed of a finite (possibly length-zero) sequence of characters from $\Sigma$. The length of a string $A$ is denoted by $|A|$. When $1 \leq i \leq |A|$, $A_i$ denotes the $i$th character of $A$. The notation $A_{i..h}$, where $i \leq h$, denotes the substring of $A$ that begins at character $A_i$ and ends at character $A_h$. Hence $A = A_{1..|A|}$. String $A$ is a subsequence of string $B$ if $B$ can be transformed into $A$ by deleting zero or more characters from it.

Let $ed(A, B)$ denote the edit distance between strings $A$ and $B$. For conveniency, the length of $A$ is $m$ and the length of $B$ is $n$ throughout the paper, and we also assume that $m \leq n$. In general, $ed(A, B)$ is defined as the minimum number of edit operations that are needed in transforming $A$ into $B$, or vice versa. In this paper we concentrate specifically on Levenshtein distance (denoted by $ed_L(A, B)$), which allows a single edit operation to insert, delete or substitute a single character. But the methods are applicable also to indel distance (denoted by $ed_{id}(A, B)$), which differs only in that it does not allow substitutions.

M. Consens and G. Navarro (Eds.): SPIRE 2005, LNCS 3772, pp. 256–266, 2005.

Indel distance is interesting for example because it is related to $llcs(A, B)$, the length of the longest common subsequence between $A$ and $B$, by the formula $2 \times llcs(A, B) = m + n - ed_{id}(A, B)$.

If $A$ takes the role of a pattern string and $B$ the role of a text string, approximate string matching is defined as searching for those locations $j$ in $B$ where $ed(A, B_{h..j}) \leq k$ for some $h \leq j$. Here $k$ is a predetermined error threshold.

An interesting variation of string comparison/matching, allowing *transposition invariance*, was proposed recently by Lemström and Ukkonen in [8] in the context of music comparison and retrieval. If musical pieces are stored as sequences of note pitches and we want to find a melody pattern $p$ (ie. a string whose characters are note pitch values) from a music database, it may be natural not to care about the overall pitch at which $p$ is found. This leads to the basic idea of transposition invariant matching: to allow $A$ to match at any pitch. Transposition invariance has also other possible uses. Mäkinen et al. [13] mention several, such as time series comparison.

A transposition is represented as an integer $t$. Characters $A_i$ and $B_j$ match under transposition $t$ if $A_i + t = B_j$. With the finite integer alphabet $\Sigma$, the set of all possible transpositions is $\mathbb{T} = \{b - a \mid a, b \in \Sigma\}$. Let $A + t$ denote string $A$ that has been shifted by transposition $t$. That is, the $i$th character of $A + t$ is $A_i + t$. The notation $ed^t(A, B)$ denotes the transposition invariant edit distance between $A$ and $B$. It requires us to find the minimum distance over all transpositions, ie. $ed^t(A, B) = \min\{ed(A + t, B) \mid t \in \mathbb{T}\}$. Then the task of transposition invariant approximate matching is to find locations $j$ in $B$ where $ed^t(A, B_{h..j}) \leq k$ for some $h \leq j$.

At the moment there is no algorithm that is able to compute $ed^t(A, B)$ more efficiently in the worse case than the approach of simply computing the distances $ed(A + t, B)$ separately for each possible transposition $t$. Hence the current nontrivial solutions, including the ones we propose, are concentrated on how to compute each distance $ed(A + t, B)$ efficiently [12,13], and on building heuristics on how to quickly discard possible transpositions from further consideration [9].

We will assume for the remaining part if the paper that the notion of edit distance refers to the Levenshtein distance (e.g. $ed(A, B) = ed_L(A, B)$).

## 2  Dynamic Programming

The classic $O(mn)$ solution [19] for computing $ed(A, B)$ is to fill an $(m + 1) \times (n + 1)$ dynamic programming table $D$, where the cell $D[i, j]$ will eventually hold the value $ed(A_{1..i}, B_{1..j})$. Under Levenshtein distance it works as follows. First the boundary values of $D$ are initialized by setting $D[i, 0] = i$ and $D[0, j] = j$ for $i \in 0, \ldots, m$ and $j \in 1, \ldots, n$. Then the remaining cells are filled recursively so that $D[i, j] = D[i - 1, j - 1]$ if $A_i = B_j$, and otherwise $D[i, j] = 1 + \min\{D[i - 1, j - 1], D[i - 1, j], D[i, j - 1]\}$. The recurrence for indel distance is very similar.

The dynamic programming method can be modified to conduct approximate string matching by changing the boundary initialization rule $D[0, j] = j$ into $D[0, j] = 0$ [16].

|   | S | T | A | I | R |
|---|---|---|---|---|---|
|   | 0 | 1 | 2 | 3 | 4 | 5 |
| S 1 | 0 | 1 |   |   |   |
| P 2 | 1 | 1 |   |   |   |
| I 3 |   |   | 2 |   |   |
| R 4 |   |   |   |   |   |
| E 5 |   |   |   |   |   |

|   | S | T | A | I | R |
|---|---|---|---|---|---|
|   | 0 | 1 | 2 | 3 | 4 | 5 |
| S 1 | 0 | 1 |   |   |   |
| P 2 | 1 | 1 | 2 |   |   |
| I 3 |   |   | 2 | 2 |   |
| R 4 |   |   |   |   | 2 |
| E 5 |   |   |   |   |   |

**Fig. 1.** Assume the values $L[1, -1] = 2$, $L[1, 0] = 2$, and $L[1, 1] = 1$ are already known. (*Left*) When computing the value $L[2, 0]$, we have $q = 0$, $s = \max\{L[1, -1], L[1, 0] + 1, L[1, 1] + 1\} = 3$, and $lcp(s + 1, q + s + 1) = lcp(4, 4) = 0$. Hence $L[2, 0] = 3$. (*Right*) When computing the value $L[2, 1]$, we have $q = 1$, $s = \max\{L[1, 0], L[1, 1] + 1, L[1, 2] + 1\} = 2$, and $lcp(s, q + s) = lcp(3, 4) = 2$. Hence $L[2, 1] = 4$.

### 2.1   Greedy Filling Order

Let diagonal $q$ be the up-left to low-right diagonal in $D$ whose cells $D[i, j]$ satisfy $j - i = q$. Ukkonen [17] proposed a greedy algorithm for computing edit distance. The cells in $D$ are filled in the order of increasing distance values $0, 1, \dots$. The algorithm is based on the well-known facts that the values along a diagonal $q$ are non-decreasing, and that moving from one diagonal to another costs one operation. Let $L[r, q]$ denote the lowest row on diagonal $q$ of $D$ that has a value less or equal to $r$, and define $L[r, q] = -1$ if such row does not exist. Thus $L[r, q] = \max\{i \mid D[i, i + q] \leq r \vee i = -1\}$. The greedy method can be used as follows under Levenshtein distance. First the values $L[r, q]$ are initialized with the value $-1$. Once the values $L[r - 1, q]$ are known, the values $L[r, q]$ can be computed using the rule $L[r, q] = \min\{m, s + lcp(s + 1, q + s + 1)\}$, where $s = \max\{L[r - 1, q - 1], L[r - 1, q] + 1, L[r - 1, q + 1] + 1\}$, and $lcp(i, j)$ is the length of the longest common prefix between $A_{i..m}$ and $B_{j..n}$. Fig. 1 shows an example. The value $lcp(i, j)$ can be computed in constant time by using the method of Chang and Lawler [1], which requires $O(n)$ time preprocessing. Hence the greedy algorithm is able to compute $d = ed(A, B)$ in $O(n + d^2)$ time, as at most $O(d)$ diagonals are processed and a single diagonal involves $O(d)$ computations.

## 3   Sparse Dynamic Programming

Sparse dynamic programming concentrates only on the matching points $D[i, j]$ where $A_i = B_j$. Let $\mathbb{M}(t) = \{(i, j) \mid A_i + t = B_j\}$ be the set of matching points under transposition $t$. A single set $\mathbb{M}(t)$ can be represented in linear space and generated in $O(n \log n)$ time [5]. By following [13], the sets $\mathbb{M}(t)$ can be computed for all relevant transpositions in $O(\sigma + mn)$ time. There are $|\mathbb{M}(t)| = O(mn)$

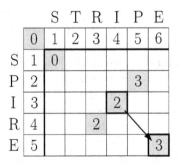

| | | S | T | R | I | P | E |
|---|---|---|---|---|---|---|---|
| | 0 | 1 | 2 | 3 | 4 | 5 | 6 |
| S | 1 | 0 | 1 | 2 | 3 | 4 | 5 |
| P | 2 | 1 | 1 | 2 | 3 | 3 | 4 |
| I | 3 | 2 | 2 | 2 | 2 | 3 | 4 |
| R | 4 | 3 | 3 | 2 | 3 | 3 | 4 |
| E | 5 | 4 | 4 | 3 | 3 | 4 | 3 |

| | | S | T | R | I | P | E |
|---|---|---|---|---|---|---|---|
| | 0 | 1 | 2 | 3 | 4 | 5 | 6 |
| S | 1 | 0 | | | | | |
| P | 2 | | | | | 3 | |
| I | 3 | | | 2 | | | |
| R | 4 | | 2 | | | | |
| E | 5 | | | | | | 3 |

**Fig. 2.** (*Left*) A completely filled dynamic programming matrix for computing $ed($"SPIRE","STRIPE"$)$. (*Right*) The same computation using sparse dynamic programming. Only the cell $(0,0)$ and the match points are considered (shown in grey). When the cell $(i,j) = (5,6)$ is computed, the preceding cells $(i',j')$ are $(0,0)$, $(1,1)$, $(2,5)$, $(3,4)$, and $(4,3)$. The corresponding values $D[i',j'] + \max\{i - i', j - j'\}$ are 6, 5, 3, 2, and 3, respectively. The minimum value 2 corresponds to $(i',j') = (3,4)$. This leads into setting $D[5,6] = D[3,4] + 2 - 1 = 3$.

matching points under a given transposition $t$. The overall number of matches under all relevant[1] transpositions is of this same complexity: $\sum_{t \in \mathbb{T}} |\mathbb{M}(t)| = mn$, as each point is a match point for exactly one value of $t$. This observation was abstracted in [13] into the following Lemma, which reduces the solution of transposition invariant matching into finding an efficient sparse dynamic programming algorithm for the basic case without transposition invariance.

**Lemma 1 ([13]).** *If distance $ed(A, B)$ can be computed in $O(g(|\mathbb{M}(t)|)f(m,n))$ time, where $g$ is a convex (concave up) increasing function, then the distance $ed^t(A, B)$ can be computed in $O(g(mn)f(m,n))$ time.*

For edit distance, a sparse recurrence can be derived in a straightforward manner from the corresponding dynamic programming recurrence. Let the notation $(i', j') \prec (i, j)$ mean that $i' < i$ and $j' < j$, and we also say that in this case $(i', j')$ *precedes* $(i, j)$. The following sparse scheme for Levenshtein distance is adapted from Galil and Park [4]. First we initialize $D[0,0] = 0$. Then each value $D[i,j]$ where $A_i = B_j$ is computed recursively by setting $D[i,j] = \min\{D[i',j'] + \max\{i - i', j - j'\} \mid (i',j') \in \mathbb{M}(t) \cup (0,0) \wedge (i',j') \prec (i,j)\} - 1$. Fig. 2 shows an example. After computing the values at all matching points, the possibly still uncomputed values $D[m,j]$ can be computed with $O(n)$ extra cost. Approximate string matching is achieved by adding the value $i$ as one possible choice in the minimum clause. The decisive factor for efficiency of sparse dynamic programming is how the preceding points $(i', j')$ that give minimal distances at $D[i,j]$ are found.

Galil and Park [4], by following the framework of Eppstein et al. [2], discussed a scheme that is able to compute the distances $D[i,j]$ for all $(i,j) \in \mathbb{M}(t)$ in over-

---

[1] A transposition is relevant if it leads into at least one match between $A$ and $B$.

all time of $O(|\mathbb{M}(t)| \log \log(\min\{|\mathbb{M}(t)|, mn/|\mathbb{M}(t)|\}))$. They process the points in row-wise manner for increasing $i$, and within each row for increasing $j$, and maintain an *owner list*. The point $(i', j')$ is the owner of $(i, j)$ if it results in the minimum distance at $(i, j)$ in the sparse dynamic programming recurrence. On row $i$, the owner list contains the column indices of the owners of the points $(i, j)$ on row $i$. Hence its size is $O(n)$, and all its key values are integers in the range $1, \ldots, n$. The list of owners at each point $(i, j)$ can be updated by doing an amortized constant number of insert, delete and lookup operations on a priority queue. This results in the overall cost $O(|\mathbb{M}(t)| \log \log(\min\{|\mathbb{M}(t)|, mn/|\mathbb{M}(t)|\}))$ if the priority queue is implemented using Johnson's data structure [6]. When the values stored in the priority queue are integers in the range $0, \ldots, z$, this data structure facilitates a homogenous sequence of $r \leq z$ insertions/deletions/lookups (all of the same type) in $O(r \log \log(z/r))$ time.

Mäkinen et al. [13] proposed to use two-dimensional range minimum queries in finding the minimum point $(i', j')$. They achieved $O(|\mathbb{M}(t)| \log \log m)$ time for indel distance and $O(|\mathbb{M}(t)| \log m \log \log m)$ time for Levenshtein distance by using the data structure of Gabow et al. [3]. With the exception of computing indel distance, Mäkinen et al. resorted to processing $B$ in segments of $O(m)$ match points within the matrix $D$ (distance), or $O(m)$ characters (approximate matching), in order to achieve the preceding time boundaries. They achieved also $O(|\mathbb{M}(t)| \log \log m)$ in all cases by applying the segmenting techniques to the approach of Eppstein et al. [2].

We note here that the above-mentioned segmenting technique is not necessary for achieving $O(|\mathbb{M}(t)| \log \log m)$: the method of Eppstein et al. does not internally rely on the typical assumption that $m \leq n$. If we switch the roles of the string pair so that $A$ becomes $B$ and vice versa, then that method leads directly into the run time $O(|\mathbb{M}(t)| \log \log m)$: now the owner list contains integers in the range $1, \ldots, m$, in which case each operation on Johnson's data structure takes $O(\log \log m)$ time. Taking this into account changes their time bound into $O(|\mathbb{M}(t)| \log \log(\min\{m, |\mathbb{M}(t)|, mn/|\mathbb{M}(t)|\}))$.

## 4    Restricting the Computation

Now we are ready to present our technique for restricting the computation with transposition invariant edit distance. The first building block is the following Lemma that is essentially similar to the idea of so-called counting filter [7,14].

**Lemma 2.** *Let $A$ and $B$ be two strings and $D$ be a corresponding dynamic programming table that has been filled as described in section 2. The condition $D[i, j] \leq k$ can hold only if the substring $B_{j-h+1..j}$, where $h = \min\{i, j\}$, matches at least $i - k$ characters of $A_{1..i}$.*

Using the preceding lemma, we get the main rule for restricting the computation.

**Lemma 3.** *Let $c > 1$ be a constant, and $ck$ and $j$ be positive integers that fulfill the conditions $k < ck \leq m$ and $1 \leq j \leq n$. There exists at most $O(k)$ different transpositions $t$ for which $D^t[ck, j] \leq k$.*

*Proof.* By Lemma 2, the length-$ck$ prefix $A_{1..ck}$ can match a substring ending at $B_j$ with at most $k$ errors only if the substring $B_{j-h+1..j}$, where $h = \min\{ck, j\}$, matches at least $ck - k$ characters of $A_{1..i}$. The corresponding $ck \times h$ submatrix of $D$, spanning rows $1, \ldots, ck$ and columns $j - h + 1, \ldots, j$, contains at most $c^2 k^2$ character-pairs $(A_i, B_j)$. Since $A_i + t = B_j$ for only one transposition $t$, there can be at most $c^2 k^2 / (ck - k) = k(c^2 / (c - 1)) = O(k)$ different transpositions that have at least $ck - k$ matches. □

The value $c^2 / (c - 1)$ in Lemma 3 gets its minimum value, 4, when we choose $c = 2$. In the following section we propose how to use this Lemma.

## 5   The Main Algorithm

Assume first that we wish to compute the accurate value $ed^t(A, B)$ only if it is at most $k$. We can use Lemma 3 in solving this problem efficiently. The first step is to conduct sparse dynamic programming on a restricted part of $D$.

**Step 1: Partial Sparse Dynamic Programming.** Consider a given relevant transposition $t$. In the first step we compute $ed(A_{1..2k} + t, B)$ by using sparse dynamic programming. During (or after) this process we compute the values $D[2k, j]$ for $j = 1, \ldots, n$ and record the values that are $\leq k$. This contains several elements. The first is an ordered list $matchPos$ that contains the pair $(j, D[2k, j])$ if $D[2k, j] \leq k$. The $matchPos$ list is in ascending order according to $j$. The positions $j$ are also recorded into $k + 1$ ordered lists $matchPos(p)$ for $p = 0, \ldots, k$. The list $matchPos(p)$ contains in ascending order all positions $j$ where $D[2k, j] = p$.

At this point we have all relevant information of $D$ until row $2k$. Fig. 3 illustrates.

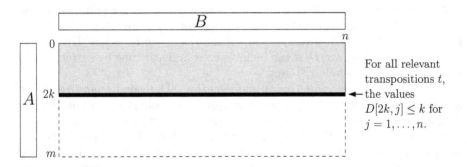

**Fig. 3.** For each relevant transposition, the values $D[2k, j] \leq k$ are recorded for $j = 1, \ldots, n$ by using sparse dynamic programming within rows $0, \ldots, 2k$ of the dynamic programming matrix (shaded area)

**Step 2: Match Extension.** Let *consPos* be an ordered list of the positions and lengths of maximal groups of consecutive values $D[2k, j] \leq k$. The pair $(x, y)$ appears in *consPos* if and only if $D[2k, j] \leq k$ for $j = x, \ldots, x + y - 1$ and it is not true that $D[2k, x - 1] \leq k$ or $D[2k, x + y] \leq k$. We do not actually store this complete list, as it can be recovered during a single linear time traversal of the list *matchPos*. We start the traversal from the beginning of *matchPos*, and every time an item $(x, y)$ is recovered, the traversal is suspended until the following checking phase is completed for that $(x, y)$. Fig. 4a illustrates. Then we recover the next item $(x, y)$, and so on until the *matchPos* has been completely traversed.

The final checking step applies a variation of a greedy edit distance algorithm [17] (Section 2.1) over each group of consecutive potentially match-seeding positions. This stage checks which, if any, of the cells $D[2k, j]$ recorded in *matchPos* array can be extended to a full match between $A$ and a prefix of $B$ within edit distance $k$. This is done separately for each item $(x, y)$ in *consPos*. The consecutive cells $D[2k, x], \ldots, D[2k, x + y - 1]$ that correspond to $(x, y)$ lie on the diagonals $x - 2k, \ldots, x - 2k + y - 1$ in $D$. It is enough to consider these diagonals using the greedy approach. Other type of diagonals are already known to contain a value larger than $k$, and values along a diagonal never decrease. The basic greedy algorithm needs to be modified in order to handle the fact that values on row $2k$ do not originally follow the greedy computing order.

For simplicity, we consider a $(m - 2k + 1) \times y$ submatrix $D'$ of $D$ in which $D'[i, j]$ is equal to $D[2k + i, x + j]$. We use the prime symbol to mean that a previously used construct addresses via $D'$ instead of $D$. For example the value $L'[r, q]$ equals $L[r, x - 2k + q] - 2k$ (see Section 2.1). We also use the notation $m' = m - 2k$.

We use two auxiliary arrays of size $y = O(n)$ in our greedy algorithm. The first array is called *initDist*, and it initially contains the values $initDist[j] = D'[0, j] = D[2k, x + j]$ for $j = 0, \ldots, y - 1$. These values are set by re-traversing the items of *matchPos* that were included in recovering the currently processed item $(x, y)$ of *consPos*. The second array, called *sortDist*, contains the indices $0, \ldots, y - 1$ sorted according to the value of the corresponding entry in *initDist*. Fig. 4b illustrates. This way $initDist[sortDist[z]] \leq initDist[sortDist[z + 1]]$ for $z = 0..y - 2$. Also *sortDist* can be initialized in linear time by extracting the values from the *matchPos(p)* lists.

We also use values *readyCount* and *currDist*. The value *readyCount* tells how many diagonals have been completely processed, and it is first initialized to 0. The value *currDist* tells the distance values that will be expanded/added next in $L'[r, q]$. That is, we will next update values of form $L'[currDist, q]$ among the diagonals that are still not completely processed. Initially *currDist* is equal to the minimum value in $initDist + 1$, and this can be first computed as $initDist[sortDist[0]] + 1$.

First all values $L'[r, q]$ are initialized by setting $L'[r, q] = -1$ for $q = 0, \ldots, y - 1$ and $r = 0, \ldots, k$. Then the list *initDist* is traversed for $j = 0, \ldots, y - 1$ and at each $j$ we set $L'[initDist[j], j] = lcp(2k + 1, x + j + 1)$. This lets the current

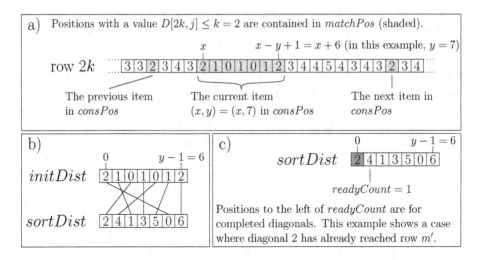

**Fig. 4.** Examples of the used variables and arrays

value $D'[0, j] = D[2k, x + j]$ propagate along the diagonal $j$ of $D'$ as long as there are consecutive matching characters, if any. This in part makes the values $L'[r, q]$ more compatible with the basic greedy algorithm.

Before beginning a new round of iterations, the greedy algorithm starts from position $z = readyCount$ of $sortDist$. The positions $readyCount, \ldots, y - 1$ of $sortDist$ always refer to the still unprocessed diagonals in ascending order of their last computed value (highest $r$ for which $L'[r, q]$ is computed along the corresponding diagonal $q$). Fig. 4c illustrates. At the beginning of each iteration, at position $z$, the algorithm checks if $z < y$ (did we already process all diagonals?) and $initDist[sortDist[z]] < currDist$ (did we already process all diagonals that may get new values $currDist$?). If the check is successful, then the diagonal that corresponds to $sortDist[z]$ is processed, as described soon, and the value of $z$ is incremented. If the check fails, the algorithm begins a new iteration if $readyCount < y$. The value $currDist$ is incremented before the next iteration.

Let us use the shorthand $q = sortDist[z]$ in the following. Processing diagonal $q$ begins by checking whether the value $L'[r-1, q]$ has been properly propagated to the two neighboring diagonals. If $r > 0$, $q > 0$ and $L'[r, q-1] < L'[r-1, q]+1$, we set $L'[r, q-1] = \min\{m', u' + lcp(2k + u' + 1, x + q + u' + 1)\}$, where $u' = L'[r-1, q]+1$. In similar way, if $r > 0$, $q < y-1$ and $L'[r, q+1] < L'[r-1, q]$, we set $L'[r, q+1] = \min\{m', u' + lcp(2k+u'+1, x+q+u'+1)\}$, where $u' = L'[r-1, q]$. This removes possible anomalies present due to the initial setting that is incompatible with the basic greedy algorithm. The cost of this extra processing is $O(1)$ per processed diagonal, but there could be more efficient ways in practice if one for example records which diagonals no longer need to be checked like this. We also note that this processing can lead into finding a match. But we let the match be discovered and handled sometime later by the following step (the same is true for the initial extension of the values from row $2k$).

Finally we update the value $L[r, q]$ itself. Because we do know that all diagonals have already been processed at least up to the value $currDist-1$, we may use the formula from section 2.1 as such, and set $L'[r, q] = \min\{m', s' + lcp(2k + s' + 1, x+q+s'+1)\}$, where $s' = \max\{L'[r-1, q-1], L'[r-1, q]+1, L'[r-1, q+1]+1\}$. If the new value is $L'[r, q] = m'$, we record the match, interchange the values $sortDist[z]$ and $sortDist[readyCount]$ in $sortDist$, and increment $readyCount$. The matches can be recorded into a size-$y$ array so that the occurrences can be reported in the end in $O(y)$ time. After recording value $L'[r, q]$, we also check whether the value affects the neighboring diagonals.

As a final note, we would like to note how to compute the $lcp$-values efficiently. A naive way would require us to do separate preprocessing for $A + t$ for each transposition $t$. But this can be avoided by doing the preprocessing for $\overline{A}$ and $\overline{B}$, where $|\overline{A}| = m - 1$, $|\overline{B}| = n - 1$, $\overline{A}_i = A_{i+1} - A_i$ for $i = 1, \ldots, m - 1$, and $\overline{B}_j = B_{j+1} - B_j$ for $j = 1, \ldots, n - 1$. Let $lcpT(i, j, t)$ be the value of $lcp(i, j)$ under transposition $t$. Now $lcpT(i, j) = 0$ if $A_i + t \neq B_j$, and otherwise $lcpT(i, j) = 1 + \overline{lcp}(i, j)$, where $\overline{lcp}(i, j)$ is as $lcp(i, j)$ but for $\overline{A}$ and $\overline{B}$ instead of $A$ and $B$.

## 6    Analysis

The time complexity of the algorithm described in the preceding section is as follows. The first stage of sparse computation takes a total time of $O(|\mathbb{M}(t)| \log \log k)$ over all relevant transpositions. These computations produce a total of $O(kn)$ match points for further checking. The initialization phase before the greedy algorithm takes linear time in the number of checked matches, that is, $O(1)$ per diagonal. The greedy algorithm spends $O(k)$ time per diagonal (ie. match point), which makes the checking cost $O(k^2n)$ time. This complexity dominates the overall time.

In order to conduct approximate string matching, we only need to change the stage of sparse dynamic programming. This does not change the cost, and hence we have $O(k^2n)$ time for approximate string matching.

The $O(k^2n)$ procedure for thresholded edit distance computation can be made $O(n + k^3)$ by considering only an $O(k)$ diagonal band in $D$. And this then enables us to obtain $O(n + d^3)$ time for computing $d = ed^t(A, B)$. We first do the computation with a limit $k = 1$, and then double $k$ until the resulting computation manages to find a value $ed^t(A, B) \leq k$. The last round spends $O((2d)^3) = O(d^3)$ time, and the total time is bounded by $O(d^3) \times \sum_{h=0}^{\infty}(1/2^h)^3 = O(d^3) \times O(1) = O(d^3)$. The term $O(n)$ comes from preprocessing (for example, reading the input strings).

The space requirements for sparse dynamic programming are typically dominated by the sets $\mathbb{M}(t)$, but a lower limit is the size of the input strings. When we restrict the matching sets to contain matches only within the area of $D$ that participates in sparse dynamic programming, we use $O(n + k^2)$ in thresholded and $O(n+d^2)$ space in edit distance computation. The space required in approximate string matching is $O(nk)$, but by imitating [13], this can be diminished to

$O(n+k^2)$ by processing $B$ in overlapping segments of length-$O(k)$. The space requirements for the second phase of extending the matches are of the same order: the dominating factor is the table $L'[r,q]$, whose size in each scenario happens to have the same asymptotic limit as the corresponding total size of the match sets (minus the basis $O(n)$).

# 7   Conclusion

Transposition invariant string matching is a relatively new and interesting problem proposed by Lemström and Ukkonen [8]. Previously Mäkinen et al. [12,13] have given $O(mn \log \log m)$ and $O(dn \log \log m)$ algorithms for transposition invariant edit distance computation, where $d$ is the transposition invariant distance between $A$ and $B$. The same authors also gave an $O(mn \log \log m)$ algorithm for transposition invariant approximate string matching. In the same work, Mäkinen et al. also stated the challenge to develop error-dependent algorithms for transposition invariant string matching. In this paper we have given an initial answer to that challenge by presenting a basic scheme for constructing transposition invariant algorithms that depend on $d$ or $k$. We then introduced an $O(n + d^3)$ algorithm for transposition invariant edit distance and an $O(k^2n)$ algorithm for transposition invariant approximate string matching. To the best of our knowledge, these are the first such error-dependent algorithms for this problem.

# References

1. Chang, W. I., and Lawler, E. L. Sublinear approximate string matching and biological applications. *Algorithmica*, 12:327–344, 1994.
2. Eppstein, D., Galil, Z., Giancarlo, R., and Italiano, G. F. Sparse dynamic programming I: linear cost functions. *Journal of ACM*, 39(3):519–545, 1992.
3. Gabow, H. N., Bentley, J. L., and Tarjan, R. E. Scaling and related techniques for geometry problems In *Proc. 16th ACM Symposium on Theory of Computing (STOC'84)*, 135–143, 1984.
4. Galil, Z., and Park, K. Dynamic programming with convexity, concavity and sparsity. *Theoretical Computer Science*, 92:49–76, 1992.
5. Hirschberg, D. S. Algorithms for the longest common subsequence problem *Journal of ACM*, 24:664–675, 1977.
6. Johnson, D. B. A priority queue in which initialization and queue operations take $O(\text{loglog } D)$ time. *Mathematical Systems Theory*, 15:295–309, 1982.
7. Jokinen, P., Tarhio, J., and Ukkonen, E. A comparison of approximate string matching algorithms. *Software Practice & Experience*, 26(12):1439–1458, 1996.
8. Lemström, K., and Ukkonen, E. Including interval encoding into edit distance based music comparison and retrieval. In *Proc. Symposium on Creative & Cultural Aspects and Applications of AI & Cognitive Science (AISB 2000)*, 53–60, 2000.
9. Lemström, K., Navarro, G., and Pinzon, Y. Practical algorithms for transposition-invariant string-matching To appear in *Journal of Discrete Algorithms*
10. Landau, G. M., and Vishkin, U. Fast parallel and serial approximate string matching *Journal of Algorithms*, 10:157–169, 1989.

11. Levenshtein, V. I. Binary codes capable of correcting spurious insertions and deletions of ones (original in Russian). *Russian Problemy Peredachi Informatsii 1*, 12–25, 1965.
12. Mäkinen, V, Navarro, G., and Ukkonen, E. Algorithms for transposition invariant string matching. In *Proc. 20th International Symposium on Theoretical Aspects of Computer Science (STACS'03)"*, LNCS 2607, 191–202, 2003.
13. Mäkinen, V, Navarro, G., and Ukkonen, E. Transposition invariant string matching. To appear in *Journal of Algorithms*.
14. Navarro, G. Multiple approximate string matching by counting. In *Proc. 4th South American Workshop on String Processing (WSP'97)*, 125–139, 1997.
15. Navarro, G. A guided tour to approximate string matching. *ACM Computing Surveys*, 33(1):31–88, 2001.
16. Sellers, P. The theory and computation of evolutionary distances: pattern recognition. *Journal of Algorithms*, 1:359–373, 1980.
17. Ukkonen, E. Algorithms for approximate string matching *Information and Control*, 64:100–118, 1985.
18. van Emde Boas, P. Preserving order in a forest in less than logarithmic time and linear space. *Information Processing Letters*, 6:80–82, 1977.
19. Wagner, R., and Fisher, M. The string-to-string correction problem. *Journal of ACM*, 21(1):168–173, 1974.

# Fast Plagiarism Detection System

Maxim Mozgovoy[1], Kimmo Fredriksson[1,*], Daniel White[2], Mike Joy[2],
and Erkki Sutinen[1]

[1] Department of Computer Science, University of Joensuu,
PO Box 111, FIN–80101 Joensuu, Finland
{Maxim.Mozgovoy, Kimmo.Fredriksson, Erkki.Sutinen}@cs.joensuu.fi
[2] Department of Computer Science, University of Warwick, Coventry CV4 7AL, U.K.
{D.R.White, M.S.Joy}@warwick.ac.uk

*Introduction.* The large class sizes typical for an undergraduate programming course mean that it is nearly impossible for a human marker to accurately detect plagiarism, particularly if some attempt has been made to hide the copying. While it would be desirable to be able to detect all possible code transformations we believe that there is a minimum level of acceptable performance for the application of detecting student plagiarism. It would be useful if the detector operated at a level that meant for a piece of work to *fool* the algorithm would require that the student spent a large amount of time on the assignment and had a good enough understanding to do the work without plagiarising.

*Previous Work.* Modern plagiarism detectors, such as Sherlock [3], JPlag [5] and MOSS [6] use a tokenization technique to improve detection. These detectors work by pre-processing code to remove white-space and comments before converting the file into a tokenized string. The main advantage of such an approach is that it negates all lexical changes and a good token set can also reduce the efficacy of many structural changes. For example, a typical tokenization scheme might involve replacing all identifiers with the <IDT> token, all numbers by <VALUE> and any loops by generic <BEGIN_LOOP>...<END_LOOP> tokens. Our algorithm also makes use of tokenised versions of the input files and we use suffix arrays [4] as our index data structure to enable efficient comparisons.

While all the above-mentioned systems use different algorithms to each other, the core idea is the same: a many-to-many comparison of all files submitted for an assignment should produce a list sorted by some similarity score that can then be used to determine which pairs are most likely to contain plagiarism. A naïve implementation of this comparison, such as that used by Sherlock or JPlag, results in $O(f(n)N^2)$ complexity where $N$ is the size (number of files) of the collection, and $f(n)$ is the time to make the comparison between one pair of files of length $n$. Without loss of detection quality, our method achieves $O(N(n + N))$ average time by using indexing techniques based on suffix arrays. If the index structure becomes too large, it can be moved from primary memory to secondary data storage without significant loss of efficiency [2].

The approach we describe can be also used to find similar code fragments in a large software system. In this case the importance of fast algorithm is especially

---

* Supported by the Academy of Finland, grant 202281.

M. Consens and G. Navarro (Eds.): SPIRE 2005, LNCS 3772, pp. 267–270, 2005.

---

**Algorithm 1.** Compare a File Against an Existing Collection

---

1    $p = 1$ // the first token of $Q$
2    WHILE $p \leq q - \gamma + 1$
3        find $Q[p...p + \gamma - 1]$ from the suffix array
4        IF $Q[p...p + \gamma - 1]$ was found
5            UpdateRepository
6            $p = p + \gamma$
7        ELSE
8            $p = p + 1$
9    FOR EVERY file $F_i$ in the collection
10    $Similarity(Q, F_i) = MatchedTokens(F_i)/q$

---

high due to large file collection size. The Dup tool [1] uses parametrized suffix trees to solve this task, but the algorithms are relatively complex compared to our approach.

*Algorithms and Complexity.* Our proposed system is based on an index structure built over the entire file collection. Before the index is built, all the files in the collection are tokenized. This is a simple parsing problem, and can be solved in linear time. For each of the $N$ files in the collection, The output of the tokenizer for a file $F_i$ is a string of $n_i$ tokens. The total number of tokens is denoted by $n = \sum n_i$.

We use suffix array as an index structure. A suffix array is a lexicographically sorted array of all suffixes of a given string [4]. The suffix array for the whole document collection is of size $O(n)$. We consider the total memory requirements to be acceptable for modern hardware. A suffix array allows us to rapidly find a file (or files), containing any given substring. This is achieved with a binary search, and requires $O(m + \log_2 n)$ time on average, where $m$ is the length of the substring (it is also possible to make this the worst case complexity, see [4]). The array can be constructed in time $O(n \log n)$, assuming atomic comparison of two tokens.

Algorithm 1 is intended for finding all files within the collection's index that are similar to a given query file. It tries to find the substrings of the tokenised query file, $Q[1..q]$, in the suffix array, where $q$ is the number of tokens. Matching substrings are recorded and each match contributes to the similarity score. The algorithm takes contiguous non-overlapping token substrings of length $\gamma$ from the query file and searches all the matching substrings from the index. These matches are recorded into a 'repository'. This phase also includes a sanity check as overlapping matches are not allowed.

The similarity between the file $Q$ being tested and any file $F_i$ in the collection is just a number of tokens matched in the collection file divided by the total number of tokens in the test file (so it is a value between 0 and 1), i.e.

$$Similarity(Q, F_i) = MatchedTokens(F_i)/q,$$

In Algorithm 2, we encounter two types of collisions. The first one appears when more than one match is found in the same file. If several matches that are found correspond to the same indexed file, these matches are extended to

**Algorithm 2.** Update the Repository

| | |
|---|---|
| 1 | Let $S$ be the set of matches of $Q[p...p + \gamma - 1]$ |
| 2 | IF some of the strings in $S$ are found in the same file /* collision of type 1 */ |
| 3 |    leave only the longest one |
| 4 | FOR every string $M$ from the remaining list $S$ |
| 5 |   IF $M$ doesn't intersect with any repository element |
| 6 |     insert $M$ to the repository |
| 7 |   ELSE IF $M$ is longer than any conflicting rep. element /* collision of type 2 */ |
| 8 |     remove all conflicting repository elements |
| 9 |     insert $M$ to the repository |

$\Gamma$ tokens, $\Gamma \geq \gamma$, such that only one of the original matches survives for each indexed file. Therefore, for each file in the index, the algorithm finds all matching substrings that are longer than other matching substrings and whose lengths are at least $\gamma$ tokens. The second one is the reverse of the first problem: we should not allow the situation when two different places in the input file correspond to the same place in some collection file. To resolve the difficulty we use 'longest wins' heuristics. We sum the lengths of all the previous matches that intersect with the current one, and if the current match is longer, we use it to replace the intersecting previous matches.

The complexity of Algorithm 1 is highly dependent on the value of the $\gamma$ parameter. Line 3 of Algorithm 1 takes $O(\gamma + \log n)$ average time, where $n$ is the total number of tokens in the collection (assuming atomic token comparisons). If we make the simplifying assumption that two randomly picked tokens match each other (independently) with fixed probability $p$, then on average we obtain $np^\gamma$ matches for substrings of length $\gamma$. If $Q$ was found, we call Algorithm 2. Its total complexity is, on average, at most $O((q/\gamma \cdot np^\gamma)^2)$. To keep the total average complexity of Algorithm 1 to at most $O(q(\gamma + \log n))$, it is enough that $\gamma = \Omega(\log_{1/p} n)$. This results in $O(q \log n)$ total average time. Since we require that $\gamma = \Omega(\log n)$, and may adjust $\gamma$ to tune the quality of the detection results, we state the time bound as $O(q\gamma)$. Finally, the scores for each file can be computed in $O(N)$ time. To summarize, the total average complexity of Algorithm 1 can be made $O(q(\gamma + \log n) + N) = O(q\gamma + N)$. The $O(\gamma + \log n)$ factors can be easily reduced to $O(1)$ (worst case) using suffix trees [7] with suffix links, instead of suffix arrays. This would result in $O(q + N)$ total time.

Note that we have excluded the tokenization of $Q$ and that we have considered the number of tokens rather than the number of characters. However, the tokenization is a simple linear time process, and the number of tokens depends linearly on the file length.

To compare every file against each other, we can just run Algorithm 1 for every file in our collection. After that, every file pair gets two scores: one when file $a$ is compared to file $b$ and one when the reverse comparison happens, as the comparison is not symmetric. We can use the average of these scores as a final score for this pair.

Summing up the cost of this procedure for all the $N$ files in the collection, we obtain a total complexity of $O(n\gamma + N^2)$, including the time to build the suffix array index structure. With suffix trees this can be made $O(n + N^2)$.

*Evaluation of the System.* It is not feasible in the nearest future to compare our system's results with a human expert's opinion on real-world datasets as a human would not have the time to conduct a thorough comparison of every possible file pair. However, we can examine the reports that are produced by different plagiarism detection software when used on the same dataset. The systems used for the analysis include MOSS [6], JPlag [5] and Sherlock [3]. Every system printed a report about the same real collection, consisting of 220 undergraduate student's Java programs.

The simple approach (to consider only detection or rejection) allows us to organize a 'voting' experiment. Let $S_i$ be the number of 'jury' systems (MOSS, JPlag and Sherlock), which marked file $i$ as suspicious. If $S_i \geq 2$, we should expect our system to mark this file as well. If $S_i < 2$, the file should, in general, remain unmarked.

For the test set consisting of 155 files marked by at least one program, our system agreed with the 'jury' in 115 cases (and, correspondingly, disagreed in 40 cases). This result is more conformist than the results obtained when the same experiment was run on the other 3 tested systems. Each system was tested while the other three acted as jury.

*Conclusions.* We have developed a new fast algorithm for plagiarism detection. Our method is based on indexing the code database with a suffix array, which allows rapid retrieval of blocks of code that are similar to the query file. This idea makes rapid pairwise file comparison possible. Evaluation shows that this algorithm's quality is not worse than the quality of existing widely used methods, while its speed performance is much higher. For the all-against-all problem our method achieves $O(\gamma n)$ (with suffix arrays) or $O(n)$ (with suffix trees) average time for the comparison phase. Traditional methods, such as JPlag, need at least $O((n/N)^2 N^2) = O(n^2)$ average time for the same task. In addition, computing the similarity matrix takes $O(N^2)$ additional time, and this cannot be improved, as it is also the size of the output.

# References

1. B. S. Baker. Parameterized Duplication in Strings: Algorithms and an Application to Software Maintenance. *SIAM Journal on Computing*, 26(5):1343–1362, 1997.
2. D. Clark and J. Ian Munro. Efficient suffix trees on secondary storage. *Proceedings of the seventh annual ACM-SIAM symposium on Discrete algorithms*, 1996.
3. M. S. Joy and M. Luck. Plagiarism in programming assignments. *IEEE Transactions on Education*, 42(2):129–133, 1999.
4. U. Manber and G. Myers. Suffix arrays: a new method for on-line string searches. In *Proceedings of SODA '90*, 319–327. SIAM, 1990.
5. L. Prechelt, G. Malpohl, and M. Phlippsen. JPlag: Finding plagiarisms among a set of programs. Technical report, Fakultat for Informatik, Universitat Karlsruhe, 2000. http://page.mi.fu-berlin.de/~prechelt/Biblio/jplagTR.pdf.
6. S. Schleimer, D. S. Wilkerson, and A. Aiken. Winnowing: local algorithms for document fingerprinting. In *Proceedings of SIGMOD '03*, 76–85. ACM Press, 2003.
7. E. Ukkonen. On-line construction of suffix trees. *Algorithmica*, 14:249–260, 1995.

# A Model for Information Retrieval Based on Possibilistic Networks

Asma H. Brini, Mohand Boughanem, and Didier Dubois

IRIT, 118, route de Narbonne, Cedex 4, Toulouse, France

**Abstract.** This paper proposes a model for Information Retrieval (IR) based on possibilistic directed networks. Relations documents-terms and query-terms are modeled through possibility and necessity measures rather than a probability measure. The relevance value for the document given the query is measured by two degrees: the necessity and the possibility. More precisely, the user's query triggers a propagation process to retrieve necessarily or at least possibly relevant documents. The possibility degree is convenient to filter documents out from the response (retrieved documents) and the necessity degree is useful for document relevance confirmation. Separating these notions may account for the imprecision pervading the retrieval process. Moreover, an improved weighting of terms in a query not present in the document is introduced. Experiments carried out on a sub-collection of CLEF, namely LeMonde 1994, a French newspapers collection, showed the effectiveness of the model.

## 1 Introduction

The Information Retrieval (IR) process consists in selecting among a large collection a set of documents that are relevant to a user's query. The set of retrieved documents in answer to a query does not usually correspond to the set of documents that are relevant to the user need. For an efficient Information Retrieval System (IRS) these two sets must be equal as often as possible. The relevance of a document to a query is usually interpreted by most of IR models, vector space [14], probabilistic [12][13][18], inference and belief networks [20][11][17], as a score computed by summing the inner products of term weights in the documents and query representations.

Whatever the used model, the response to a user need is a list of documents ranked according to a relevance value. Many approaches consider term weights as probability of relevance. In such models the incompleteness of information is not considered when representing or evaluating documents given a query. Notions of certainty or possibility are not distinguished in this relevance computing. Yet, the rough nature of document descriptions (a multi set of terms) and of the query description (a list of terms) is hardly compatible with the high precision of relevance values obtained by current methods. The aim of this work is to propose an IR model based on possibilistic networks. Instead of using a unique

M. Consens and G. Navarro (Eds.): SPIRE 2005, LNCS 3772, pp. 271–282, 2005.

relevance value, we propose a possibilistic approach for computing relevance. The relevance of a document to a given query is measured using two values i.e. the necessity and the possibility of relevance. The possibility of relevance is meant to eliminate irrelevant documents (weak plausibility). The necessity of relevance focuses attention on what looks very relevant.

We briefly discuss the use of Bayesian networks in Information Retrieval in section 2. In section 3 we present a general possibilistic approach for IR. We separate reasons for rejecting a document as irrelevant from reasons for selecting it by means of two evaluations: possibility and necessity. This approach is a significant extension of a previous attempt based on possibilistic networks [3]. This extension results from difficulties to find an efficient way of querying the system. It is too restrictive (and demanding) to aggregate query terms by an *AND* operator when the only information we have is a set of terms. Thus, the idea is to aggregate query terms by conjunction or disjunction operators according to different aggregation methods when no information is given about the logical description of the query. To provide for such a flexibility, a query node is required in the model architecture. We discuss in the latter section the experiments we carried out, showing the importance of weighting schemes we use and by comparing our approach by existing known models on a realistic benchmark.

## 2   Related Works

We discuss in this section the use of Bayesian networks [9][7] in IR, with a view to later comparing it to our model. Bayesian Nets (BNs) [7] provide an efficient tool for storing and reasoning from large probability distributions involving many discrete variables. When probability measures depend on a subjective view, probabilities do not necessarily interpret relative frequencies (related to chance events) but account for degrees of belief (conditional or not). BNs have been used in IR since 1990. The well known IR models using BNs are Inference Networks (INs) and Belief Networks. INs are used in INQUERY system [20] and their efficiency is related to distinct IR approaches and their combination in one model. This system evaluates the belief in a document with respect to a query, and a list of weighted documents is retrieved. Belief Networks [11][17] have been used to "model knowledge derived from past queries and combine it with the vector space model" [11]. The ranking of a document is based on the similarity between document $d_j$ and query $Q$, computing the probability $P(d_j = 1/Q = 1)$. $Q = 1$ and $d_j = 1$ means respectively $Q$ activated and $d_j$ activated. Recent researchers [4] [5], designed the Bayesian Network Retrieval Model, with a flexible topology that can take into account term relationships as well as document relationships.

The meaning of document and query representations for all these models and relevant document retrieval is identical. For these models a unique degree of relevance is computed and generally weights given to arcs when term nodes are instantiated are based on a combination of $tf - idf$. However, the model we propose provides a different meaning to document and query representations as well as to the selection of a document given a query. One way to solve our key issue can be given by the use of possibilistic networks.

## 3   The Possibilistic Model

One main original idea behind our possibilistic model concerns the relevance interpretation. Instead of using a unique relevance value of a document with respect to a query, we propose a possibilistic approach [6]. A possibility distribution $\pi$ is a mapping from $U$ to $[0,1]$. $\pi(u)$ evaluates the plausibility that $u$ is the actual value of some variable to which $\pi$ is attached. $\pi(u) = 0$ means that $u$ is impossible but $\pi(u) = 1$ only indicates a lack of surprise about $u$. A proposition $A$ is evaluated by its degree of possibility $\Pi(A) = max_{u \in A} \pi(u)$ and its degree of necessity (or certainty) $N(A) = 1 - \Pi(\overline{A})$ where $\overline{A}$ is the complement of $A$ [6].

### 3.1   Model Architecture

Our approach is based on possibilistic directed networks [1][2], where relations between documents, query and term nodes are quantified by possibility and necessity measures. The proposed network architecture appears on Figure (1). From a qualitative point of view, nodes in the graphical component represent query, index terms and documents and the graph reflects the (in)dependence relations existing between nodes. Document and query nodes have binary domains. A document $D_j$ is invoked or not, taking its values in the domain $\{d_j, \overline{d_j}\}$. The activation of a document node, i.e. $D_j = d_j$ (resp. $\overline{d_j}$) means that a document is relevant or not. A query $Q$ takes its values in the domain $\{q, \overline{q}\}$. As only the positive query instantiation is of interest, we consider $Q = q$ only, and denote it as $Q$. The domain of an index term node $T_i$, is $\{t_i, \overline{t_i}\}$. $(T_i = t_i)$ means a term $t_i$ is present in the object (document or query) and thus is *representative* of the object to a certain degree. A *non-representative* term, denoted by $\overline{t_i}$, is a term absent from the object.

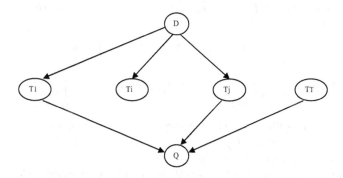

**Fig. 1.** Model architecture

Let $\mathcal{T}(D_j)$ (resp. $\mathcal{T}(Q)$) be the set of terms indexed in document $D_j$ (resp. in the query). The query expresses a request for documents containing some terms but excluding other terms. Arcs are directed from document node to index term nodes defining dependence relations existing between index terms and documents. The values taken by index term nodes depend on the document node (parent) instantiation. The query instantiation only gives evidence to propagate through invoked terms, thus arcs are directed from term to query nodes. The terms appearing in the user query form the parent set of $Q$ in the graph. There is an instantiation of the parent set $Par(Q)$ of the query $Q$ that represents the query in its most demanding (conjunctive) form. Let $\theta^Q$ be such an instantiated vector. Any instance of the parent set of $Q$ is denoted $\theta$. We show, later in this section, how values are assigned to arcs. For simplicity a query is supposed to contain positive terms only.

## 3.2   Evaluation Process

In this model, the propagation process is similar to the probabilistic Bayesian propagation [1][2]. The query evaluation consists in the propagation of new evidence through activated arcs to retrieve relevant documents. Our model should be able to infer propositions like:

- It is plausible to a certain degree that the document is relevant for the user need, denoted by $\Pi(d_j \mid Q)$
- It is almost certain (in possibilistic sense) that the document is relevant to the query, denoted by $N(d_j \mid Q)$

The first kind of proposition is meant to eliminate irrelevant documents (weak plausibility). The second answer focuses attention on what looks very relevant. Under a possibilistic approach, given the query, we are thus interested in retrieving necessarily or at least possibly relevant documents. Thus, the propagation process evaluates the following quantities

$$\Pi(d_j \mid Q) = \frac{\Pi(Q \wedge d_j)}{\Pi(Q)}, \quad N(d_j \mid Q) = 1 - \Pi(\overline{d_j} \mid Q) \text{ where } \Pi(\overline{d_j} \mid Q) = \frac{\Pi(Q \wedge \overline{d_j})}{\Pi(Q)}$$

The possibility of $Q$ is $\Pi(Q) = max(\Pi(Q \wedge d_j), \Pi(Q \wedge \overline{d_j}))$ so that $\Pi(d_j \mid Q) = min(1, \frac{\Pi(Q \wedge d_j)}{\Pi(Q \wedge \overline{d_j})})$ [6][1].

We are interested in defining $\Pi(Q \wedge D_j)$. Given the model architecture, it is of the form:

$$\max_{\theta}(\Pi(Q \mid \theta) \cdot \prod_{T_i \in \mathcal{T}(Q) \wedge \mathcal{T}(D_j)} \Pi(\theta_i \mid D_j) \cdot \Pi(D_j) \cdot \prod_{T_k \in \mathcal{T}(Q) \setminus \mathcal{T}(D_j)} \Pi(\theta_k)) \quad (1)$$

for $\theta$ being the possible instances of the parent set of $Q$, $\theta_i$ is the instance of $T_i$ in $\theta$. This is computed for $D_j \in \{d_j, \overline{d_j}\}$. Note that terms $T_i \in \mathcal{T}(D_j) \setminus \mathcal{T}(Q)$ are not involved in this computation.

The top retrieved documents are those having a necessary relevance value greater than 0, and the set of possibly relevant documents are retrieved as a second choice.

# 4   Query Aggregation

The possibility of the query given the index terms depend on query interpretation. Several interpretations exist, whereby query terms ($\Pi(Q \mid \theta)$) are defined as expressing *conjunction, disjunction...* or, like in Bayesian probabilistic networks, by *sum* and *weighted sum* as proposed for example in the works of Turtle [20]. The basic idea is that for any instantiation $\theta$, the conditional possibility $\Pi(Q \mid \theta)$ is specified by some aggregation function merging elementary possibilistic likelihood functions $\Pi(Q \mid \theta_i)$. Each $\Pi(Q \mid \theta_i)$ is the weight of instance $\theta_i$ in view of its conformity with the instantiation of $T_i$ in the query (in $\theta^Q$). We do not consider relations that may exist between terms even if the use of networks would make it possible. Hence, it is difficult (space and time consuming) to store all possible query term configurations or to compute them when the query is submitted to the system. A reasonable organization is to let each query term bear a weight and to compute the weight of joint terms in the query. When the user does not give any information on the aggregation operators to be used, the only available evidence one can use is the importance of each query term in the collection. This evidence is available for single terms that form the query. We give in what follows different manners to aggregate query terms.

## 4.1   Conjunctive, Disjunctive and Quantified Aggregations

For a Boolean *AND* query, the evaluation process searches documents containing all query terms. Then, $\Pi(Q \mid \theta_i) = 1$ if $\theta_i = \theta_i^Q$, and 0 otherwise. The possibility of the query $Q$ given an instance $\theta$ of all its parents, is given by $\Pi(Q \mid \theta)$, where $\Pi(Q \mid \theta) = 1$ if $\forall T_i \in Par(Q)$ $\theta_i = \theta_i^Q$ means that the term $T_i$ in $\theta$ is instantiated as in the query. Generally this interpretation of the query is too demanding.

For a Boolean *OR* query, the document is already somewhat relevant if there exists a query term in it. The final document relevance should increase with the number of present query terms. The pure disjunctive query is handled by changing $\forall$ into $\exists$ in the conjunctive query. But this interpretation is too weak to discriminate among documents.

Assume a query is considered satisfied by a document if they have at least $K$ common terms. Consider an increasing function, $f(\frac{K(\theta)}{n})$, where $K(\theta)$ is the number of terms in the query instantiated like in a given configuration $\theta$ of $Par(Q)$, given that the query contains $n$ terms. It is supposed that $f(0) = 0$ and $f(1) = 1$. $f$ is a fuzzy quantifier [22]. For instance, $f(i/n) = 1$ if $i \geq \frac{K(\theta)}{n}$, and 0 otherwise, requires that at least $K$ terms in the query are in conformity with $\theta$. But more generally $f$ can be a non-Boolean function.

The quantifier approach to computing the possibility of the query $Q$ given an instance $\theta$ of all its parents, is given by:

$$\Pi(Q \mid \theta) = f(\frac{K(\theta)}{n}) \tag{2}$$

## 4.2   Noisy OR

In general, we may assume that the conditional possibilities $\Pi(Q \mid \theta_i)$ are not Boolean-valued, but depend on suitable evaluations of terms $t_i$. A possible query term combinations can be "noisy-Or" [9] based. It means that $\Pi(Q \mid \theta)$ is evaluated in terms of conditional possibilities of the form $\Pi(Q \mid t_i \wedge_{k \neq i} \overline{t_k})$ using a probabilistic sum. The primitive terms in a noisy OR are $\Pi(Q \mid t_i \wedge_{k \neq i} \overline{t_k}) = \frac{idf_i}{N} = nidf_i$, denoted $1 - q_i$ for simplicity. Then

$$\Pi(Q \mid \theta) = 0 \text{ if } \not\exists \text{ i s.t. } \theta_i = \theta_i^Q \tag{3}$$
$$= \frac{1 - \prod_{i:t_i=\theta_i=\theta_i^Q} q_i}{1 - \prod_{T_k \in Par(Q)} q_k} \text{ otherwise}$$

Only positive terms in the query configuration appear on the numerator. The more query terms present in the document with the same positive instantiation as in the query is, the higher the relevance of the document will be[1].

## 5   Arc Values

In the first part of this section, we define term-document arc values depending on term instantiations. Then we propose a weighting scheme for root nodes. We also weight query-term arcs to aggregate query terms. For the proposed approach we give a weight to prior document possibility, not a uniform one like in inference network model [20] but based on its length.

### 5.1   Document-Term Arcs

To evaluate the possibility and necessity of a document relevance we need to express and define relevance represented by arcs in the network. Our approach tries to distinguish between terms which are possibly representative of documents (whose absence rules out a document) and those which are necessarily representative of documents, i.e. terms which suffice to characterize documents.

*Postulate 1*: A term is all the more possibly representative of a document as it appears frequently in that document;

*Postulate 2*: A term is all the more necessarily representative of a document as it appears more frequently in that document and it appears fewer times in the whole collection.

According to Postulate 1, $\Pi(t_i/d_j)$ can be estimated from the frequency $tf$:

$$\Pi(t_i/d_j) = nf_{t_{ij}} \tag{4}$$

where $nf_{t_{ij}}$ is normalized term frequency, $nf_{t_{ij}} = \frac{tf_{ij}}{max_{\forall t_k \in d_j}(tf_{kj})}$.

---

[1] We assume Closed World Assumption (CWA): $\Pi(Q \mid t_i) = \Pi(Q \mid t_i \wedge_{k \neq i} \overline{t_k})$.

A term weight 0 means that a term is not compatible with the document. If it is equal to 1, then the term is possibly representative or relevant to describe the document. Here, "representative" should not be necessarily understood in the general sense, but only as "useful to retrieve this document in the collection". If a term is representative of a document in the general sense, it may not be of much help to retrieve a document. Namely, for a document in a collection devoted to fuzzy sets, the word "fuzzy" is very representative, but it is potentially useless as it does not characterize it among other documents in the same area. Note that the possibility degree is normalized (its maximum is 1). A term not appearing in a document is not compatible with it, and if it appears with a maximal frequency it is considered as a possible candidate to represent it. A discriminant term in a collection is a term which appears (often) in few documents of the whole collection. We assume that a discriminant term is a term which is necessarily representative of a document thus certainly contributes to selecting a document. We define the necessary relevance degree, $\phi_{ij}$, of term $i$ to represent a document $j$ as a weight of the form:

$$\phi_{ij} = \mu_1 \left(\frac{N}{n_i}\right) * \mu_2 \left(nf_{t_{ij}}\right) \tag{5}$$

where $*$: product operator and $\mu_1, \mu_2$: normalization functions. For instance, $\mu_1$ logarithmic function, $\mu_2$ identity function, and then $\phi_{ij} = \frac{log \frac{N}{n_i}}{log(N)} \cdot ntf_{ij}$ This degree of necessarily relevance shows the necessity for a term to imply a document and thus works to retrieve a document by:

$$N(t_i \rightarrow d_j) = \phi_{ij} \tag{6}$$

Since, $\Pi(\overline{d_j}) = 1$ a priori, $\Pi(t_i \mid \overline{d_j}) = \Pi(t_i \wedge \overline{d_j}) = 1 - N(t_i \rightarrow d_j) = \phi_{ij}$, while $\Pi(\overline{t_i} \mid \overline{d_j}) = 1$. In table 1, we summarize the conditional possibilities of term instantiations given the document instantiations.

**Table 1.** Conditional possibility table $\Pi(T_i \mid D_j)$

|          | $d_j$          | $\neg d_j$       |
|----------|----------------|------------------|
| $t_i$    | $nf_{t_{ij}}$  | $1 - \phi ij$    |
| $\neg t_i$ | 1            | 1                |

## 5.2   Root Terms

Weights assigned to terms are mostly the result of a frequentist view because no other information is available. Several works in the literature focus on the definition and on the valuation of the term importance among a collection of documents [12] [18] [8] [10]. Those problems are dealt with, using semantic [21] or a statistical [14][19], [8][23], or probabilistic [10] point of views.

For our approach, when computing the relevance degree to a document given a query, weights must be assigned not only to common terms between the document and the query but also to terms that are present in the query and absent from the document. To be sure to pick up the relevant set of documents, terms must have a discrimination power. The more important is the discrimination power assigned to a term the more efficiently this term helps in the retrieval of documents. This power depends on the distribution of the term among the collection and this distribution is quantified by the density of this term in documents or by the importance of terms across the collection. The less peaked is the density distribution, the less discriminant is the term. We define a new discriminative factor based on entropy, denoted by $df_i$ for a term $i$ in a collection. It improves over the usual $idf$. The notion of entropy was firstly proposed to evaluate how peaked is a density [16]. The weighting scheme aims to maximize the entropy of the density of the term across the collection.

$$df_i = -\sum_j p_{ij} \, log \, p_{ij}; \tag{7}$$

$$p_{ij} = \frac{\frac{tf_{ij}}{l_j}}{\sum_k \frac{tf_{ik}}{l_k}}$$

The lower the $df_i$ value is, the more picked the density distribution, and the more interesting the term for retrieving some documents. Thus:

$$\forall t_i \notin \mathcal{T}(D_j), \ \Pi(\theta_i) = 1 \text{ if } \theta_i^Q = \overline{t_i}$$

$$= \frac{df_i}{max_{k \in T} df_k} = ndf_i \text{ otherwise} \tag{8}$$

where $T$ is the set of terms in the collection.

The more discrimination power the term has, the less documents $d_j$ not containing it are relevant to the query and the lower is $\Pi(Q \wedge d_j)$. It is clear that two documents with same $idf's$ may have different $df$ values. $idf$ is less discriminant than $df$.

## 5.3   Prior Possibility of Documents

In absence of information, the *a priori* possibility on a document node is uniform $(\Pi(d_j) = \Pi(\overline{d_j}) = 1)$. Actually, we can obtain information on a document given the importance of its terms, its length etc. This knowledge can be given for instance, by the user, the user profile etc. Hence, for example, if we are interested in retrieving long documents, we define the prior possibility of activating a document $D_j = d_j$, $\Pi(d_j) = \frac{l_j}{max_{k=1,..,N} l_k} = nl_{d_j}$ where $l_j$ is the length in frequency of document $d_j$; $l_j = \sum_i tf_{ij}$. The shorter the document is, the less possibly relevant it is. Besides, $\Pi(\overline{d_j}) = 1$.

## 6   Experiments and Results

The experiments were undertaken on the dataset *Le Monde*. The aim of these experiments is to evaluate the reliability of the proposed approach based on different weighting schemes specifically one considering the term discriminative power based on entropy $ndf$ and the second one the known $nidf$ and its evaluation process, i.e. its evaluation of documents given a query, based on two measures of relevance. The results obtained are discussed below and then compared to OKAPI's weighting scheme.

## 6.1   The Dataset Collection

Experiments are carried out on a sub-collection of CLEF, namely LeMonde 1994, a French newspapers collection (44013 documents and 34 queries, 154 MB). For each query, 1000 top documents are retrieved. We give the results evaluation by means of $P5, P10, ...$ i.e. the precision at point 5, which is the ratio of relevant documents among the 5 top retrieved documents, among the 10 ones, etc.

## 6.2   Parameters

Our model is based on two measures that evaluate two kinds of relevance i.e. the necessary and the possible one. The trust in necessarily relevant documents is greater than in possibly relevant documents. If less than 1000 necessary documents are retrieved we complete to 1000 by adding possibly relevant documents.

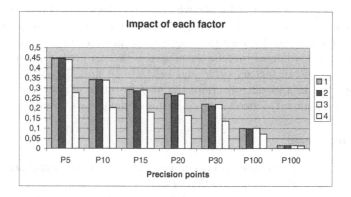

**Fig. 2.** Weights impact

As shown in sections above and by formula 1, different pieces of information are used for the document evaluation given the query: the distribution of terms inside a collection, the a priori possibility of documents, the importance of query term absent from documents ($ndf$ factor). The tuning of parameters is described in table 2. *Yes* indicates the parameter is used and *No* that it is not used in the computations. Figure 2 shows the impact on precision points of each piece of information. As example, in case 1, the normalized length($nl_d$), the normalized entropy ($ndf' = \frac{ndf}{max_{t \in T(Q)} ndf_t}$) and the Noisy Or are used for the document evaluation given the query. The $ndf'$ as shown in section 5.2 is normalized given

**Table 2.** Parameters tuning

|   | $nl_d$ | $ndf'$ | Noisy Or | $ndf$ |
|---|---|---|---|---|
| 1 | Yes | Yes | Yes | No |
| 2 | Yes | No | Yes | Yes |
| 3 | Yes | Yes | No | No |
| 4 | No | Yes | Yes | No |

all terms of the collection and a second time given the query terms. When not $ndf$ nor $ndf'$ factors are considered inside the computations the results decrease strongly (about 90% less for average precision)[2]. It is the case especially for query terms having a high potential discrimination power between documents, i.e., terms which have high density in few documents of the collection. In a such case, there are no necessary documents for an important number of queries. This $ndf$ factor strongly decreases the relevance of documents not containing "interesting" terms. In the case of removing the documents length, from the propagation process, the number of necessary relevant documents increases. This is because short documents relevance grows up as they contain not interesting terms: documents with higher $ndf$ (for root terms) than $nidf$ (for present terms). When the Noisy or is not considered, weights affeted to query terms present in documents equal 1.

The average precision is higher when $nidf$ is kept out computations than when $ndf$ is removed from the propagation computations. In our model, the $df$ factor is used once to decrease the relevance of documents not containing "interesting" terms, whereas $nidf$ factor is used for terms present in the document under concern. Both factors ($nidf$ and $ndf$) try to find the extent to which a term is specific in a given collection, the $nidf$ in terms of presence/absence (of a term in a document) whereas $ndf$ in terms of density distribution.

### 6.3   Comparison

Figure 3 shows the precision points comparison between our approach ($Pi$-nets) and the probabilistic approach ($OKAPI$).

The comparisons is between the BM25 weighting scheme (OKAPI) [12][13] and our approach. We can note from figure above (figure 3) that precision points are better for our approach for any points of precision. It improves average precision by 7.99% compared to OKAPI system.

## 7   Conclusion

This paper presents a new IR approach based on possibility theory and a new term discrimination index based on entropy. In a general way, the possibility measure is convenient to filter out documents (or index terms from the set of

---

[2] This result does not appear in the figure 2.

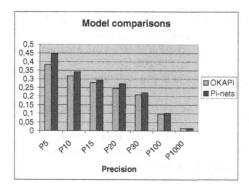

**Fig. 3.** Models comparison

representative terms of documents) whereas necessity captures document relevance (or index terms representativeness). The originality of the proposed approach is due to the use of information about the distribution of terms across the collection (example: use of $df$), and a new way of indexing documents by separating different kinds of information. The first experiments carried out on LeMonde are promising. Other experiments on the benchmark $WT10G$ of TREC are also promising as we seem to obtain in most cases better average precision than known models.

## Acknowledgments

Thank to Luis de Campos for helpful discussions on the original possibilistic model.

## References

1. S. Benferhat, D. Dubois, L.Garcia, H. Prade: Possibilistic logic bases and possibilistic graphs. In Proc. of the 15th Conference on Uncertainty in Artificial Intelligence, (1999) 57-64.
2. C. Borgelt, J. Gebhardt and R. Kruse: Possibilistic graphical models. Computational Intelligence in Data Mining, CISM Courses and Lectures 408, Springer, Wien, (2000) 51-68.
3. A.H. Brini and M. Boughanem and D. Dubois: Towards a possibilistic approach for Information Retrieval. Proc. EUROFUSE Data and Knowledge Engineering, Warsaw, (2004) 92-102
4. L.M. Campos, J.M. Fernandez-Luna, J.F. Huete. The BNR Model: Foundations and Performance of Bayesian Network-based Retrieval Model. JASIST, (2003) 54(4),302-313.
5. F. Crestani, L.M. de Campos, J.M. Fernandez-Luna, J.F. Huete: A Multi-layered Bayesian Network Model for Structured Document Retrieval, Proc. ECSQARU LNAI 2711, Springer, (2003) 74-86.

6. D. Dubois and H. Prade: Possibility theory. Plenum, (1988)
7. F.V. Jensen. Bayesian Networks and Decision Graphs, Springer, 2000.
8. H.P. Luhn: The automatic creation of literature abstracts. IBM Journal of Research and Development, (1958), 2, 159-165.
9. J. Pearl: Probabilistic Reasoning in Intelligent Systems. Morgan Kaufmann San Mateo, Ca, (1988)
10. M.E. Maron: Automatic indexing: an experimental enquiry. Journal of the ACM, (1961), 8, 404-417
11. B. Ribeiro-Neto, I. Silva, R. Muntz: A Belief Network Model for IR. In Proc. Of the 19th ACM SIGIR Conf. on Research and Development in Information Retrieval, (1996), 253-260, Zurich, Switzerland.
12. S.E. Robertson and S. Walker: Some simple effective approximations to the 2-poisson model for probabilistic weighted retrieval, Proc. of the 17th Ann. Inter. ACM SIGIR Conf. on Research and Dev. in Information Retrieval, Springer-Verlag, (1994) 232-241
13. S.E. Robertson, S. Walker, S. Jones, M.M Hancock-Beaulieu and M. Gatford: Okapi at TREC-3. Proc. of the 3rd Text REtrieval Conference (TREC-3), NIST Special Publication (1995) 109-126
14. G. Salton: The Smart retrieval system-experiments. In Automatic Document Processing, Prentice Hall Inc, (1971)
15. G. Salton and C. Buckley: Term-weighting approaches in automatic text retrieval. IPM, (1988) VOL 24, 513-523
16. C. E. Shannon: The mathematical theory of communication. Bell System Technical Journal, (1948) Vol. 27, 379-423 and 623- 656.
17. Ilmerio Silva, Berthier Ribeiro-Neto, Pavel Calado, Edleno Moura, and Nivio Ziviani: Link-Based and Content-Based Evidential Information in a Belief Network Model. In ACM SIGIR 23rd Int. Conference on Information Retrieval, Athens, Greece, (2000), 96-103.
18. K. Sparck Jones, S. Walker and S.E. Robertson: A probabilistic model of information retrieval: development and comparative experiments, Parts 1 & 2. IPM, (2000) VOL 36 779-808,809-840
19. K. Sparck Jones: A statistical interpretation of term specificity and its application in retrieval. Journal of Doc, (1972) VOL 28 111-121
20. H.R. Turtle and W.B. Croft: Inference networks for document retrieval. In Proc. 13th Int.Conf. on Research and Development in Information Retrieval. (1990) 1-24
21. K. Van Rijsbergen: A Theoretical basis for the use of co-occurrence data in information retrieval. In Jour. of Doc. (1977) 33 106-119.
22. R. R. Yager and H. Legind Larsen: Retrieving information by fuzzification of queries. Int. Jour. of Intelligent Inf. Systems, (1993) Vol. 2(4).
23. H.P. Zipf: Human behaviour and the principle of least effort. Addison-Wesley, Cambridge, Massachusetts (1949).

# Comparison of Representations of Multiple Evidence Using a Functional Framework for IR

Ilmério R. Silva, João N. Souza, and Luciene C. Oliveira

School of Computer Science,
Federal University of Uberlândia,
Uberlândia - MG, Brazil
{ilmerio, nunes}@facom.ufu.br, luciene@pos.facom.ufu.br

**Abstract.** The combination of sources of evidence is an important subject of research in information retrieval and can be a good strategy for improving the quality of rankings. Another active research topic is modeling and is one of the central tasks in the development of information retrieval systems. In this paper, we analyze the combination of multiple evidence using a functional framework, presenting two case studies of the use of the framework to combine multiple evidence in contexts bayesian belief networks and in the vector space model. This framework is a meta-theory that represents IR models in a unique common language, allowing the representation, formulation and comparison of these models without the need to carry out experiments. We show that the combination of multiple evidence in the bayesian belief network can be carried at in of several ways, being that each form corresponds to a similarity function in the vector model. The analysis of this correspondence is made through the functional framework. We show that the framework allows us to design new models and helps designers to modify these models to extend them with new evidence sources.

**Keywords:** Combination of Multiple Evidence, Functional Framework, Information Retrieval Models.

## 1 Introduction

Modeling is one of the central and most active research topics in Information Retrieval (IR). The implementation of any new idea to improve quality and accuracy of an information retrieval system usually requires the first step of modeling. The modeling task is complex, and also important, in modern IR systems, such as search engines on the web and enterprise search systems. In these cases, it is common to have more than one source of evidence available to be exploited by the model in the task of providing answers to a given query. This abundance of evidence sources certainly offers an opportunity to the development of better systems, but also poses a challenge for who is in charge of developing an IR model.

When searching for documents on the Web, the text of the documents may be insufficient to provide good results. A recent popular approach for improving

M. Consens and G. Navarro (Eds.): SPIRE 2005, LNCS 3772, pp. 283–294, 2005.

search effectiveness is to use alternative sources of evidence information. IR models which combine evidence sources is a important subject of research in IR [3,11,10,13]. We analyze two case studies for the combination of multiple sources evidence in the bayesian belief network model and in the classic vector space model using a powerful tool for developing of new IR models, called functional framework. We show that the combination of multiple evidence in the bayesian belief network can be can be carried at in of several ways, being that each form corresponds to a similarity function in the vector model. The analysis of this correspondence is made through the functional framework. This framework is a meta-theory for IR models which help in producing new IR models. The meta-theory is useful not only for devising new models, but also for representing previously proposed models and helps designers to modify these models to extend them with new sources of evidence. Moreover, the functional framework allows the analysis of different models characteristics to fit inside the same plan of representation and also allows the comparison of IR models through the use of functions.

The main motivation behind this work is in the application of the framework to combine multiple evidence and show that we can design new models equivalent to other models using different modeling semantic. We present two forms of modeling in two different environments, bayesian modeling and vector modeling, for combining evidence. The functional framework allows to make the passage among different types of modeling with ease, due to its capacity of abstraction and expressibility supplying a functional environment. This passage of the models for the functional framework facilitates the design and comparison between them therefore they are represented in a same language.

The rest of the paper is organized as follows. In section 2 we show the related works to the functional framework and the combination of evidence in models IR. We explain the basics concepts of the functional framework for IR in section 3. In sections 4 and 5 we analysis two case studies to combine multiple evidence in the bayesian network and vector models using the functional framework. Finally in section 6, we discuss the potential advantages of using the functional framework for creating IR models and combination of multiple source of evidence, summarize our contributions and discuss future research directions.

## 2    Related Works

There are some works on combination of content-based and link-based pieces of evidence in a single IR model. In paper [10] the use of belief networks to represent and to combine link-based and content-based information is proposed. [6] extends the model proposed in [10] presenting a generalization of this model to combine evidence sources in belief network. In [11] the link and content combination in the context of the vector space model is presented. Here we present two case studies for the use of a meta-theory, functional framework, to combine some evidence in the context of the belief network and vector space model. In the first case study, we use the model proposed in [6] generalizing the operator used in the composition of documents and design a model correspondent in the vector model. This model designed uses the same semantics of the model proposed in [11], but it is generalized for multiple evidence. In the second case study, we use the vector model to combine multiple evidence found in the first case study modifying the similarity function for the cosine similarity and then we design a new corresponding belief network model through the use of functional framework.

There is also a lot of works on meta-theories for IR: formal models [1,4,5,8], logic-based meta-models [2,7,12,14] and probability-based meta-models [9]. The functional meta-theory uses a different approach of previous meta-theories, based on functions. One advantage in relation to the other meta-theories is the capacity of the framework to represent the classic models, models that combine evidence and all the models that can be expressed by an algorithm.

Our work presents a analysis of the combination of multiple evidence using the functional framework. It differs from the others for a series of factors. We show an application of the functional meta-theory for generic combination of evidence pieces in two different contexts and in two distinct ways. This combination can be made in various forms in any modeling context. A tool to help designers in the accomplishment of this task is the functional framework, which unifies models. Moreover, we discuss the capacity of the framework to represent, design, combine IR models and compare similar models or not without using experimentation.

## 3   Fundamentals of the Functional Framework for IR

We present here the main definitions of the functional framework. All components and definitions are based on functions.

### 3.1   Representation of Models

To represent IR models in the functional framework we define functional term, weight function, functional objects, similarity function between two functional objects, and functional model.

**Definition 1.** *Functional Term. A functional term is a function whose semantics relates to set of index terms. A functional term $f$ is denoted by $f(k_l, ..., k_s)$, where $k_l, ..., k_s$ are index terms.*

Let $K = \{k_1, \ldots, k_t\}$ be a set of index terms and $\rho(K)$ be a set of subsets of $K$ called power set. For example, the function $syn : \mathbf{K} \longrightarrow \rho(\mathbf{K})$ is the synonymous function such that given a term returns a set of terms, this is the set of synonyms of that term. The function $syn(k_i) = \{k_{i1}, \ldots, k_{is}\}$ returns the set of synonyms of the term $k_i$.

**Definition 2.** *Weight Function. A weight function is a function whose result is the weight of the term in a document or in a query. Let $C = \{d_1, \ldots, d_z\}$*

be a collection of documents, $K = \{k_1, \ldots, k_t\}$ be an index terms set in $C$, and $q$ a query. The weight function $g : K \times \{C \cup \{q\}\} \longrightarrow \mathbb{R}$ is such that $g(k_i, d_j)$ returns the weight associated with the pair $(k_i, d_j)$ and $g(k_i, q)$ returns the weight associated with the pair $(k_i, q)$.

To simplify we use the following notation. Let $g_j : K \longrightarrow \mathbb{R}$ an unary function that returns the weights of a term in the document $d_j$. This function returns the weight associated with the pair $(k_i, d_j)$. Analogously, let $g_q : K \longrightarrow \mathbb{R}$ an unary function that returns the weights of a term in the query $q$. This function returns the weight associated with the pair $(k_i, q)$. The weight functions $g_j$ and $g_q$ are functional terms.

**Definition 3.** *Functional Objects. Functional objects are functional documents and functional queries. These objects are represented by a set of functional terms. A functional document $df_j$ is represented by a set of functional terms that relate the index terms in the document $d_j$. A functional query $qf$ is represented by a set of functional terms that relate terms in the query $q$.*

**Definition 4.** *Similarity Function between two Functional Objects. Given a set of functional objects $O = \{df_1, \ldots, df_n, qf_1, qf_2, \ldots, qf_m\}$, the similarity is a function $\Delta : O \times O \longrightarrow \mathbb{R}$ which assigns a positive real number $\Delta(of_j, of_i)$ for every pair $(of_j, of_i)$, where $\{of_j, of_i\} \subseteq O$, satisfying the following properties (or axioms):*

1. $0 \leq \Delta(\boldsymbol{of}_j, \boldsymbol{of}_i) \leq 1$ *(normalization)*
2. $\Delta(\boldsymbol{of}_j, \boldsymbol{of}_j) = 1$ *(reflexivity)*
3. $\Delta(\boldsymbol{of}_j, \boldsymbol{of}_i) = \Delta(\boldsymbol{of}_i, \boldsymbol{of}_j)$ *(symmetry)*

The similarity function relates functional terms of the functional objects. Notice that in this case the similarity function does not necessarily denote a function of distance or metric (the property of the triangular inequality is not mandatory). The vector model based on cosine similarity, for example, does not satisfy the property of the triangular inequality, satisfying only the properties of symmetry, reflexivity and normalization. The normalization property is important for the combination of the models, the reflexivity property is important because for two identical functional objects need have the greatest possible similarity and the symmetry property is important for documents clustering.

**Definition 5.** *Functional Model. A functional model is defined by the tuple*

$$\Psi = \langle D, Q, T, \Delta \rangle$$

*where $D$ is a set of functional documents $\{df_1, \ldots, df_n\}$; $Q$ is a set of functional queries $\{qf_1, \ldots, qf_m\}$; $T$ is a set of functional terms of the functional documents and queries $\{g_1, \ldots, g_v\}$; and $\Delta$ is the similarity function, with the three properties: normalization, reflexivity and symmetry.*

The functional terms are extracted from functional documents and queries. In general, functional queries are functional ad hoc queries, where pre-computation cannot be anticipated. Then, without loss of representation power, and to simplify the notation, we use of the following form: a functional model is represented by a functional document collection, a functional query unitary set and a similarity function. It is denoted by $\Psi = \langle \{df_1, \ldots, df_n\}, \{qf\}, \Delta \rangle$, where $\Delta$ is a similarity function over pairs of functional documents or functional documents and queries of a collection.

## 3.2   Comparison of Models

After carrying out the representation of the IR models in the functional framework we can compare them. We define one relation of comparison between models: relation equivalence that will be shown as follows.

**Definition 6.** *Functional Models Equivalence. Two functional models*
$\Psi_a = \langle \{df_{a1}, \ldots, df_{an}\}, \{qf_a\}, \Delta_a \rangle$ *and*  $\Psi_b = \langle \{df_{b1}, \ldots, df_{bn}\}, \{qf_b\}, \Delta_b \rangle$ *are equivalents if and only if $\forall$ functional query $qf$ there exist an bijective function:*
$\phi : \{df_{a1}, \ldots, df_{an}\} \rightarrow \{df_{b1}, \ldots, df_{bm}\}$ *such that if $\phi(df_{ai}) = df_{bi}$ and $\phi(df_{ak}) = df_{bk}$ then the two conditions below must be satisfied:*

1. $\Delta_a(qf, df_{ai}) = \Delta_a(qf, df_{ak}) \Leftrightarrow \Delta_b(qf, df_{bi}) = \Delta_b(qf, df_{bk})$
2. $\Delta_a(qf, df_{ai}) > \Delta_a(qf, df_{ak}) \Leftrightarrow \Delta_b(qf, df_{bi}) > \Delta_b(qf, df_{bk})$

These conditions guarantee that the models $\Psi_a$ and $\Psi_b$ are equivalent if and only if they generate the same ranking. To represent a model in the functional framework means that its similarity and form of representation of documents and queries are translated into another language. The objective of this representation is to have the model in the formalism of the functional framework. The comparison of equivalence between the models is important for reutilization of code or choice of implementation of a model and a better understanding of the model semantics.

# 4   Combination of Multiple Evidence Using the Functional Framework – From Bayesian to Vector Model

In this section we discuss a first case study to combine multiple evidence in the belief network model and in the vector space model with the utilization of the functional framework. We propose a belief network model to combine multiple evidence witch is a extension of the belief network model proposed in [6]. We represent this belief network model in the functional framework, find the corresponding vector model to combine multiple evidence, pass this vector model to the functional framework and verify that they are equivalent by their construction.

## 4.1   Belief Network Model to Combine Multiple Evidence Sources

The belief network model can be used to combine multiple evidence sources, such as the text of links content and the information comes from the link analysis between documents of the collection.

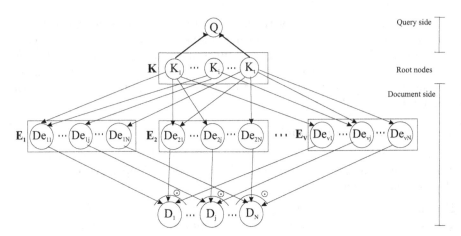

**Fig. 1.** Generic belief network model to combine multiple evidence sources

We propose in this work a generic model of belief network to combine multiple evidence sources. This model is an extension of the belief network model proposed in [6]. The difference is that we generalize the operator used in the composition of documents. Figure 1 illustrates this belief network generalized to combine multiple evidence.

In the bayesian network of Figure 1, the node $Q$ models the user query and the nodes set $\mathbf{K}$ models the set of keywords in the documents collection. The sets of nodes $\mathbf{E}_1, \ldots, \mathbf{E}_v$ represent $v$ evidence modeled in the network. To represent a new evidence source $e_i$ in this network, new nodes $De_{i,j}$ are associated with each document $D_j$ in retrieval set for the query $Q$. The nodes set $\mathbf{K}$ is used to model the occurrence of keywords in query $Q$ that induces values of belief in each one of the nodes of the sets $\mathbf{E}_1, \ldots, \mathbf{E}_v$. The node $D_j$ represents the combination of all the evidence modeled.

The ranking of a document is computed as the probability $P(dj|q)$, as follows:

$$P(d_j|q) = \eta \sum_{\forall \mathbf{k}} P(d_j|\mathbf{k}) \times P(q|\mathbf{k}) \times P(\mathbf{k}) \tag{1}$$

where $\eta$ is a normalizing constant. Details on the derivation of this expression can be seen in [10]. However, the conditional probability $P(dj|\mathbf{k})$ now depends of multiple pieces of evidence, combined through the operator $\odot$, that can be disjunctive, conjunctive and noisy-OR operators.

For the disjunctive operator, this is accomplished as follows:

$$P(d_j|\mathbf{k}) = 1 - (1 - P(de_{1j}|\mathbf{k})) \times (1 - P(de_{2j}|\mathbf{k})) \times \cdots \times (1 - P(de_{vj}|\mathbf{k})) \tag{2}$$

where $P(de_{ij}|\mathbf{k})$ is the value calculated for each evidence $E_i$ in relation to the document $d_j$ that we denote here as $E_{ij}$. $E_{ij}$ can be, for instance, the weight of content part of the document $d_j$, computed by the classic vector model, or the degree of hub and authority of the document $d_j$. And $P(q|\mathbf{k})$ is defined by:

$$P(q|\mathbf{k}) = \begin{cases} 1, & \text{if } \mathbf{q} = \mathbf{k} \\ 0, & \text{otherwise} \end{cases} \tag{3}$$

Substituting each $P(de_{ij}|\mathbf{k})$ for $E_{ij}$ in Eq.(2), and Eq.(2) and (3) into Eq.(1), defining the *a priori* probability $P(\mathbf{k})$ as constant and considering that the constant $\eta$ does not influence in the final result of ranking, we can define the *similarity function* as:

$$sim(d_j, q) = 1 - (1 - E_{1j})(1 - E_{2j}) \ldots (1 - E_{vj}) \tag{4}$$

Observe that any evidence $e_i$ can be ignored, attributing $E_{ij} = 0$. This similarity function does not satisfy the symmetry property.

For the conjunctive operator, we have the multiplication of the values of each evidence as shown in the following function: $sim(d_j|q) = E_{1j} \times E_{2j} \cdots \times E_{vj}$. Notice that if for any evidence $e_i$, $E_{ij} = 0$, then $sim(d_j|q) = 0$, ignoring all the other evidence. For this the conjunctive operator is not utilized very much in practice.

The combination model using the disjunctive and conjunctive operator does not make *a priori* assumption about the importance of each evidence source. The probabilities to be combined depend only on the characteristics of the algorithms and on the parameters used. However, the model can be modified to allow the insertion of weights which can be accomplished by the use of the noisy-OR operator. Thus, we have the following equation for the similarity function:

$$sim(d_j|q) = 1 - (1 - W_1 \times E_{1j})(1 - W_2 \times E_{2j}) \ldots (1 - W_v \times E_{vj}) \tag{5}$$

where $W_1 \ldots W_v$ are the weights given to the multiple evidence $e_1, \ldots, e_v$, respectively. These weights can be defined by user, can depend or not on the query or can be automatically calculated.

### 4.2  Functional Belief Network Models to Combine Multiple Evidence Sources

We show here the representation of the belief network model using the disjunctive operator presented previously. To represent the generic belief network model to combine multiple evidence sources using the disjunctive operator in the functional framework we define the functional model $\Psi_{ng} = \langle \{df_{ng_1}, ..., df_{ng_n}\}, \{qf_{ng}\}, \Delta_{ng} \rangle$. The bayesian model with multiple evidence can be represented in functional framework by:

- $df_{ng_j} = \{g_{e1_j}, g_{e2_j}, \ldots, g_{ev_j}\}$, where $g_{e1_j} = R_{j,q}$ is function calculated by cosine of the vector model and $g_{e2_j} \ldots g_{ev_j}$ are functions that define values for the evidence $e_2, \ldots, e_v$ associated with the document $d_j$, respectively.
- $qf_{ng} = \{g_{e1_q}, g_{e2_q}, \ldots, g_{ev_q}\}$, where $g_{e1_q} = 1$ and other functional terms are defined in an analogous form to the functional documents
- Similarity function is given by

$$\Delta_{ng}(df_{ng_j}, qf_{ng}) = 1 - (1 - R_{j,q})(1 - g_{e2_j} g_{e2_q}) \ldots (1 - g_{ev_j} g_{ev_q}) \tag{6}$$

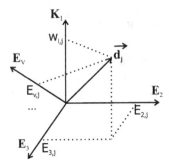

**Fig. 2.** Generic Vector Model for combination of multiple evidence sources

The Eq. (4) does not satisfy the symmetry property and Eq. (6) subsumes Eq.(4) setting $g_{e2_q} = \ldots = g_{ev_q} = 1$. Notice that we modified the original function similarity Eq.(4) to satisfy the properties necessary for a similarity function in our framework Eq. (6). Observe that in this case we have the noisy-OR operator introduced here. Then, to represent the belief network using the disjunctive operator we need to modify the similarity function for the operator noisy-OR to satisfy the properties of the functional framework. This alteration can be carried out due to the high power abstraction of functions. It is important, for example, to work with the clustering of documents and in this case we have to calculate the similarity between two documents.

### 4.3   Vector Model to Combine Multiple Evidence Sources

We define a vector space model extended to combine multiple evidence corresponding to the previous bayesian network. In the vector classic space model the set of index terms $\{k_i | 1 \leq i \leq t\}$, forms the axes of the vector space. The documents and queries are represented as vectors in this space: $d_j = (w_{1j}, w_{2j}, \ldots, w_{tj})$ and $q = (w_{1q}, w_{2q}, \ldots, w_{tq})$, respectively.

We propose in this work a model that combines information of multiple evidence sources through an extension of the vector space model. For this, we extend the vector space adding $v - 1$ new axes, where $v - 1$ is the number of new evidence. Figure 2 shows this vector model to combine of multiple evidence sources.

In this case, the equation of the similarity function is:

$$sim(d_j, q) = 1 - (1 - R_{jq})(1 - E_{2q} \times E_{2j}) \ldots (1 - E_{2v} \times E_{vj}) \qquad (7)$$

where $R_{jq}$ is calculated by cosine of the vector model, $E_{2q}, \ldots, E_{vq}$ are the values of each evidence $e_2, \ldots, e_v$ associated to query $q$ and $E_{2j}, \ldots, E_{vj}$ are the values of each evidence $e_2, \ldots, e_v$ associated to document $d_j$, respectively.

### 4.4   Functional Vector Model to Combine Multiple Evidence Sources

We represent the vector model to combine multiple evidence sources presented previously in the functional framework. To represent the generic vector model for combination of multiple evidence in the functional framework we define the functional model $\Psi_{vg} = \langle \{df_{vg_1}, ..., df_{vg_n}\}, \{qf_{vg}\}, \Delta_{vg} \rangle$. The vector model $\Psi_{vg}$ with multiple evidence sources can be represented in the functional framework by:

- $df_{vg_j} = \{g_{e1_j}, g_{e2_j}, \ldots, g_{ev_j}\}$, where $g_{e1_j} = R_{j,q}$ is function calculated by cosine of the vector model and $g_{e2_j} \ldots g_{ev_j}$ are functions that define values for the evidence $e_2, \ldots, e_v$ associated with the document $d_j$, respectively.
- $qf_{vg} = \{g_{e1_q}, g_{e2_q}, \ldots, g_{ev_q}\}$, where $g_{e1_q} = 1$ and other functional terms are defined in an analogous form to the functional documents
- Similarity function is given by

$$\Delta_{vg}(df_{vg_j}, qf_{vg}) = 1 - (1 - R_{jq})(1 - g_{e2_j}g_{e2_q}) \ldots (1 - g_{ev_j}g_{ev_q}) \quad (8)$$

We design a vector model to combine multiple evidence equivalent to the belief network model to combine multiple evidence. We verify that the functional models corresponding to them are equivalent by its construction. They possess the same similarity function, then the two following properties of equivalence are satisfied. The models $\Psi_{vg}$ and $\Psi_{ng}$ are equivalents, because the models generate the same ranking.

## 5 Combination of Multiple Evidence Using the Functional Framework – From Vector to Bayesian Model

In this section we present a second case study to combine multiple evidence sources in the belief network model and in the vector space model with the utilization of the functional framework. We propose another form to combine evidence sources in the vector model. We represent this vector model extended in the functional framework, find the corresponding belief network model to combine multiple evidence, represent this belief network model in the functional framework and verify that they are equivalent by its construction.

### 5.1 Vector Model to Combine Multiple Evidence Sources

There are some ways to combine multiple evidence sources in the vector model, modifying the similarity function. Another form to extend the vector model to combine evidence sources is use the cosine similarity function.

The vector modeling is the same presented in section 4.3. The model proposed here also is an extension of the vector space model through the addition of $v - 1$ new axes, where $v - 1$ is the number of new evidence. This vector model extended to combine multiple evidence is presented in Figure 2.

The *similarity function* is defined by:

$$sim(d_j, q) = \frac{\sum_{i=1}^{t} w_{i,j} \cdot w_{i,q} + E_{2j}E_{2q} + \cdots + E_{vj}E_{vq}}{\sqrt{\sum_{i=1}^{t} w_{i,j}^2 + E_{2j}^2 + \cdots + E_{vj}^2} \times \sqrt{\sum_{i=1}^{t} w_{i,q}^2 + E_{2q}^2 + \cdots + E_{vq}^2}} \quad (9)$$

where $w_{i,j}$ is the weight of the term $k_i$ in the document $d_j$, $w_{i,q}$ is the weight of the term $k_i$ in the query $q$, $E_{2j}, \ldots, E_{vj}$ are the values of each evidence $e_2, \ldots, e_v$ associated to document $d_j$ and $E_{2q}, \ldots, E_{vq}$ are the values of each evidence $e_2, \ldots, e_v$ associated to query $q$, respectively.

## 5.2 Functional Vector Model to Combine Multiple Evidence Sources

We represent the vector model to combine multiple evidence sources presented previously in the functional framework. To represent the generic vector model to combine of multiple evidence sources in the functional framework we define the functional model $\Psi_{vc} = \langle \{df_{vc_1}, ..., df_{vc_n}\}, \{qf_{vc}\}, \Delta_{vc} \rangle$. The vector model with multiple evidence sources can be represented in functional framework by:

- $df_{vc_j} = \{g_j, g_{e2_j}, \ldots, g_{ev_j}\}$, where $g_j(k_i) = w_{i,j}$ is function that defines the weight of the terms in the document and $g_{e2_j} \ldots g_{ev_j}$ are functions that define values associated to evidence $e_1, \ldots, e_v$,respectively.
- $qf_{vc} = \{g_q, g_{e2_q}, \ldots, g_{ev_q}\}$, where functional terms are defined in an analogous form to the functional documents
- Similarity function is given by

$$\Delta_{vc}(df_{vc_j}, qf_{vc}) = \frac{\sum_{i=1}^{t} g_j(k_i) \cdot g_q(k_i) + g_{e2_j} g_{e2_q} + \cdots + g_{ev_j} g_{ev_q}}{\sqrt{\sum_{i=1}^{t} g_j(k_i)^2 + g_{e2_j}^2 + \cdots + g_{ev_j}^2} \times \sqrt{\sum_{i=1}^{t} g_q(k_i)^2 + g_{e2_q}^2 + \cdots + g_{ev_q}^2}} \quad (10)$$

## 5.3 Belief Network Model to Combine Multiple Evidence Sources

We design a belief network model to combine multiple evidence sources equivalent to the vector model to combine multiple evidence sources using cosine similarity function. The resulting network is shown in Figure 3.

In this bayesian network model, the ranking is computed as follows:

$$sim(d_j, q) = \frac{\sum_{i=1}^{t} w_{i,j} \cdot w_{i,q} + E_{2j} E_{2q} + \cdots + E_{vj} E_{vq}}{\sqrt{\sum_{i=1}^{t} w_{i,j}^2 + E_{2j}^2 + \cdots + E_{vj}^2} \times \sqrt{\sum_{i=1}^{t} w_{i,q}^2 + E_{2q}^2 + \cdots + E_{vq}^2}} \quad (11)$$

The derivation of this equation is made in a similar form to that presented in [10] for Eq.(1) making the necessary substitutions for modeling the cosine function.

## 5.4 Functional Belief Network Models to Combine Multiple Evidence Sources

We show here the representation of the belief network models presented previously. To represent the second bayesian network model in the functional framework we define the functional model $\Psi_{nc} = \langle \{df_{nc_1}, ..., df_{nc_n}\}, \{qf_{nc}\}, \Delta_{nc} \rangle$. The second bayesian model with multiple evidence sources can be represented in functional framework by:

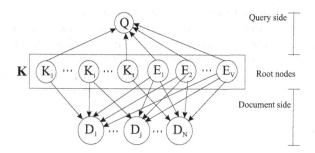

**Fig. 3.** Generic model of belief network to combine multiple evidence sources

– $df_{nc_j} = \{g_j, g_{e2_j}, \ldots, g_{ev_j}\}$, where $g_j(k_i) = w_{i,j}$, is function that defines the weight of the terms in the document and $g_{e2_j} \ldots g_{ev_j}$ are functions that define values associated to evidence $e_1, \ldots, e_v$, respectively.

– $qf_{nc} = \{g_q, g_{e2_q}, \ldots, g_{ev_q}\}$, where functional terms are defined in an analogous form to the functional documents

– Similarity function is given by

$$\Delta_{nc}(df_{nc_j}, qf_{nc}) = \frac{\sum_{i=1}^{t} g_j(k_i) \cdot g_q(k_i) + g_{e2_j} g_{e2_q} + \cdots + g_{ev_j} g_{ev_q}}{\sqrt{\sum_{i=1}^{t} g_j(k_i)^2 + g_{e2_j}^2 + \cdots + g_{ev_j}^2} \times \sqrt{\sum_{i=1}^{t} g_q(k_i)^2 + g_{e2_q}^2 + \cdots + g_{ev_q}^2}} \quad (12)$$

We propose a belief network model to combine multiple evidence equivalent to the vector model to combine multiple evidence with the cosine similarity function. We verify that the functional models $\Psi_{vc}$ and $\Psi_{nc}$ are equivalent by its construction.

## 6   Conclusions and Future Works

The functional framework allows the analysis of different models using different modeling semantic. This framework is a simple, powerful, and flexible tool that defines a level of abstraction to the representation, comparison, combination and design of IR models. This allows one to work with theoretical and practical applications, making it practical in the sense of implementation. Moreover, this framework can to be used to generalize all the IR models that can be expressed by algorithms because it is based on functions.

As seen in literature the combination of evidence can improve the ranking quality. We present case studies using the bayesian and vector modeling for combination of the multiple evidence sources, but other approaches can be used. One of our contributions is the proposal of the use of the functional framework as unifying of IR models and application of this framework to combine evidence sources. The functional framework can be used for combination of multiple evidence in several ways and helps to formulate new models and combine them.

Future work will include the design of other models for combining multiple evidence sources using other modeling semantics and to verify through experiments which of the models of the case studies possess the best quality. Moreover, other future work will be to make the comparison between other IR models and the study of new models which are equivalent to the existing models, simpler and of easier implementation and with more semantics than the existing models. In addition, an interesting area of research will be to study the characteristics of the models, for example, which define properties that the models must have so that they possess more precision or recall than others.

# References

1. R. Baeza-Yates and B. Ribeiro-Neto. *Modern Information Retrieval.* Addison Wesley, 1999.
2. F. Crestani and M. Lalmas. Logic and uncertainty in information retrieval. In *ESSIR*, pages 179–206, 2000.
3. M. A. Pinheiro de Cristo, P. Calado, M. Silveira, I. Silva, R. R. Muntz, and B. A. Ribeiro-Neto. Bayesian belief networks for ir. *International Journal of Approximate Reasoning*, 34(2-3):163–179, 2003.
4. S. Dominich. A unified mathematical definition of classical information retrieval. *Jornal of the American Society for Information Science*, 51(7):614–624, 2000.
5. S. Dominich. On applying formal grammar and languages, and deduction to information retrieval modelling. In *Proceedings of the ACM SIGIR MF/IR*, pages 37–41, 2001.
6. E. M. Abinader Júnior. Combinação e avaliação de múltiplas fontes de evidências para recuperação de documento na web. Master's thesis, Universidade Federal do Amazonas, Instituto de Ciências Exatas, Amazonas,Manaus, 2004.
7. M. Lalmas and P. D. Bruza. The use of logic in information retrieval modeling. *Knowledge Engineering Review. In press.*, pages 13(3):263–295, 1998.
8. A. Montejo Rez. Formal models for ir: a review and a proposal for keyword assignment. In *Workshop on Mathematical/Formal Methods in Information Retrieval.* ACM-SIGIR, 2003.
9. B. Ribeiro(Ribeiro-Neto) and R. Muntz. A belief network model for ir. In *Proc. of the 19th ACM SIGIR Conference on Research and Development in Information Retrieval*, pages 253–260, Zurich, Switzerland, 1996.
10. I. Silva, B. Ribeiro-Neto, P. Calado, E. Moura, and N. Ziviani. Link-based and content-based evidential information in a belief network model. In *Proceedings of the 23rd Annual International ACM SIGIR Conference on Research and Development in Information Retrieval*, pages 96–103, Athens, Greece, July 2000.
11. I. Silva, J. Souza, R. Moura, and Ribeiro-Neto. Informação de Links no Modelo Vetorial Usando uma Estrutura Funcional. In *Anais do 18th Simpósio Brasileiro de Banco de Dados*, pages 170–184, Manaus, AM, Brasil, 2003.
12. D.W. Song, K.F. Wong, P.D. Bruza, and Cheng C.H. Towards a common-sense aboutness theory for information retrieval modeling. In *In Proceedings of the FourthWorld Multiconference on Systemics, Cybernetics and Informatics (SCI 2000)*, pages 23–26, Orlando, Florida (USA), July 2000.
13. T. Westerveld, W. Kraaij, and D. Hiemstra. Retrieving Web pages using content, links, URLs and anchors. In *The Tenth Text REtrieval Conference (TREC-2001)*, pages 663–672, Gaithersburg, Maryland, USA, November 2001.
14. K.F. Wong, D. Song, P. Bruza, and C.H. Cheng. Application of aboutness to functional benchmarking in information retrieval. *ACM Trans. Inf. Syst.*, 19(4):337–370, 2001.

# Deriving TF-IDF as a Fisher Kernel

Charles Elkan

Department of Computer Science and Engineering,
University of California, San Diego
elkan@cs.ucsd.edu
http://www.cs.ucsd.edu/users/elkan/

**Abstract.** The Dirichlet compound multinomial (DCM) distribution
has recently been shown to be a good model for documents because
it captures the phenomenon of word burstiness, unlike standard models such as the multinomial distribution. This paper investigates the
DCM Fisher kernel, a function for comparing documents derived from
the DCM. We show that the DCM Fisher kernel has components that
are similar to the term frequency (TF) and inverse document frequency
(IDF) factors of the standard TF-IDF method for representing documents. Experiments show that the DCM Fisher kernel performs better
than alternative kernels for nearest-neighbor document classification, but
that the TF-IDF representation still performs best.

## 1 Introduction

A fundamental property of text documents, regardless of language, is that if a
word occurs once, it is likely that the same word will occur again. This phenomenon is called burstiness [3]. Unfortunately, standard probabilistic models
for documents, in particular multinomial distributions, do not allow for burstiness, since they assume that each word in a document is generated independently.
These models are therefore incorrect in a fundamental way. In recent research, an
alternative distribution has been proposed called the Dirichlet compound multinomial (DCM) [7]. This distribution can capture the phenomenon of burstiness.
Experimentally, DCM models lead to significantly better classification accuracy
than multinomial models on standard document collections [7].

In this paper, we derive the Fisher kernel for the DCM distribution. A Fisher
kernel is a function that measures the similarity of two data items not in iso-
lation, but rather in the context provided by a probability distribution. For
documents, a Fisher kernel measures how much two members of a collection
are similar taking into account a whole corpus as background information. We
show that the Fisher kernel based on the DCM has a mathematical form related
to the well-known TF-IDF representation for documents [1]. This demonstration is a new approach towards explaining why the TF-IDF heuristic is justified
and why it is so successful experimentally. We provide experimental results for
nearest neighbor classification for seven different kernel functions, that is for

M. Consens and G. Navarro (Eds.): SPIRE 2005, LNCS 3772, pp. 295–300, 2005.

seven different document representations: TF-IDF, the DCM Fisher kernel, the multinomial Fisher kernel, the Bhattacharyya kernel [6], and $L_0$, $L_1$, and $L_2$ normalized representations.

## 2     The Dirichlet Compound Multinomial Distribution

Throughout this paper, we assume the so-called "bag of words" representation for documents. In this representation, a document $x$ is a vector of counts $\langle x_1, \ldots, x_w, \ldots, x_W \rangle$ where $x_w$ is the number of appearances of word $w$ and $W$ is the vocabulary size. The DCM distribution is

$$p(x) = \frac{n!}{\prod_{w=1}^{W} x_w!} \frac{\Gamma(s)}{\Gamma(s+n)} \prod_{w=1}^{W} \frac{\Gamma(x_w + \alpha_w)}{\Gamma(\alpha_w)}$$

where the length of the document is $n = \sum_w x_w$ and $s = \sum_w \alpha_w$ is the sum of the DCM parameters [2] [7]. Like a multinomial, formally a DCM is a distribution over alternative count vectors of the same length $n$. Since different lengths give rise to different distributions, but a corpus always contains documents of different lengths, we assume that a corpus is modeled by a family of DCMs that all have the same parameter values $\alpha_w$.

Given a set $D$ of documents, there is no closed-form expression for the maximum likelihood $\alpha_w$ parameter values. However, these can be approximated closely as

$$\alpha_w = \frac{\sum_{d \in D} I(x_{dw} \geq 1)}{\sum_{d \in D} \Psi(s + n_d) - \Psi(s)}$$

where $x_{dw}$ is the count of word $w$ in document $d$, $n_d$ is the length of document $d$, and $\Psi(\cdot)$ is the digamma function (proof to be published elsewhere).

For typical document sets $s$ and $n_d$ are both in the hundreds, so $\Psi(s + n_d) - \Psi(s)$ is around one for each document, and therefore

$$\alpha_w \approx \frac{1}{|D|} \sum_{d \in D} I(x_{dw} \geq 1) \tag{1}$$

where $|D|$ is the number of documents in the collection. Since $x_{dw} = 0$ for most documents $d$ and most words $w$, $\alpha_w \ll 1$ for most words. For example, for a DCM trained on one class of newsgroup articles, the average $\alpha_w$ is 0.004. Of the 59,826 parameters, 99% are below 0.1, only 17 are above 0.5, and only 5 are above 1.0.

## 3     The Fisher Kernel for the DCM

In general, a kernel function $k(x, y)$ is a way of measuring the resemblance between two data items $x$ and $y$. Standard kernel functions are scalar products in some space of alternative representations of data items, that is $k(x, y) = s(x) \cdot s(y)$ where $s(x)$ is a re-representation of $x$.

For documents, one common approach is to re-represent a count vector by $L_2$ normalization: $s(x) = x/||x||_2$ where $||x||_2 = \sqrt{\sum_w x_w^2}$. This yields what is called cosine similarity, since $k(x, y) = x \cdot y/||x||_2||y||_2$ is the cosine of the angle between the vectors $x$ and $y$. Intuitively, this re-representation is unsatisfying for at least two reasons: (a) repeated appearances of one word in the same document are of decreasing informativeness–a consequence of the burstiness phenomenon, and (b) words that appear across a large number of different documents are less informative. TF-IDF (term frequency-inverse document frequency) representations were proposed to address these concerns several decades ago [1] [9]. Most commonly, each term frequency $x_w$ (i.e. each word count) is (a) log-transformed and (b) multiplied by the log of the inverse of the number of documents that word $w$ appears in. This specific version of TF-IDF is

$$\text{TF-IDF}(x_w) = \log(x_w + 1) \cdot \log \frac{|D|}{\sum_{d \in D} I(x_{dw} > 0)}.$$

Typically (and in our experiments below) the TF-IDF representation is then $L_2$ normalized.

The normalized TF-IDF representation and the corresponding kernel are among the best approaches for retrieving documents relevant to a query, and for categorizing documents into classes. However, no compelling theoretical reason for preferring TF-IDF to other heuristic representations is known [9]. Here, we show that a representation similar to TF-IDF arises naturally from the DCM.

A high-level motivation for TF-IDF is that it incorporates knowledge about the distribution of all documents into the similarity measure for individual documents. Given a probability distribution $p(x)$, the Fisher kernel measures the similarity of $x$ and $y$ in the context of this distribution: $k(x, y) = s(x)^T H s(y)$ where $s(x) = \nabla x$ is the Fisher score vector for $x$, i.e. the vector of partial derivatives of the log-likelihood $l(x) = \log p(x)$ with respect to the parameters $\alpha_w$, and $H$ is the Hessian of second partial derivatives of $l(x)$ with respect to the parameters [4] [5]. With this definition, $k(x, y)$ is invariant to changes in the parameterization of $p$. However, $H$ is usually approximated by the identity matrix, and in this case the Fisher kernel is different for different parameterizations.

For the DCM, the partial derivative of the log-likelihood is

$$\frac{\partial l(x)}{\partial \alpha_w} = \Psi(s) - \Psi(s + n) + \Psi(x_w + \alpha_w) - \Psi(\alpha_w). \tag{2}$$

The Fisher kernel $k(x, y)$ is then the scalar product of the partial derivative vectors for $x$ and $y$.

Asymptotic values for the digamma function give insight into these score vectors. For $z \geq 1$, $\Psi(z)$ is close to $\log(z - 0.5)$ with the difference tending to zero as $z$ tends to infinity. Similarly, $\Psi(z)$ is close to $-1/z + \Psi(1)$ for $z \ll 1$, where $\Psi(1) \approx -0.577$, with the difference tending to zero as $z$ tends to zero from above. As mentioned in Section 2, for a typical corpus $\alpha_w \ll 1$ for most words $w$. Therefore, Equation (2) can be approximated as

$$\frac{\partial l(x)}{\partial \alpha_w} \approx \Psi(s) - \Psi(s + n) + I(x_w \geq 1)[\log(x_w - 0.5) + 1/\alpha_w - \Psi(1)].$$

In this form the Fisher score is clearly related to TF-IDF. First, given a document, the term $\Psi(s) - \Psi(s + n)$ is the same for all words $w$ and as explained in Section 2, it is typically around minus one for all documents. Therefore, it has little influence on the ranking of which documents $y$ are closest to a document $x$, i.e. which $y$ give the smallest $k(x, y)$ values. Second, the term $\log(x_w - 0.5)$ is a log transform of term frequency. Finally, Equation (1) says that $1/\alpha_w \approx |D|/\sum_{d \in D} I(x_{dw} \geq 1)$ which is precisely inverse document frequency.

## 4   Experiments

In this section, we examine the performance of nine methods for document classification. Two methods are Bayesian classifiers based on training multinomial and DCM models. Seven methods are $k$-nearest neighbor classifiers. Of these, three use different $L_p$ normalizations of documents $s(x) = x/||x||_p$ for $p = 0, 1, 2$. One nearest neighbor method uses the TF-IDF representation with $L_2$ normalization. Finally, three nearest neighbor methods use theoretically motivated kernels: the Fisher DCM kernel and two that are representative of those proposed in other recent research. The Bhattacharyya kernel uses the representation $s(x) = \langle \sqrt{x_1/||x||_1}, \ldots, \sqrt{x_W/||x||_1} \rangle$, following the experiments of [6]. The Fisher kernel based on the multinomial distribution uses the representation $s(x) = \langle x_1/\hat{\theta}_1, \ldots, x_W/\hat{\theta}_W \rangle$ where $\hat{\theta}_w = \sum_{d \in D} x_{dw}/\sum_{d \in D} n_d$ is the maximum likelihood parameter value for word $w$ for the multinomial distribution fitted to the given document collection.

Bayesian classification uses Bayes' rule and a different DCM or multinomial model learned from the training documents in each class. However, classification using a Fisher kernel uses just one DCM or multinomial model learned from the entire collection of training documents.

We use two standard document collections called industry sector and 20 newsgroups. Documents are tokenized, stop words removed, and count vectors extracted using the Rainbow toolbox [8]. The industry sector[1] collection contains 9555 documents distributed in 104 classes. It has a vocabulary of 55,055 words, and each document contains on average 606 words. The data are split into halves for training and testing. The 20 newsgroups[2] collection contains 18,828 documents belonging to 20 classes. This collection has a vocabulary of 61,298 words with an average document length of 116 words. The data are split into 80/20 fractions for training and testing.

Table 1 shows classification accuracy averaged over ten splits of each document collection. Nearest neighbor results are for $k = 3$ neighbors. Not surprisingly, the TF-IDF kernel performs best. Both Fisher kernel methods perform well, particularly on the industry sector collection, which has only a small number of documents per class. The DCM Fisher kernel performs slightly better than the multinomial Fisher kernel.

---

[1] http://www.cs.umass.edu/~mccallum/code-data.html
[2] http://people.csail.mit.edu/people/jrennie/20Newsgroups

**Table 1.** Accuracy averaged over ten random splits for different classifiers

| Collection | M | DCM | L0 | L1 | L2 | TF-IDF | Bhattacharyya | Fisher-DCM | Fisher-M |
|---|---|---|---|---|---|---|---|---|---|
| 20 news | 0.843 | 0.845 | 0.606 | 0.675 | 0.778 | 0.828 | 0.744 | 0.677 | 0.665 |
| industry | 0.791 | 0.795 | 0.624 | 0.017 | 0.567 | 0.881 | 0.254 | 0.761 | 0.735 |

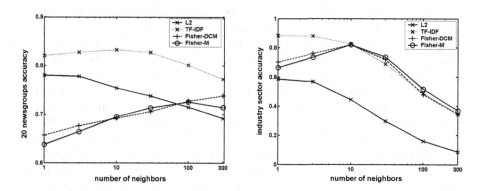

**Fig. 1.** Average accuracy scores for increasing numbers of nearest neighbors for 20 newsgroups (left) and industry sector (right)

Given the theoretical arguments in favor of the DCM over the standard multinomial model, it is surprising that a Bayesian classifier using multinomial models performs so well. We have three explanations for this. First, our multinomial model uses additive smoothing with constant 0.01 instead of with constant 1.0, which is the standard Laplace smoothing, but performs considerably worse. Second, both the 20 newsgroups and industry sector collections consist of relatively short documents, in which burstiness is less apparent than in longer documents. Third, it is well-known that Bayesian classifiers can be highly accurate even when they use models that produce inaccurate probabilities, since the ordering of the probabilities may still be correct.

Figure 1 shows how four of the $k$-nearest neighbor methods perform as $k$ varies. Both Fisher kernel methods benefit from using many neighbors on the 20 newsgroups collection, while the performance of cosine similarity decreases as more neighbors are used. This fact possibly indicates that cosine similarity can identify neighbors correctly only if they are very close, whereas the Fisher kernel methods and TF-IDF can pick out not-so-close neighbors well also. For each of the four methods, the optimum value of $k$ is smaller on the industry sector collection. This is perhaps because each class has fewer members in this collection, so each document has fewer genuine neighbors.

## 5   Discussion

Although the TF-IDF representation for documents is widely used, its origin is heuristic and it does not have a convincing theoretical basis [9]. However, TF-

IDF implicitly contains an important insight: the similarity of two documents (or two data items in general) should be a function not just of the documents themselves, but also of the context of other documents in which they lie.

Fisher kernels are a general implementation of this idea of exploiting background context when computing the degree of similarity of two data items. Above, we have derived and investigated the Fisher kernel induced by the Dirichlet compound multinomial (DCM) distribution. We have shown that the expression for the DCM Fisher kernel contains components similar to the log-term-frequency and inverse-document-frequency components of TF-IDF. We have also shown experimentally that nearest neighbor classifiers based on the DCM Fisher kernel perform well, although not as well as TF-IDF-based classifiers.

We are excited about continuing the research of this paper in three directions. First, we want to experiment with collections of longer documents, where we expect the superiority of the DCM over the multinomial to be greater. Second, we want to use the DCM Fisher kernel in an SVM classifier, since SVMs are generally rather more accurate than nearest neighbor methods. Third, given that TF-IDF remains the best known representation for documents, can we find a new probability distribution whose Fisher kernel is even more similar to TF-IDF?

**Acknowledgments.** David Kauchak and Rasmus Madsen assisted with the experimental part of this paper.

# References

1. Akiko Aizawa. An information-theoretic perspective of tf-idf measures. *Information Processing and Management*, 39(1):45–65, 2003.
2. N. Balakrishnan, Norman L. Johnson, and Samuel Kotz. *Discrete Multivariate Distributions*. New York: John Wiley and Sons Inc., 1997.
3. Kenneth W. Church and William A. Gale. Poisson mixtures. *Natural Language Engineering*, 1(2):163–190, 1995.
4. Thomas Hofmann. Learning the similarity of documents: An information-geometric approach to document retrieval and categorization. In *Proceedings of NIPS*, pages 914–920, 2000.
5. Tommi S. Jaakkola and David Haussler. Exploiting generative models in discriminative classifiers. In *Proceedings of NIPS*, pages 487–493, 1999.
6. Tony Jebara and Risi Kondor. Bhattacharyya and expected likelihood kernels. In *Proceedings of the Conference on Learning Theory (COLT)*, pages 57–73, 2003.
7. Rasmus E. Madsen, David Kauchak, and Charles Elkan. Modeling word burstiness using the Dirichlet distribution. To appear in *Proceedings of ICML*, 2005.
8. Andrew K. McCallum. *Bow: A toolkit for statistical language modeling, text retrieval, classification and clustering.* www.cs.cmu.edu/~mccallum/bow, 1996.
9. Stephen Robertson. Understanding inverse document frequency: On theoretical arguments for IDF. *Journal of Documentation*, 60(5):503–520, 2004.

# Utilizing Dynamically Updated Estimates in Solving the Longest Common Subsequence Problem

Lasse Bergroth[1,2]

[1] TUCS – Turku Centre for Computer Science,
Lemminkäisenkatu 14–18 A, 20520 Turku, Finland
[2] Turku University, Department of Information Technology, Programming techniques,
Ylhäistentie 2, 24130 Salo, Finland
bergroth@it.utu.fi

**Abstract.** The running time of longest common subsequence (lcs) algorithms is shown to be dependent of several parameters. To such parameters belong e. g. the size of the input alphabet, the distribution of the characters in the input strings and the degree of similarity between the strings. Therefore it is very difficult to establish an lcs algorithm that could be efficient enough for all relevant problem instances. As a consequence of that fact, many of those algorithms are planned to be applied only on a restricted set of all possible inputs. Some of them are besides quite tricky to implement.

In order to speed up the running time of lcs algorithms in common, one of the most crucial prerequisities is that preliminary information about the input strings could be utilized. In addition, this information should be available after a reasonably quick preprocessing phase. One informative a priori -value to calculate is a lower bound estimate for the length of the lcs. However, the obtained lower bound might not be as accurate as desired and thus no appreciable advantages of the preprocessing can be drawn.

In this paper, a straightforward method for updating dynamically the lower bound value for the lcs is presented. The purpose is to refine the estimate gradually to prune more effectively the search space of the used exact lcs algorithm. Furthermore, simulation tests for the new presented method will be performed in order to convince us of the benefits of it.

**Keywords:** Longest common subsequence, string algorithms, heuristic algorithms.

## 1 Introduction

There exist several practical applications where comparing the contents of two strings is of essential importance. For instance, in molecular biology it is important to estimate the similarity of two DNA or protein sequences. Especially for pre-selection purposes, those biological sequences can be treated as strings from an appropriate input alphabet. The degree of similarity can be measured by counting the maximal number of identical symbols existing in both input strings in the same order. Collecting these identical symbols and concatenating them produces (one of) the

M. Consens and G. Navarro (Eds.): SPIRE 2005, LNCS 3772, pp. 301–314, 2005.

longest common subsequence(s) of the strings the length of which numerically describes the similarity between the strings. In order to specify some further application fields, especially text and image compression and version maintenance related problems in computer science are worth mentioning.

In this paper, the longest common subsequence problem of exactly two input strings $X[1..m]$ and $Y[1..n]$ will be considered. Without loss of generality it can be assumed that $m \leq n$. The elements of the input strings are taken from the input alphabet denoted by $\Sigma$, which consists of $\sigma$ different symbols. A *subsequence* $S[1..s]$ $(0 \leq s \leq m)$ of $X$ can be obtained by deleting arbitrarily $m$-$s$ symbols from $X$. Further, if $S$ is also a subsequence of $Y$, then $S$ is a *common subsequence* of $X$ and $Y$, denoted briefly by $cs(X,Y)$. The *longest common subsequence* of $X$ and $Y$, abbreviated by $lcs(X,Y)$ (or solely *lcs*) is the $cs(X,Y)$ having maximal length, which will be denoted by $r$. The longest common subsequence need not be unique. That means that there may be several subsequences satisfying the lcs criterion for the actual problem instance. As well it is possible that the same longest common subsequence can be collected from different positions of the input strings. The following example clarifies this.

**Example 1:** $X[1..11]=$ '*aadddbcdacd*', $Y[1..11]=$ '*cdacbddbaab*', $\Sigma=\{$ *a, b, c, d* $\}$, $\sigma=4$
$\rightarrow r = 5$, $lcs(X,Y) =$ '*addba*', '*dddba*' or '*cdacd*'

In example 1, three different sequences satisfying the lcs criterion can be found. Also it can be noticed that there exist several ways to build the sequence '*addba*' by alternating, which pair of symbols '*d*' lying in positions 3, 4 and 5 in $X$ will be chosen to the $lcs(X,Y)$. Also the first character of the lcs, '*a*', can be selected freely from either of the positions 1 or 2 of $X$.

The lcs problem has actually two variants. Sometimes it is enough that only the length of the lcs, $r$, would be required, whereas in some applications the sequence itself has to be produced. The former variant will be called *r*-variant and the latter *lcs*-variant. Basically, every algorithm calculating $r$ only can be modified to solve lcs also by introducing additional bookkeeping that records the algorithm progression. After $r$ is known, the lcs can be constructed by backtracking the selections made. In this paper, the lcs-variant will be emphasized. To be able to understand the lcs problem properly, some additional definitions have to be declared.

The very first manageable approach to solve the lcs problem is from the year 1974. The method was based on dynamic programming technique [1]. That means that each character lying in the input string $X$ is compared with characters from each position of $Y$. This leads to calculation of an lcs for all possible prefixes of $X$ and $Y$. Let us denote by $r(i,j)$ the length of the lcs for the prefix pair $X[1..i]$ and $Y[1..j]$, where $0 \leq i \leq m$ and $0 \leq j \leq n$. The following recursive rule defines the connection between the length of the lcs of two arbitrary prefix pairs, $r(i,j)$, and the immediately shorter ones, where at least one of the prefixes of $X$ and $Y$ has shortened exactly by one character:

$$r(i,j) = \begin{cases} 0 & \text{if } i = 0 \text{ or } j = 0 \\ r(i\text{-}1, j\text{-}1) + 1 & \text{if } i \neq 0 \text{ and } j \neq 0 \text{ and } X[i] = Y[j] \\ \max\{ r(i\text{-}1, j), r(i, j\text{-}1) \} & \text{if } i \neq 0 \text{ and } j \neq 0 \text{ and } X[i] \neq Y[j] \end{cases}$$

Considering the recursive rule it can easily be realized that $r(i,j)$ may be incremented from its earlier value only in such index pairs $(i,j)$ where two same characters are found. Those index pairs are called *matches*. A match residing in the index pair $(i,j)$ is assigned to a certain *class k*, where $k$ is equivalent with the value $r(i,j)$. The $k$'th class contains all the matches having the $r(i,j)$ values $k$. In order to get the solution to the original problem by using the recursive rule, the value $r(m,n)$ has to be determined. Before being able to calculate that value, all respective values for shorter prefix pairs between $X$ and $Y$ have to be calculated. That means that this approach always performs $mn$ comparisons regardless of the properties of the input strings. The calculation procedure can be illustrated by filling from top to bottom and from left to right all the cells of a *matrix*, whose row indices refer to the positions of the input string $X$ and column indices analogically to the positions of the input string $Y$. The 0'th row and column are needed to the initialization of the process. The numbers in each cell denote the appropriate $r(i,j)$-values.

The fact that only those matrix cells containing a match can have contribution to $r$ – and thus the pioneer lcs algorithm does a lot of excessive work – was detected in 1977 [2, 3, 4]. That means that all the non-matching index pairs can be skipped over. To find exactly the matches without applying a linear scan repeatedly on one of the input strings, it is possible to construct a case-supporting data structure called *matchlist*, which contains ordered information about the positions where the next (previous) instance of each symbol of the input alphabet can be found after (before) the current index position. When discarding the non-matching index pairs, quite a lot of excessive work can be avoided. Even though the construction cost of the matchlist is linear to $n$, the methods basing on that technique can still be inefficient in numerous situations. When the alphabet size is small, there may exist a great number of matching index pairs, and the improvements in the running time and space complexity are thus only marginal. Most algorithms using matchlists process one row (or column) at a time. Also it is possible to search for all the matches belonging to one class at a time. Those methods are suitable for the problem instances where lcs is relatively short [2, 5, 6, 7].

In 1984, it was realized [7] that there also exist matches which need not be considered at all. It was proven that a match $(i,j)$ belonging to a class $k$ ($1 \leq k \leq m$) is important only, if there does not exist any other match $(i',j')$ belonging to the same class $k$ so that $i' = i$ and $j' < j$ or alternatively $i' < i$ and $j' = j$. If such a match $(i',j')$ cannot be found, the match $(i,j)$ is called a *dominant match*. The lcs problem can be solved by qualifying only all the dominant matches. Auxiliary data structures, such as *closest-matrices*, assist effectively in separating all the dominant matches from the set of all matches [8, 9]. Unfortunately, the cost of building a data structure supporting direct access during the preprocessing is $O(n\sigma)$, which means that an increase in the alphabet size quickly demolishes the advantages gained by the smarter processing [5, 8, 9].

A valuable observation contributing to the establishment of this paper was discovered in 1990, when Rick in his technical report [9] proved that if $r$ is known beforehand, it might be possible to discard even some of the dominant matches. Let us assume to be known that $r = k + h$, and $k$-1 is the length of the lcs found so far.

Then it would be unnecessary to register a dominant match of class $k$ residing in index pair $(i,j)$ where $\min\{\ m\text{-}i,\ n\text{-}j\ \} < h$. The reason for rejecting such matches is that the amount of symbols left either in $X$ following the $i$'th index or in $Y$ following the $j$'th index is insufficient, if a cs of length $k + h$ should be able to be constructed. The remaining dominant matches which are potential candidates to be selected to the lcs are called *minimal witnesses*. Recently, in 2003, it was realized that even though $r$ is not known exactly, a good approximation for it enables us to get rid of the majority of unnecessary dominant matches [10]. In this paper will be shown in addition that the estimate for $r$ need necessarily not be extremely accurate immediately after the preprocessing phase. In contrary, it can be dynamically refined during the running time of the applied exact lcs algorithm.

To clarify the concepts, the next example illustrates a matrix derived from the same strings as in example 1. The values in its cells present the $r(i,j)$ values for each prefix combination. The dominant matches are surrounded by boxes and the non-dominant matches are encircled. The broken line, called a *contour*, separates two adjacent areas having different $r(i,j)$ values. Example 2 is a visualization of the first example, where, for instance, value 4 at $(6,11)$ represents $r(X[1..6], Y[1..11])$. There is also a match at that index pair. Because there is a 4'th class match also in the position $(6,8)$, the match at $(6,11)$ must be a non-dominant match. In contrary, $(6,8)$ is a dominant match, because no cs of length 4 can be found when a shorter prefix from either $X$ or $Y$ is selected while another prefix is kept unchanged. If $r = 5$ is known in advance, a dominant match of class 3 at $(4,7)$ is a minimal witness, because there are enough characters (at least two) in both input strings to construct a lcs of length 5 through it. Conversely, the dominant match of the 2'nd class at $(2,9)$ does not fulfil the criterion of a minimal witness.

|    |     | 0 | 1 | 2 | 3 | 4 | 5 | 6 | 7 | 8 | 9 | 10 | 11 |
|----|-----|---|---|---|---|---|---|---|---|---|---|----|----|
|    | Y   | ø | c | d | a | c | b | d | d | b | a | a  | b  |
| X  |     |   |   |   |   |   |   |   |   |   |   |    |    |
| 0  | ø   | 0 | 0 | 0 | 0 | 0 | 0 | 0 | 0 | 0 | 0 | 0 | 0 |
| 1  | a   | 0 | 0 | 0 | [1] | 1 | 1 | 1 | 1 | 1 | (1) | (1) | 1 |
| 2  | a   | 0 | 0 | 0 | (1) | 1 | 1 | 1 | 1 | 1 | [2] | (2) | 2 |
| 3  | d   | 0 | 0 | [1] | 1 | 1 | 1 | [2] | (2) | 2 | 2 | 2 | 2 |
| 4  | d   | 0 | 0 | (1) | 1 | 1 | 1 | (2) | [3] | 3 | 3 | 3 | 3 |
| 5  | d   | 0 | 0 | (1) | 1 | 1 | 1 | (2) | (3) | 3 | 3 | 3 | 3 |
| 6  | b   | 0 | 0 | 1 | 1 | 1 | [2] | 2 | 3 | [4] | 4 | 4 | (4) |
| 7  | c   | 0 | [1] | 1 | 1 | [2] | 2 | 2 | 3 | 4 | 4 | 4 | 4 |
| 8  | d   | 0 | 1 | [2] | 2 | 2 | 2 | [3] | (3) | 4 | 4 | 4 | 4 |
| 9  | a   | 0 | 1 | 2 | [3] | 3 | 3 | 3 | 3 | 4 | [5] | (5) | 5 |
| 10 | c   | 0 | (1) | 2 | 3 | [4] | 4 | 4 | 4 | 4 | 5 | 5 | 5 |
| 11 | d   | 0 | 1 | (2) | 3 | 4 | 4 | [5] | (5) | 5 | 5 | 5 | 5 |

**Example 2:** *Graphical illustration of a matrix established by dynamic programming for input strings X = 'aadddbcdacd' and Y ='cdacbddbaab'*

The majority of the lcs algorithms follows either of the two principal processing paradigms: either one matrix row at a time or one class at a time. But there is still one approach that could be applied – advancing one *diagonal* at a time. Almost all diagonal-wise processing algorithms have originally been developed for calculating the *edit distance* between the input strings [11, 12, 13]. The edit distance problem is closely connected with the lcs problem [5]. However, there exists one pure lcs algorithm which processes the input strings one diagonal at a time [14].

In this paper, the most common approach for processing – one row at a time – is taken into consideration. According to the results in [5], the exact lcs algorithm of Kuo and Cross [15], briefly *KC*, seems to have very competitive running times for different types of input strings. The processing method of that algorithm is quite simple, and its auxiliary data structures can be constructed in $O(m)$ time. For these reasons, that algorithm was selected for refinement. The motivation of this paper is to show that despite the beforehand effectiveness of the KC algorithm its running time can still remarkably be reduced. In the next section, the original KC algorithm is described. In the third section, the idea of providing dynamically updated lower bound estimates to the disposal of KC algorithm is presented. The purpose of the fourth section is to present the practical impact and to demonstrate that the presented idea of embedding the dynamic lower bound information can be applied analogically on any row-wise lcs algorithm. Finally, the last section is reserved for conclusions and discussion.

# 2   The Original KC Algorithm

The row-wise algorithms process one symbol of the shorter input string at a time, and that symbol will be compared against symbols residing in the longer input string. Depending on the degree of the refinement of the algorithm, either all matches or only dominant matches for the appropriate symbol will be sought. The length of the lcs can increase only by one during one execution round in the outermost loop of the algorithm.

The KC algorithm uses matchlists for all different symbols of $X$ and searches for all the matching symbols in $Y$. As another auxiliary data structure an array denoted by *MinYPrefix*[0..m] is needed. Each position $l$ ($0 \leq l \leq m$) of that array contains information, where the $l$'th contour crosses the current row $i$. The value $p$ for *MinYPrefix*[$l$] can in other words be interpreted, how long is the shortest prefix of $Y$ to form a cs of length $l$ with $X[1..i]$. If no prefix of $Y$ fulfils that requirement, then the value $p$ is set to $n + 1$, which has the meaning 'undefined'. On the other hand, the 0'th index of that array is initialized to zero and it remains the same during the calculation. All the not undefined values of *MinYPrefix* array are in an increasing order, because the contours never intersect each other. The value in some position $l$ of *MinYPrefix* is updated to a lower one when processing the $i$'th row, if fewer symbols of $Y$ are needed to construct a cs of length $l$ with $X[1..i]$ than before with $X[1..i-1]$. In other case, *MinYPrefix*[$l$] will be kept unchanged on the $i$'th round of the outermost algorithm loop.

The rows will be processed from left to right. Each time when a match on a row $i$ at a position $j$ is detected, the KC algorithm will determine to which class it belongs. That can be found most effectively by performing a binary search on the array *MinYPrefix* and searching for an index $l$ of it for which is valid *MinYPrefix*[$l$-1] $< j \le$ *MinYPrefix*[$l$]. If the equality on the right hand side is met, the match under consideration is not dominant and need not be registered. When, instead, the value $j$ is strictly smaller than *MinYPrefix*[$l$], a new dominant match of the class $l$ is found and the value *MinYPrefix*[$l$] will be updated to $j$. After processing the $m$'th row, $r$ can be found as the highest index of *MinYPrefix* containing a not undefined value.

**Table 1.** Contents of the array MinYPrefix[0..m] after processing of each row

| | | | | | | | | | | | | |
|---|---|---|---|---|---|---|---|---|---|---|---|---|
| | | | | | | **MinYPrefix** | | | | | | |
| **Row** | 0 | 1 | 2 | 3 | 4 | 5 | 6 | 7 | 8 | 9 | 10 | 11 |
| 0 | 0 | 12 | 12 | 12 | 12 | 12 | 12 | 12 | 12 | 12 | 12 | 12 |
| 1 | 0 | *3* | 12 | 12 | 12 | 12 | 12 | 12 | 12 | 12 | 12 | 12 |
| 2 | 0 | 3 | *9* | 12 | 12 | 12 | 12 | 12 | 12 | 12 | 12 | 12 |
| 3 | 0 | *2* | *6* | 12 | 12 | 12 | 12 | 12 | 12 | 12 | 12 | 12 |
| 4 | 0 | 2 | 6 | *7* | 12 | 12 | 12 | 12 | 12 | 12 | 12 | 12 |
| 5 | 0 | 2 | 6 | 7 | 12 | 12 | 12 | 12 | 12 | 12 | 12 | 12 |
| 6 | 0 | 2 | *5* | 7 | *8* | 12 | 12 | 12 | 12 | 12 | 12 | 12 |
| 7 | 0 | *1* | *4* | 7 | 8 | 12 | 12 | 12 | 12 | 12 | 12 | 12 |
| 8 | 0 | 1 | *2* | *6* | 8 | 12 | 12 | 12 | 12 | 12 | 12 | 12 |
| 9 | 0 | 1 | 2 | *3* | 8 | *9* | 12 | 12 | 12 | 12 | 12 | 12 |
| 10 | 0 | 1 | 2 | 3 | *4* | 9 | 12 | 12 | 12 | 12 | 12 | 12 |
| 11 | 0 | 1 | 2 | 3 | 4 | *6* | 12 | 12 | 12 | 12 | 12 | 12 |

It is worth mentioning that if only $r$ is needed, the previous values of *MinYPrefix* can be discarded totally. However, when at least one instance for the lcs is needed, all the previously found dominant matches should be kept in memory so that the sequence can be found by backtracking the dominant matches. The maintenance of the lcs-path information is nonetheless skipped here because of its insignificance for the theoretic running time.

Let us consider again the input strings in example 2. For those strings, the contents of the *MinYPrefix*-array would be the following after processing of each row. All the updates after the initialization are marked with bold italic.

In figure 1 below, the original algorithm of Kuo and Cross is listed. The time complexity of the algorithm is $O(|M| + m(r + \log m))$, where $M$ is the number of matches between the input strings. The latter term describes the time needed for checking the contents of *MinYPrefix* and performing the binary search on that array. The extraordinary handling of indices for recovering one of the lcs-sequences is omitted here. The original KC algorithm will be abbreviated by *KC-ORIG* in the simulation test results.

---

```
      begin
(1)     for i := 1 to m do MinYPrefix[i] := n + 1;
(2)     MinYPrefix[0] := 0;
(3)     r := 0;
(4)     for i := 1 to m do
                /* Update the array values for row i. */
(5)        for each match j on the i'th row do  /* Scan all the matches on the row i. */
(6)           Find the value k for which MinYPrefix[k] < j ≤ MinYPrefix[k+1]
(7)           if j < MinYPrefix[k+1] then
(8)              MinYPrefix[k+1] := j;  /* The contour k is shifted to the left. */
(9)              if k = r then r := r + 1;  /* r increases by one. */
           end  /* for each match j */
        end  /* for i */
(10)    return r;
      end;
```

---

Fig. 1. The formal description of the original KC algorithm

## 3   Modifications to the Original KC Algorithm

We have now considered the behavior of the original KC algorithm with such granularity that we are able to propose on it some modifications, which would be relevant to improve the practical running time of that algorithm. Two suggestions how to intensify the functionality of the original method will be presented. Finally, the algorithm *KC-DYN* will be presented as a sum of the gradual enhancements to the original KC method.

### 3.1   Utilizing a Static a Priori Information Concerning the Length of the Lcs

The starting point for innovations of this paper is utilization of a *heuristic lower bound* for the lcs. When the input strings are long, solving the lcs problem by using an exact algorithm directly may take an unbearably long time. In some applications it is enough that we get *a reliable approximation for r*. In such situations, a lower bound for the lcs can give us a guard criterion, whether a more accurate investigation of the strings is needed or not. For those purposes, heuristic approximation algorithms for the lcs have been developed [16, 17, 18].

The idea of a lower bound heuristic is to relax the original problem. There exist several methods to get the lower bound for $r$ (abbreviated *lbr*). This can be done, for instance, by remapping the original symbols onto a smaller alphabet [17]. Some heuristics divide the original problem to smaller subproblems and combine their results so that legal cs for the original problem can be extracted [17]. Some methods have their foundations simply on the symbol frequency information concerning the input strings [16, 17]. The fourth approach to calculate a lower bound is based on a greedy selection of matches. The principle of them is to process one class at a time, but only one locally viewed dominant match is selected from each class. The heuristic called *BestNext* uses this kind of a processing manner selecting such a dominant match that enables construction of the longest possible cs after the appropriate selection [17, 18]. For instance, if we had input strings of length 10, and the dominant matches of class 1 resided at (1,8), (3,5) and (7,1), the selection of (3,5) would enable constructing an lcs of length 6, because there would be five symbols left in $Y$ and seven in $X$ after that selection. Clearly, when choosing either of the other candidates, one of the input strings would exhaust earlier. So the point (3,5) would be locally the most economical selection. The BestNext heuristic performs in practice reliably and very fast [17, 18], which contributed to the selection of that lower bound method for this research work. The time complexity of the BestNext heuristic can be described with the expression $O(n + \sigma + \sigma \cdot lbr)$.

If a reliable *lbr* is available, that can be used for pruning the search space of the KC algorithm. The static variant of this technique has earlier been applied on a diagonal-wise lcs method [10]. Let us assume that we were processing the $i$'th row using the KC algorithm, and *lbr* in the actual problem instance were $k + h$. Further, we will suppose that a match of class $k$ would be found in an index-pair $(i, j)$. If there are now fewer than $h$ columns left in $Y$ after the $j$'th index of it, the match can be regarded as non-significant (even if the match were dominant, it cannot be a minimal witness), because it could not belong any more to a cs of length *lbr* or longer. When we detect the first match like that on any row, all the remaining matches on that row can be discarded as uninformative. For the same reason, if we detect a match of class $k$ on the $i$'th row, and there are not at least $h$ rows left, all the matches belonging to classes $1..k$ can be omitted on the following rows. In addition, the number of omitted classes increases by one from $k$ on every following row for the same reason. It is evident that the longer the estimated *lbr* value is, the more efficiently search space of the KC algorithm can be cut. The running time of the KC algorithm embedded by a static heuristic lower bound estimate is $O(|M| + m(r + \log m) + (n + \sigma + \sigma \cdot lbr))$, but the simulation tests showed that it practically never performs worse than the original KC algorithm. The KC algorithm refined with a static lower bound calculation will be denoted by *KC-STAT*.

## 3.2  Updating the Lower Bound Dynamically

When the properties of heuristic lower bound algorithms are studied accurately, it is easy to realize that they unfortunately contain some weak points. For instance, if a very fast but a rough lower bound method is used, e.g. utilizing solely the frequency of the most common symbol in both input strings, the lower bound can be very poor – especially, if the symbol distribution is even. Also the quite a reliable BestNext algorithm fails to find a good estimate for $r$, if the lcs does not lie near the main diagonal of the matrix presenting the index pairs of the input strings.

Therefore updating the lower bound could be worth trying. The recalculation of the lower bound could be done by following some specific guideline – e.g. after processing a predefined fixed amount of rows. When recalculating the lower bound, the sequences $X$ and $Y$ are both cut by a certain amount of characters from their beginning. The length of the newly calculated $lbr$ for the suffixes $X[i+1..m]$, $Y[i+1..n]$ (denoted by $lbr_{suff}$) of the original strings can be added to the already calculated exact lcs of the prefixes of $X[1..i]$ and $Y[1..i]$ (denoted by $r_{pref}$). The lcs of the appropriate prefixes can be found by searching for the highest index $q$ for which $MinYPrefix[q] \leq i$. If the sum $r_{pref} + lbr_{suff}$ now exceeds the previously calculated $lbr$ value, a new better estimate for $r$ is obtained, and therefore the value $lbr$ will be updated. The upgraded value for $lbr$ will immediately be taken into consideration. The procedure of recalculating the lower bound for suffixes of decreasing length will be repeated again after the beforehand defined amount of rounds in the outer loop of the algorithm. Every time we get an improved approximation for $r$, the lower bound will be adjusted upwards according to the new estimate. If the approximation remains unchanged or deteriorates, clearly no updates will be made at that moment.

The following example 3 clarifies the idea of dynamic updates. The boxes denote again dominant matches in the problem, non-dominant matches are not marked. Before applying the exact KC algorithm, $lbr$ is calculated for the whole problem. The matches found by the BestNext lower bound heuristics are along the dotted line beginning at the index pair $(3,2)$ and ending at $(16,15)$, the value for $lbr$ is 9, whereas $r$ for this problem is 11. Because the BestNext algorithm maximizes the area for remaining matches, it does not find the optimal choice $(1,4)$ as its first step, but instead the match at $(3,2)$ is preferred. After that selection it is impossible to find the run of contiguous matches in $(1,4)$, $(2,5)$, $(3,6)$, $(4,7)$ and $(5,8)$, because the heuristic never does any kind of backtracking due its greedy method of advancing. When the run of the KC algorithm starts, it is no more necessary to record any matches which cannot belong to an lcs of the length 9 or longer. For this reason, there is no need to scan e.g. the 1'st row after the 8'th position of $Y$.

Let us suppose that after the 8'th row the new value for $lbr$ is calculated, and 8 characters from the beginning of both input strings will be discarded. The BestNext heuristic will find now that $lbr$ is 6 for the suffixes $X[9..16]$, $Y[9..16]$. The matches belonging to the new lower bound are also combined with each other with a dotted line. When the data structure *MinYPrefix* is then scanned after the 8'th row, it will be revealed that $r = 5$ for the prefixes $X[1..8]$, $Y[1..8]$. Sum of the new *lbr* and the current $r$ values is thus 11 and it will be registered as the updated value for *lbr*. This means that we can restrict the search from now on only on those matches which can be members of an lcs of length at least 11. Due to this, all the dominant matches which lie under the 8'th row and are marked with dotted borderlines can be dropped out of consideration. The matches chosen by the heuristics are marked with italic, and the matches belonging to $r$ are emboldened and underlined. In the test runs, the KC-variant containing dynamic updating of *lbr* is abbreviated by KC-DYN.

|   |   | 0 | 1 | 2 | 3 | 4 | 5 | 6 | 7 | 8 | 9 | 10 | 11 | 12 | 13 | 14 | 15 | 16 |
|---|---|---|---|---|---|---|---|---|---|---|---|----|----|----|----|----|----|----|
|   | Y | ø | e | a | f | b | d | a | c | e | b | c | d | f | a | d | b | b |
| **X** |   |   |   |   |   |   |   |   |   |   |   |   |   |   |   |   |   |   |
| 0 | ø | 0 | 0 | 0 | 0 | 0 | 0 | 0 | 0 | 0 | 0 | 0 | 0 | 0 | 0 | 0 | 0 | 0 |
| 1 | b | 0 | 0 | 0 | 0 | **1** | 1 | 1 | 1 | 1 | 1 | 1 | 1 | 1 | 1 | 1 | 1 | 1 |
| 2 | d | 0 | 0 | 0 | 0 | 1 | **2** | 2 | 2 | 2 | 2 | 2 | 2 | 2 | 2 | 2 | 2 | 2 |
| 3 | a | 0 | 0 | *1* | 1 | 1 | 2 | **3** | 3 | 3 | 3 | 3 | 3 | 3 | 3 | 3 | 3 | 3 |
| 4 | c | 0 | 0 | 1 | 1 | 1 | 2 | 3 | **4** | 4 | 4 | 4 | 4 | 4 | 4 | 4 | 4 | 4 |
| 5 | e | 0 | 1 | 1 | 1 | 1 | 2 | 3 | 4 | **5** | 5 | 5 | 5 | 5 | 5 | 5 | 5 | 5 |
| 6 | f | 0 | 1 | 1 | 2 | 2 | 2 | 3 | 4 | 5 | 5 | 5 | 5 | 6 | 6 | 6 | 6 | 6 |
| 7 | f | 0 | 1 | 1 | 2 | 2 | 2 | 3 | 4 | 5 | 5 | 5 | 5 | 6 | 6 | 6 | 6 | 6 |
| 8 | f | 0 | 1 | 1 | 2 | 2 | 2 | 3 | 4 | 5 | 5 | 5 | 5 | 6 | 6 | 6 | 6 | 6 |
| 9 | b | 0 | 1 | 1 | 2 | *3* | 3 | 3 | 4 | 5 | **6** | 6 | 6 | 6 | 6 | 6 | *7* | 7 |
| 10 | a | 0 | 1 | *2* | 2 | 3 | 3 | *4* | 4 | 5 | 6 | 6 | 6 | 6 | *7* | 7 | 7 | 7 |
| 11 | d | 0 | 1 | 2 | 2 | 3 | *4* | 4 | 4 | 5 | 6 | 6 | **7** | 7 | 7 | *8* | 8 | 8 |
| 12 | f | 0 | 1 | 2 | *3* | 3 | 4 | 4 | 4 | 5 | 6 | 6 | 7 | **8** | 8 | 8 | 8 | 8 |
| 13 | f | 0 | 1 | 2 | 3 | 3 | 4 | 4 | 4 | 5 | 6 | 6 | 7 | 8 | 8 | 8 | 8 | 8 |
| 14 | a | 0 | 1 | 2 | 3 | 3 | 4 | *5* | 5 | 5 | 6 | 6 | 7 | 8 | **9** | 9 | 9 | 9 |
| 15 | d | 0 | 1 | 2 | 3 | 3 | 4 | 5 | 5 | 5 | 6 | 6 | 7 | 8 | 9 | **10** | 10 | 10 |
| 16 | b | 0 | 1 | 2 | 3 | *4* | 4 | 5 | 5 | 5 | 6 | 6 | 7 | 8 | 9 | 10 | **11** | 11 |

**Example 3:** *Graphical illustration of the technique of dynamic updates for lbr.*

The refined version KC-DYN of the exact algorithm is listed below.

---

    **begin**
(1)    *lbr* := Calculate lower bound for *r* using the BestNext heuristic.
(2)    *repeat* := value, how often the lower bound for *r* will be calculated
(3)    **for** *i* := 1 **to** *m* **do** *MinYPrefix*[*i*] := *n* + 1;
(4)    *MinYPrefix*[0] := 0;
(5)    *r* := 0;
(6)    **for** *i* := 1 **to** *m* **do**
        /* Update the array values for relevant classes on the row *i*. */
(7)      *minClass* := max{ 1, *lbr* + *i* - *m* }
(8)      **for** each match *j* in at least *minClass* on the *i*'th row **do**
        /* Scan all the relevant matches on the row *i*. */
(9)        Find the value *k* for which *MinYPrefix*[*k*] < *j* ≤ *MinYPrefix*[*k*+1]
(10)      **if** *n* - *j* < *lbr* - *k* - 1/* Too near to the rightmost column? */ **then exit**;
(11)      **if** *j* < *MinYPrefix*[*k*+1] **then**
(12)        *MinYPrefix*[*k*+1] := *j*; /* The contour *k* is shifted to the left. */
(13)        **if** *k* = *r* **then** *r* := *r* + 1; /* *r* increases by one. */
(14)      **end**; /* **for** each match *j* */
(15)      **if** (*row* **mod** *repeat* = 0) **then**
        **begin**
(16)        *newlbr* := Calculate lower bound for *r*(*X*[*i*+1..*m*], *Y*[*i*+1..*n*]) using *BestNext*;
(17)        *lbr* := max{ *lbr*, *r*(*X*[1..*i*], *Y*[1..*i*]) + *newlbr* }
        **end**; /* **for** *i* */
(18) **return** *r*
**end**;

---

**Fig. 2.** The formal description of the KC-DYN algorithm

## 4 Practical Impact

In order to validate and verify the practical usefulness of the ideas presented in the previous section, various simulation tests were performed. All the compared algorithms were written in the C programming language. When compiling the programs, gcc version 3.2 with optimization –O3 was used. The tests were run on a Pentium IV (1.8 GHz) under Red Hat Linux 8.0 3.2-7. The space allocation and deallocation for data structures were planned carefully and fairly. The original KC version was implemented following exactly the algorithm description of its authors. Tests were performed for a skew (Zipfian) character distribution. The length of the input strings was held fixed — 12000 characters for both input strings. For each type of test case, 10 repetitions were made. The recalculation interval for the value *lbr* was fixed to 1000 rows.

When considering figure 3, it can be noticed that the original KC algorithm is clearly slower than its both refined variants. The difference between the two heuristic

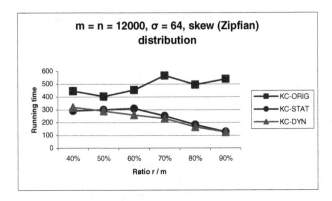

**Fig. 3.** Running times, when m = n = 12000, σ = 64, Zipfian character distribution, the ratio r / m ranges from 40 % to 90 %

variants is conversely rather small. This can be clarified by the method used for generating the input strings. Both *X* and *Y* were filled following the same selected symbol distribution. So it is evident that the lcs path lies quite near the main diagonal and no bigger distortions are expectable. When the ratio *r / m* increases the heuristic preprocessing notably speeds up the exact algorithm. It is also valuable to realize that using heuristic preprocessing – either static or dynamic – the performance of the original algorithm never gets worse, although the calculated lower bounds are not very tight.

If we let the alphabet size increase from 64 to 256, there are no radical differences in the results, as we can see in figure 4. Because the amount of matches decreases while the alphabet size increases, it is no wonder that the running times get fastened. This is a remarkable advantage, when long input strings have to be taken into consideration.

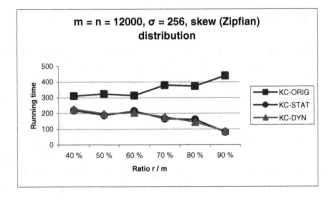

**Fig. 4.** Comparison of the variants of KC: original, static and dynamic, with m = n = 12000, σ = 256 and Zipfian character distribution

# 5  Conclusions

In this paper, methods for improving the performance of the lcs algorithm of Kuo and Cross by utilizing static and dynamic lower bound estimates have been presented. It was demonstrated that the usability of KC is greatly enhanced, when those intelligent additional properties have been embedded into it. The test results showed that even though the input strings may be long, it is not unavoidable to use linear space algorithm which usually perform slowly if not only $r$ but also one of the lcs instances has to be recovered. Because the time and space complexities of KC depend indeed on the amount of matches, it is quite acceptable to believe that the presented techniques also remarkably reduce the bookkeeping for retrieving the lcs.

To the class of the newest and the most sophisticated linear lcs algorithms belong e.g. the algorithms of Rick [19] and Goeman & Clausen [20]. It is worth mentioning that even those two methods need a preprocessing phase the space complexity of which is $O(n\sigma)$. When the size of input alphabet increases strongly, the possibilities for using direct access methods (closest-matrices etc.) when scanning the input strings deteriorate undoubtedly. As a core conclusion of this paper can be expressed that the utilization of heuristic methods is a key for improving any original lcs method, and embedding heuristic preprocessing in the lcs algorithms is still a very fruitful field of research in the future.

# References

1. Wagner, R. A. & Fischer, M. J.: The string to string correction problem, Journal of the Association for Computing Machinery, Vol. 21, nr 1, pages 168-173, 1974
2. Hirschberg, Daniel S.: Algorithms for the Longest Common Subsequence problem, Journal of the  Association for Computing Machinery, Vol. 24, nr 4, pages 664-675, October 1977
3. Hunt, James W. & Szymanski, Thomas G.: A Fast Algorithm for Computing Longest Common Subsequences, Communications of the ACM, Vol. 20, nr 5, pages 350-353, may 1977
4. Mukhopadhyay, Amar: A Fast Algorithm for the Longest-Common-Subsequence Problem, Information Sciences 20, pages 69-82, Elsevier North Holland Inc., 1980
5. Bergroth, L & Hakonen H & Raita T: A Survey of Longest Common Subsequence Algorithms, Proceedings of SPIRE 2000, A Coruña, Spain, 2000, pages 39 to 47
6. Chin, Francis Y. L. & Poon, C. K.: A Fast Algorithm for Computing Longest Common Subsequences of Small Alphabet Size, Journal of Information Processing, Vol. 13 nr 4, pages 463-469, 1990
7. Hsu, W. J. & Du, M. W.: New Algorithms for the LCS Problem, Journal of Computer and System Sciences 29, pages 133-152, 1984
8. Apostolico, A. & Guerra, C.: The Longest Common Subsequence Problem Revisited, Algorithmica (1987) 2: pages 315-336, Springer-Verlag
9. Rick, Claus: New Algorithms for the Longest Common Subsequence Problem, Institut für Informatik der Universität Bonn, Research Report No. 85123-Cs, October 1994
10. Bergroth, L. & Hakonen, H. & Väisänen, J: New Refinement Techniques for Longest Common Subsequence Algorithms, Proceedings of  SPIRE 2003, Manaus, Brazil, October 2003, pages 287 - 303

11. Miller, Webb & Myers, Eugene W.: A File Comparison Program, Software – Practice and Experience, Vol. 15(11), pages 1025-1040, November 1985
12. Myers, Eugene W.: An O(ND) Difference Algorithm and Its Variations, Algorithmica (1986) 1: pages 251 – 266, Springer-Verlag
13. Wu, Sun & Manber, Udi & Myers, Gene & Miller, Webb: An O(NP) Sequence Comparison Algorithm, Information Processing Letter 35 (1990), North-Holland, pages 317-323
14. Nakatsu, Narao & Kambayashi, Yahiko & Yajima, Shuzo: A Longest Common Subsequence Algorithm Suitable for Similar Text Strings, Acta Informatica 18, pages 171-179, Springer-Verlag 1982
15. Kuo, Shufen & Cross, George R.: An Improved Algorithm to Find the Length of the Longest Common Subsequence of Two Strings, ACM SIGIR Forum, Spring / Summer 1989, Vol. 23, No. 3-4, pages 89-99
16. Chin, F. & Poon, C. K: Performance Analysis of Some Simple Heuristics for Longest Common Subsequences, Algorithmica, 12: 293-311
17. Bergroth, L. & Hakonen H. & Raita T.: New Approximation Algorithms for Longest Common Subsequences, Proceedings of SPIRE 1998, Santa Cruz de la Sierra, Bolivia, September 1998
18. Johtela, T. & Smed, J. & Hakonen, H. & Raita, T.: An Efficient Heuristic for the LCS Problem, Third South American Workshop on String Processing, WSP'96, Recife, Brazil, August 1996, pp. 126-140
19. Rick, Claus: Simple and Fast Linear Space Computation of Longest Common Subsequences. Information Processing Letters 75(6): 275-281 (2000)
20. Goeman, H. & Clausen, M.: A New Practical Linear Space Algorithm for the Longest Common Subsequence Problem, Proceedings of the Prague Stringology Club Workshop '99

# Computing Similarity of Run-Length Encoded Strings with Affine Gap Penalty*

Jin Wook Kim[1], Amihood Amir[2], Gad M. Landau[3], and Kunsoo Park[1]

[1] School of Computer Science and Engineering,
Seoul National University
{jwkim, kpark}@theory.snu.ac.kr
[2] Department of Computer Science,
Bar-Ilan University and Georgia Tech
amir@cs.biu.ac.il
[3] Department of Computer Science,
University of Haifa and Polytechnic University
landau@cs.haifa.ac.il

**Abstract.** The problem of computing similarity of two run-length encoded strings has been studied for various scoring metrics. Many algorithms have been developed for the longest common subsequence metric and some algorithms for the Levenshtein distance metric and the weighted edit distance metric. In this paper we consider similarity based on the affine gap penalty metric which is a more general and rather complicated scoring metric than the weighted edit distance. To compute similarity in this model efficiently, we convert the problem to a path problem on a directed acyclic graph and use some properties of maximum paths in this graph. We present an $O(nm' + n'm)$ time algorithm for computing similarity of two run-length encoded strings in the affine gap penalty model, where $n'$ and $m'$ are the lengths of given two run-length encoded strings, and $n$ and $m$ are the decoded lengths of given two strings, respectively.

## 1  Introduction

A string $S$ is *run-length encoded* if it is described as an ordered sequence of pairs $(\sigma, i)$, often denoted "$\sigma^i$", each consisting of an alphabet symbols, $\sigma$, and an integer, $i$ [2]. Each pair corresponds to a *run* in $S$, consisting of $i$ consecutive occurrences of $\sigma$. Let $A$ and $B$ be two strings with lengths $n$ and $m$, respectively. Let $A'$ and $B'$ be two run-length encoded strings of $A$ and $B$, and $n'$ and $m'$ be the lengths of $A'$ and $B'$, respectively.

The problem of computing similarity of two run-length encoded strings, $A'$ and $B'$, has been studied for various scoring metrics. For the longest common subsequence metric, Bunke and Csirik [3] presented an $O(nm' + n'm)$ time algorithm, while Apostolico, Landau, and Skiena [1] gave an $O(n'm' \log(n'm'))$ time

---

\* This work was supported by FPR05A2-341 of 21C Frontier Functional Proteomics Project from Korean Ministry of Science & Technology.

M. Consens and G. Navarro (Eds.): SPIRE 2005, LNCS 3772, pp. 315–326, 2005.

algorithm and Mitchell [13] obtained an $O((d + n' + m') \log(d + n' + m'))$ time algorithm, where $d$ is the number of matches of compressed characters. Mäkinen, Navarro, and Ukkonen [12] conjectured an $O(n'm')$ time algorithm on average without proof.

For the Levenshtein distance metric, Arbell, Landau, and Mitchell [2] and Mäkinen, Navarro, and Ukkonen [11] presented $O(nm' + n'm)$ time algorithms, independently. Mäkinen, Navarro, and Ukkonen [11] posed as an open problem the challenge of extending these results to more general scoring metrics. Crochemore, Landau, and Ziv-Ukelson [5,4] and Mäkinen, Navarro, and Ukkonen [12] gave $O(nm' + n'm)$ time algorithms for the weighted edit distance metric using techniques completely different from each other.

In this paper we consider similarity based on the affine gap penalty metric. The affine gap penalty metric is a more general and rather complicated scoring metric than the weighted edit distance. To compute similarity in this model efficiently, we convert the problem to a path problem on a directed acyclic graph and use some properties of maximum paths in this graph. It is not necessary to build the graph explicitly since we come up with recurrences using the properties of the graph.

We present an $O(nm' + n'm)$ time algorithm for computing similarity of two run-length encoded strings in the affine gap penalty model, where $n'$ and $m'$ are the lengths of given two run-length encoded strings, and $n$ and $m$ are the decoded lengths of given two strings, respectively. This result shows that we successfully extended comparison of run-length encoded strings to a more general scoring metric.

## 2   Preliminaries

We first give some definitions and notations that will be used in this paper. A string is concatenations of zero or more characters from an alphabet $\Sigma$. A space is denoted by $\Delta \notin \Sigma$; we regard $\Delta$ as a character for convenience. The length of a string $A$ is denoted by $|A|$. Let $a_i$ denote $i$th character of a string $A$ and $A[i..j]$ denote a substring $a_i a_{i+1} \ldots a_j$ of $A$. When a string $\alpha$ is a substring of a string $A$, we denote it by $\alpha \prec A$. Given two strings $A = a_1 a_2 \ldots a_n$ and $B = b_1 b_2 \ldots b_m$, an alignment of $A$ and $B$ is $A^* = a_1^* a_2^* \ldots a_l^*$ and $B^* = b_1^* b_2^* \ldots b_l^*$ constructed by inserting zero or more $\Delta$s into $A$ and $B$ so that each $a_i^*$ maps to $b_i^*$ for $1 \leq i \leq l$. There are three kinds of mappings in $a^*$ and $b^*$ according to the characters of $a_i^*$ and $b_i^*$.

- match : $a_i^* = b_i^* \neq \Delta$,
- mismatch : $(a_i^* \neq b_i^*)$ and $(a_i^*, b_i^* \neq \Delta)$,
- insertion or deletion (indel for short) : either $a_i^*$ or $b_i^*$ is $\Delta$.

Note that we do not allow the case of $a_i^* = b_i^* = \Delta$.

### 2.1   Global Alignments

Given two strings $A$ and $B$, an optimal global alignment of $A$ and $B$ is an alignment of $A$ and $B$ that has the highest similarity. We denote the similarity of an optimal global alignment by $SG(A, B)$.

A well-known algorithm to find an optimal alignment was given by Smith and Waterman [14], and Gotoh [7]. Given two strings $A$ and $B$ where $|A| = n$ and $|B| = m$, the algorithm computes $SG(A, B)$ using a dynamic programming table (called the $H$ table) of size $(n + 1)(m + 1)$. Let $H_{ij}$ for $0 \leq i \leq n$ and $0 \leq j \leq m$ denote $SG(A[1..i], B[1..j])$. Then, $H_{ij}$ can be computed by the following recurrence:

$$
\begin{aligned}
H_{i,0} &= -g_i, \; H_{0,j} = -g_j \quad && \text{for } 0 \leq i \leq n, 0 \leq j \leq m \\
H_{ij} &= \max\{H_{i-1,j-1} + s(a_i, b_j), C_{ij}, R_{ij}\} \quad && \text{for } 1 \leq i \leq n, 1 \leq j \leq m
\end{aligned}
\tag{1}
$$

where

$$
\begin{aligned}
C_{0,j} &= R_{i,0} = -\infty \quad && \text{for } 0 \leq i \leq n, 0 \leq j \leq m \\
C_{ij} &= \max\{H_{i-1,j} - g_1, C_{i-1,j} - \mu\} \quad && \text{for } 1 \leq i \leq n, 1 \leq j \leq m \\
R_{ij} &= \max\{H_{i,j-1} - g_1, R_{i,j-1} - \mu\} \quad && \text{for } 1 \leq i \leq n, 1 \leq j \leq m
\end{aligned}
\tag{2}
$$

and $s(a_i, b_j)$ is the similarity score between elements $a_i$ and $b_j$ such that $s(a_i, b_j) = 1$ if $a_i = b_j$ and $s(a_i, b_j) = -\delta$ if $a_i \neq b_j$, and $g_k$ is the gap penalty for an indel of $k \geq 1$ bases such that $g_k = \gamma + k\mu$ where $\delta$, $\gamma$, and $\mu$ are non-negative constants. Then the value $H_{nm}$ is $SG(A, B)$ and it is computed in $O(nm)$ time.

## 2.2   Gap Penalty Models [8]

We defined the gap penalty $g_k$ as $g_k = \gamma + k\mu$ where $\gamma$ and $\mu$ are non-negative constants. This is called the *affine gap penalty model*, where $\gamma$ is the gap initiation penalty and $\mu$ is the gap extension penalty. We define $g_0 = 0$. When there is no gap initiation penalty, i.e., $g_k = k\mu$, it is called the *linear gap penalty model*.
    The problem we consider in this paper is follows.

*Problem 1.* Let $A$ and $B$ be two strings, and let $A'$ and $B'$ be run-length encoded strings of $A$ and $B$, respectively. Given $A'$ and $B'$, compute $SG(A, B)$ with affine gap penalty.

## 2.3   Black and White Blocks [2]

We divide the $H$ table into submatrices, which called "blocks". A *block* is a submatrix $H_{i_1..i_2, j_1..j_2}$ consisting of two runs - one of $A$ and one of $B$. Thus, by definition, the $H$ table is divided into exactly $n'm'$ blocks where $n'$ and $m'$ are the run-length encoded lengths of $A$ and $B$, respectively. The blocks are of two types: *black blocks*, corresponding to pairs of identical letters $a_{i_1} = b_{j_1}$, and *white blocks*, corresponding to pairs of distinct letters $a_{i_1} \neq b_{j_1}$.
    For a same block, there exists only one kind of similarity score $s(a_i, b_j)$. In a black block, every $a_i$ is equal to every $b_j$ and thus we use only 1 for $s(a_i, b_j)$. In a white block, every $a_i$ is different from every $b_j$ and thus we use only $-\delta$ for $s(a_i, b_j)$.

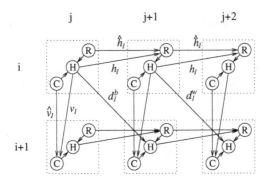

**Fig. 1.** An alignment graph for $a_{i+1} = g$ and $b_{j+1}b_{j+2} = gt$

## 2.4   Dependency of Elements

The computation of similarity can be viewed as a path problem on a directed acyclic graph called an *alignment graph* [9]. See Figure 1. At each position $(i, j)$ for $0 \leq i \leq n$ and $0 \leq j \leq m$, there are three kinds of vertices: an *H-vertex*, a *C-vertex* and an *R-vertex*. An alignment graph has the following edges:

1. $h_1$ : a horizontal edge from an *H*-vertex at $(i, j)$ to an *R*-vertex at $(i, j+1)$. The edge weight $|h_1|$ is $-\gamma - \mu$.
2. $\hat{h}_1$ : a horizontal edge from an *R*-vertex at $(i, j)$ to an *R*-vertex at $(i, j+1)$. $|\hat{h}_1| = -\mu$.
3. $v_1$ : a vertical edge from an *H*-vertex at $(i, j)$ to a *C*-vertex at $(i+1, j)$. $|v_1| = -\gamma - \mu$.
4. $\hat{v}_1$ : a vertical edge from a *C*-vertex at $(i, j)$ to a *C*-vertex at $(i+1, j)$. $|\hat{v}_1| = -\mu$.
5. $d_1$ : a diagonal edge from an *H*-vertex at $(i, j)$ to an *H*-vertex at $(i+1, j+1)$. There are two kinds of diagonal edges: $d_1^b$ when $a_{i+1} = b_{j+1}$ and $d_1^w$ when $a_{i+1} \neq b_{j+1}$. $|d_1^b| = 1$ and $|d_1^w| = -\delta$.
6. Edges at $(i, j)$ from an *R*-vertex to an *H*-vertex and a *C*-vertex to an *H*-vertex. The edge weights are 0.

The edges from 1 to 4 are defined from recurrence (2) and the edges from 5 to 6 are defined from recurrence (1). Since $R_{ij}$ is the maximum of $H_{i,j-1} - g_1$ and $R_{i,j-1} - \mu$ in recurrence (2), we define an edge $h_1$ from an *H*-vertex to an *R*-vertex with edge weight $-\gamma - \mu$ and define an edge $\hat{h}_1$ from an *R*-vertex to an *R*-vertex with $-\mu$. The other edges are defined similarly.

We can define a *path* $\langle \cdot \rangle$ from a vertex to a vertex. A horizontal path $\langle \hat{h}_i \rangle$ for $i > 1$ is defined as $i$ consecutive $\hat{h}_1$ edges, i.e., $\langle \hat{h}_1 \ldots \hat{h}_1 \rangle$ and a horizontal path $\langle h_i \rangle$ is defined as $\langle h_1 \hat{h}_{i-1} \rangle$. Vertical paths $\langle \hat{v}_i \rangle$ and $\langle v_i \rangle$ are defined similarly. A diagonal path $\langle d_i \rangle$ is defined as $i$ consecutive $d_1$ edges. A path $P$ from $(k, l)$ to $(i, j)$ is a sequence of edges from a vertex at $(k, l)$ to a vertex at $(i, j)$. For example, $\langle h_2 d_1 v_1 \rangle$ is a path from an *H*-vertex at $(i, j)$ to a *C*-vertex (or an *H*-vertex) at $(i + 2, j + 3)$. Let $|\langle \cdot \rangle|$ denote a path weight of $\langle \cdot \rangle$ which is the sum of all edge weights in the path. For example, the path weight of $\langle h_2 d_1^w v_1 \rangle$ is $|\langle h_2 d_1^w v_1 \rangle| = -\gamma - 2\mu - \delta - \gamma - \mu$.

We can merge two paths or divide a path, denoted by $\langle\alpha\rangle\langle\beta\rangle \leftrightarrow \langle\alpha\beta\rangle$, if the path weights are the same. For example, $\langle d_a h_b\rangle\langle h_c v_d\rangle \leftrightarrow \langle d_a h_{b+c} v_d\rangle$, $\langle d_a h_b\rangle\langle h_c v_d\rangle \leftrightarrow \langle d_a h_b h_c v_d\rangle$. However, for the following cases, the path weights are changed: $|\langle h_a h_b\rangle| \leq |\langle h_{a+b}\rangle|$, $|\langle v_a v_b\rangle| \leq |\langle v_{a+b}\rangle|$.

We can exchange the order of adjacent two edges in a path. If $\langle h_a v_b d_c\rangle$ is a path from $(k, l)$ to $(i, j)$, then $\langle v_b h_a d_c\rangle$ is also a path from $(k, l)$ to $(i, j)$ and the path weights are the same. However, an exchange of edge $d$ can cause the change of a path weight because $d$ depends on the match/mismatch of the position. Since $|d_1^w| < |d_1^b|$, $|\langle v_a d_b^w\rangle| \leq |\langle d_b v_a\rangle|$ and $|\langle v_a d_b^b\rangle| \geq |\langle d_b v_a\rangle|$ at any time.

We also define a *maximum path* from $(k, l)$ to $(i, j)$ which is a path that has the maximum path weight among all paths from $(k, l)$ to $(i, j)$. A maximum path from an H-vertex to another H-vertex will be called an *HH-mp*. Similarly, we will use a *CC-mp*, an *RR-mp*, an *HC-mp*, etc. Each maximum path has some restrictions: An $Hx$-mp cannot start with $\hat{v}$ or $\hat{h}$ ($x$ is don't care symbol). A $Cx$-mp and an $Rx$-mp must start with $\hat{v}$ and $\hat{h}$, respectively. An $xC$-mp must end with $v$ or $\hat{v}$ and an $xR$-mp must end with $h$ or $\hat{h}$.

From recurrence 1, we can get a relation between $H_{ij}$ and its previously defined entries.

**Lemma 1.** *Let $P$ be an HH-mp from $(k, l)$ to $(i, j)$. Then $H_{ij} \geq H_{kl} + |P|$.*

Note that the symmetric versions of Lemma 1 hold for an *HH*-mp, an *RR*-mp, an *HC*-mp, etc.

Now we consider a maximum path in one block. Every maximum path in one block consists of a permutation of $d_i$, $h_j$ and $v_k$. We know that $|\langle h_a h_b\rangle| \leq |\langle h_{a+b}\rangle|$ and the order exchange of $d$ does not change the path weight because there is only one kind of edge $d$ in one block. Thus, a path that consists of a permutation of $d_i$, $h_j$ and $v_k$ is a maximum path.

The number of diagonal edges in a maximum path depends on the weight of $d$ and that of $v$ and $h$. Let $\langle h_{k-t} d_t v_{k-t}\rangle$ be a path from $(i, j)$ to $(i + k, j + k)$ for $0 \leq t \leq k$. Then the path weight is $|\langle h_{k-t} d_t v_{k-t}\rangle| = -2g_{k-t} + t|d_1| = -2\gamma\lceil(k-t)/k\rceil - 2k\mu + (2\mu + |d_1|)t$ since $g_{k-t} = -\gamma - (k-t)\mu$ if $t < k$; it is 0 if $t = k$. The term $-2\gamma\lceil(k-t)/k\rceil$ has a maximum value when $t = k$ and the term $(2\mu + |d_1|)t$ has a maximum value when $t = k$ for $2\mu + |d_1| \geq 0$ and $t = 0$ for $2\mu + |d_1| < 0$. Thus, we compare $|\langle d_k\rangle|$ with $|\langle h_k v_k\rangle|$ and select the greater one for an *HH*-mp from $(i, j)$ to $(i + k, j + k)$. To determine an *HH*-mp from $(i, j)$ to $(i + k + s, j + k)$ for $s > 0$, we compare $|\langle d_k\rangle|$ with $|\langle h_k \hat{v}_k\rangle|$ and select

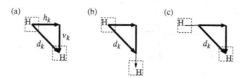

**Fig. 2.** Three cases for edge selection. $|\langle d_k\rangle|$ is compared with (a) $|\langle h_k v_k\rangle|$, (b) $|\langle h_k \hat{v}_k\rangle|$, (c) $|\langle \hat{h}_k v_k\rangle|$.

the greater one, i.e., one of $\langle d_k v_s \rangle$ and $\langle h_k v_{k+s} \rangle$ is an $HH$-mp. To determine an $HH$-mp from $(i, j)$ to $(i + k, j + k + s)$, we compare $|\langle d_k \rangle|$ with $|\langle \hat{h}_k v_k \rangle|$. See Figure 2.

For a black block, $|\langle d_k^b \rangle| = k > 0$ and $|\langle h_k \rangle| = |\langle v_k \rangle| \leq 0$ (also $|\langle \hat{h}_k \rangle| = |\langle \hat{v}_k \rangle| \leq 0$). Thus we get the following proposition.

**Proposition 1.** *Given a black block, we must maximize the number of diagonal edges in a path.*

Proposition 1 does not hold for a white block, because the similarity score for mismatch, $-\delta$, is also less than or equal to 0.

## 3   Algorithm

In this section we present an algorithm that computes the similarity between two run-length encoded strings with affine gap penalty.

The outline of the algorithm is the same as that for the LCS [3], the Levenshtein distance [2,11] and the weighted edit distance [5,12]. Given two run-length encoded strings $A'$ and $B'$, we compute blocks from left to right and from top to bottom. For each block, we compute the bottom row from left to right and the rightmost column from top to bottom. See Figure 3.

Given a block $H_{i+1..i+p,j+1..j+q}$, our goal is to compute the value of $C_{i+p,j+l}$, $R_{i+p,j+l}$ and $H_{i+p,j+l}$ for $1 \leq l \leq q$ and $C_{i+k,j+q}$, $R_{i+k,j+q}$ and $H_{i+k,j+q}$ for $1 \leq k \leq p$ in $O(p + q)$ time using $C_{i+k,j}$, $R_{i+k,j}$ and $H_{i+k,j}$ for $0 \leq k \leq p$ and $C_{i,j+l}$, $R_{i,j+l}$ and $H_{i,j+l}$ for $0 \leq l \leq q$.

We present two algorithms, one for a white block and another for a black block. For each block, we first present how to compute the values of $C$ and $R$, and then show how to compute the values of $H$.

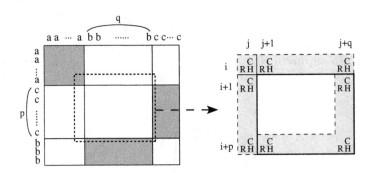

**Fig. 3.** *H table* for $a^r c^p b^t$ and $a^s b^q c^u$ is divided into 9 blocks which consist of 3 black blocks and 6 white blocks. For one of the white blocks, $H_{i+1..i+p,j+1..j+q}$, we only need to compute $H_{i+p,j+1..j+q}$ and $H_{i+1..i+p,j+q}$ from $H_{i..i+p,j}$ and $H_{i,j..j+q}$.

## 3.1  White Blocks

We give an algorithm for a white block. We only show how to compute the values of the elements on the bottom row of the block. Computing the elements on the rightmost column is done similarly.

**Computing $C_{i+p,j+l}$.** To compute the value of $C_{i+p,j+l}$ for $1 \leq l \leq q$, we need $R_{i+k,j}$ for $1 \leq k \leq p$, $C_{i,j+s}$ for $1 \leq s \leq l$, $H_{i+k,j}$ for $1 \leq k \leq p$ and $H_{i,j+s}$ for $0 \leq s \leq l$. Since there are various ways from each element to $C_{i+p,j+l}$, we give some lemmas to select *essential paths*, i.e., the paths that must be considered to compute $C_{i+p,j+l}$.

**Lemma 2.** *Let $H_{i+k,j+l}$ for $1 \leq k \leq p$ be an element within a white block and $P_1$ be a CH-mp from $(i, j + l - s)$ to $(i + k, j + l)$ for $0 \leq s < l$. Then, there exists an element $C_{i,j+l}$ such that $C_{i,j+l-s} + |P_1| \leq C_{i,j+l} + |P_2|$ where $P_2$ is a CH-mp from $(i, j + l)$ to $(i + k, j + l)$ or exists $H_{i,j+l-t}$ for $0 \leq t < s$ such that $C_{i,j+l-s} + |P_1| \leq H_{i,j+l-t} + |P_3|$ where $P_3$ is an HH-mp from $(i, j + l - t)$ to $(i + k, j + l)$.*

**Lemma 3.** *Let $P_1$ be a CC-mp from $(i, j + l - s)$ to $(i + p, j + l)$ for $0 \leq s < l$. Then, there exists an element $C_{i,j+l}$ such that $C_{i,j+l-s} + |P_1| \leq C_{i,j+l} + |P_2|$ where $P_2$ is a CC-mp from $(i, j + l)$ to $(i + p, j + l)$ or exists an element $H_{i,j+l-t}$ for $0 \leq t < s$ such that $C_{i,j+l-s} + |P_1| \leq H_{i,j+l-t} + |P_3|$ where $P_3$ is an HC-mp from $(i, j + l - t)$ to $(i + p, j + l)$*

**Lemma 4.** *Let $P_1$ be an RC-mp from $(i+k, j)$ to $(i+p, j+l)$ for $1 \leq k \leq p-1$. If $-\delta > -2\mu$, there exists an element $R_{i+p-1,j}$ such that $R_{i+k,j} + |P_1| \leq R_{i+p-1,j} + |P_2|$ where $P_2$ is an RC-mp from $(i + p - 1, j)$ to $(i + p, j + l)$ or exists $H_{i+t,j}$ for $k < t \leq p-1$ such that $R_{i+k,j} + |P_1| \leq H_{i+t,j} + |P_3|$ where $P_3$ is an HC-mp from $(i + t, j)$ to $(i + p, j + l)$.*

*If $-\delta \leq -2\mu$, the RC-mp from $(i + k, j)$ to $(i + p, j + l)$ for $1 \leq k \leq p - 1$ is $\langle \hat{h}_l v_{p-k} \rangle$ and it is an essential path for every $1 \leq k \leq p - 1$.*

**Lemma 5.** *Let $P_1$ be an HC-mp from $(i, j + l - s)$ to $(i + p, j + l)$ for $0 \leq s \leq l$. Then, there exists an element $H_{i,j+l-t}$ for $0 \leq t \leq \min\{l, p - 1\}$ such that $H_{i,j+l-s} + |P_1| \leq H_{i,j+l-t} + |P_2|$ where $P_2$ is an HC-mp from $(i, j + l - t)$ to $(i + p, j + l)$.*

By the lemmas above, we can select essential paths from the row of $C$ on top of the block, the column of $R$ to the left of the block, and the row of $H$ on top of the block to $C_{i+p,j+l}$. The maximum paths from the column of $H$ to the left of the block to $C_{i+p,j+l}$, i.e., the HC-mps from $(i + k, j)$ to $(i + p, j + l)$ for $1 \leq k \leq p - 1$, are all essential paths. From these, we derive that the value of $C_{i+p,j+l}$ is the maximum of the following. See Figure 4.

(i) $\max_{1 \leq s \leq p-1} \{R_{i+s,j} - g_{p-s}\} - l\mu$
(ii) $C_{i,j+l} - p\mu$

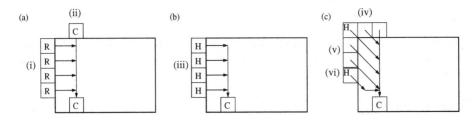

**Fig. 4.** Computing $C_{i+p,j+l}$ in a white block. (a) formulas (i) and (ii), (b) formula (iii), (c) formulas (iv), (v) and (vi).

(iii) $\max_{1\leq s\leq p-1}\{H_{i+s,j} - g_{p-s}\} - g_l$

(iv) $\max_{0\leq s\leq \min\{l,p-1\}}\{H_{i,j+l-s} - s\delta - g_{p-s}\}$

(v) $\max_{1\leq s\leq p-1-l}\{H_{i+s,j} - g_{p-s-l}\} - l\delta$ when $l < p-1$

(vi) $\max_{1\leq s\leq \min\{l-1,p-2\}}\{H_{i+p-1-s,j} - g_{l-s} - s\delta\} - g_1$ when $l \geq 2$.

The value of each formula can be computed in $O(p)$ time (of course, (ii) in constant time) and the maximum of them is computed in constant time. Thus we need $O(p)$ time to compute the value of $C_{i+p,j+l}$.

Computing all the values of $C$ of the bottom row needs $O(pq)$ time using the above result. However, since we compute the bottom row from left to right, i.e., $l$ is increased from 1 to $q$, we can reduce the time complexity to $O(p+q)$ using some properties of the recurrences that two adjacent entries are very similar.

First, consider (i) $\max_{1\leq s\leq p-1}\{R_{i+s,j} - g_{p-s}\} - l\mu$ and (iii) $\max_{1\leq s\leq p-1}\{H_{i+s,j} - g_{p-s}\} - g_l$. The index $s$ of the maximum value in (i) and that in (iii) do not depend on $l$. Hence we compute (i) and (iii) for $l = 1$ in $O(p)$ time and then get the maximum value for $l \geq 2$ in constant time by adding $-(l-1)\mu$.

Second, consider (iv) $\max_{0\leq s\leq \min\{l,p-1\}}\{H_{i,j+l-s} - s\delta - g_{p-s}\}$. The range of the column index for $H$ in (iv) is $j$ to $j+l$ for $1 \leq l < p$ and $j+l-p+1$ to $j+l$ for $l \geq p$. As $l$ increases, the range is increased by one till $l < p$ and then the position of the range is shifted to the right by one. See Figures 5(a) and 5(c). Each time $l$ increases, value $-\delta + \mu$ is added to all the rest elements. It is almost the same as the recurrence for $C$ in Case 2 of [10]. Thus, using MQUEUE [10], we can get the maximum value in amortized constant time. We can use a deque with heap order [6] to get worst-case constant time.

Third, consider (v) $\max_{1\leq s\leq p-1-l}\{H_{i+s,j} - g_{p-s-l}\} - l\delta$ when $l < p-1$. The range of the row index for $H$ in (v) is $i+1$ to $i+p-1-l$ for $1 \leq l < p-1$. That is, the range is decreased by one till $l < p-1$. See Figure 5(a). Hence we make a stack with heap order for $l = 1$ in $O(p)$ time and then get the maximum value for $l \geq 2$ one by one in constant time by popping one element, getting the maximum value of the stack and adding $(l-1)(-\delta + \mu)$ to it.

Last, consider (vi) $\max_{1\leq s\leq \min\{l-1,p-2\}}\{H_{i+p-1-s,j} - g_{l-s} - s\delta\} - g_1$ when $l \geq 2$. The range of the row index for $H$ in (vi) is $i+p-l$ to $i+p-2$ for $2 \leq l < p-1$ and $i+1$ to $i+p-2$ for $l \geq p-1$. As $l$ increases, the range is increased by one till $l < p-1$ and then the index $s$ of the maximum value does not depend

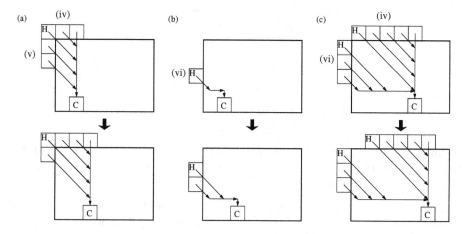

**Fig. 5.** The changes from $C_{i+p,j+l}$ to $C_{i+p,j+l+1}$ when (a) (b) $l < p-1$ and (c) $l \geq p-1$

on $l$ for $l \geq p - 1$. See Figures 5(b) and 5(c). Thus, we can get the maximum value for $l = 2$ in constant time and then get the maximum value till $l < p - 1$ one by one in constant time by adding $-\mu$ to the previous maximum value and comparing it with a new element. We also get the maximum value for $l \geq p - 1$ in constant time by adding $-(l - p + 2)\mu$ to the maximum value for $l = p - 2$.

From above all, we compute (i) and (iii) in $O(p+q)$ time, (ii), (iv) and (vi) in $O(q)$ time, and (v) in $O(p)$ time. Therefore, we compute $C_{i+p,j+l}$ for $1 \leq l \leq q$ in $O(p + q)$ time.

**Computing $R_{i+p,j+l}$.** To compute the value of $R_{i+p,j+l}$ for $1 \leq l \leq q$, we need $R_{i+p,j+l-1}$ and $H_{i+p,j+l-1}$ by recurrence (2). Since we know the values of $R_{i+p,j}$ and $H_{i+p,j}$ and we compute the value of the elements from left to right, we have no problem to compute $R_{i+p,j+l}$ and it takes $O(1)$ time. Therefore, we compute all the values of $R$ of the bottom row in $O(q)$ time.

**Computing $H_{i+p,j+l}$.** To compute the value of $H_{i+p,j+l}$ for $1 \leq l \leq q$, we need $C_{i+p,j+l}$, $R_{i+p,j+l}$ and $H_{i+p-1,j+l-1}$. Since we know the values of $C_{i+p,j+l}$ and $R_{i+p,j+l}$, we need only compute the diagonal incoming value.

**Lemma 6.** *Let $P_1$ be a RH-mp from $(i + k, j)$ to $(i + p - 1, j + l - 1)$ for $1 \leq k \leq p - 1$. Then, $R_{i+k,j} + |P_1| + |\langle d_1^w \rangle| \leq R_{i+p,j+l}$.*

**Lemma 7.** *Let $P_1$ be a CH-mp from $(i, j + s)$ to $(i + p - 1, j + l - 1)$ for $1 \leq s \leq l - 1$. Then, $C_{i,j+s} + |P_1| + |\langle d_1^w \rangle| \leq C_{i+p,j+l}$.*

**Lemma 8.** *Let $P_1$ be a HH-mp from $(i + k, j)$ to $(i + p - 1, j + l - 1)$ for $1 \leq k \leq p-1$. Then, $H_{i+k,j} + |P_1| + |\langle d_1^w \rangle| \leq R_{i+p,j+l}$ or $H_{i+k,j} + |P_1| + |\langle d_1^w \rangle| \leq H_{i+p-l,j} + |\langle d_l^w \rangle|$ when $l \leq p$.*

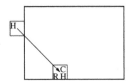

**Fig. 6.** Computing $H_{i+p,j+l}$ in a white block.

**Lemma 9.** *Let $P_1$ be a HH-mp from $(i, j + s)$ to $(i + p - 1, j + l - 1)$ for $1 \leq s \leq l - 1$. Then, $H_{i,j+s} + |P_1| + |\langle d_1^w \rangle| \leq C_{i+p,j+l}$ or $H_{i,j+s} + |P_1| + |\langle d_1^w \rangle| \leq H_{i,j+l-p} + |\langle d_l^w \rangle|$ when $l > p$.*

By the lemmas above, we derive that the value of $H_{i+p,j+l}$ is the maximum of the following: (i) $R_{i+p,j+l}$, (ii) $C_{i+p,j+l}$, (iii) $H_{i+p-l,j} - l\delta$ when $l \leq p$, (iv) $H_{i,j+l-p} - p\delta$ when $l > p$. See Figure 6.

Since each value of (i), (ii), (iii) and (iv) is computed in constant time, we can compute all the values of $H$ of the bottom row in $O(q)$ time.

**Analysis.** Given a white block with $p$ rows and $q$ columns, the bottom row of the block is computed in $O(p + q)$ time. The values of $C$ of the bottom row are computed in $O(p + q)$ time and the values of $R$ and $H$ of the bottom row are computed in $O(q)$ time.

The rightmost column of the block is also computed in $O(p + q)$ time and thus the similarity of the white block can be computed in $O(p + q)$ time.

## 3.2   Black Blocks

We give an algorithm for a black block. As white blocks, we only show how to compute the values of the elements on the bottom row of the block.

**Computing $C_{i+p,j+l}$.** To compute the value of $C_{i+p,j+l}$ for $1 \leq l \leq q$, we need $R_{i+k,j}$ for $1 \leq k \leq p$, $C_{i,j+s}$ for $1 \leq s \leq l$, $H_{i+k,j}$ for $1 \leq k \leq p$ and $H_{i,j+s}$ for $0 \leq s \leq l$. We give two lemmas for a black block to select essential paths.

**Lemma 10.** *Let $P_1$ be a CC-mp from $(i, j + l - s)$ to $(i + p, j + l)$ for $1 \leq s < l$. Then, there exists an element $H_{i,j+l-s}$ such that $C_{i,j+l-s} + |P_1| \leq H_{i,j+l-s} + |P_2|$ where $P_2$ is an HC-mp from $(i, j + l - s)$ to $(i + p, j + l)$*

**Lemma 11.** *Let $P_1$ be an HC-mp from $(i, j + l - s)$ to $(i + p, j + l)$ for $0 \leq s \leq l$. Then, there exists an element $H_{i,j+l-t}$ for $0 \leq t \leq \min\{l, p - 1\}$ such that $H_{i,j+l-s} + |P_1| \leq H_{i,j+l-t} + |P_2|$ where $P_2$ is an HC-mp from $(i, j + l - t)$ to $(i + p, j + l)$.*

By Lemmas 10, 11 and Proposition 1, we can select essential paths from the row of $C$ on top of the block and the row of $H$ on top of the block to $C_{i+p,j+l}$. The maximum paths from the column of $R$ to the left of the block and the column of $H$ to the left of the block to $C_{i+p,j+l}$ are all essential paths. From these, we derive that the value of $C_{i+p,j+l}$ is the maximum of the following. See Figure 7. (Hence we need $O(p)$ time to compute $C_{i+p,j+l}$.)

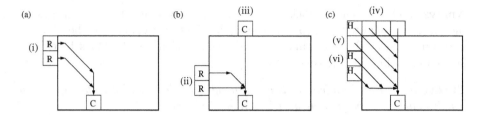

**Fig. 7.** Computing $C_{i+p,j+l}$ in a black block. (a) formula (i), (b) formulas (ii) and (iii), (c) formulas (iv), (v) and (vi).

(i) $\max_{1\le s\le p-l}\{R_{i+s,j} - g_{p-s-l+1}\} - \mu + (l-1)$ for $l \le p-1$
(ii) $\max_{0\le s\le \min\{l-2,p-2\}}\{R_{i+p-1-s,j} - (l-s)\mu + s\} - g_1$ when $l \ge 2$
(iii) $C_{i,j+l} - p\mu$
(iv) $\max_{0\le s\le \min\{l,p-1\}}\{H_{i,j+l-s} + s - g_{p-s}\}$
(v) $\max_{1\le s\le p-1-l}\{H_{i+s,j} - g_{p-s-l}\} + l$ when $l < p-1$
(vi) $\max_{1\le s\le \min\{l-1,p-2\}}\{H_{i+p-1-s,j} - g_{l-s} + s\} - g_1$ when $l \ge 2$.

We can compute all the values of $C$ of the bottom row in $O(p+q)$ time. Recurrences (iii), (iv), (v) and (vi) are essentially the same as recurrences (ii), (iv), (v) and (vi) of a white block, and (i) and (ii) are similar to (v) and (vi), respectively.

**Computing $R_{i+p,j+l}$.** Computing $R_{i+p,j+l}$ for $1 \le l \le q$ in a black block is the same as in a white block. We can compute $R_{i+p,j+l}$ by recurrence (2) and it takes $O(1)$ time. Therefore, we compute all the values of $R$ of the bottom row in $O(q)$ time.

**Computing $H_{i+p,j+l}$.** To compute the value of $H_{i+p,j+l}$ for $1 \le l \le q$, we need $C_{i+p,j+l}$, $R_{i+p,j+l}$ and $H_{i+p-1,j+l-1}$. Since we know the values of $C_{i+p,j+l}$ and $R_{i+p,j+l}$, we need only compute the diagonal incoming value.

To compute $H_{i+p,j+l}$, we need more terms than that in a white block. Because Lemmas 6 and 7 do not hold for a black block, we need to compute paths from $R_{i+k,j}$ for $1 \le k \le p-1$ and $C_{i,j+s}$ for $1 \le s \le l-1$.

**Lemma 12.** *Let $P_1$ be a RH-mp from $(i+k,j)$ to $(i+p-1,j+l-1)$ for $1 \le k \le p-1$. Then, $R_{i+k,j} + |P_1| + |\langle d_1^b\rangle| \le H_{i+p-l,j} + |\langle d_l^b\rangle|$ when $p \ge l$ and $R_{i+k,j} + |P_1| + |\langle d_1^b\rangle| \le H_{i,j+l-p} + |\langle d_p^b\rangle|$ when $l > p$.*

**Lemma 13.** *Let $P_1$ be a CH-mp from $(i,j+s)$ to $(i+p-1,j+l-1)$ for $1 \le s \le l-1$. Then, $C_{i,j+s} + |P_1| + |\langle d_1^b\rangle| \le H_{i+p-l,j} + |\langle d_l^b\rangle|$ when $p \ge l$ and $C_{i,j+s} + |P_1| + |\langle d_1^b\rangle| \le H_{i,j+l-p} + |\langle d_p^b\rangle|$ when $l > p$.*

By Lemmas 12, 13, 8, 9 and Proposition 1, we derive that the value of $H_{i+p,j+l}$ is the maximum one of the followings: (i) $R_{i+p,j+l}$, (ii) $C_{i+p,j+l}$, (iii) $H_{i+p-l,j} + l$ when $p \ge l$, (iv) $H_{i,j+l-p} + p$ when $l > p$.

Since each value of (i), (ii), (iii) and (iv) is computed in constant time, we can compute all the values of $H$ of the bottom row in $O(q)$ time.

**Analysis.** Given a black block with $p$ rows and $q$ columns, the bottom row of the block is computed in $O(p + q)$ time. The rightmost column of the block is also computed in $O(p + q)$ time and thus the similarity of the black block can be computed in $O(p + q)$ time.

**Theorem 1.** *The similarity of two run-length encoded strings in the affine gap penalty model can be computed in $O(nm' + n'm)$ time.*

# References

1. A. Apostolico, G. M. Landau, and S. Skiena. Matching for Run Length Encoded Strings. *Journal of Complexity*, 15(1):4–16, 1999
2. O. Arbell, G. M. Landau, and J. Mitchell. Edit Distance of Run-Length Encoded Strings. *Information Processing Letters*, 83(6):307–314, 2002
3. H. Bunke and H. Csirik. An Improved Algorithm for Computing the Edit Distance of Run Length Coded Strings. *Information Processing Letters*, 54:93–96, 1995
4. M. Crochemore, G. M. Landau, B. Schieber, and M. Ziv-Ukelson. Re-Use Dynamic Programming for Sequence Alignment: An Algorithmic Toolkit. *String Algorithmices*, NATO Book series, KCL Press, 2004
5. M. Crochemore, G. M. Landau, and M. Ziv-Ukelson. A Subquadratic Sequence Alignment Algorithm for Unrestricted Scoring Matrices. *SIAM Journal on Computing*, 32(6):1654–1673, 2003
6. H. Gajewska and R. E. Tarjan. Deques with Heap Order. *Information Processing Letters*, 22:197–200, 1986
7. O. Gotoh. An Improved Algorithm for Matching Biological Sequences. *Journal of Molecular Biology*, 162:705–708, 1982
8. D. Gusfield. *Algorithms on Strings, Trees, and Sequences*, Cambridge University Press, 1997
9. X. Huang and W. Miller. A Time-Efficient, Linear-Space Local Similarity Algorithm. *Advances in Applied Mathematics*, 12:337–357, 1991
10. J. W. Kim and K. Park. An Efficient Local Alignment Algorithm for Masked Sequences. *In Proc. 10th COCOON*, LNCS, 3106:440-449, 2004
11. V. Mäkinen, G. Navarro, and E. Ukkonen. Approximate Matching of Run-Length Compressed Strings, *In Proc. 12th CPM*, LNCS, 2089:31–49, 2001
12. V. Mäkinen, G. Navarro, and E. Ukkonen. Approximate Matching of Run-Length Compressed Strings, *Algorithmica*, 35:347–369, 2003
13. J. Mitchell. A Geometric Shortest Path Problem, with Application to Computing a Longest Common Subsequence in Run-Length Encoded Strings. Technical Report, Dept. of Applied Mathematics, SUNY Stony Brook, 1997
14. T. F. Smith and M. S. Waterman. Identification of Common Molecular Subsequences. *Journal of Molecular Biology*, 147:195–197, 1981

# $L_1$ Pattern Matching Lower Bound*

Ohad Lipsky[1] and Ely Porat[2]

[1] Department of Computer Science, Bar-Ilan University, Ramat-Gan 52900, Israel
Tel: +972 3 531-8408
ohadlipsky@yahoo.com
[2] Department of Computer Science, Bar-Ilan University, 52900 Ramat-Gan, Israel
Tel: 972 3 531-7620
porately@cs.biu.ac.il

**Abstract.** Let a text string $T = t_0, \ldots, t_{n-1}$ and a pattern string $P = p_0, \ldots, p_{m-1}$ $t_i, p_j \in \mathbb{N}$ be given. In The *Approximate Pattern Matching in the $L_1$ metric* problem ($L_1$-matching for short) the output is, for every text location $i$, the $L_1$ distance between the pattern and the length $m$ substring of the text starting at $i$, i.e. $\Sigma_{j=0}^{m-1}|t_{i+j} - p_j|$. The *Less Than Matching* problem is that of finding all locations $i$ of $T$ where $t_{i+j} \geq p_j$ $j = 0, \ldots, m-1$. The *String Matching with Mismatches* problem is that of finding the number of mismatches between the pattern and every length $m$ substring of the text. For the three above problems, the fastest known deterministic solution is $O(n\sqrt{m \log m})$ time.

In this paper we show that the latter two problems can be linearly reduced to the problem of $L_1$-matching.

## 1   Introduction

*Approximate matching* is one of the fundamental problems in pattern matching. In approximate matching one defines a distance function between strings, and seeks for the distance between the pattern and every length $m$ substring of the text. In the problem of *String Matching with Mismatches* [6] the hamming distance is used as distance function. Almost two decades ago Abrahamson [1] showed an $O(n\sqrt{m \log m})$ time algorithm for the problem, which is the fastest known so far. Amir, Lewenstein and Porat [3] gave an $O(n\sqrt{k \log k})$ time algorithm for the case that we can discard locations with distance greater than a given bound $k$. However, in the general case $k = m$ and it is not better than Abrahamson's result. More than that, Indyk [9] showed that improvement over the Abrahamson algorithm to $O(nm^c)$ time will yield an $O(n^{2+2c})$ time for boolean matrix multiplication.

The number of mismatches is an important measure as part of an *edit distance* between typed strings. Advances in Multimedia, Digital Libraries and Computational Biology have shown that a much more generalized theoretical basis of string matching could be of tremendous benefit [16,15]. In computer vision, for example, it does not make sense to say that a pattern pixel with a close

---

* Partially supported by GIF Young Scientists Program grant 2055-1168.6/2002.

M. Consens and G. Navarro (Eds.): SPIRE 2005, LNCS 3772, pp. 327–330, 2005.

grey level to the text pixel should generate the same error penalty as, say, a white pattern pixel being matched to a black text pixel. Similarly in biology, the energy level for bonding different proteins is different, suggesting that not all "mismatches" should be counted equally. In various other applications, such as earthquake prediction [14], stock market analysis [13], and music retrieval [17], the distance measure used is the Minkowsky $L_1$ norm. The $L_1$ norm is of particular importance also due to the fact that strings can be embedded in the $L_1$ space and the distance between their associated vectors approximates the *edit distance with moves* of the strings [8]. The problem of $L_1$ matching was solved in $O(n\sqrt{m}\log m)$ time in [12,4,7]. The solution of [12] was further extended to solve closest pair problems in [10]; it was also the base for other pattern matching algorithms in [11]. We believe it is hard to show a better time for the $L_1$ matching.

**Our Contribution:** We show here that *String Matching with Mismatches* can be linearly reduced to $L_1$ matching. We further show that the problem of *Less Than Matching* can also be linearly reduced to $L_1$ matching.

The problem of *Less Than Matching* was presented in [2] and solved in $O(n\sqrt{m}\log m)$. It is still open problem whether that time is optimal.

## 2    Preliminaries

Let $A = a_0,\ldots,a_{m-1}$ and $B = b_0,\ldots,b_{m-1}$ be two string of numbers. The $L_1$ *distance* between $A$ and $B$, denoted by $L1(A,B)$ is defined by $L1(A,B) = \sum_{j=0}^{m-1}|a_j - b_j|$.

Formally define the **Approximate Pattern Matching in the $L_1$ metric** problem:

*Input:* text $T = t_0,\ldots,t_{n-1}$, pattern $P = p_0,\ldots,p_{m-1}$,where $t_i, p_j \in \mathbb{N} \forall i,j$.
*Output:* $l_0, l_1,\ldots,l_{n-m}$ where $l_i = L1(t_i\cdots t_{i+m-1}, P)$ For every $i \in [0, n-m]$.

The problem of **String Matching with Mismatches** is the problem of counting the mismatches between the pattern and the text, for every alignment possible. Formally:

*Input:* Text $T = t_0, t_1,\ldots,t_{n-1}$ and pattern $P = p_0, p_1,\ldots,p_{m-1}$, where $t_i, p_j \in \Sigma$.
*Output:* $h_0,\ldots,h_{n-m}$ where $h_i = \sum_{j=0}^{m-1} neq(t_{i+j}, p_j)$ for every $i \in [0, n-m]$ and where

$$neq(x,y) = \begin{cases} 1, & x \neq y \\ 0, & x = y \end{cases}$$

The problem of **Less than Matching** is defined as:

*Input:* text $T = t_0,\ldots,t_{n-1}$, pattern $P = p_0,\ldots,p_{m-1}$ where $t_i, p_j \in \mathbb{N}$.
*Output:* All locations $i$ in $T$ $i \in [0, n-m]$ s.t. $t_{i+j} \geq p_j$ $\forall j = 0,\ldots,m-1$

## 3   Reduction from String Matching with Mismatches

We use the following key observation:

**Lemma 1.** *For any pair of integers $x, y$*

$$|x - y + 1| + |x - y - 1| - 2|x - y| = \begin{cases} 0, & x \neq y \\ 2, & x = y \end{cases}$$

**The Reduction:** The input is a text $T = t_0 t_1 \ldots t_{n-1}$ and a pattern $P = p_0 p_1 \ldots p_{m-1}$ both string of symbols from some alphabet $\Sigma$. We can assume that $\Sigma = \{1, 2, \ldots, |\Sigma|\}$. If not, just change $T$ and $P$ using any bijective function from $\Sigma$ to $\{1, 2, \ldots, |\Sigma|\}$. This can be done in linear time.

1. Construct $P^-$ to be $p_0 - 1, p_1 - 1, \cdots, p_{m-1} - 1$
   and $P^+$ to be $p_0 + 1, p_1 + 1, \cdots, p_{m-1} + 1$.
2. Let
   $l_0, l_1, \ldots, l_{n-m}$ be the output of $L_1$ matching of $T$ and $P$,
   $l_0^+, l_1^+ \ldots, l_{n-m}^+$ be the output of $L_1$ matching of $T$ and $P^+$
   and $l_0^-, l_1^- \ldots, l_{n-m}^-$ be the output of $L_1$ matching of $T$ and $P^-$.
3. For every $i \in [0, n - m]$:
   $m_i \leftarrow \frac{1}{2}(l_i^+ + l_i^- - 2l_i)$.
   $h_i \leftarrow m - m_i$.

It follows from the observation that $m_i$ equals to the number of matches at location $i$ and then clearly $h_i$ is the number of mismatches at location $i$. The reduction is linear.

## 4   Reduction from Less Than Matching

**Lemma 2.** *For any pair of integers $x, y$*

$$|x - y + 1| - |x - y| = \begin{cases} +1, & x \geq y \\ -1, & x < y \end{cases}$$

**The Reduction:** The input is text $T = t_0 t_1 \ldots t_{n-1}$ and pattern $P = p_0 p_1 \ldots p_{m-1}$ both strings of natural numbers. Take the following steps:

1. Run $L_1$ matching with $T$ and $P$, output will be $l_0, \ldots, l_{n-m}$.
2. Run $L_1$ matching with $T^+$ and $P$, where $T^+ = t_0 + 1, t_1 + 1, \ldots, t_{n-1} + 1$.
   The output will be $l_0^+, l_1^+, \ldots, l_{n-m}^+$.
3. For every $i \in [0, n - m]$:
   If $l_i^+ - l_i = m$ then output location $i$ as a *match*.

Clearly, the reduction is linear. At location $i$, every pair of numbers $t_{i+j}$ and $p_j$ can either add 1 to $l_i^+ - l_i$ or subtract 1 from $l_i^+ - l_i$, it follows that if $l_i^+ - l_i = m$ all pairs added 1, which mean that for all $j = 0, \ldots, m - 1$ it holds that $t_{i+j} \geq p_j$, which mean we have a match. It is easily seen that if there is a match at location $i$ it will hold that $l_i^+ - l_i = m$.

# References

1. Karl R. Abrahamson. Generalized string matching. *SIAM J. Comput.*, 16(6):1039–1051, 1987.
2. Amihood Amir and Martin Farach. Efficient 2-dimensional approximate matching of half-rectangular figures. In *Information and Computation*, pages 118(1):1–11, 1995.
3. Amihood Amir, Moshe Lewenstein, and Ely Porat. Faster algorithms for string matching with k mismatches. *J. Algorithms*, 50(2):257–275, 2004.
4. Amihood Amir, Ohad Lipsky, Ely Porat, and Julia Umanski. Approximate matching in the $l_1$ metric. In Apostolico et al. [5], pages 91–103.
5. Alberto Apostolico, Maxime Crochemore, and Kunsoo Park, editors. *Combinatorial Pattern Matching, 16th Annual Symposium, CPM 2005, Jeju Island, Korea, June 19-22, 2005, Proceedings*, volume 3537 of *Lecture Notes in Computer Science*. Springer, 2005.
6. Alberto Apostolico and Zvi Galil, editors. *Combinatorial Algorithms on Words*. Springer-Verlag New York, Inc., Secaucus, NJ, USA, 1985.
7. Peter Clifford, Raphaël Clifford, and Costas S. Iliopoulos. Faster algorithms for delta, gamma-matching and related problems. In Apostolico et al. [5], pages 68–78.
8. Graham Cormode and S. Muthukrishnan. The string edit distance matching problem with moves. In *SODA*, pages 667–676, 2002.
9. Piotr Indyk. Private communications. 1999.
10. Piotr Indyk, Moshe Lewenstein, Ohad Lipsky, and Ely Porat. Closest pair problems in very high dimensions. In *ICALP*, volume 3142 of *Lecture Notes in Computer Science*, pages 782–792. Springer, 2004.
11. Piotr Indyk, Ohad Lipsky, and Ely Porat. Approximate translation matching, manuscript. 2004.
12. Ohad Lipsky. Efficient distance computations. Master's thesis, Bar-Ilan University, Department of Computer Science, 2003.
13. E. Maasoumi and J. Racine. Entropy and predictability of stock market returns. In *Journal of Econometrics*, pages 107(1):291–312, 3, 2002.
14. L. Malagnini, R. B. Herman, and M. Di Bona. ground motion scaling in the apenines (italy). In *Bull. Seism. Soc. Am.*, pages 90:1062–1081, 2000.
15. M. V. Olson. A time to sequence. In *Science, 270*, pages 394–396, 1995.
16. Alex Pentland. Invited talk. nsf institutional infrastructure workshop. 1992.
17. Ilya Shmulevich, O. Yli-Harja, E. Coyle, D. Povel, and K. Lemstrom. Perceptual issues in music pattern recognition - complexity of rhythm and key fining. 1999.

# Approximate Matching in the $L_\infty$ Metric[*]

Ohad Lipsky[1] and Ely Porat[2]

[1] Department of Computer Science, Bar-Ilan University, Ramat-Gan 52900, Israel
Tel: +972 3 531-8408
ohadlipsky@yahoo.com
[2] Department of Computer Science, Bar-Ilan University, 52900 Ramat-Gan, Israel
Tel: 972 3 531-7620
porately@cs.biu.ac.il

**Abstract.** Let a text $T = t_0, \ldots, t_{n-1}$ and a pattern $P = p_0, \ldots, p_{m-1}$, strings of natural numbers, be given. In the *Approximate Matching in the $L_\infty$ metric* problem the output is, for every text location $i$, the $L_\infty$ distance between the pattern and the length $m$ substring of the text starting at $i$, i.e. $Max_{j=0}^{m-1}|t_{i+j} - p_j|$. We consider the *Approximate $k - L_\infty$ distance* problem. Given text $T$ and pattern $P$ as before, and a natural number $k$ the output of the problem is the $L_\infty$ distance of the pattern from the text only at locations $i$ in the text where the distance is bounded by $k$. For the locations where the distance exceeds $k$ the output is $\phi$. We show an algorithm that solves this problem in $O(n(k + \log(\min(m, |\Sigma|))) \log m)$ time.

## 1 Introduction

One of the classical approximate pattern matching problems is the *String matching with Mismatches* [3]. The input is a text string $T = t_0, \ldots, t_{n-1}$ and pattern string $P = p_0, \ldots, p_{m-1}$ and the output is the number of mismatches between the pattern and every length $m$ substring of the text. The number of mismatches is a very intuitive way for comparing strings, motivated by typing errors. However, this measure is not completely suitable in many fields. in computer vision, for example, one might allow the pattern pixel to differ from text pixel within some tolerated distance, without causing a mismatch. In music information retrieval [9] one might consider half tone changes as allowed, and seek for large distortions only. In various other applications the $L_\infty$ Minkowsky norm is in use. The problem of $L_\infty$ matching was discussed by Lipsky [7], and in recent papers [4,1]. In [7,1] an $O(n|\Sigma| \log(m + |\Sigma|))$ time algorithm is presented for the $L_\infty$ matching problem. Their technique was also extended to closest pair problems [5] and to some other pattern matching problems [6]. In [4] They showed how to find all locations for which the $L_\infty$ distance is less than some given $\delta$ in $O(\delta n \log m)$ time. We present a simpler solution, that not only finds the locations it matches, but also computes the exact $L_\infty$ distance at those locations. Let $T^i$ denote the length $m$ substring of the text starting at $i$.

---

[*] Partially supported by GIF Young Scientists Program grant 2055-1168.6/2002.

M. Consens and G. Navarro (Eds.): SPIRE 2005, LNCS 3772, pp. 331–334, 2005.

## 2    The Algorithm

**Algorithm Outline**

1. Discard all locations with distance $> 2k$.
2. Reduce the alphabet to an alphabet of size $2k$.
3. Run the $L_\infty$ algorithm with the reduced alphabet.
4. Take distances as cyclic.

We discuss each step in detail.

**Discarding Locations with too High a Distance:** We use here an approximation method for $L_\infty$ which approximates the $L_\infty$ values up to a factor of $1 \pm \epsilon$ [8]. We use $\epsilon = \frac{1}{4}$. Then we discard every location with distance greater than $\frac{5}{4}k$. This ensures us that location with distance greater than $\frac{5}{3}k$ are discarded for sure, and location with distance less than $k$ are kept for sure. The locations with distance between $k$ and $\frac{5}{3}k$ might be discarded, and might be kept. In the next steps we will exploit the fact that while there are locations with distance greater than $k$, since they are bounded by $5k/3$ it will only add a constant factor. The time needed for this step is $O(n \log m \log |\Sigma|)$.

**Reducing the Alphabet:** This is done simply by taking each number modulo $4k$, i.e. use $T' = t_0 \bmod 4k, t_1 \bmod 4k, \ldots, t_{n-1} \bmod 4k$ and $P' = p_0 \bmod 4k$, $p_1 \bmod 4k, \ldots p_{m-1} \bmod 4k$ instead of the original $T$ and $P$. The idea behind this step is the following lemma:

**Lemma 1.** *For any integers* $x, y, k$ *s.t.* $|x - y| < 2k$

$$|x \bmod 4k - y \bmod 4k| = |x - y| \operatorname{or} |x \bmod 4k - y \bmod 4k| = 4k - |x - y|$$

**Corollary 1.** *For any integers* $x, y, k$ *s.t.* $|x - y| < 2k$

$$|x - y| = |x \bmod 4k - y \bmod 4k| \operatorname{or} |x - y| = 4k - |x \bmod 4k - y \bmod 4k|$$

This step is executed in linear time.

**Run $L_\infty$ Algorithm with Reduced Alphabet:** We simply use the algorithm from the next section on $P'$ and $T'$, the alphabet size of $T'$ and $P'$ is bounded by $4k$. This step will take $O(nk \log(m+k))$ time. Let $l_0, \ldots, l_{n-m}$ be the output.

**Take Distances as Cyclic:** Following from the corollary we know that for each non discarded location $i$ either the $L_\infty$ distance of the pattern from the length $m$ substring of the text starting at location $i$ equals $l_i$ or it equals $4k - l_i$. Since location $i$ is not discarded we know that the output should be less than $2k$, and therefore only one of the two options fit. This step is executed in linear time.

## 3   $L_\infty$ Matching $O(n|\Sigma|\log(m+|\Sigma|))$ Algorithm

The method in this algorithm is encoding the text and the pattern in such a way that in a single convolution, and a linear time pass on the convolution result we compute the output.

**Key Idea:** We look at one text number, $t$, and one pattern number $p$. We encode both of them to a $|\Sigma|$ long binary strings. The encoding of $t$ is all 0's except the $t$-th bit which is 1, and similarly with $p$, which is encoded to all 0's except the $p$-th bit. Let $c(i)$ denote the encoded $i$. Now, we start by $c(p)$ aligned below $c(t)$ and start at position $-|\Sigma|$ (where $c(t)$ fixed to start at position 1). We move $c(p)$ to the right till both 1-bits are one below the other. At this position, the distance between the starting position of $c(t)$ and the starting position of $c(p)$ equals to the difference $|t-p|$, an example is given in Figure 1. If we look at $r = c(t) \otimes c(p)$ we will have either $r[-|t-p|] = 1$ or $r[|t-p|] = 1$. Extending this idea to encoding strings of numbers requires adding leading (or tracing) zeros between the encoded numbers.

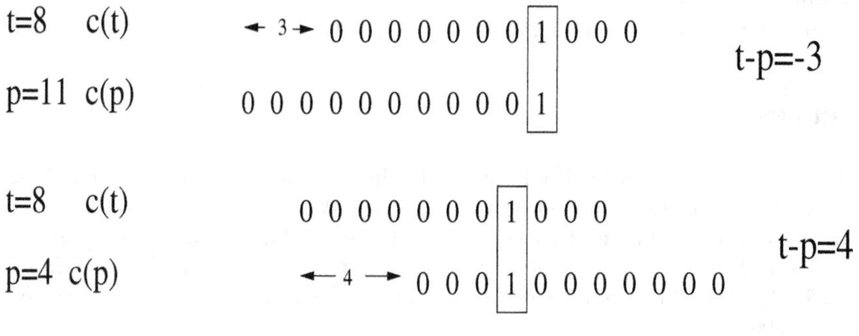

t=8   c(t)          ← 3 →  0 0 0 0 0 0 0 1 0 0 0

p=11  c(p)          0 0 0 0 0 0 0 0 0 0 1                    t-p=-3

t=8   c(t)          0 0 0 0 0 0 0 1 0 0 0

p=4   c(p)          ← 4 →  0 0 0 1 0 0 0 0 0 0 0              t-p=4

**Fig. 1.** $c(p)$ moved below $c(t)$ till the 1-bits are aligned

In detail: first, define $\chi_{\neq 0}(x) = 1$ if $x \neq 0$ and 0 otherwise. Next, define For every $x \in \Sigma = \{1, \ldots, n\}$, $c^t(x) = c^t(x)_1, \ldots, c^t(x)_{2|\Sigma|}$ where $c^t(x)_i = 1$ if $i = |\Sigma| + x$ and 0 otherwise. Similarly define $c^p(x) = c^p(x)_1, \ldots, c^p(x)_{2|\Sigma|}$ where $c^p(x)_i = 1$ if $i = x$ and 0 otherwise.

**Algorithm Steps**

1. Construct $c^t(T) = c^t(t_1) \cdots c^t(t_n)$
2. Construct $c^p(P) = c^p(p_1) \cdots c^p(p_m)$
3. Compute $R = c^t(T) \otimes c^p(P)$
4. For $i = 1, \ldots, n - m + 1$
   $$O[i] \leftarrow \max_{s=-|\Sigma|}^{|\Sigma|} \chi_{\neq 0}(R[(2i-1)|\Sigma| + 1 + s])|s|$$

*Claim.* At the end of the algorithm $O[i] = \max_{j=1}^{m} |t_{i+j-1} - p_j|$.

**Proof:** First, we show that $O[i] \geq |t_{i+j-1} - p_j|$ for every $j \in \{1, \ldots, m\}$. In order to see that, it is enough to show that for every $j \in \{1, \ldots, m\}$ it holds that $R[(2i-1)|\Sigma| + 1 + t_{i+j-1} - p_j] \neq 0$ (since, then, the value of $|t_{i+j-1} - p_j|$ is one of the values for the $max$ taken in step 4). Now, since $R[(2i-1)|\Sigma| + 1 + t_{i+j-1} - p_j] = \sum_{k=1}^{2m|\Sigma|} c^t(T)_{(2i-1)|\Sigma|+1+t_{i+j-1}-p_j+k-1} c^p(P)_k$ and for $k = p_j$ we have $c^t(T)_{(2i-1)|\Sigma|+t_{i+j-1}} = 1$ and $c^p(P)_{p_j} = 1$ (from the way we defined the encoding) it holds that $R[(2i-1)|\Sigma| + 1 + t_{i+j-1} - p_j] \neq 0$.

We have left to show that $O[i] = |t_{i+j-1} - p_j|$ for some $j \in \{1, \ldots, m\}$. Let $s_m$ be the value for which $O[i] = \chi_{\neq 0}(R[(2i-1)|\Sigma|+1+s_m])|s_m|$. We can assume that $R[(2i-1)|\Sigma|+1+s_m] \neq 0$ (otherwise $O[i] = 0$ and $0 \leq \max_{j=1}^m |t_{i+j-1} - p_j|$ and our proof is done). The fact that $R[(2i-1)|\Sigma|+1+s_m] \neq 0$ implies that for some $j' \in \{1, \ldots, m\}, k' \in \{0, \ldots, 2|\Sigma| - 1\}$ we have $c^t(T)_{(2i-1)|\Sigma|+1+s_m+2j'|\Sigma|+k'-1} = c^t(T)_{(2(i+j')-1)|\Sigma|+s_m+k'} = 1$ and $c^p(P)_{2j'|\Sigma|+k'} = 1$, which in turn implies that for $j'$ it holds that $|t_{i+j'-1} - p_{j'}| = s_m$. This complete our proof. $\blacksquare$

**Time:** The time needed to convolve 2 strings of size $n|\Sigma|$ and $m|\Sigma|$ is $O(n|\Sigma| \log n)$. The computation of $O[i]$ takes $2|\Sigma|$ steps , and $i = 1, 2, \ldots, n - m + 1$ so this step takes $O(n|\Sigma|)$. Both steps together take $O(n|\Sigma| \log n)$. We can slightly improve the time by using the technique of cutting the text into $n/m$ overlapping segments, each of length $2m$ to a total time of $O(n|\Sigma| \log(m + |\Sigma|))$.

# References

1. Amihood Amir, Ohad Lipsky, Ely Porat, and Julia Umanski. Approximate matching in the $l_1$ metric. In Apostolico et al. [2], pages 91–103.
2. Alberto Apostolico, Maxime Crochemore, and Kunsoo Park, editors. *Combinatorial Pattern Matching, 16th Annual Symposium, CPM 2005, Jeju Island, Korea, June 19-22, 2005, Proceedings*, volume 3537 of *Lecture Notes in Computer Science*. Springer, 2005.
3. Alberto Apostolico and Zvi Galil, editors. *Combinatorial Algorithms on Words*. Springer-Verlag New York, Inc., Secaucus, NJ, USA, 1985.
4. Peter Clifford, Raphaël Clifford, and Costas S. Iliopoulos. Faster algorithms for delta, gamma-matching and related problems. In Apostolico et al. [2], pages 68–78.
5. Piotr Indyk, Moshe Lewenstein, Ohad Lipsky, and Ely Porat. Closest pair problems in very high dimensions. In *ICALP*, volume 3142 of *Lecture Notes in Computer Science*, pages 782–792. Springer, 2004.
6. Piotr Indyk, Ohad Lipsky, and Ely Porat. Approximate translation matching, manuscript. 2004.
7. Ohad Lipsky. Efficient distance computations. Master's thesis, Bar-Ilan University, Department of Computer Science, 2003.
8. Ohad Lipsky and Ely Porat. Approximate $l_\infty$ matching. Technical report, Department of Computer Science, 2003.
9. Ilya Shmulevich, O. Yli-Harja, E. Coyle, D. Povel, and K. Lemstrom. Perceptual issues in music pattern recognition - complexity of rhythm and key fining. 1999.

# An Edit Distance Between RNA Stem-Loops

Valentin Guignon[1,*], Cedric Chauve[2,**], and Sylvie Hamel[3,**]

[1] Programme de Bioinformatique, Université de Montréal,
Pav. André-Aisenstadt, CP 6128 succ. Centre-ville, Montréal (QC), H3C 3J7, Canada
[2] LaCIM, Département d'Informatique, Université du Québec à Montréal,
C.P. 8888 succ. Centre-Ville, Montréal (QC), H3C 3P8, Canada
[3] LBIT, DIRO, Université de Montréal,
Pav. André-Aisenstadt, CP 6128 succ. Centre-ville, Montréal (QC), H3C 3J7, Canada
`guignonv@yahoo.fr`, `chauve@lacim.uqam.ca`, `sylvie.hamel@umontreal.ca`

**Abstract.** We introduce the notion of conservative edit distance and mapping between two RNA stem-loops. We show that unlike the general edit distance between RNA secondary structures, the conservative edit distance can be computed in polynomial time and space, and we describe an algorithm for this problem. We show how this algorithm can be used in the more general problem of complete RNA secondary structures comparison.

## 1 Introduction

In this paper, we address a classical problem in bioinformatics, the comparison of two RNA secondary structures, and we describe a new algorithm to compute an edit distance and a mapping between two RNA secondary structures. The importance and functional variety of the several types of known RNA molecules, especially non-coding RNAs like transfer RNAs (tRNAs), ribosomal RNAs (rRNAs), untranslated regions (UTRs), small nuclear RNA (snRNAs) for example, is a strong motivation for their study [4], in particular, in the comparative approach that relates combinatorial similarity between molecules to functional similarity. Several algorithms exist to compare RNA secondary structures, most of them based on the encoding of RNA secondary structures by ordered trees, followed by the computation of an edit distance and a mapping between these trees. This approach was first used in [9,10], and is part of the popular Vienna RNA Package [7]. However, the set of edit operations considered in these works appears to be too limited, as it does not contain some edit operations that correspond to natural evolutionary events for RNA structures, a problem that is discussed in [1]. Recently, Jiang *et al.* introduced a new edit distance model [8] that is more realistic, as it contains a broader set of edit operations, but also less tractable algorithmically, as computing the edit distance between two RNA secondary structures in this model is NP-hard [3].

---

* Supported by a scholarship of the Génome Québec program "Comparative and integrative bioinformatics".
** Supported by grants from NSERC and FQRNT.

M. Consens and G. Navarro (Eds.): SPIRE 2005, LNCS 3772, pp. 335–347, 2005.
© Springer-Verlag Berlin Heidelberg 2005

In the present work, we consider the set of edit operations defined in [8]. Our main result is that, when the compared RNA structures are simple enough and the set of allowed mappings is restricted, the distance and mapping computation becomes tractable, and even easy. More precisely, we consider the computation of a *conservative mapping* between two *stem-loops*; a precise statement of the problem is given in Section 2, but intuitively, a mapping is said to be conservative if it describes only evolutionary events involving bases that are closely located in terms of secondary structure. We show that such a computation can be done efficiently with a dynamic programming algorithm that is an extension of the classical algorithm computing an edit distance between strings. The motivations for considering such a restricted notion of distance and mapping are the following. First, every RNA secondary structure can be decomposed into a sequence of stem-loops and stem-loop-like substructures, that can be handled by our algorithm. For example, several families of non-coding RNAs, like tRNAs, snoRNAs H/ACA or miRNAs precursors, have a secondary structure mostly composed of a few stem-loops. Second, two closely related, from the evolutionary point of view, RNA structures – not limited to stem-loops – should share several stem-loops that are very similar and whose comparison will be well represented by a conservative mapping. Moreover, if two compared RNA structures share only some similar motifs that contain stem-loops, our approach based on the decomposition in stem-loop-like substructures and their pairwise comparisons can highlight such locally conserved motifs, a problem that received some attention recently [2,6].

This paper is organized as follows. In Section 2, we state precisely the problem we address, namely the computation of a conservative distance and mapping between stem-loops. In Section 3, we describe and analyze a dynamic programming algorithm that solves this problem. In Section 4, we show how the stem-loops comparison can be used in a very simple way to compare complete RNA secondary structures and we illustrate this approach with the comparison of two RNAse P RNA.

## 2    Edit Distance Between Stem-Loops

*Tree Representation of Stem-Loops.* RNA primary structure is generally represented by a string over the four letters alphabet $\{A, C, G, U\}$. This primary structure folds back onto itself to form the secondary structure, that is a planar structure containing unpaired bases from $\{A, C, G, U\}$ and base pairs, that are ordered pairs of bases, the most common being $AU, UA, CG, GC, GU, UG$. This secondary structure can be represented by an ordered tree[1], where, for a given RNA secondary structure, each base pair is represented by an internal node labeled by this base pair and each unpaired base by a leaf, labeled by this base (see Figure 1). In such a representation, a multi-loop corresponds to an internal node having several children that are internal nodes. We call *stem-loop* an RNA

---

[1] Depending on the level of comparison, an RNA can be represented by different trees, see Allali and Sagot [1] for example.

secondary structure without multi-loop, which implies that the tree representing
a stem-loop is *linear*: each internal node has at most one child that is also an
internal node. Note that linear trees are quite similar to strings, as a string can
be seen as a linear tree with only one leaf.

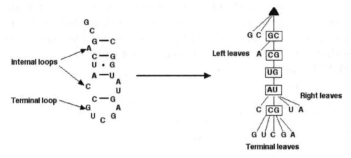

**Fig. 1.** A stem-loop and its tree representation

Given a linear tree representing a stem-loop, leaves can be of three kinds: *left
leaves*, *right leaves* and *terminal leaves*. The leaves representing unpaired bases
of the terminal loop are the terminal leaves, the leaves representing unpaired
bases located on the left (resp. right) of the stem-loop are the left leaves (resp.
right leaves). In Figure 1, the terminal leaves are $G, U, C, G, A$, the left leaves
are $G, C, A, C$ and the right leaves $U, A$.

*Edit Operations.* Let $T_1$ and $T_2$ be two linear trees representing two stem-loops.
Edit operations represent evolutionary events acting on secondary structures.
The set of edit operations we describe now is the same that was described in
[8][2], augmented of additional constraints on some of these operations. These
operations are illustrated in Figure 2.

We define an *edit operation between $T_1$ and $T_2$* as a couple $(a, b)$, where $a$
(resp. $b$) can be an internal node or a leaf of $T_1$ (resp. of $T_2$), or a pair of leaves
of $T_1$ (resp. of $T_2$) that have the same parent, or the symbol $-$ (but both $a$ and
$b$ can not be $-$).

If $a$ and $b$ are both internal nodes or both leaves, the operation $(a, b)$ is a
*relabeling*: the node $a$ changes its label to become $b$. If $b = -$, $(a, b)$ is a *deletion*:
the node $a$ is removed from $T_1$. If $a = -$, $(a, b)$ is an *insertion*, the symmetric
operation of a deletion. We add the following constraint on the relabeling operation: if $a$ and $b$ are both leaves, respectively of $T_1$ and $T_2$, then either one of
these two leaves is a terminal leaf, or both of them are left leaves, or both are
right leaves. We call these three operations the *simple operations*.

If $a$ is an internal node and $b$ a leaf, $(a, b)$ is called an *altering* – the internal
node $a$ is replaced by the leaf $b$ in $T_1$ –, and if $a$ is a leaf and $b$ an internal
node, it is called a *completion*, which is the symmetric operation of an altering.

---

[2] Note that in [8] RNA secondary structures are represented by *non-crossing arc-
annotated sequences*, but such sequences are naturally equivalent to ordered trees.

Finally, $(a, b)$ is called an *arc-breaking* if $a$ is an internal node and $b$ is a pair of leaves, and an *arc-creation* if $a$ is a pair of leaves and $b$ is an internal node. These last four operations, that represent evolutionary events that act on base pairs, were introduced in [8]. We call them *complex operations*. The fact that these operations represent evolutionary events on base pairs that transform a stem-loop into another stem-loop impose some implicit conditions on the nodes of a linear tree that are involved. In particular, an arc-creation can only involve two leaves that have the same parent, and these two leaves can not be both left leaves or right leaves.

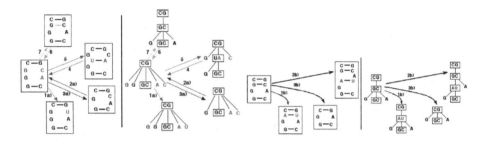

**Fig. 2.** Illustration of edit operations: relabeling (1a and b), insertion (2a and b), deletion (3a and b), altering and completion (4 and 5), arc-breaking and arc-creation (6 and 7) on stem-loops and trees

*Conservative Mapping and Edit Distance.* Let $T_1$ and $T_2$ be two linear trees and $S = s_1, \ldots, s_n$ a sequence of edit operations between $T_1$ and $T_2$ such that their successive application transforms $T_1$ into $T_2$. Such a sequence of edit operations is called a *conservative edit sequence between $T_1$ and $T_2$*. We shall here notice that the definition of a conservative edit sequence is less general than the general definition of edit sequence, used in [8], as we impose that all the operations of this sequence have to be described on $T_1$ and $T_2$. This, for example, forbids that a base that was unpaired by an arc-breaking could later be involved in a arc-creation or a completion. This is why we call conservative such an edit sequence, and we will justify later in this section why we consider such restrictions.

The edit operations of a conservative edit sequence between $T_1$ and $T_2$, other than insertions and deletions, naturally induce, by their definition, a *mapping* between nodes of $T_1$ and $T_2$, where a leaf of a tree can be mapped to a leaf or an internal node of the other tree, and an internal node can be mapped to an internal node, a leaf, or a pair of leaves. This mapping highlights the bases that are common, up to relabeling, between the stem-loops represented by $T_1$ and $T_2$. We call such a mapping a *conservative mapping*.

Finally, we associate to each edit operation $(a, b)$ a *cost* denoted $\delta(a, b)$. If $(a, b)$ is a relabeling, where the nodes $a$ of $T_1$ and $b$ of $T_2$ have the same label, then $\delta(a, b) = 0$. Note that a complex operation $(a, b)$, depending on the label of the nodes in $a$ and $b$, can also imply a relabeling. Hence, the cost of an operation depends of its nature and of the labels of the involved nodes. The cost associated

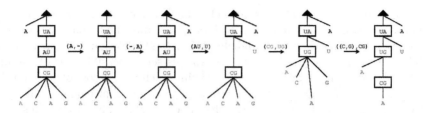

**Fig. 3.** A conservative edit sequence between two linear trees

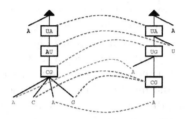

**Fig. 4.** The conservative mapping corresponding to the edit sequence of Figure 3

to an edit sequence between $T_1$ and $T_2$ is the sum of the individual cost of each operations in the sequence. The conservative edit distance is the minimal cost of a conservative edit sequence that transforms $T_1$ into $T_2$.

We describe in this work an algorithm that computes a conservative mapping and the corresponding edit distance between two linear trees $T_1$ and $T_2$.

*Discussion on Various Distances.* Several RNA comparison algorithms have been defined based on different subsets of the set of edit operations we defined above. We illustrate now, through a simple example on real data, the influence of the choice of the set of allowed edit operations on the comparison of two structures. Given the two micro-RNAs (miRNAs) precursors of Figure 5, we describe two possible sequences of edit operations that transform the first structure into the second one, based on different sets of edit operations.

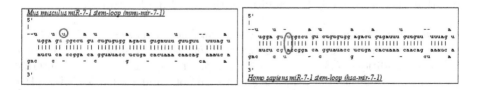

**Fig. 5.** Two miRNAs (mouse and human)

If we consider only the simple operations of relabeling, insertion and deletion, with cost 0.75 (resp. 1.25) for the insertion or deletion of an unpaired base (resp. a base pair) and 0.25 for the relabeling of a base[3], a possible optimal scenario to transform the mouse miRNA precursor into the human one contains 4 operations, for a cost of 3: relabeling $(UA, UG)$, deletion $(U, -)$, insertion $(-, U)$ and insertion $(-, UA)$. If we add to this set of possible edit operations the operations of arc-creation and arc-breaking, each with a cost of 0.5, and the altering and completion operations, with a cost of 1 each for example, then the following scenario, that is conservative, has a better score of 2.25 and seems more plausible, from the evolutionary point of view than the previous one: insertion $(-, U)$ (below), relabeling $(UA, UG)$ and completion $(U, UA)$.

This example is a good illustration of why we believe that the set of all edit operations we described above should be considered when comparing RNA secondary structures. However, the problem of computing the general edit distance between RNA secondary structures is NP-hard [3]. And even in the case of the comparison of two stem-loops, it is not known if the general edit distance can be computed in polynomial time. As we will see in the next section, in the case of a conservative distance and mapping between stem-loops, the problem can be solved in polynomial time, due to the similarity between this problem and the problem of computing an edit distance between strings.

Finally, one can see that the restrictions that we impose to define a conservative mapping prevent the evolutionary scenarios corresponding to such mappings to create a base pair between bases that are not closely located in the secondary structure. Hence, if considering only conservative mapping is a strong combinatorial restriction, it should not prevent to obtain a pertinent distance and mapping between stem-loops that are close from an evolutionary point of view, which is our goal in this work. Our experiments on real data, miRNAs precursors (not shown) and RNAse P RNA (Section 4) seem to confirm this intuition.

## 3    A Dynamic Programming Algorithm

We now describe a dynamic programming algorithm that computes the conservative distance between two stem-loops, by using a unique two-dimensional dynamic programming table. Through all this section, we use distance and mapping respectively for conservative distance and conservative mapping. We recall that a depth-first prefix traversal (DFP) of an ordered tree is a traversal of the tree that visits recursively the children of the root from left to right.

*Indexing Pairs, Predecessor and Successor of a Node.* An ordered pair $I = (x, y)$ of nodes of a linear tree $T$ is called an *indexing pair* if it satisfies one the five following conditions: (1) $x$ is an internal node and $y = x$, (2) $x$ is an internal node and $y$ a right leaf of $x$, (3) $y$ is an internal node and $x$ a left leaf of $y$, (4) $x$ and $y$ are respectively a left leaf and a right leaf and they have the same parent, or (5) $x$ and $y$ are terminal leaves, and $x$ is located to the left of $y$.

An indexing pair $(x, y)$ of $T$ defines a subtree of $T$, denoted by $T_{(x,y)}$, in the following way: $T_{(x,y)}$ is the tree obtained from $T$ by removing all the nodes

---

[3] The scores we use here are the same we use in Section 4, where they are discussed.

visited between $x$ and $y$ during a DFP traversal of $T$ (if $x = y$, this corresponds to removing from $T$ all the nodes other than $x$ in the subtree rooted in $x$).

We define the *predecessor* of a node $x$, $p(x)$, as its immediate left sibling if $x$ is not the leftmost child of its parent, and its parent otherwise. Symmetrically, the *successor* of a node $x$, $s(x)$, is its immediate right sibling if $x$ is not the rightmost child of its parent, and its parent otherwise. Note that for an internal node $x$ that is the only child of its parent $y$, $p(x) = s(x) = y$. According to the previous definitions, the root $r$ of a tree does not have a predecessor, neither a successor, so we define them formally by $p(r) = s(r) = \emptyset$.

Finally, an indexing pair $(x, y)$ is said to be *terminal* if $x$ (resp. $y$) is a terminal leaf and $y = s(x)$ (resp. $x = p(y)$).

*A Dynamic Programming Algorithm.* We can now define the dynamic programming table that we use to compute the edit distance between two stem-loops. This table, denoted $D$, is a two-dimensional table indexed by pairs $(I, J)$ such that $I$ is either an indexing pair of $T_1$ or $I = \emptyset$, and $J$ is either an indexing pair of $T_2$ or $\emptyset$. The cell $D[I, J]$ of this table contains the edit distance between the two linear trees $T_I$ and $T_J$.

It follows immediately from the definition of indexing pairs that we can define the edit distance between $T_1$ and $T_2$ in terms of $D[I, J]$. Indeed, if we denote by $F_1$ and $F_2$ the sets of terminal indexing pairs respectively of $T_1$ and $T_2$, we have:

$$d(T_1, T_2) = \min_{(x,y) \in F_1, (u,v) \in F_2} \{D[(x, y), (u, v)]\}. \tag{1}$$

To compute the table $D$, we use a dynamic programming algorithm, based on the following equations. First, we initialize the table

$$\begin{cases} D[\emptyset, \emptyset] = 0, \\ D[(x, y), \emptyset] = \sum_{a \text{ node of } T_{1(x,y)}} \delta(a, -), \text{ for all indexing pairs } (x, y) \text{ of } T_1, \\ D[\emptyset, (u, v)] = \sum_{b \text{ node of } T_{2(u,v)}} \delta(-, b), \text{ for all indexing pairs } (u, v) \text{ of } T_2. \end{cases} \tag{2}$$

The general case is composed of 4 sub-cases. In the following equations, we denote by (R) an equation corresponding to a relabeling event, (I) an insertion, (D) a deletion, (AC) an arc-creation, (AB) an arc-breaking, (C) a completion and (A) an altering.

1. If $x = y$ and $u = v$ ($x$ and $u$ are internal nodes),

$$D[(x, y), (u, v)] = \min \left\{ \begin{array}{ll} D[(p(x), s(y)), (p(u), s(v))] + \delta(x, u), & (R) \\ D[(p(x), s(y)), (u, v)] + \delta(x, -), & (D) \\ D[(x, y), (p(u), s(v))] + \delta(-, u), & (I) \end{array} \right\}, \tag{3}$$

where $(p(x), s(y)) = \emptyset$ if $x$ is the root of $T_1$ and $(p(u), s(v)) = \emptyset$ if $u$ is the root of $T_2$.

2. If $x \neq y$ and $u \neq v$,

$$D[(x,y),(u,v)] = \min \left\{ \begin{array}{ll} D[(\mathrm{p}(x),y),(u,v)] + \delta(x,-), & \text{(D)} \\ D[(x,\mathrm{s}(y)),(u,v)] + \delta(y,-), & \text{(D)} \\ D[(x,y),(\mathrm{p}(u),v)] + \delta(-,u), & \text{(I)} \\ D[(x,y),(u,\mathrm{s}(v))] + \delta(-,v), & \text{(I)} \\ D[(\mathrm{p}(x),y),(\mathrm{p}(u),v)] + \delta(x,u), & \text{(R)} \\ D[(x,\mathrm{s}(y)),(u,\mathrm{s}(v))] + \delta(y,v), & \text{(R)} \end{array} \right\}. \qquad (4)$$

Note that in the above equation, some of the 6 terms of the form $D[I, J] + \delta(\dots)$ can be undefined. This can happen if $I$ and/or $J$ is neither $\emptyset$, nor an indexing pair: for example if $x$ is an internal node and $y$ a right leaf of $x$, then $(\mathrm{p}(x), y)$ is not an indexing pair of nodes of $T_1$. In such a case, the function min will not take into account these undefined terms.

3. If $x = y$, and $u \neq v$

$$D[(x,y),(u,v)] = \min \left\{ \begin{array}{ll} D[(\mathrm{p}(x),\mathrm{s}(y)),(u,v)] + \delta(x,-), & \text{(D)} \\ D[(x,y),(\mathrm{p}(u),v)] + \delta(-,u), & \text{(I)} \\ D[(x,y),(u,\mathrm{s}(v))] + \delta(-,v), & \text{(I)} \\ D[(\mathrm{p}(x),\mathrm{s}(y)),(\mathrm{p}(u),v)] + \delta(x,u), & \text{(A)} \\ D[(\mathrm{p}(x),\mathrm{s}(y)),(u,\mathrm{s}(v))] + \delta(x,v), & \text{(A)} \\ D[(\mathrm{p}(x),\mathrm{s}(y)),(\mathrm{p}(u),\mathrm{s}(v))] + \delta(x,(u,v)) & \text{(AB)} \end{array} \right\},$$

$$(5)$$

where the same remark as in sub-case 2, about possibly undefined terms, applies.

4. If $x \neq y$ and $u = v$,

$$D[(x,y),(u,v)] = \min \left\{ \begin{array}{ll} D[(x,y),(\mathrm{p}(u),\mathrm{s}(v))] + \delta(-,u), & \text{(I)} \\ D[(\mathrm{p}(x),y),(u,v)] + \delta(x,-), & \text{(D)} \\ D[(x,\mathrm{s}(y)),(u,v)] + \delta(y,-), & \text{(D)} \\ D[(\mathrm{p}(x),y),(\mathrm{p}(u),\mathrm{s}(v))] + \delta(x,u), & \text{(C)} \\ D[(x,\mathrm{s}(y)),(\mathrm{p}(u),\mathrm{s}(v))] + \delta(y,u), & \text{(C)} \\ D[(\mathrm{p}(x),\mathrm{s}(y)),(\mathrm{p}(u),\mathrm{s}(v))] + \delta((x,y),u) & \text{(AC)} \end{array} \right\},$$

$$(6)$$

where again the same remark as in sub-case 2, about possibly undefined terms, applies.

We now describe the algorithm to fill all the cells of the table $D$. An indexing pair $(u, v)$ of a tree $T$ is said to be *ancestral* for the indexing pair $(w, z)$ of $T$ if $(u, v) = \emptyset$ or $u$ (resp. $v$) is not visited after $w$ (resp. before $z$) during a DFP traversal of $T$. It follows from this definition and from the equations above that, in order to compute the table $D$, we have to enumerate all the couples $(I, J)$ of indexing pairs of $T_1$ and $T_2$ in a way that preserves the ancestral order for $I$ and $J$: $D[I, J]$ will be computed after all the cells $D[I', J']$ where $I'$ is ancestral for $I$ and $J'$ is ancestral for $J$. Such an enumeration scheme is easy to design for a given tree, based on parallel depth-first prefix and postfix traversals of this tree, and can be performed in time that is linear in the number of indexing pairs for this tree. Given $D$, a mapping is a path in this table computed with the classical *backtracking* method used to compute the alignment of two strings.

*Complexity Analysis.* The space complexity of this algorithm is given by the size of the table $D$, i.e the number of couples $(I, J)$ where $I$ is an indexing pair of $T_1$ and $J$ an indexing pair of $T_2$. Let $\text{ind}(T_1)$ and $\text{ind}(T_2)$ denote respectively the number of indexing pairs of $T_1$ and $T_2$: the table $D$ contains $\Theta(\text{ind}(T_1) \times \text{ind}(T_2))$ cells.

As the enumeration of all indexing pairs of the trees $T_1$ and $T_2$ respecting the ancestral relation can be performed in time linear in the number of such pairs for each tree, the initialization of the table (equation (2)) can be computed in $O(\text{ind}(T_1) \times \text{ind}(T_2))$ time. Moreover, filling one cell of the table, using the dynamic programming equations (3), (4), (5) and (6) can be done in constant time, since testing if a pair of nodes is indexing takes a constant time. Note also that the predecessor and successor of every node of a tree can easily be computed, prior to the computation of the table $D$, in linear time during a DFP traversal of this tree. Finally, once $D$ has been filled, computing the edit distance using equation (1) can be done by visiting the cells indexed by pairs of terminal indexing pairs, and so in $O(\text{ind}(T_1) \times \text{ind}(T_2))$ time. This leads to the result that the time complexity for computing the conservative edit distance between $T_1$ and $T_2$ is $\Theta(\text{ind}(T_1) \times \text{ind}(T_2))$ time. It follows from the similarity between our algorithm and the string edit distance algorithm that computing a mapping from $D$ asks for the same time, that is $\Theta(\text{ind}(T_1) \times \text{ind}(T_2))$.

Let $n_1$ be the number of nodes of $T_1$, $m_1$ be the number of internal nodes of $T_1$, $\{x_1, \ldots, x_{m_1}\}$ these internal nodes, $\ell_i$ and $r_i$ the number of left and right leaves of $x_i$, for $i = 1, \ldots, m - 1$, and $t_1$ the number of terminal leaves. The number $\text{ind}(T_1)$ of indexing pairs in $T_1$ is exactly

$$m_1 + (t_1(t_1 - 1)/2) + 2t_1 + \sum_{i=1}^{m_1-1} ((\ell_i + 1) \times (r_i + 1) - 1), \qquad (7)$$

where these four terms correspond respectively to the number of indexing pairs formed by two occurrences of the same internal node, those formed by two terminal leaves, those formed by a terminal leaf and $x_{m_1}$ and, finally, those formed with at least one non terminal leaf.

Hence, $\text{ind}(T_1) \in O(n_1^2)$, and, if we denote by $n_2$ is the number of nodes in $T_2$, the overall distance and mapping algorithm has a worst-case time complexity in $O(n_1^2 \times n_2^2)$. However, it is interesting to remark that, if $T_1$ is a tree representing a stem-loop with few unpaired bases, or small loops (internal loops and the terminal loop), then $\text{ind}(T_1)$ is closer to $n_1$ than to $n_1^2$. Hence, when comparing stem-loops with such characteristics in terms of unpaired bases and loops, the algorithm asks for a time that is, in practice, in only quadratic.

## 4   Comparison of Complete Secondary Structures

In this section, we describe a simple method that allows the comparison of two complete RNA secondary structures $R_1$ and $R_2$, based on the stem-loops comparison algorithm of the previous section. This method has three phases: (1)

decomposition of the two RNA structures into two sequences of stem-loops sub-structures, (2) independent pairwise comparisons between the stem-loops of $R_1$ and the stem-loops of $R_2$, and (3) finally, an alignment of these two sequences of stem-loops using the distances computed during the phase (2).

Given an RNA secondary structure $R$, if one removes all the unpaired bases belonging to multi-loops, one obtains a set of substructures with no multi-loops. Even if these substructures are not all stem-loops under the classical definition of the term, due to the fact that some of them do not have a terminal loop, we call them stem-loops, as the algorithm we described in Section 3 does not need any major modification to handle stem-loops that do not have a terminal loop. This set of substructures is naturally ordered by the sequence of bases that forms the primary structure of $R$, as illustrated in Figure 6.

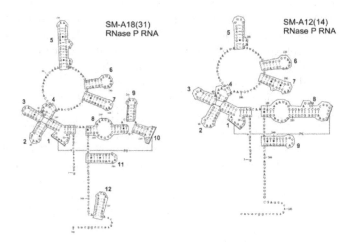

**Fig. 6.** Decomposition of two RNase P RNA into stem loops. Numbers indicate the order on each of the two sets of stem-loops.

Now, let $R_1^1, \ldots, R_1^k$ and $R_2^1, \ldots, R_2^\ell$ be the two sequences of stem-loops given by the decompositions of two complete RNA secondary structures $R_1$ and $R_2$. We use a table $P$, indexed by pairs of integers belonging to $\{0, \ldots, k\} \times \{0, \ldots, \ell\}$ where $P[i,j]$ is the distance between $R_1^i$ and $R_2^j$ – with $R_1^0 = R_2^0$ being the empty stem-loop –, computed using the algorithm of Section 3. The table of Figure 7 corresponds to the pairwise comparisons of the stem-loops of Figure 6, with the following costs: 0.25 for the relabeling of a single base, 0.4 for the relabeling of the two bases of a base pair, 0.75 (resp. 1.25) for the deletion and the insertion of a leaf (resp. an internal node), 0.5 for an arc-breaking and an arc-creation, and 1 for a completion and an altering. These costs were chosen in such a way that no edit operation can be replaced, for a smaller cost, by a sequence of other edit operations. Moreover, the results we present below did not differ when alternative cost schemes, that had the same property, were used.

**Stem-loops Distances**
SM-A18(31) RNase P RNA

SM-A12(14) RNase P RNA

| | - | 1 | 2 | 3 | 4 | 5 | 6 | 7 | 8 | 9 |
|---|---|---|---|---|---|---|---|---|---|---|
| - | 0 | 26 | 15 | 15 | 19 | 22 | 19 | 24 | 52 | 20 |
| 1 | 27 | 2,8 | 13,45 | 16,25 | 13,65 | 15,6 | 13,05 | 16,25 | 25,7 | 14,15 |
| 2 | 15 | 12,95 | 0,8 | 4,2 | 10,1 | 7,9 | 5,35 | 8,65 | 33 | 6,95 |
| 3 | 15 | 14,05 | 3,65 | 1,3 | 9,7 | 7,4 | 6,45 | 8,3 | 32,75 | 5,8 |
| 4 | 19 | 11,85 | 9,85 | 9,45 | 0,25 | 12,95 | 9,7 | 12,6 | 32,45 | 10,55 |
| 5 | 37 | 21,55 | 18,8 | 18,65 | 19,95 | 13 | 16,35 | 13,4 | 25,8 | 16,1 |
| 6 | 19 | 13,3 | 4,95 | 5,9 | 9,7 | 7,5 | 0 | 8,25 | 30,4 | 4,8 |
| 7 | 23 | 15,25 | 9,2 | 7,55 | 12,2 | 7,15 | 7,7 | 2,7 | 29,4 | 5,1 |
| 8 | 33 | 11,95 | 17,8 | 19 | 15,2 | 15,7 | 15,5 | 16,85 | 18,4 | 16,95 |
| 9 | 16 | 13,6 | 3,7 | 3,2 | 9,15 | 7,45 | 4,55 | 7,8 | 32,65 | 4,55 |
| 10 | 16 | 9,95 | 4,6 | 5,7 | 8,25 | 8,95 | 4,5 | 9,95 | 31,4 | 6,4 |
| 11 | 20 | 13,4 | 6,85 | 6,05 | 10,55 | 5,65 | 5,05 | 5,05 | 30,55 | 2,35 |
| 12 | 20 | 13,3 | 6,05 | 6,45 | 11,35 | 5,4 | 5,25 | 5,75 | 30,7 | 3,3 |

**Fig. 7.** Pairwise distances between the stem-loops of Figure 6

Finally, we apply the classical string global alignment algorithm (see [5] for example) to these two sequences of stem-loops, using the table $P$ to define the cost of the insertion or deletion of a given stem-loop, and the score of a matching between two stem-loops. The resulting dynamic programming table is given in Figure 8, where marked cells describe the alignment of stem-loops obtained by backtracking. To obtain from this table a mapping, one can use the classical backtracking method, both in the table of the alignment of the sequences of stem-loops and in the tables of the pairwise alignments of stem-loops.

**Global Alignment**
SM-A18(31) RNase P RNA

SM-A12(14) RNase P RNA

| | - | 1 | 2 | 3 | 4 | 5 | 6 | 7 | 8 | 9 |
|---|---|---|---|---|---|---|---|---|---|---|
| - | 0 | 26 | 41 | 56 | 75 | 97 | 116 | 140 | 192 | 212 |
| 1 | 27 | 2,8 | 17,8 | 32,8 | 51,8 | 73,8 | 92,8 | 116,8 | 165,7 | 185,7 |
| 2 | 42 | 17,8 | 3,6 | 18,6 | 37,6 | 59,6 | 78,6 | 101,5 | 149,8 | 169,8 |
| 3 | 57 | 32,8 | 18,6 | 4,9 | 23,9 | 45 | 64 | 86,9 | 134,2 | 154,2 |
| 4 | 76 | 51,8 | 37,6 | 23,9 | 5,15 | 27,15 | 46,15 | 70,15 | 119,4 | 139,4 |
| 5 | 113 | 88,8 | 70,6 | 56,25 | 42,15 | 18,15 | 37,15 | 59,55 | 95,95 | 116 |
| 6 | 132 | 107,8 | 89,6 | 75,25 | 61,15 | 37,15 | 18,15 | 42,15 | 89,95 | 100,8 |
| 7 | 155 | 130,8 | 112,6 | 97,15 | 84,15 | 60,15 | 41,15 | 20,85 | 71,55 | 91,55 |
| 8 | 188 | 163,8 | 145,6 | 130,2 | 112,4 | 93,15 | 74,15 | 53,85 | 39,25 | 59,25 |
| 9 | 204 | 179,8 | 161,6 | 146,2 | 128,4 | 109,2 | 90,15 | 69,85 | 55,25 | 43,8 |
| 10 | 220 | 195,8 | 177,6 | 162,2 | 144,4 | 125,2 | 106,2 | 85,85 | 71,25 | 59,8 |
| 11 | 240 | 215,8 | 197,6 | 182,2 | 164,4 | 145,2 | 126,2 | 105,9 | 91,25 | 73,6 |
| 12 | 260 | 235,8 | 217,6 | 202,2 | 184,4 | 165,2 | 146,2 | 125,9 | 111,3 | 93,6 |

**Fig. 8.** Alignment of the two sequences of stem-loops of Figure 6 using the table of Figure 7: marked cells indicate an optimal stem-loops alignment

As it appears on Figure 8, this algorithm, applied on the two quite similar RNAse P RNAs of Figure 6 gives a good result, even if the stem-loops 8, 9 and 10 of the RNAse P RNA SM-A18(31) show that this method is sensitive to the insertion of a stem-loop into another stem-loop. However, the comparisons of the other stem-loops that were very similar compensated this problem.

If $n_1$ is the number of bases of $R_1$ and $n_2$ the number of bases of $R_2$, it follows immediately from the complexity of comparing two stem-loops that the comparison of $R_1$ and $R_2$ is performed in $O(n_1^2 \times n_2^2)$ in the worst-case time. However, the low number of unpaired bases in the two sets of stem-loops of our example makes that the effective time complexity was only quadratic.

It is also interesting to notice that all the different variants of the alignment of strings can be used with our method. For example, if one wants to discover clusters of close stem-loops that are similar in $R_1$ and $R_2$, that is local motifs, one just has to use the algorithm for local alignment of strings instead of the global string alignment algorithm.

## 5   Conclusion

We described in this paper an efficient algorithm for the comparison of stem-loops, intended to give good results for stem-loops that are evolutionary close. By imposing some restrictions on the set of possible mappings that are considered, we were able to use the complete set of edit operations defined in [8]. Moreover, we sketched a method that allows to use the stem-loops comparison algorithm as a basis for the comparison of complete RNA secondary structures. Experimental results suggest that this approach gives interesting results.

This work raises several interesting algorithmical questions. First, it would be interesting to see at which point the fact to consider stem-loops makes easier the edit distance computation: is computing the general edit distance of [8] between stem-loops NP-hard ? And if this is the case, are there definitions of some mapping, less restrictive than conservative mappings, that allow a polynomial time computation. From a preliminary work on this question, it seems that it is possible to relax the locality of interactions between bases imposed in a conservative mapping and that one can consider interactions between bases that do not belong to the same internal loop. However, this makes the computation more time-consuming, at least in practice.

It would also be important to understand more deeply the influence of the cost scheme of edit operations on the final result, as it was done in string algorithms. It is for example possible that some cost schemes allow to compute in polynomial time the general edit distance.

The most interesting question concerns the way to use the comparison of stem-loops in the comparison of complete secondary structures. We used here a simple method based on the alignment of strings, that has the good property to be efficient in terms of computing time. In [1], Allali and Sagot introduced the notion of *multilevel RNA structure comparison*, that considers several levels of representation of an RNA structure into trees. The method we described in Section 4 follows this principle in fact, but does not consider the high level architecture of this structure as it considers the stem-loops in a simple sequence. It then would be interesting to combine our algorithm with the multi-level approach of [1].

# References

1. J. Allali and M.-F. Sagot. A new distance for high level RNA secondary structure comparison. *IEEE/ACM Transactions on Computational Biology and Bioinformatics*, 2(1):4–14. 2005.
2. R. Backofen and S. Will. Local sequence-structure motifs in RNA. *J. Bioinform. Comput. Biol.*, 2(4):681–698. 2004.
3. G. Blin, G. Fertin and C. Sinoquet. RNA sequences and the EDIT(NESTED, NESTED) problem. Report RR-IRIN-03.07, IRIN (Nantes, France). 2003.
4. J. Couzin. Breakthrough of the year: small RNAs make big splash. *Science*, 298:2296–2297. 2002.
5. D. Gusfield *Algorithms on strings, trees and sequences.* Cambridge University Press. 1997.
6. M. Höchsmann, T. Töller, R. Giegerich and S. Kurtz. Local similarity in RNA secondary structures. In *2nd IEEE Computer Society Bioinformatics Conference (CSB 2003)*, pages 159–169, IEEE Computer Society. 2003.
7. I.L Hofacker. Vienna RNA secondary structure server. *Nucleic Acids Res.*, 31(13):3429–3431. 2003.
8. T. Jiang, G. Lin, B. Ma and K. Zhang. A general edit distance between RNA structures. *J. Comput. Biol.*, 9(2):371–388. 2002.
9. B.A. Shapiro and K. Zhang. Comparing multiple RNA secondary structures using tree comparisons. *Comput. Appl. Biosci.*, 6(4):309–318. 1988.
10. K. Zhang and D. Shasha. Simple fast algorithms for the editing distance between trees and related problems. *SIAM J. Comput.*, 18(6):1245–1262. 1989.

# A Multiple Graph Layers Model with Application to RNA Secondary Structures Comparison

Julien Allali[1] and Marie-France Sagot[2]

[1] Institut Gaspard-Monge, Université de Marne-la-Vallée, Cité Descartes,
Champs-sur-Marne, 77454 Marne-la-Vallée Cedex 2, France
allali@univ-mlv.fr
[2] Inria Rhône-Alpes et LBBE, Université Claude Bernard, Lyon I, 43 Bd du 11
Novembre 1918, 69622 Villeurbanne cedex, France and King's College, London, UK
Marie-France.Sagot@inria.fr

**Abstract.** We introduce a new data structure, called MiGaL for "Multiple Graph Layers", that is composed of various graphs linked together by relations of abstraction/refinement. The new structure is useful for representing information that can be described at different levels of abstraction, each level corresponding to a graph. We then propose an algorithm for comparing two MiGaLs. The algorithm performs a step-by-step comparison starting with the most "abstract" level. The result of the comparison at a given step is communicated to the next step using a special colouring scheme. MiGaLs represent a very natural model for comparing RNA secondary structures that may be seen at different levels of detail, going from the sequence of nucleotides, single or paired with another to participate in a helix, to the network of multiple loops that is believed to represent the most conserved part of RNAs having similar function. We therefore show how to use MiGaLs to very efficiently compare two RNAs of any size at different levels of detail simultaneously.

**Keywords:** Graph layers, graph comparison, edit distance, RNA, secondary structure.

## 1 Introduction

We introduce in this paper a new data structure, called *MiGaL* for *"Multiple Graph Layers"*, that is composed of various graphs linked together by relations of abstraction/refinement. The new structure is useful for representing information that can be described at different levels of abstraction, each level corresponding to a graph. Similar structures have already been used for modelling a spatial environment [4,5] or plant architectures [8,6]. It could also be used for analysing web pages or source codes. MiGaL is a more general structure in that it can model graphs besides trees, and edges can be added between the different layers.

After giving a formal presentation of the MiGaL structure, we propose an algorithm for comparing two MiGaLs. The algorithm performs a step-by-step

M. Consens and G. Navarro (Eds.): SPIRE 2005, LNCS 3772, pp. 348–359, 2005.

comparison starting with the most "abstract" level. The result of the comparison at a given step is communicated to the next step using a special colouring scheme.

We then present an application of MiGaL to the analysis of RNA secondary structures. RNAs have different functions in a cell, these being strongly related to the spatial fold adopted by the molecule. It is therefore meaningful to compare such folds in order to infer or understand function. For RNAs, the comparison is usually done by considering the secondary structure which, in the case of RNAs, provides a planar view of the fold. A secondary structure may be seen at different levels of detail, going from the sequence of nucleotides, single or paired with another to participate in a helix, to the network of multiple loops that is believed to correspond the most conserved part of RNAs having similar function.

MiGaLs represent therefore a natural way of modelling such secondary structures. To perform the comparison at each different level, an edit distance algorithm is applied to trees that are rooted and ordered (reflecting the orientation of the RNA molecule). We use for this the algorithm particularly adapted to RNAs that was introduced in a recent work by Allali *et al.* [1]. In this paper, we show how to optimise it using the colouring scheme of the algorithm for comparing two MiGaLs introduced earlier in the paper in order to compare RNAs at different levels of detail simultaneously. We show that our method leads to quite satisfying results when applied to the analysis of RNA secondary structures. Furthermore, we show that this new algorithm for comparing RNAs allows to address the *scattering effect* problem mentioned in the literature [1]. Finally, the divide strategy used as the different levels are considered in turn allows also to very efficiently compare RNAs of any size.

## 2 Multiple Graph Layers

We present in this section the MiGaL data structure. After a formal definition, we provide an algorithm to compare such structures.

### 2.1 Definition

From now on, we adopt the following notations. Given a set $S$, we denote by $|S|$ its cardinality. Let $G(V, E)$ be a graph with vertex set $V(G) = \{v_1, \ldots, v_{|V(G)|}\}$ and edge set $E(G) = \{e_1, \ldots, e_{|E(G)|}\}$.

We start by defining the new data structure. It can be described as a layered sequence of graphs, with each graph linked to its neighbours by two applications, one of which corresponds to an abstraction from the previous graph, the other a refinement leading to the next graph. The formal definition of the MultI-GrAphLayers data structure is as follows.

**Definition 1 (Definition of a *MiGaL*).** *A MiGaL structure $M(G, R)$ of size $|M|$ is defined as a layered sequence $G = G_1, \ldots, G_{|M|}$ of $|M|$ graphs and a sequence $R = \alpha_1, \ldots, \alpha_{|M|-1}$ of $|M| - 1$ applications called refinements. Each refinement $\alpha_i$ is an application from $V(G_i)$ to $\mathcal{P}(V(G_{i+1}))$, that is, each vertex in $V(G_i)$ has a subset of $V(G_{i+1})$ as image. The inverse application of $\alpha_i$, denoted*

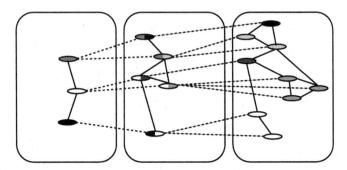

**Fig. 1.** Example of a *MiGaL* structure of size three defined by three graphs and two refinements

by $\beta_i$, is a surjection called an abstraction *that maps every vertex in $V(G_{i+1})$ to a vertex of $V(G_i)$. In addition, $M(G, R)$ satisfies the three conditions:*

1. *For each vertex $v \in V(G_i)$, the subgraph induced by $\alpha_i(v)$ is connected and not empty.*
2. *For each edge $(u, v) \in E(G_i)$, there exists at least one vertex of $\alpha_i(u)$ connected to a vertex of $\alpha_i(v)$.*
3. *For each pair of unconnected vertices $u, v \in V(G_i)$, there is no edge between a vertex of $\alpha_i(u)$ and a vertex of $\alpha_i(v)$.*

Figure 1 illustrates this definition. From now on, we refer to the first graph of a MiGaL as the *top* graph and to the last one as the *bottom* graph.

This structure is clearly useful to encode data at different levels of detail. A comparable data structure called *Multihierarchical Graph* has been used in [4,5] to represent the environment (towns, buildings, rooms, ...) and drive robots in space. Godin *et al.* in [8,6] use a similar structure, that they call *quotiented trees*, to model and study plants. As compared to the latter, MiGaLs can represent graphs and not just trees, and contrary to Multihierarchical Graphs, MiGaLs can have edges between nodes belonging to different layers.

## 2.2   Top-Down Comparison Algorithm

In this section, we present an algorithm to compare two MiGaLs. The algorithm performs a top-down traversal of the structure. We assume both MiGaLs have the same number of layers, and compare them layer by layer starting with the graph at the top. The result of the comparison at a given layer is transmitted to the next layer by colouring vertices and edges of the graph and using the refinement application. We then compare the two graphs of the next layer taking into account such colouring. The process continues until the last layer is reached (bottom graph).

This approach assumes the existence of an algorithm allowing to compare two graphs of a same layer. This algorithm clearly depends on the data that is modelled using a MiGaL. In what follows, we consider a black box that is able to produce an *extended mapping* between two graphs where by an extended mapping is meant the following:

**Definition 2 (Extended mapping).** *Given two graphs $G_1$ and $G_2$, we define an extended mapping $\mathcal{M}$ between them as a set of couples of vertex sets of the two graphs: $\mathcal{M} = \{(S_1, S_2)/S_1 \subset V(G_1) \text{ and } S_2 \subset V(G_2)\}$ such that for all $(S_1, S_2) \in \mathcal{M}$:*

- *$S_1$ is a connected component of $G_1$*
- *$S_2$ is a connected component of $G_2$*
- *$\forall(S_1', S_2') \in \mathcal{M}, S_1 \cap S_1' = \emptyset$ or $S_2 \cap S_2' = \emptyset$*

We define a colour-partitioned graph $G$ with vertex set $V(G)$ as a graph fitted with an application $C_G : S \subset V(G) \rightarrow \mathbb{N}^+$ which gives a colour to each vertex of $S \subset V(G)$ such that subgraphs defined by vertices of the same colour are connected. For convenience, uncoloured vertices are given the colour 0. The vertices with colour 0 are therefore elements of $V(G) \setminus S$.

We now extend the definition of an *extended mapping* to colour-partitioned graphs as follows.

**Definition 3 (Colour-constrained extended mapping).** *Given two colour-partitioned graphs $G_1$ and $G_2$, a colour constrained extended mapping $\mathcal{M}$ between them is defined as an extended mapping which further satisfies the following conditions:*

- *$\forall(S_1, S_2) \in \mathcal{M}$, every vertex of $S_1$ has the same colour $c$ and every vertex of $S_2$ also has this colour $c$.*
- *There is no vertex of $G_1$ or $G_2$ coloured with 0 that is involved in $\mathcal{M}$.*

Now that we have defined a mapping on colour-partitioned graphs, we give a general algorithm for comparing two MiGaLs, $M(G, R)$ and $M'(G', R')$, having the same number of layers. This number is denoted by $|M| = |M'|$. The algorithm assumes again that we have a black box $B$ that computes a colour-constrained mapping between two colour-partitioned graphs.

The initialisation step colours the vertices of $G_1$ and $G_1'$ with 1.

The algorithm is then divided into $|M|$ steps. For each layer $i$ from 1 to $|M|$, we compute the mapping $\mathcal{M}_i$ between $G_i$ and $G_i'$ using $B$ and (except for the last step) colour the vertices of $G_{i+1}$ and $G_{i+1}'$ such that vertices associated by the mapping have the same colour and vertices absent from $\mathcal{M}_i$ are coloured with 0. The result of the algorithm is the set of all mappings between each pair of layers.

The pseudo-code of this algorithm is shown in Figure 2. Let $b$ be the time complexity of algorithm $B$; the time required to compare two MiGaLs of size $n$ is $O(n * b)$.

It is important to notice that the main idea of this algorithm is to compute mappings from the top to the bottom layer. Each layer is compared using the mapping of the previous one without reconsidering the choices implied by this mapping.

**Compare**$(M(G,R), M'(G',R'))$
1.    $colour = 2$
2.    Set all vertices of $G_1$ and $G'_1$ to 1.
3.    for each layer $i$ from 1 to $|M|$
4.        $\mathcal{M} = B(G_i, G'_i)$
5.        set the color of $V(G_{i+1})$ and $V'(G'_{i+1})$ to 0
6.        for each couple of sets $(u,v)$ in $\mathcal{M}$
7.            for all nodes $n$ in $u$
8.                set the colour of nodes in $\alpha(n)$ to $colour$
9.            for all nodes $m$ in $v$
10.                set the colour of nodes in $\alpha'(m)$ to $colour$
11.            set $colour$ to $colour + 1$
12.    return $\mathcal{M}_{1...|M|}$

**Fig. 2.** Algorithm for the comparison of two MiGaL structures using an external algorithm $B$ that computes a colour-constrained mapping between two colour-partitioned graphs

We now present a practical application of the MiGaL structure and of the comparison algorithm to the study of RNA secondary structures. In particular, we show, in this case, how to significantly improve the complexity of algorithm $B$ by taking advantage of the node-colouring.

## 3   RNA-MiGaL

RNAs are one of the most important molecules in the cell. They are composed by a succession of nucleotides named A,C,G and U (also referred to as bases). Inside a cell, RNAs do not keep a linear form but instead fold in space. The fold is given by the set of interactions between nucleotides. An RNA can be described at three different levels, respectively called its primary, secondary and tertiary structures. The primary structure refers to the sequence of nucleotides. The secondary structure is composed of the list of base pairs that participate in an helix (see below). The tertiary structure corresponds to all interactions (base pairings) in the RNA that is, to its 3D structure.

The function of an RNA (be it the well known ribosomal and transfert RNAs or the more recently discovered snoRNAs, microRNAs, etc.) is strongly linked to the shape adopted by the RNA in space. It is accepted that two RNAs that have the same function will have closely similar secondary structures but not necessarily similar primary structures. Considering this, it is fundamental to have efficient algorithms to compare RNAs from the point of view of their secondary structures.

Various structural elements can be distinguished in the secondary structure of an RNA: *helices* which correspond to consecutive base pairs, *hairpin-loops* which are unpaired bases at the end of an helix, *internal-loops* defined by the

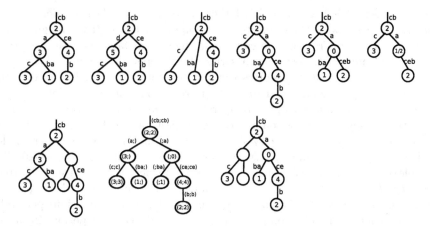

**Fig. 3.** Edit operations, edit distance and alignment on rooted ordered labelled (on nodes and edges) trees. On the first line, the first tree is edited by a substitution of $a-3$ to $d-5$. Then the node 5 is deleted and after that the node 0 is inserted. From this point, if operations have unit cost (1 for mismatch, deletion, insertion), the edit cost is 3 (the edit distance is 2 because the substitution is useless). The last two trees on the first line give an example of edge fusion (edges $ce$ and $b$, the new label is constructed by concatenation) and node fusion (nodes 0 and 1, the new label is computed using mean). The second line shows how to align the first tree of the first line with the fourth. We see here that the alignment cost is four (two insertions and two deletions).

unpaired bases between two helices and called *bulges* when one side is empty, *multi-loops* which are the meeting point of at least three helices and, finally, *stems* that are series of helices, internal-loops and bulges.

Many approaches have been used for modelling secondary structures. Mainly, three codings has been proposed to represent RNA secondary structures: rooted ordered trees [16]; arc annotated sequences [3]; 2-intervals [14].

Here we focus on the tree model. We represent secondary structures by rooted ordered trees because the root correspond to the beginning and the end of the molecule and the order between the children of a node corresponds to the orientation of the sequence. In fact, we can use different tree models depending on the information we want to encode. In [16], Zuker and Sankoff use trees where internal nodes code for base pairs and leaves for free nucleotides. This tree can be compacted [7] into a homeomorphically irreducible tree where internal nodes correspond to helices and leaves correspond to fragments of unpaired bases. Shapiro in [12] uses a tree where edges code for helices and nodes are labelled on $\{M, B, I, H, R\}$ (multi-loop, budge, internal-loop, hairpin-loop and the root). We can also use a more abstract tree where edges code for stems and nodes for multi-loops and hairpin-loops, or just for multi-loops.

To compare rooted ordered trees, there exist essentially two methods. The first is the edit distance [11,13,15] based on three edit operations. A *substitution* changes the label of a node. A *deletion* removes a node of the tree, its children are then re-attached to the node's father preserving the relative order between nodes. An *insertion* is the opposite operation of a deletion. If we assign a score to each of these operations, we can define the edit distance between two trees as the minimum of the score of a series of edit operations (sum of the score of each operation) that transforms the first tree into the second one. Recently, Allali *et al.* [1] extended this distance by introducing four new operations: *node fusion*, *edge fusion* and the opposite operations of *node split* and *edge split*. These operations allow to address some of the limitations of the classical edit distance when comparing RNA secondary structures using high level trees (where nodes and edges code for secondary structure elements only, not for nucleotides). The complexity of the classical edit distance is $O(n^4)$ where $n$ is the size of the trees and $O((2d)^l n^4)$ for the algorithm with node and edge fusions where $d$ is the degree of the trees and $l$ the maximum number of consecutive fusions per node.

The second method for comparing rooted ordered trees is by performing a tree alignment [10,9]. In this case, the goal is to insert blank labelled nodes in the two trees such that they become isomorphic. The score of the alignment corresponds to the sum of the scores of the association of node labels (the blank character assumes the role of insertions and deletions in sequence alignment). The optimal score, maximal or minimal depending on the scoring scheme adopted, is then sought. A tree alignment can be expressed as an edit distance computation where insertions must precede all deletions. We do not give further details here on alignments as we only use edit distance later in the paper.

We now suggest a new modelling of RNA secondary structures based on the MiGaL data structure introduced earlier in the paper. We thus call RNA-MiGaL a MiGaL structure composed by four layers, each modelled by a rooted ordered tree. The next section is dedicated to a description of RNA-MiGaL.

### 3.1   The Four Layers Definition

The four layers contained in an RNA-MiGaL correspond to the secondary structure of an RNA observed at different levels of detail. Thus, each layer is modelled by a rooted ordered labelled tree. In the bottom layer, nodes and leaves code for nucleotides while the top layer encodes the network of multi-loops of an RNA. This choice has been dictated by two assumptions. The first, already mentioned, is that structure is more important than sequence. Thus we introduce information on nucleotides at the bottom layer only which will be treated last by the comparison algorithm. The second is that the network of multi-loops can be considered as the skeleton of the secondary structure of an RNA. RNAs of the same family should thus have strongly conserved multi-loop networks. For this reason, the top layer of an RNA-MiGaL is a tree that codes for the multi-loop network. Intermediate layers correspond to the structure encoded using stems or helices.

We provide below a summary of the layers and of the meaning of a node and a leaf in each layer:

– The tree of layer 1 corresponds to the multi-loop network. The nodes encode multi-loops and the edges encode stems. In the nodes, we store the number of helices connected by the multi-loop. In the edges, we store the number of nucleotides contained in the stem.

- Layer 2 consists in the structure defined by the stems. The internal nodes represent multi-loops, leaves represent hairpin-loops and edges encode stems. In the edges, we store the number of base pairs and the number of unpaired bases contained in the stem. In the internal nodes and leaves, we store the number of unpaired bases contained in the multi-loops and hairpin-loops.
- The tree of layer 3 encodes secondary structure elements: nodes encode hairpin-loops, multi-loops, internal-loops and bulges and store the number of unpaired bases of the corresponding element. The edges represent helices and store the number of base pairs of the helices.
- The last tree models the RNA primary structure. The internal nodes thus represent base pairs and leaves unpaired bases. Both store the names of the corresponding bases.

### 3.2   RNA-MiGaL Comparison

The problem now is to compare two secondary structures using RNA-MiGaLs. To do so, we use edit distance with fusions and the algorithm described in [1] to compare the pairs of trees of layers 1, 2 and 3. To take into account the colour of the nodes and edges, we just have to test if two nodes or edges have the same colour to allow them to be fusioned (in a tree) or substituted (between the trees). The trees of layer 4 are compared using the classical edit distance as fusions make no sense for nucleotides.

Figures 4 and 5 show the result of the comparison using RNA-MiGaLs between two Group I Intron RNAs retrieved from [2]. The left RNA is found in *Acanthamoeba griffini* and the right one in *Chlorella sorokiniana*.

The time complexity required to compare two trees of a same layer is $O((2d)^l n^4)$ for the first 3 layers and $O(n^4)$ for the last layer with $n$ the size of the trees. Since we colour nodes while we progress in the comparison, and, once trees are

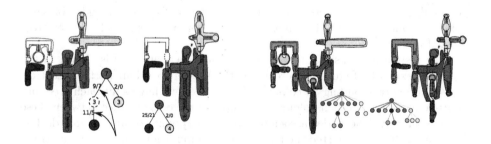

**Fig. 4.** Result of the comparison of two Group I Introns. On the left, the result of the comparison of the trees of layer 1; on the right, the result of the comparison of the trees of layer 2.

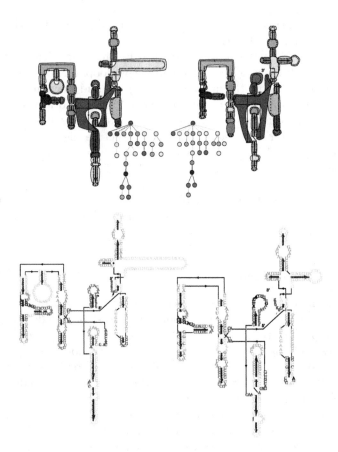

**Fig. 5.** Result of the comparison of two Group I Introns. On the top, the comparison of the trees of layer 3; on the bottom, the result of the comparison of the trees of layer 4.

coloured, substitutions and fusions can only be performed on nodes of same colour, we may wonder if it is not possible to compare separately the subtrees defined by each colour.

The response is positive but it requires some attention as is shown in the following example. Figure 6 presents on the top left a colour-partitioned tree, with white corresponding to colour 0. In the box, we have split each tree into subtrees according to their colour and computed the edit distance between these trees separately. The nodes associated by the mapping are linked with dashed lines. We have then reported the result of the computations on the original trees. The problem is pointed to by the bold dashed lines in the tree at the bottom. We have two couples of nodes associated via two different computations. Inside each computation, the relative order of the nodes is respected by the edition (we work on ordered trees). However, on the final tree, the associations do not respect the order between the nodes.

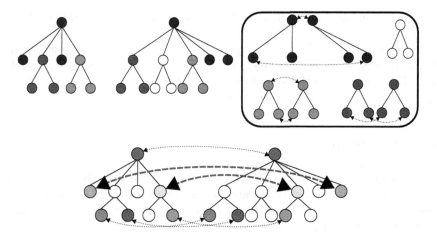

**Fig. 6.** Problem about splitting tree according to colour to optimize edition computation

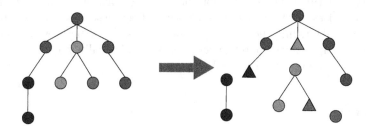

**Fig. 7.** Splitting the left tree into subtrees according to the colours of the nodes and addition of anchors (represented by triangles)

We therefore have to modify the algorithm such as to take into account the presence of a subtree of a colour $x$ hanging from an edge of a subtree of colour $y \neq x$. To do so, we introduce anchors during the splitting step. When we split trees into subtrees according to their colour, we add anchors to the subtrees. These anchors represent the subtrees of another colour that hang from an edge of the subtree being considered as shown in Figure 7. The anchors have the same colour as the subtrees they stand for. We then have to modify the edit distance scoring scheme such that:

- deleting the anchor of colour $c$ costs the deletion of the subtree corresponding to the anchor;
- an anchor of colour $c$ can only match with an anchor of the same colour.

Finally, we order the edit computations according to the dependence implied by the colouring scheme, that is we consider the subtree from the leaves to the root.

The consequence on the complexity is important as the time required to compare two trees of a given layer has a time (and space) complexity which depends on the size of the biggest subtree and not on the size of the whole trees anymore.

In the worst-case, the time complexity is $0((2d)^l \times n^4)$ for each level but in the case of RNAs, on average, the tree at each layer contains a number of subtree that is proportional to the size of the tree, and the observed time complexity is close to linear except for the first layer.

## 4   Conclusion

We introduced in this paper a new data structure and an algorithm for comparing such structures. Both are generic and may be applied to very different types of problems. They are particularly interesting for comparing objects that may be described at different levels of abstraction.

RNA secondary structures are one example of such objects and we therefore showed how to optimise the generic algorithm, in particular its special colouring scheme for going from one level of abstraction to the next, to compare two RNA secondary structures. The results obtained are very satisfying. In particular, the algorithm addresses the so-called scattering effect described in the literature [1]. The new algorithm allows to compare large sized structures in a fast way while the multiple layer approach represents a biologically very natural way of modelling and analysing with RNAs.

## References

1. Julien Allali and Marie-France Sagot. A new distance for high level rna secondary structure comparison. *IEEE/ACM Trans. Comput. Biol. Bioinformatics*, 2(1):3–14, 2005.
2. J. J Cannone, S. Subramanian, M. N. Schnare, J. R. Collett, L. M. D'Souza, Y. Du, B . Feng, N. Lin, L. V. Madabusi, K. M. Muller, N. Pande, Z. Shang, N. Yu, and R. R. Gutell. The comparative RNA web (CRW) site: an online database of comparative sequence and structure infor mation for ribosomal, intron, and other RNAs. *BMC Bioinformatics*, 3(1), 2002.
3. P. A. Evans. Finding common subsequence with arcs and pseudoknots. In *Proceedings of the 10th Annual Symposium on Combinatorial Pattern Matching (CPM'99)*, number 1645 in LNCS, pages 270–280, 1999.
4. Juan A. Fernandez and Javier Gonzalez. Hierarchical graph search for mobile robot path planning. In *ICRA*, pages 656–661, 1998.
5. Juan-Antonio Fernández-Madrigal and Javier González. Multihierarchical graph search. *IEEE Trans. Pattern Anal. Mach. Intell.*, 24(1):103–113, 2002.
6. P. Ferraro and C. Godin. An edit distance between quotiented trees. *Algorithmica*, 36:1–39, 2003.
7. W. Fontana, D. A. M. Konings, P. F. Stadler, and P. Schuster. Statistics of RNA secondary structures. *Biopolymers*, 33:1389–1404, 1993.
8. C. Godin and Y. Caraglio. A multiscale model of plant topological structures. *Journal of theoretical biology*, 191:1–46, 1998.

9. Matthias Höchsmann, Thomas Töller, Robert Giegerich, and Stefan Kurtz. Local similarity in RNA secondary structures. In *Proceedings of the IEEE Computer Society Conference on Bioinformatics*, page 159. IEEE Computer Society, 2003.

10. Tao Jiang, Lusheng Wang, and Kaizhong Zhang. Alignment of trees - an alternative to tree edit. In *Proceedings of the 5th Annual Symposium on Combinatorial Pattern Matching*, pages 75–86. Springer-Verlag, 1994.

11. S. M. Selkow. The tree-to-tree editing problem. *Inform. Process. Lett.*, 6(6):184–186, 1977.

12. B. Shapiro. An algorithm for multiple RNA secondary structures. *Comput. Appl. Biosci.*, 4(3):387–393, 1988.

13. Kuo-Chung Tai. The tree-to-tree correction problem. *J. ACM*, 26(3):422–433, 1979.

14. S. Vialette. Pattern matching problems over 2-interval sets. In *Proceedings of the 13th Annual Symposium on Combinatorial Pattern Matching*, pages 53–63. Springer-Verlag, 2002.

15. K. Zhang and D. Shasha. Simple fast algorithms for the editing distance between trees and related problems. *SIAM J. Comput.*, 18(6):1245–1262, 1989.

16. M. Zuker and D. Sankoff. RNA secondary structures and their prediction. *Bull. Math. Biol.*, 46:591–621, 1984.

# Normalized Similarity of RNA Sequences

Rolf Backofen[1], Danny Hermelin[2], Gad M. Landau[3], and Oren Weimann[2]

[1] Institute of Computer Science, Friedrich-Schiller Universität Jena,
Jena Center for Bioinformatics, Germany
backofen@inf.uni-jena.de

[2] Department of Computer Science, University of Haifa, Israel
danny@cri.haifa.ac.il, oweimann@cs.haifa.ac.il

[3] Department of Computer Science, University of Haifa, Haifa - Israel and
Department of Computer and Information Science,
Polytechnic University, New York, USA
landau@cs.haifa.ac.il

**Abstract.** We introduce a normalized version of the LCS metric as a new local similarity measure for comparing two RNAs. An $\mathcal{O}(n^2 m \lg m)$ time algorithm is presented for computing the maximum normalized score of two RNA sequences, where $n$ and $m$ are the lengths of the sequences and $n \leq m$. This algorithm has the same time complexity as the currently best known global LCS algorithm.

## 1 Introduction

Sequence comparison is an extensively studied topic with many applications, especially in biology. One commonly used metric is *longest common subsequence* (LCS) [2,10,11] which measures the longest subsequence of symbols that appears in both input sequences. While the LCS metric is a suitable metric for global comparison, in many real-life applications one is often interested in finding local regions of high similarity [16]. One approach for transforming the global LCS metric into a local version, is to calculate the *normalized longest common subsequence* [3,7]. Here, one divides the LCS score of two substrings by the sum of their lengths. This approach overcomes various weaknesses that are inherent in the standard local similarity algorithm [16].

In RNA sequences, as in other biological applications, it is not sufficient to perform pure sequence-based comparisons without respecting the underlying semantics of the sequences. RNAs are polymers consisting of four nucleotides A,C,G and U which are connected linearly via a backbone. In addition, the complementary nucleotides A—U, G—C and G—U can form bonds, which define the secondary structure of the RNA. In recent years, RNA sequences gained increasing interest due to numerous discoveries of biological functions which are associated with them. Consequently, research on small RNAs has been elected as the scientific breakthrough of the year 2002 by the readers of Science [6].

One major challenge of this research is to find common patterns in RNAs, since they suggest functional similarities. For this purpose, one has to investigate

M. Consens and G. Navarro (Eds.): SPIRE 2005, LNCS 3772, pp. 360–369, 2005.

not only sequential features, but also structural features for the following reasons. First, a major fraction of the function of an RNA is determined by its secondary structure [14]. Second, it is known that RNA structure is often more conserved than the sequence during evolution [5].

One promising approach for comparing RNAs while considering their sequence and secondary structure is to use appropriate variants of LCS-like metrics. This has been widely studied in the literature. One variant which is known under the term "*longest common subsequence for arc-annotated sequences*" (LAPCS), was first introduced by Evans [8], and then later extensively studied in [1,9,12]. However, the major downfall of this variant is that a base-pair, *i.e.* hydrogen bond, is not regarded as a whole entity. For example, in comparison of two RNAs, one nucleotide in a base-pair can be matched in the LCS, while the other nucleotide unmatched. In this paper we adopt a different variant introduced by Zhang [17] which treats base-pairs as a whole. This method is closer to the spirit of the comparative analysis method currently being used in the analysis of RNA secondary structures, either manually or automatically.

All known approaches for transferring the LCS metric to a metric of comparing RNA sequences have been restricted to global comparisons so far. However, as in almost all biological applications, global similarity is inferior to local similarity when comparing RNA sequences. In this paper, we consider a local variant of the LCS metric. Specifically, we consider the local normalized LCS metric for RNA sequences which measures the highest LCS scoring consecutive subsequences divided by their length. The advantages of the normalized approach in the context of strings [3,7] also apply for RNAs. We present an $\mathcal{O}(n^2 m \lg m)$ time algorithm for this problem which is conceptually inspired by the algorithm given in [7]. Its time complexity origins in the global LCS algorithm presented in [13], which is used as a preprocessing procedure in the algorithm. Therefore, our algorithm can compute the local normalized LCS score at the cost of computing the global LCS score.

This paper is organized as follows. In the following section we introduce definitions and terminology which will be used throughout the paper. In Section 3, we describe methods which provide the basis of our algorithm. The algorithm itself is later presented in Section 4.

## 2 Preliminaries

### 2.1 RNA Sequences

An RNA sequence $\mathcal{R}$ is an ordered pair $(S, P)$, where $S = s_1 \cdots s_{|S|}$ is a string over the alphabet $\Sigma = \{A, C, G, U\}$, and $P \subseteq \{1, \ldots, |S|\} \times \{1, \ldots, |S|\}$ is the set of hydrogen bonds between bases of $\mathcal{R}$ (*i.e.* the secondary structure), such that $\forall (i, i') \in P : i < i'$. Any base in $\mathcal{R}$ can bond with at most one other base, therefore we have $\forall (i_1, i'_1), (i_2, i'_2) \in P, i_1 = i_2 \Leftrightarrow i'_1 = i'_2$. Furthermore, following Zuker [18,19], we assume a model where the bonds in $P$ are *non crossing*, *i.e.* for any $(i_1, i'_1), (i_2, i'_2) \in P$, we cannot have $i_1 < i_2 < i'_1 < i'_2$ nor $i_2 < i_1 < i'_2 < i'_1$. We refer to a bond $(i, i') \in P$ as an *arc*, where $i$ ($i'$) is referred to as the *left* (*right*) *endpoint* of the arc. Finally, we let $|\mathcal{R}|$ denote the number of nucleotides in $\mathcal{R}$, *i.e.* $|\mathcal{R}| = |S|$.

We will require a notion similar to that of a substring for RNA sequences. Therefore, for any $1 \leq i < i' \leq |S|$, we let $\mathcal{R}[i, i']$, the *consecutive subsequence* of $\mathcal{R}$, be the RNA $(S', P')$ with $S' = S_i \cdots S_{i'}$ and $P' = P \cap \{i, \ldots, i'\} \times \{i, \ldots, i'\}$. Note that arcs of $\mathcal{R}$ with one endpoint in $\mathcal{R}[i, i']$ are absent in $\mathcal{R}[i, i']$. If there are no such arcs, then $\mathcal{R}[i, i']$ is said to be *arc complete*.

## 2.2   LCS Similarity

This paper deals with comparing two RNA sequences. We denote these two RNAs by $\mathcal{R}_1 = (S_1, P_1)$ and $\mathcal{R}_2 = (S_2, P_2)$, and we set $|\mathcal{R}_1| = |S_1| = n$ and $|\mathcal{R}_2| = |S_2| = m$. Furthermore, we assume $n \leq m$.

A base in one sequence can be matched to an identical base in the other sequence if they are both non arc endpoints. In the case of arcs, given $(i, i') \in P_1$ and $(j, j') \in P_2$, we require that $(i, j)$ be matched if and only if $(i', j')$ is matched. This captures the notion of arcs as single entities. A match between two left (right) endpoints of arcs is called a *left arc match* (respectfully *right arc match*). Note that for every left arc match $(i, j)$ there exists exactly one corresponding right arc match $(i', j')$, such that $(i, i') \in P_1$ and $(j, j') \in P_2$. In this case we let $right(i, j) = (i', j')$ and $left(i', j') = (i, j)$.

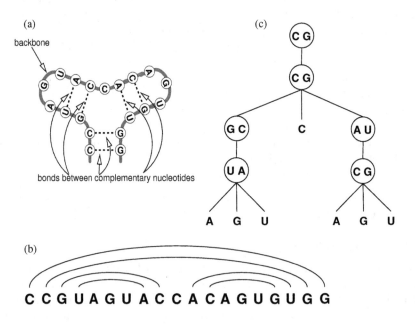

**Fig. 1.** Three different ways of viewing an RNA sequence. In (a), a schematic 2-dimensional description of an RNA folding. In (b), a linear representation of the RNA. In (c), the RNA as a rooted ordered tree.

An RNA sequence $\mathcal{R}'$ is a *common subsequence* of $\mathcal{R}_1$ and $\mathcal{R}_2$ if it can be obtained by omitting unpaired bases and arcs (along with their endpoints) in both $\mathcal{R}_1$ and $\mathcal{R}_2$. Alternatively, $\mathcal{R}'$ is a set of matches $\mathcal{M} = (i_1, j_1), \ldots, (i_p, j_p)$, such that $i_k < i_{k+1}$, $j_k < j_{k+1}$ for all $1 \leq k < p$, and $(i, j) \in \mathcal{M} \Leftrightarrow right(i, j) \in \mathcal{M}$ for any left arc match $(i, j)$. The longest common subsequence (LCS) of $\mathcal{R}_1$ and $\mathcal{R}_2$, denoted $LCS(\mathcal{R}_1, \mathcal{R}_2)$, is a common subsequence of $\mathcal{R}_1$ and $\mathcal{R}_2$ of maximum cardinality.

The non crossing formation formed by the arcs in both $\mathcal{R}_1$ and $\mathcal{R}_2$ conveniently allows representing these RNAs as trees [17]. Each arc is represented by an internal node in the tree, and each unpaired base by a leaf. The set of ordered children of an internal node which is associated with the arc $(i, i')$ is all unpaired bases $i''$ such that $i < i'' < i'$, and all arcs $(l, l')$ such that $i < l < l' < i'$ (see Fig. 1). In [15], an algorithm for tree editing was presented which was later improved in [13]. This algorithm can be used to determine the minimum number of unpaired base and arc deletions needed in order to obtain a common subsequence of $\mathcal{R}_1$ and $\mathcal{R}_2$ in $\mathcal{O}(n^2 m \lg m)$ time, and is currently the fastest (worst-case) algorithm for computing the global LCS score of two RNA sequences [4]. Furthermore, an important property of this algorithm is that it computes the score between every pair of subtrees of the two given trees. In our setting this means that $LCS(\mathcal{R}[i, i'], \mathcal{R}[j, j'])$ is computed between all pairs of arcs $(i, i') \in P_1$ and $(j, j') \in P_2$ in a single execution of this algorithm. The importance of this property will become apparent later on.

### 2.3   Normalized Similarity

We next present an extension of a local similarity metric for strings that uses normalization [7], to a metric for RNA sequences. Following this, we define the computational problem considered in this paper.

**Definition 1 (Normalized LCS score).** *The normalized LCS score of two RNA sequences $\mathcal{R}'_1$ and $\mathcal{R}'_2$ is given by*

$$\frac{|LCS(\mathcal{R}'_1, \mathcal{R}'_2)|}{|\mathcal{R}'_1| + |\mathcal{R}'_2|}.$$

The above definition is of a global nature. We therefore define the *local* normalized LCS score of two RNA sequences, $\mathcal{R}_1$ and $\mathcal{R}_2$, as the normalized LCS score of the two highest scoring consecutive subsequences of $\mathcal{R}_1$ and $\mathcal{R}_2$. More formally:

**Definition 2 (Local normalized LCS score).** *The local normalized LCS score of two RNA sequences, $\mathcal{R}_1$ and $\mathcal{R}_2$, is the maximal value of*

$$\frac{|LCS(\mathcal{R}_1[i, j], \mathcal{R}_2[i', j'])|}{|\mathcal{R}_1[i, j]| + |\mathcal{R}_2[i', j']|}$$

*where $LCS(\mathcal{R}_1[i, j], \mathcal{R}_2[i', j'])$ is also a common subsequence of $\mathcal{R}_1$ and $\mathcal{R}_2$, and $1 \leq i \leq j \leq n$, $1 \leq i' \leq j' \leq m$.*

Note the requirement of $LCS(\mathcal{R}_1[i,j], \mathcal{R}_2[i',j'])$ being a common subsequence of $\mathcal{R}_1$ and $\mathcal{R}_2$ is crucial in case either $\mathcal{R}_1[i,i']$ or $\mathcal{R}_2[j,j']$ are not arc complete. In this case, the above requirement prevents matching an endpoint of an arc whose other endpoint is absent in the consecutive subsequence, thereby ensuring that local solutions are valid also as global solutions.

Furthermore, notice that by the above definition, one single match gets the optimal normalized LCS score $(1/2)$. To solve this problem, we require that $|LCS(\mathcal{R}_1[i,j], \mathcal{R}_2[i',j'])| \geq I$, where $I$ is some integer (perhaps dependent of $n$) predefined according to the application at hand.

**Definition 3 (The local normalized LCS problem).** *Given two RNA sequences $\mathcal{R}_1 = (S_1, P_1)$ and $\mathcal{R}_2 = (S_2, P_2)$, and an integer $I$, the local normalized LCS problem asks to compute the local normalized LCS score of $\mathcal{R}_1$ and $\mathcal{R}_2$.*

## 3   Decomposing Common Subsequences

In the following section we present techniques for decomposing common subsequences of our two RNA sequences $\mathcal{R}_1 = (S_1, P_1)$ and $\mathcal{R}_2 = (S_2, P_2)$. We will be interested in a small fraction of such subsequences. These can intuitively be thought of as locally optimal common subsequences which contain exactly $k$ matches, for all $1 \leq k \leq n$. In [7], a $k$-*Chain* is defined as a common subsequence of two given strings which consists of $k$ matches. We adopt this terminology for our case as follows.

**Definition 4.** $k$-$Chain_{(i',j')}^{(i,j)}$ *is a common subsequence of $\mathcal{R}_1[i,i']$ and $\mathcal{R}_2[j,j']$ which consists of $k$ matches.*

- $k$-$Chain_{(i',j')}^{(i,j)}$ *starts at $(i,j)$ and ends at $(i',j')$.*

- *The head of $k$-$Chain_{(i',j')}^{(i,j)}$ is the first match in $k$-$Chain_{(i',j')}^{(i,j)}$.*

- *The tail of $k$-$Chain_{(i',j')}^{(i,j)}$ is the last match in $k$-$Chain_{(i',j')}^{(i,j)}$.*

- *The length of $k$-$Chain_{(i',j')}^{(i,j)}$ is the sum of $|\mathcal{R}_1[i,i']|$ and $|\mathcal{R}_2[j,j']|$, i.e. $j' - j + i' - i$.*

- *The normalized score of $k$-$Chain_{(i',j')}^{(i,j)}$ is given by $\frac{k}{j'-j+i'-i}$.*

This definition differs from the definition in [7], since $k$-$Chain_{(i',j')}^{(i,j)}$ is defined there only when $(i,j)$ and $(i',j')$ are matches. Next we define the best scoring $k$-Chain that starts at $(i,j)$.

**Definition 5.** $k$-$Chain^{(i,j)}$ *is $k$-$Chain_{(i',j')}^{(i,j)}$ with the highest normalized score (shortest length) over all $i' \leq n$ and $j' \leq m$.*

A major obstacle in constructing $k$-Chains, is that any attempt to construct $k$-$Chain^{(i,j)}$ simply by tying another match to the tail of $(k-1)$-$Chain^{(i,j)}$ will not necessarily result in the optimal $k$-Chain. We therefore take the opposite approach. From among all chains which start at $(i', j')$, $i \leq i', j \leq j$ and $(i, j) \neq (i', j')$, we choose the one that when concatenated to $(i, j)$, creates $k$-Chain$^{(i,j)}$.

In order to construct $k$-Chain$^{(i,j)}$, we distinguish between four different cases depending on $(i, j)$. Indeed, $(i, j)$ can either be a mismatch, a non arc match, a right arc match, or a left arc match. Note that if $(i, j)$ is a non arc match then we can assume that it is the head of $k$-Chain$^{(i,j)}$, since otherwise by replacing the head with $(i, j)$, we obtain a $k$-Chain with the same score. Furthermore, if $(i, j)$ is a right arc match, then $(i, j)$ cannot be the head of $k$-Chain$^{(i,j)}$, since the left arc match corresponding to $(i, j)$ is not in $k$-Chain$^{(i,j)}$, and by definition, $(i, j)$ cannot appear alone in any common subsequence.

In the following we give further details concerning each one of these four cases. Later these will provide the basis for a dynamic programming procedure which we design for solving the local normalized LCS problem.

For a pair of matches $(i, j), (i', j') \in \{1, \ldots, n\} \times \{1, \ldots, m\}$, we refer to the value $|i' - i| + |j' - j|$ as the *distance* between $(i, j)$ and $(i', j')$.

**Definition 6 ($k$-closest).** *The $k$-closest chain to $(i, j)$ is the $k$-Chain with minimum distance between its tail and $(i, j)$ among all $k$-Chains starting at $(i', j')$ with $i \leq i', j \leq j'$, and $(i', j') \neq (i, j)$.*

(a)

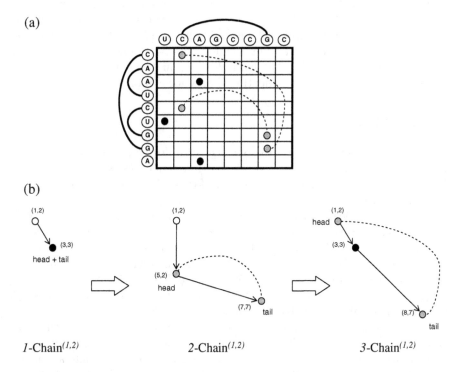

(b)

1-Chain$^{(1,2)}$          2-Chain$^{(1,2)}$          3-Chain$^{(1,2)}$

**Fig. 2.** Constructing $k$-Chains. (a) The matches of two RNA sequences. (b) The construction of 3-Chain$^{(1,2)}$.

*Case 1: $(i,j)$ is a mismatch.* In this case, by definition, $k$-Chain$^{(i,j)}$ consists of the $k$ matches of the $k$-closest chain to $(i,j)$.

*Case 2: $(i,j)$ is a non arc match.* In this case, $(i,j)$ is the head of $k$-Chain$^{(i,j)}$. Therefore, $k$-Chain$^{(i,j)}$ consists of $(i,j)$ and the $k-1$ matches of the $(k-1)$-closest chain to $(i,j)$ starting at $(i',j')$ such that $i < i'$ and $j < j'$.

*Case 3: $(i,j)$ is a right arc match.* In this case, $(i,j)$ cannot be the head of $k$-Chain$^{(i,j)}$. Furthermore, any match $(i,j')$ or $(i',j)$ is also a right arc match. Therefore, $k$-Chain$^{(i,j)}$ consists of the $k$ matches of the $k$-closest chain to $(i,j)$ starting at $(i',j')$ such that $i < i'$ and $j < j'$.

*Case 4: $(i,j)$ is a left arc match.* This is the most delicate case. Indeed, $k$-Chain$^{(i,j)}$ may or may not include $(i,j)$. If $(i,j) \notin k$-Chain$^{(i,j)}$, then the head of $k$-Chain$^{(i,j)}$ can still be $(i,j')$ or $(i',j)$ for some $i' > i$ and $j' > j$. Therefore, in this case $k$-Chain$^{(i,j)}$ consists of the $k$ matches of the $k$-closest chain to $(i,j)$.

In case $(i,j) \in k$-Chain$^{(i,j)}$, then $(i',j') = right(i,j)$ is also in $k$-Chain$^{(i,j)}$ by definition. Therefore, $k$-Chain$^{(i,j)}$ is of length at least $i' - i + j' - j$. Let $\mathcal{M} = LCS(\mathcal{R}_1[i,i'], \mathcal{R}_2[j,j'])$. From the optimality of $k$-Chain$^{(i,j)}$, it follows that in this case $k > |\mathcal{M}| - 2$, since otherwise there exists a shorter $k$-Chain which doesn't include $(i,j)$ nor $(i',j')$. If $k = |\mathcal{M}|$, then $k$-Chain$^{(i,j)}$ consists exactly of the matches in $\mathcal{M}$. If $k \geq |\mathcal{M}|$, then $k$-Chain$^{(i,j)}$ consists of all matches of $\mathcal{M}$ and all matches of $(k-|\mathcal{M}|)$-Chain$^{(i',j')}$. The case $k = |\mathcal{M}| - 1$ is eccentric. Here, $k$-Chain$^{(i,j)}$ consists of all but one match of $\mathcal{M}$ (which is neither $(i,j)$ or $right(i,j)$). In this case, $(k+1)$-Chain$^{(i,j)}$ has a higher normalized score than $k$-Chain$^{(i,j)}$, and so this case is mentioned only for completeness.

## 4    The Algorithm

We are now in position to describe our algorithm for computing the local normalized similarity score of our two given RNA sequences $\mathcal{R}_1 = (S_1, P_1)$ and $\mathcal{R}_2 = (S_2, P_2)$. Recall that we are looking for the highest normalized scoring $k$-Chain such that $k \geq I$.

Our algorithm begins in a preprocessing stage which consists of two phases. In the first phase, for each $(i,j) \in \{1,\ldots,n\} \times \{1,\ldots,m\}$, the algorithm classifies $(i,j)$ as one of the following four types: mismatch, non arc match, right arc match, or left arc match. In the second phase, the algorithm computes and stores the longest common subsequence of $\mathcal{R}_1[i,i']$ and $\mathcal{R}_2[j,j']$ for every $(i,i') \in P_1$ and $(j,j') \in P_2$. As mentioned in Section 2.2, this can be done by a single execution of the algorithm given in [13].

The second phase is the bottleneck of our algorithm. Nevertheless, this computation is necessary for efficiently constructing $k$-Chains which start at left arc matches. According to the fourth case above, if $(i,j)$ is a left arc match and $right(i,j) = (i',j')$, then the computation of $k$-Chain$^{(i,j)}$ is based on $\mathcal{M} = LCS(\mathcal{R}_1[i,i'], \mathcal{R}_2[j,j'])$. In the dynamic programming computation below, $\mathcal{M}$ is computed for any pair of arcs $(i,i') \in P_1$ and $(j,j') \in P_2$. Using [13] allows us to compute this efficiently, in some cases at the cost of computing $\mathcal{M}$ for a single pair. On the other hand, in cases where the number of nesting edges is small, one can use standard LCS algorithms.

After the preprocessing stage is complete, the algorithm computes and stores the score of $k$-Chain$^{(i,j)}$, for all $(i,j) \in \{1, \ldots, n\} \times \{1, \ldots, m\}$ and for all $1 \leq k \leq n$. More precisely, the algorithm computes and stores the lengths of all these $k$-Chains, since the score of a $k$-Chain can be derived from its length and vice versa. Let $DP^k[i, j]$ denote the length of $k$-Chain$^{(i,j)}$. For $k = 1$, the recursion of $DP^1[i, j]$ is given by:

$$DP^1[i,j] = \begin{cases} 2 & (i,j) \text{ is a non arc match,} \\ \min \begin{cases} DP^1[i+1, j+1] + 2 \\ DP^1[i, j+1] + 1 \\ DP^1[i+1, j] + 1 \end{cases} & \text{Otherwise.} \end{cases}$$

For $k > 1$, we let $(i', j') = right(i, j)$ and $\mathcal{M} = LCS(\mathcal{R}_1[i, i'], \mathcal{R}_2[j, j'])$. The recursion of $DP^k[i, j]$ is then given by:

$$DP^k[i,j] = \begin{cases} \min \begin{cases} DP^k[i+1, j+1] + 2 \\ DP^k[i, j+1] + 1 \\ DP^k[i+1, j] + 1 \end{cases} & (i,j) \text{ is a mismatch.} \\[2em] DP^{k-1}[i+1, j+1] + 2 & (i,j) \text{ is a non arc match.} \\[1em] DP^k[i+1, j+1] + 2 & (i,j) \text{ is a right arc match.} \\[1em] \min \begin{cases} DP^k[i+1, j+1] + 2 \\ DP^k[i, j+1] + 1 \\ DP^k[i+1, j] + 1 \\ j' - j + i' - i + DP^{k-|\mathcal{M}|}[i', j'] \end{cases} & \begin{array}{l} (i,j) \text{ is a left arc match} \\ \text{and } k > |\mathcal{M}|. \end{array} \\[3em] \min \begin{cases} DP^k[i+1, j+1] + 2 \\ DP^k[i, j+1] + 1 \\ DP^k[i+1, j] + 1 \\ j' - j + i' - i \end{cases} & \begin{array}{l} (i,j) \text{ is a left arc match} \\ \text{and } k \leq |\mathcal{M}|. \end{array} \end{cases}$$

The final stage of the algorithm consists of analyzing all $DP$ tables and reporting a solution. Here, the algorithm can either report the normalized score of the highest scoring $k$-Chain such that $k \geq I$, or the consecutive subsequences $\mathcal{R}_1[i, i']$ and $\mathcal{R}_2[j, j']$ that correspond to this chain. Furthermore, if required, the algorithm can be modified to report all chains with a score higher than some given threshold, e.g. 80%.

Correctness of our algorithm follows from the discussion in Section 3.

*Time complexity.* The preprocessing stage can be done in $\mathcal{O}(n^2 m \lg m)$ time using the algorithm in [13]. Furthermore, computing all $DP$ requires $\mathcal{O}(n^2 m)$ time. Hence, the total time complexity of our algorithm is $\mathcal{O}(n^2 m \lg m)$.

# Acknowledgments

We would like to thank Dennis Shasha and Kaizhong Zhang for fruitful discussions. Furthermore, we owe our gratitude to an anonymous referee for pointing out an error in the preliminary version of this paper.

# References

1. Alber J., J. Gramm, J. Guo and R. Niedermeier. Towards optimally solving the longest common subsequence problem for sequences with nested arc annotations in linear time. *Proc. of the 13th Combinatorial Pattern Matching conference (CPM 2002)*, LNCS vol. 2373, 99-114, 2002.
2. Apostolico A. and C. Guerra. The longest common subsequence problem revisited. *Algorithmica*, 2:315-336, 1987.
3. Arslan A.N., Ö. Eğecioğlu and P.A. Pevzner. A new approach to sequence alignment: normalized sequence alignment. *Bioinformatics*, 17(4):327-337, 2001.
4. Bille P. A survey on tree edit distance and related problems. *Theoretical Computer Science*, 337:217-239, 2005.
5. Chartrand P., X-H. Meng, R.H. Singer and R.M. Long. Structural elements required for the localization of ASH1 mRNA and of a green fluorescent protein reporter particle *in vivo*. *Current Biology*, 9:333-336, 1999.
6. Couzin J. Breakthrough of the year. Small RNAs make big splash. *Science*, 298(5602):2296-2297, 2002.
7. Efraty N. and G.M. Landau. Sparse normalized local alignment. *Proc. of the 15th Combinatorial Pattern Matching conference (CPM 2004)*, LNCS vol. 3109, 333-346, 2004.
8. Evans P.A. Algorithms and complexity for annotated sequence analysis. *PhD thesis, University of Alberta*, 1999.
9. Gramm J., J. Guo and R. Niedermeier. Pattern matching for arc annotated sequences. *Proc. of the 22nd Foundations of Software Technologies and Theoretical Computer Science conference (FSTTSC 2002)*, LNCS vol. 2556, 182-193, 2002.
10. Hirschberg D.S. Algorithms for the longest common subsequence problem. *Journal of the ACM*, 24(4):664-675, 1977.
11. Hunt J.W. and T.G. Szymanski. A fast algorithm for computing longest common subsequences. *Communications of the ACM*, 20(5):350-353, 1977.
12. Jiang T., G-H. Lin, B. Ma and K. Zhang. The longest common subsequence problem for arc-annotated sequences. *Proc. of the 11th Combinatorial Pattern Matching conference (CPM 2000)*, LNCS vol. 1848, 154-165, 2000.
13. Klein P.N. Computing the Edit-Distance between Unrooted Ordered Trees. *Proc. of the 6th European Symposium on Algorithms conference (ESA 1998)*, LNCS vol. 1461, 91-102, 1998.
14. Moore P.B. Structural motifs in RNA. *Annual review of biochemistry*, 68:287-300, 1999.
15. Shasha D. and K. Zhang. Simple fast algorithms for the editing distance between trees and related problems. *SIAM Journal on Computing*, 18(6):1245-1262, 1989.
16. Smith T.F. and M.S. Waterman. The identification of common molecular subsequences. *Journal of Molecular Biology* , 147:195-197, 1981.

17. Zhang K. Computing similarity between RNA secondary structures. *Proc. of the IEEE joint symposium on Intelligence and Systems conference*, 126-132, 1998.
18. Zuker M. On finding all suboptimal foldings of an RNA molecule. *Science*, 244(4900):48-52, 1989.
19. Zuker M. and P. Stiegler. Optimal computer folding of large RNA sequences using thermodynamics and auxiliary information. *Nucleic Acids Research*, 9(1):133-148, 1981.

# A Fast Algorithmic Technique for Comparing Large Phylogenetic Trees

Gabriel Valiente

Department of Software, Technical University of Catalonia, E-08034, Barcelona

**Abstract.** The comparison of rooted phylogenetic trees is essential to querying phylogenetic databases such as TreeBASE. Current comparison methods are based on either tree edit distances or common subtrees. However, a limitation of such methods is their inherent complexity. In this paper, a new distance over fully resolved phylogenetic trees, the *transposition distance*, is described which is based on a well-known bijection between perfect matchings and phylogenetic trees, and simple linear-time algorithms are presented for computing the new distance.

## 1 Introduction

The comparison of phylogenetic trees is essential to performing phylogenetic queries on databases of phylogenetic trees [4]. The main repository of published phylogenetic analyses, the TreeBASE [3] phylogenetic database, currently contains over 2,500 phylogenies with over 36,000 taxa among them. Previous work on phylogenetic tree comparison has focused on unrooted trees only, while the phylogenies stored in TreeBASE are rooted trees. The tools currently used to perform phylogenetic queries on TreeBASE are TreeSearch [5], to find trees that share a specified subtree, and TreeRank [8], to compute tree dissimilarity scores. Computation of a phylogenetic query by TreeRank takes time quadratic in the size of the trees.

In this paper, a new distance between phylogenetic trees, the transposition distance, is proposed which can be computed in time linear in the size of the trees, much faster than previous tree dissimilarity measures. Phylogenetic trees with polytomies can be fully resolved into a canonical representation, which is unique up to isomorphism, in linear time. The transposition distance between fully resolved phylogenetic trees over the same taxa is defined below as the minimum number of transpositions needed to transform the matching representation of one tree into the matching representation of the other. For fully resolved phylogenetic trees with overlapping taxa, the transposition distance can also take the number of non-common taxa and the number of contracted edges (in the topological restriction of the trees to their common taxa) into account.

Throughout this paper, by a *phylogenetic tree* we mean a *rooted phylogenetic tree*, that is, a directed finite graph $T = (V, E)$ with $V$ either empty or containing a distinguished node $r \in V$, called the *root*, such that for every other node $v \in V$ there exists one, and only one, path from the root $r$ to $v$. The *children* of a node

M. Consens and G. Navarro (Eds.): SPIRE 2005, LNCS 3772, pp. 370–375, 2005.

$v$ in a tree $T$ are those nodes $w$ such that $(v, w) \in E(T)$. The nodes without children are the *leaves* of the tree. The *height* of a node $v$ in a tree $T$, denoted by $height(v)$, is the length of a longest path from $v$ to any node in the subtree of $T$ rooted at $v$. The leaves of a phylogenetic tree, of height zero, are injectively labeled in a fixed, but arbitrary, set.

**Definition 1.** *A phylogenetic tree will be denoted by $T = (V, E)$, where $V$ is the set of nodes and $E \subset V \times V$ is the set of edges. A phylogenetic tree is said to be fully resolved if all nodes have either zero or two children. The taxon associated with a leaf node $v \in V$ will be denoted by $\ell(v)$.*

Phylogenetic analyses often produce phylogenies with polytomies, that is, phylogenetic trees that are not fully resolved. As a matter of fact, TreeBASE contains, at the time of this writing, 2,592 phylogenies over 36,593 taxa, 1,725 (66.55%) of which have polytomies and the remaining 867 (33.45%) are fully resolved phylogenetic trees over 11,950 taxa.

A phylogenetic tree with polytomies can be turned into a fully resolved phylogenetic tree in a canonical way, such that two isomorphic phylogenetic trees have exactly the same full resolution. The transformation of a phylogenetic tree with polytomies into a fully resolved phylogenetic tree can be, for instance, based on the natural correspondence between general trees and those binary trees that have a root but no right subtree [2, Sect. 2.3.2], and also on the fact that phylogenetic trees have unique leaf labels. All phylogenetic trees are assumed to be fully resolved in the rest of this paper.

## 2   Matching Representation of Phylogenetic Trees

Let us first recall the bijection between perfect matchings and fully resolved phylogenetic trees, studied by [1]. A partition of $\{1, \ldots, 2n\}$ into 2-subsets is a set of $n$ pairwise disjoint unordered pairs. Let $u \prec v$ denote that node $u$ is a predecessor of node $v$ in sorted bottom-up order [7], that is, either $height(u) < height(v)$ or $height(u) = height(v)$ and $\ell(u) < \ell(v)$.

**Definition 2.** *The* matching representation *$M(T)$ of a fully resolved phylogenetic tree $T = (V, E)$ with $n$ leaves labeled $1, \ldots, n$, is the partition of $\{1, \ldots, 2n\}$ into 2-subsets defined as follows. Let the internal nodes of $T$ be labeled as $\ell(v) = \max\{\ell(u) \mid u \prec v\} + 1$. Then, $M(T) = \{\{\ell(v), \ell(w)\} \mid (u, v), (u, w) \in E$ for some $u \in V\}$.*

Operationally, the matching representation of a fully resolved phylogenetic tree $T = (V, E)$ with $n$ leaves labeled $1, \ldots, n$, can be obtained as follows. First, the internal nodes of $T$ are labeled according to the following scheme: the parent in $T$ of the labeled nodes $v, w \in V$ with smallest $\ell(v)$ or $\ell(w)$ is assigned label $n + 1$, the one with next-smallest child label is assigned label $n + 2$, and so on. Then, $M(T) = \{\{\ell(v), \ell(w)\} \mid (u, v), (u, w) \in E$ for some $u \in V\}$.

The matching representation of a fully resolved phylogenetic tree can be obtained in time linear in the size of the trees by bottom-up tree traversal techniques [7]. The detailed pseudocode is given in Algorithm 1.

```
begin
   foreach node v of T do
      if v is a leaf node of T then
      |  set ℓ(v) to the index of ℓ(v) in L
      else
      └  ℓ(v) := 0
   i := |L|
   foreach level h of T from the leaves up to the root do
      let S be the set of nodes of T at level h, ordered by label
      foreach v ∈ S do
         let w be the parent of v in T
         if ℓ(w) = 0 and height(w) = h + 1 then
         |  i := i + 1
         └  ℓ(w) := i
   M := ∅
   foreach non-leaf node v of T do
      let x and y be the children of v in T
      └  M := M ∪ {{ℓ(x), ℓ(y)}}
   return M
end
```

**Algorithm 1: Matching representation.** Given a set $L$ and a fully resolved phylogenetic tree $T$ with leaves labeled in $L$, the algorithm computes the matching representation $M$ of $T$.

**Corollary 1.** *Let $T = (V, E)$ be a fully resolved phylogenetic tree. Then, $\{i, j\} \in M(T)$ if and only if there are sibling nodes $v, w \in V$ such that $\ell(v) = i$ and $\ell(w) = j$.*

## 3    Transposition Distance Between Phylogenetic Trees

The transformation of a phylogenetic tree into another one by means of *transpositions* in the matching representation is studied next. This operation is just a generalization of transposition in permutations to partitions into 2-subsets.

**Definition 3.** *Let $M$ be a partition of $\{1, \ldots, 2n\}$ into 2-subsets, and let $\{i, j\}$, $\{k, \ell\} \in M$. The transposition of $M$ at $j$ and $k$ is the replacement of $\{i, j\}$ by $\{i, k\}$ and $\{k, \ell\}$ by $\{j, \ell\}$ in $M$.*

It is easy to see that transpositions are sufficient to transform any two partitions of $\{1, \ldots, 2n\}$ into 2-subsets to each other.

**Proposition 1.** *Let $M_1$ and $M_2$ be partitions of $\{1, \ldots, 2n\}$ into 2-subsets. Then, there exists a set of transpositions that transform $M_1$ into $M_2$.*

*Proof.* $M_1$ and $M_2$ already agree on $M_1 \cap M_2$. Let $M_1' = M_1 \setminus M_1 \cap M_2$ and $M_2' = M_2 \setminus M_1 \cap M_2$. Now, $0 \leqslant |M_1'| = |M_2'| \leqslant n = |M_1| = |M_2|$. If $|M_1'| = 0$, then the empty set of transpositions suffices to transform $M_1$ into $M_2$. Otherwise, let $\{i,j\} \in M_1'$ and $\{i,k\} \in M_2'$. Since $\{i,j\} \notin M_1 \cap M_2$, it must be $j \neq k$. Let also $\{k,\ell\} \in M_1'$. The transposition of $M_1'$ at $j$ and $k$, makes $M_1'$ and $M_2'$ agree on $\{i,k\}$, so that the transformation of $M_1'$ into $M_2'$ can proceed with $M_1'' = M_1' \setminus \{i,k\}$ and $M_2'' = M_2' \setminus \{i,k\}$. Thus, it can be shown by induction on $|M_1'|$ that $M_1'$ can be transformed into $M_2'$ by means of transpositions and therefore, $M_1$ can also be transformed into $M_2$ by means of transpositions.    □

The previous result entails that transpositions are sufficient to transform the matching representations of any two fully resolved phylogenetic trees over the same taxa to each other.

**Corollary 2.** *Let $T_1$ and $T_2$ be fully resolved phylogenetic trees over the same taxa. Then, there exists a set of transpositions that transform $M(T_1)$ into $M(T_2)$.*

The *matching distance* $MD(T_1, T_2)$ between two fully resolved phylogenetic trees $T_1$ and $T_2$ over the same taxa, is the minimum number of transpositions needed to transform $M(T_1)$ into $M(T_2)$.

**Definition 4.** *Let $T_1$ and $T_2$ be fully resolved phylogenetic trees over the same taxa. The* matching distance *between $T_1$ and $T_2$, denoted by $MD(T_1, T_2)$, is the minimum number of transpositions needed to transform $M(T_1)$ into $M(T_2)$.*

The following result establishes a simple procedure for computing the matching distance between fully resolved phylogenetic trees over the same taxa.

**Theorem 1.** *Let $M_1$ and $M_2$ be partitions of $\{1, \ldots, 2n\}$ into 2-subsets, and let $G = (V, E)$ be the undirected graph with vertex set $V = \{1, \ldots, 2n\}$ and edge set $E = M_1 \cup M_2$. Let also $C$ be the set of connected components of $G$. Then, $MD(M_1, M_2) = |E|/2 - |C|$.*

*Proof.* Each connected component of $G$ consists of either a single 2-subset, that is, a trivial component $A \subseteq M_1 \cap M_2$, or an alternating cycle of 2-subsets $A \subseteq M_1 \cup M_2 \setminus M_1 \cap M_2$ coming in turn from $M_1$ and $M_2$, because $\{i,j\} \cap \{k,\ell\} = \emptyset$ for all $\{i,j\}, \{k,\ell\} \in M_1$ with $\{i,j\} \neq \{k,\ell\}$, and similarly for $M_2$. Now, each trivial component $A \subseteq M_1 \cap M_2$ represents the agreement of $M_1$ and $M_2$ on a 2-subset and thus, contributes $0 = |A| - 1$ to $MD(M_1, M_2)$. Each non-trivial

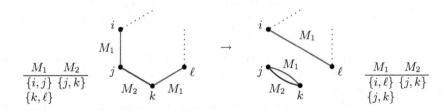

**Fig. 1.** Illustration for the proof of Theorem 1. The transposition of $M_1$ at $i$ and $k$ separates a 2-subset from the alternating cycle.

```
begin
    MD := 0
    foreach pair {i₁, j₁} of M₁ do
        let {i₂, j₂} be the pair of M₂ with i₁ = i₂
        if j₁ ≠ j₂ then
            let {k₁, ℓ₁} be the pair of M₁ with k₁ = j₂
            replace {i₁, j₁} by {i₁, k₁} in M₁
            replace {k₁, ℓ₁} by {j₁, ℓ₁} in M₁
            MD := MD + 1
    return MD
end
```

**Algorithm 2: Matching distance.** Given an integer $n$ and two partitions $M_1$ and $M_2$ of $\{1, \ldots, 2n\}$ into 2-subsets, the algorithm computes the matching distance $MD$ between $M_1$ and $M_2$.

component $A \subseteq M_1 \cup M_2 \setminus M_1 \cap M_2$, on the other hand, represents an alternating cycle of 2-subsets upon which each transposition separates a single 2-subset from $A$ and thus, also contributes $|A|/2 - 1$ to $MD(M_1, M_2)$. As a matter of fact, for each 2-subset $\{i, j\} \in A \cap M_1$, $\{j, k\} \in A \cap M_2$, and $\{k, \ell\} \in A \cap M_1$, the only possible transposition of $M_1$ is at $i$ and $k$, which separates the 2-subset $\{j, k\}$ from $A$ and decreases $|A|$ by 2, and similarly for a transposition of $M_2$. (See Fig. 1.) Therefore, $MD(M_1, M_2) = \sum_{A \in C}(|A|/2 - 1) = (\sum_{A \in C}|A|)/2 - |C| = |E|/2 - |C|$. $\quad\square$

The matching distance can thus be obtained in time linear in the size of the trees. An alternative linear-time algorithm computes the matching distance between two partitions $M_1$ and $M_2$ of $\{1, \ldots, 2n\}$ into 2-subsets while performing the actual transpositions that transform $M_1$ into $M_2$. The detailed pseudocode is given in Algorithm 2.

The *transposition distance* between two fully resolved phylogenetic trees, on the other hand, is the matching distance between the topological restriction of the trees to their common taxa.

**Definition 5.** *Let $T_1$ and $T_2$ be fully resolved phylogenetic trees, let $L_1 = \{\ell(v_1) \mid v_1 \in V_1\}$, let $L_2 = \{\ell(v_2) \mid v_2 \in V_2\}$, and let $L = L_1 \cap L_2$. Let also $T_1' = T_1|L$ and $T_2' = T_2|L$ be the topological restriction of $T_1$ and $T_2$, respectively, to their common taxa $L$. The* transposition distance *between $T_1$ and $T_2$, denoted by $TD(T_1, T_2)$, is $TD(T_1, T_2) = MD(M(T_1'), M(T_2'))$.*

The transposition distance between two fully resolved phylogenetic trees can thus be obtained in time linear in the size of the trees. The detailed pseudocode is given in Algorithm 3.

For phylogenetic trees with overlapping taxa, the transposition distance can be extended by also taking non-common taxa and contracted edges (in the topological restriction of the trees to their common taxa) into account. The transposition distance can be normalized to a value between zero and one by dividing

```
begin
    L₁ := {ℓ(v₁) | v₁ ∈ V₁}
    L₂ := {ℓ(v₂) | v₂ ∈ V₂}
    L := L₁ ∩ L₂
    T₁′ := T₁|L
    T₂′ := T₂|L
    relabel the leaves of T₁′ and T₂′ with integers {1, …, |L|}
    TD := MD(M(T₁′, L), M(T₂′, L))
    return TD
end
```

**Algorithm 3: Transposition distance.** Given two fully resolved phylogenetic trees $T_1$ and $T_2$, the algorithm computes the transposition distance $TD$ between $T_1$ and $T_2$.

by the total size of the trees, which is an upper bound on the transposition distance. Further work is needed to establish additional properties of the transposition distance, as done for instance in [6] for dissimilarity metrics over unrooted phylogenetic trees.

## Acknowledgment

The research described in this paper was partially supported by the Spanish CICYT, project GRAMMARS (TIN2004-07925-C03-01), and by the Japan Society for the Promotion of Science through Long-term Invitation Fellowship L05511 for visiting JAIST (Japan Advanced Institute of Science and Technology).

## References

1. Diaconisa, P.W., Holmes, S.P.: Matchings and phylogenetic trees. Proc. Natl. Acad. Sci. USA **95** (1998) 14600–14602
2. Knuth, D.E.: Fundamental Algorithms. 3rd edn. Volume 1 of The Art of Computer Programming. Addison-Wesley (1997)
3. Morell, V.: TreeBASE: The roots of phylogeny. Science **273** (1996) 569–570 http://www.treebase.org.
4. Page, R.D.M.: Phyloinformatics: Towards a phylogenetic database. In Wang, J.T.L., Zaki, M.J., Toivonen, H., Shasha, D.E., eds.: Data Mining in Bioinformatics. Springer-Verlag (2005) 219–241
5. Shan, H., Herbert, K.G., Piel, W.H., Shasha, D., Wang, J.T.L.: A structure-based search engine for phylogenetic databases. In: Proc. 14th Int. Conf. Scientific and Statistical Database Management, IEEE Computer Society (2002) 7–10
6. Steel, M.A., Penny, D.: Distributions of tree comparison metrics—some new results. Syst. Biol. **42** (1993) 126–141
7. Valiente, G.: Algorithms on Trees and Graphs. Springer-Verlag (2002)
8. Wang, J.T.L., Shan, H., Shasha, D., Piel, W.H.: TreeRank: A similarity measure for nearest neighbor searching in phylogenetic databases. In: Proc. 15th Int. Conf. Scientific and Statistical Database Management, IEEE Computer Society (2003) 171–180

# Practical and Optimal String Matching

Kimmo Fredriksson[1,*] and Szymon Grabowski[2]

[1] Department of Computer Science, University of Joensuu,
PO Box 111, FIN–80101 Joensuu, Finland
kfredrik@cs.joensuu.fi
[2] Technical University of Łódź, Computer Engineering Department,
Al. Politechniki 11, 90–924 Łódź, Poland
sgrabow@kis.p.lodz.pl

**Abstract.** We develop a new exact bit-parallel string matching algorithm, based on the Shift-Or algorithm (Baeza-Yates & Gonnet, 1992). Assuming that the pattern representation fits into a single computer word, this algorithm has optimal $O(n \log_\sigma m/m)$ average running time, as well as optimal $O(n)$ worst case running time, where $n$, $m$ and $\sigma$ are the sizes of the text, the pattern, and the alphabet, respectively. We also study several implementation details. The experimental results show that our algorithm is the fastest in most of the cases where it can be applied, displacing even the long-standing BNDM (Navarro & Raffinot, 2000) family of algorithms. Finally, we show how to adapt our techniques for the Shift-Add algorithm (Baeza-Yates & Gonnet, 1992), obtaining optimal time for searching under Hamming distance.

## 1 Introduction

We address the well known exact string matching problem. The problem is to search the occurrences of the pattern $P[0 \ldots m-1]$ from the text $T[0 \ldots n-1]$, where the symbols of $P$ and $T$ are taken from some finite alphabet $\Sigma$, of size $\sigma$. Numerous efficient algorithms solving the problem have been obtained. The first linear time algorithm (KMP) was given in [8], and the first sublinear average time algorithm (BM) in [2]. Many practical variants of BM family have been suggested, see e.g. [7,13]. An average optimal $O(n \log_\sigma m/m)$ time algorithm (BDM) is obtained e.g. in [4].

Recently bit-parallelism has been shown to lead to the most efficient algorithms for relatively short patterns, in practice. The first algorithm in this class was Shift-Or [1,16], which runs in time $O(n\lceil m/w \rceil)$ time, where $w$ is the number of bits in computer word (typically 32 or 64). Shift-Or is extremely simple to implement, and can be easily adapted to more complex search problems; common properties for most of the bit-parallel algorithms.

Currently, among the fastest algorithms in practice (for $m \le w$) are BNDM [11] and SBNDM [10,12]. BNDM is bit-parallel version of BDM, and SBNDM is a simplified version of BNDM. Their common feature is combining bit-parallelism

---

* Supported by the Academy of Finland, grant 202281.

M. Consens and G. Navarro (Eds.): SPIRE 2005, LNCS 3772, pp. 376–387, 2005.
© Springer-Verlag Berlin Heidelberg 2005

with skipping characters, in the manner of the BM family of algorithms [2]. In SBNDM the shift over the text is reduced, but nevertheless the algorithm is shown to be a bit faster than BNDM in practice. BNDM is optimal on average, but has quadratic worst case complexity. LNDM algorithm [5], which is based on BNDM, has also optimal linear worst case time.

The goal of this paper is to develop an algorithm that has optimal worst and average case complexities (assuming $m = O(w)$), and that in practice performs well on modern CPU architectures. Experimental results show that our algorithm is clearly the fastest in the majority of the cases it can be applied. The same techniques can be adapted to some other algorithms as well, an explicit example being the Shift-Add algorithm [1] for searching under Hamming distance.

## 2    Optimal Shift-Or

We use the following notation. The *pattern* is $P[0 \ldots m - 1]$ and the *text* is $T[0 \ldots n - 1]$. The symbols of $P$ and $T$ are taken from some finite alphabet $\Sigma$, of size $\sigma$. A machine word has $w$ bits, numbered from the least significant bit to the most significant bit. We use C–like notation for the bit-wise operations of words; & is bit-wise **and**, | is **or**, $\wedge$ is **xor**, $\sim$ negates all bits, $<<$ is shift to left, and $>>$ shift to right, both with zero padding. For brevity, we make the assumption that $m \leq w$, unless explicitly stated otherwise.

### 2.1    Standard Shift-Or

The standard Shift-Or automaton is constructed as follows. The automaton has states $0, 1, \ldots, m$. The state 0 is the initial state, state $m$ is the final (accepting) state, and for $i = 0, \ldots, m - 1$ there is a transition from the state $i$ to the state $i + 1$ for character $P[i]$. In addition, there is a transition for every $c \in \Sigma$ from and to the initial state, which makes the automaton nondeterministic.

The preprocessing algorithm builds a table $B$, having one bit-mask entry for each $c \in \Sigma$. For $0 \leq i \leq m - 1$, the mask $B[c]$ has $i$th bit set to 0, iff $P[i] = c$. These correspond to the transitions of the implicit automaton. That is, if the bit $i$ in $B[c]$ is 0, then there is a transition from the state $i$ to the state $i + 1$ with character $c$.

We also need a bit-vector $D$ for the states of the automaton. The $i$th bit of the state vector is set to 0, iff the state $i$ is active. Initially each bit is set to 1. For each text symbol $c$ the vector is updated by $D \leftarrow (D << 1) \mid B[c]$. This simulates all the possible transitions of the nondeterministic automaton in a single step. If after the update the $m$th bit of $d$ is zero, then there is an occurrence of $P$. Alg. 1 gives the code. If $m \leq w$, then the algorithm runs in time $O(n)$.

### 2.2    Average Optimal Shift-Or

We now show how to skip text characters with Shift-Or. Our algorithm takes a parameter $q$, and from the original pattern we generate a set $\mathcal{P}$ of $q$ new patterns $\mathcal{P} = \{P^0, \ldots, P^{q-1}\}$, each of length $m' = \lfloor m/q \rfloor$, as follows:

**Alg. 1.** Shift-Or($T, n, P, m$).

```
1        for i ← 0 to σ − 1 do B[i] ← ~0
2        for i ← 0 to m − 1 do B[P[i]] ← B[P[i]] & ~(1 << i)
3        D ← ~0; mm ← 1 << (m − 1); i ← 0
4        while i < n do
5            D ← (D << 1) | B[T[i]]
6            if (D & mm) ≠ mm then report match
7            i ← i + 1
```

$$P^j[i] = P[j + iq], \quad j = 0 \ldots q - 1, \quad i = 0 \ldots \lfloor m/q \rfloor - 1.$$

In other words, we generate $q$ different alignments of the original pattern $P$, each alignment containing only every $q$th character. The total length of the patterns $P^j$ is $q\lfloor m/q \rfloor \leq m$. For example, if $P = $ abcdef and $q = 3$, then $P^0 = $ ad, $P^1 = $ be and $P^2 = $ cf.

Assume now that $P$ occurs at $T[i..i + m − 1]$. From the definition of $P^j$ it directly follows that

$$P^j[h] = T[i + j + hq], \quad j = i \mod q, \quad h = 0 \ldots m' - 1.$$

This means that we can use the set $\mathcal{P}$ as a filter for the pattern $P$, and that the filter needs only to scan every $q$th character of $T$. Fig. 1 illustrates.

The set of patterns can be searched simultaneously using the Shift-Or algorithm, as long as $qm' \leq w$. All the patterns are preprocessed together, as if they were concatenated. For our example pattern, $P = $ abcdef, we effectively preprocess a pattern $P' = P^0 P^1 P^2 = $ adbecf. Alg. 2 gives the code for preprocessing and filtering algorithms. If the pattern $P^j$ matches, then the $(j + 1)m'$-th bit in $D$ is zero. This is detected with $(D \ \& \ mm) \neq mm$, where $mm$ has every $(j + 1)m'$-th bit set to 1. These bits have also to be cleared in $D$ before the shift operation $(D \ \& \ \sim mm)$, to correctly initialize the first bit corresponding to each of the successive patterns.

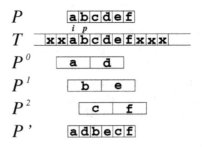

**Fig. 1.** An example. Assume that $P = $ abcdef occurs at text position $T[i \ldots i+m−1]$, and that $q = 3$. The current text position is $p = 10$, and $T[p] = $ b. The next character the algorithm reads is $T[p + q] = T[13] = $ e. This triggers a match of $P^{p \mod q} = P^1$, and the text area $T[p − 1..p − 1 + m − 1] = T[i \ldots i + m − 1]$ is verified.

**Alg. 2.** Average-Optimal-Shift-Or$(T, n, P, m, q)$.

```
1        for i ← 0 to σ − 1 do B[i] ← ∼0
2        h ← 0; mm ← 0
3        for j ← 0 to q − 1 do
4            for i ← 0 to ⌊m/q⌋ − 1 do
5                B[P[iq + j]] ← B[P[iq + j]] & ∼(1 << h)
6                h ← h + 1
7            mm ← mm | (1 << (h − 1))
8        D ← ∼0; i ← 0
9        while i < n do
10           D ← ((D & ∼mm) << 1) | B[T[i]]
11           if (D & mm) ≠ mm then Verify(T, i, n, P, m, q, D, mm)
12           i ← i + q
```

**Alg. 3.** Verify$(T, i, n, P, m, q, D, mm)$.

```
1        D ← (D & mm) ∧ mm
2        while D ≠ 0 do
3            s ← ⌊log₂(D)⌋
4            c ← −(⌊m/q⌋ − 1) q − ⌊s/⌊m/q⌋⌋
5            if P[0 . . . m − 1] = T[i + c . . . i + c + m − 1] then report match
6            D ← D & ∼(1 << s)
```

Whenever an occurrence of $P^j$ is found in the text, we must verify if $P$ also occurs, with the corresponding alignment. To efficiently detect which patterns in $\mathcal{P}$ match, we first set $D \leftarrow (D \mathbin{\&} mm) \wedge mm$, i.e. the $(j + 1)m'$-th bit in $D$ is now one if $P^j$ matches, and all other bits are zero. Now $s \leftarrow \lfloor \log_2(D) \rfloor$ gives the index of the highest bit set in $D$, and therefore $j$ is $\lfloor s/m' \rfloor$, which is our alignment offset, see Fig. 1. The corresponding text position is then verified. Finally, we clear the bit $s$ in $D$. This is repeated until $D$ becomes zero, indicating that there are no more matches. Note that computing $\lfloor \log_2(x) \rfloor$ can be done very efficiently in modern computers, e.g. by casting $x$ to real number, and extracting the exponent from the standardized floating point representation. Alg. 3 gives the verification code.

The filtering time of Alg. 2 is $O(n/q)$. The filter searches the exact matches of $q$ patterns, each of length $\lfloor m/q \rfloor$. Assuming that each character occurs with probability $1/\sigma$, the probability that $P^j$ occurs in a given text position is $(1/\sigma)^{\lfloor m/q \rfloor}$. A brute force verification cost is in the worst case $O(m)$ (but only $O(1)$ on average). To keep the total time at most $O(n/q)$ on average, we select $q$ so that $n/q = mn/\sigma^{m/q}$, i.e. $q = O(m/\log_\sigma m)$. The total average time is therefore $O(n \log_\sigma m/m)$, which is optimal [17].

*Note:* As a historical remark we note that the approximate string matching algorithms in [15,14] have some resemblance to our algorithm. These techniques can be used for exact matching as well, but even for this special case they still differ significantly from our method.

## 2.3   Handling Longer Patterns

If $qm' > w$, we must use more computer words, and the running time must be multiplied by $O(\lceil qm'/w \rceil) = O(\lceil m/w \rceil)$, i.e. the average time becomes $O(n \log_\sigma m/w)$.

However, the trick used in [12] to make BNDM work with $m > w$ can be applied to our algorithms too. The idea is to partition the pattern into $r = \lfloor m/h \rfloor$ consecutive parts. The length of each part is now $h = \lfloor (m-1)/w \rfloor + 1$. All the $h$ characters of each part are then superimposed into a single 'supercharacter'. The resulting $r$ supercharacters are then concatenated to form a single pattern of length $r$. This pattern fits into a single computer word, and it can be searched by reading only every $h$th character of the text. This turns any algorithm, where it is applied to, into a filter, so the potential matches must be verified. See [12] for more details. This technique permits long patterns for the average optimal Shift-Or as well. The result is an algorithm with $O(n \log_{\sigma/h} m/m)$ time on average. This is not optimal any more, but for $\sigma \gg h$ should work quite well.

## 2.4   Linear Worst Case Time

The worst case running time of Alg. 2 is $O(mn)$. However, the verification algorithm is easy to combine with standard Shift-Or, so that the verifications take at most $O(n)$ total time. This is done as follows. Whenever we must verify a pattern occurrence, we do it with Shift-Or. The last text position verified is saved in a variable, as well as the state vector $D$ (for plain Shift-Or). If the next verification area overlaps with the previous, we restore the Shift-Or search state from the previous verification. Otherwise, if the next verification area starts after the previous ended, we reinitialize the Shift-Or search state. The verification algorithm then reads every text character at most once, and therefore the time is at most $O(n)$ (or $O(n\lceil m/w \rceil)$ for long patterns). However, if the verification time becomes an issue, the filter does not work well, and one could use plain Shift-Or just as well.

## 2.5   Implementation

In modern pipelined CPUs branching is costly. In Alg. 1 there are two conditionals in the search code; first to detect the matches, and the second to check the end of the input. A simple way to avoid these to some degree is to unroll the line 5, i.e. repeat the code

$$D \leftarrow (D << 1) \mid B[T[i]]$$

inline several, say $U$, times (with increasing offsets for the variable $i$). This means that the bit $m - 1$ of $D$, indicating on occurrence, will be overflowed due to the repeated shifts, and hence in line 6 we must detect if any of the bits $m - 1..m + U - 1$ is zero. This means that we need $U - 1$ extra bits, and the pattern length is therefore limited to $m \le w - U + 1$.

The second optimization involves detecting the matches. Line 6 in Alg. 1 involves a variable $mm$. This can be avoided if the bit vectors are aligned so

**Alg. 4.** Fast-Shift-Or$(T, n, P, m)$.

```
1      for i ← 0 to σ − 1 do B[i] ← ((1 << m) − 1) << (w − U − m)
2      for i ← 0 to m − 1 do B[P[i]] ← B[P[i]] & ∼(1 << (w − U − m + i))
3      D ← ∼0; mm ← 1 << (m − 1); i ← 0
4      while i < n do
5          for r ← 0 to U − 1 do D ← (D << 1) | B[T[i + r]]
6          if ∼D >> (w − U) ≠ 0 then report matches
7          i ← i + U
```

**Alg. 5.** Fast-Average-Optimal-Shift-Or$(T, n, P, m, q)$.

```
1      for i ← 0 to σ − 1 do B[i] ← ∼0
2      h ← 0; mm ← 0
3      for j ← 0 to q − 1 do
4          for i ← 0 to ⌊m/q⌋ − 1 do
5              B[P[iq + j]] ← B[P[iq + j]] & ∼(1 << h)
6              h ← h + 1
7          for r ← 0 to U − 1 do
8              mm ← mm | (1 << (h − 1))
9              h ← h + 1
10         h ← h − 1
11     D ← ∼mm; i ← 0
12     while i < n do
13         for r ← 0 to U − 1 do D ← (D << 1) | B[T[i + rq]]
14         if (D & mm) ≠ mm then verify(T, i, n, P, m, q, U, D)
15         D ← D & ∼mm
16         i ← i + Uq
```

that the highest bit is in position $w - U + 1$, instead of in position $m + U - 1$. This means that the matches can be detected with $\sim D >> (w - U) \neq 0$, which is efficient if $U$ is constant.

These two simple optimizations (shown in Alg. 4) give about $2 - 5\times$ speed-up for standard Shift-Or (Alg. 1), depending on the architecture. The line 5 in Alg. 4 is automatically inlined by compilers, for small constant $U$. Altough the speed-up is considerable, note that this can depend on the architecture.

Unrolling speeds-up also the Optimal Shift-Or, but the second optimization cannot be applied in this case, since the bit positions indicating the matches are not consecutive. The unrolling technique uses $U - 1$ extra bits per pattern, so we need $q(U - 1 + \lfloor m/q \rfloor)$ bits in total, which is $O(m(U + \log_\sigma m)/\log_\sigma m)$ with the optimal $q$. Alg. 5 gives the code.

Finally, observe that while unrolling is well suited to Shift-Or, the benefits are negligible e.g. for BNDM algorithm, since the more complex control logic cannot be avoided.

## 3   Optimal Shift-Add

Shift-Add [1] is a bit-parallel algorithm for approximate searching under Hamming distance, i.e. it allows at most $k$ mismatches of pattern characters in the occurrences. Shift-Add is very similar to Shift-Or. Shift-Or reserves only one bit per pattern character in the state vector $D$. If some bit is 0 in the vector, it means that the corresponding pattern prefix matches with 0 mismatches the current text position, while bit 1 means that the prefix matches with one or more mismatches. This is possible to extend to allow $k$ mismatches by reserving $\ell = \lceil \log_2(k+1) \rceil + 1$ bits for each character, and replacing the OR operation with addition operation [1].

   More precisely, the $i$th $\ell$-bit field in $B[c]$ is $\ell$ bit binary number 0, if the $i$th character of $P$ matches the character $c$, and 1 otherwise. Then we can accumulate the mismatches as

$$D \leftarrow (D << \ell) + B[T[i]].$$

If the $m$th field of $D$ has a value less than $k+1$, the pattern matches with at most $k$ mismatches. Note that since the pattern length is $m$, the number of mismatches can also be $m$, but we have allocated only $\ell = O(\log_2 k)$ bits for the counters. This means that the counters can overflow. The solution is to store the highest bits of the fields in a separate computer word $o$, and keep the corresponding bits cleared in $D$:

$$D \leftarrow (D << \ell) + B[T[i]]$$
$$o \leftarrow (o << \ell) \mid (D \mathbin{\&} om)$$
$$D \leftarrow D \mathbin{\&} \sim om$$

The bit mask $om$ has bit one in the highest bit position of each $\ell$-bit field, and zeros elsewhere. Note that if $o$ has bit one in some field, the corresponding counter has reached at least value $k+1$, and hence clearing this bit from $D$ does not cause any problems. There is an occurrence of the pattern whenever

$$(D + o) \mathbin{\&} mm < (k+1) << ((m-1)\ell),$$

i.e. when the highest field is less than $k+1$. The bit mask $mm$ selects the $m$th field. Shift-Add clearly works in $O(n)$ time, if $m(\lceil \log_2(k+1) \rceil + 1) \leq w$.

   Our method of skipping text characters with Shift-Or clearly works with Shift-Add as well. The pattern is again splitted to $q$ partitions. If some of our $q$ patterns occur with at most $k$ mismatches, then we verify if the whole pattern occurs with at most $k$ mismatches. Note that this is different from most of the other pattern partitioning based approaches, that partition the pattern into $q$ pieces, and then search the pieces with $\lfloor k/q \rfloor$ errors. This latter approach leads to $O(nk \log_\sigma m/m)$ average time in general, and works for $k = O(m/\log_\sigma m)$. This time is not optimal, whereas our approach leads to $O(n(k + \log_\sigma m)/m)$ optimal average time, see below.

   Adapting the Shift-Add algorithm to multiple patterns requires some modifications on the preprocessing and searching algorithms. The problem is how to detect the matches of several patterns in parallel. This is solved by initializing the counters to $2^{\ell-1} - (k+1)$, instead of to zero. This trick has been used before, e.g. in [3]. This ensures that the overflow bit is activated immediately when the counter reaches a value $k + 1$, and is therefore easy to detect for all patterns in parallel. This could be implemented explicitly, by setting the first field in $D$ of each pattern to this value after the shift operation. Instead, we add $2^{\ell-1} - (k+1)$ to all fields of the $B[c]$ vectors that correspond to the first character of each of the patterns. This ensures that the counters in $D$ get correctly initialized, assuming the first counters of each pattern were zero before the addition. This zeroing is done explicitly with a bit mask. Alg. 6 gives the code.

---

**Alg. 6.** Average-Optimal-Shift-Add$(T, n, P, m, q, k)$.

```
1      ℓ ← ⌈log₂(k + 1)⌉ + 1
2      iv ← 0
3      for i ← 0 to m − 1 do iv ← iv | (1 << (iℓ))
4      for i ← 0 to σ − 1 do B[i] ← iv
5      iv ← (1 << (ℓ − 1)) − (k + 1)
6      h ← 0; mm ← 0; hm ← 0; om ← 0
7      for j ← 0 to q − 1 do
8          for i ← 0 to ⌊m/q⌋ − 1 do
9              B[P[iq + j]] ← B[P[iq + j]] ^ (1 << h)
10             h ← h + ℓ
11         hm ← hm | (((1 << ℓ) − 1) << (h − ℓ))
12         mm ← mm | (1 << (h − 1))
13         iv ← iv | (iv << h)
14     for i ← 0 to σ − 1 do B[i] ← B[i] + iv
15     for i ← 0 to ⌊m/q⌋q − 1 do om ← om | (1 << (((i + 1)ℓ) − 1))
16     D ← 0; o ← om; i ← 0
17     while i < n do
18         D ← (D << ℓ) + B[T[i]]
19         o ← (o << ℓ) | (D & om)
20         D ← D & ∼hm & ∼om
21         if (o & mm) ≠ mm then Verify(T, i, n, P, m, q, k, o, mm)
22         o ← o & ∼hm
23         i ← i + q
```

---

The probability of a match of our $\lfloor m/q \rfloor$ length pattern piece with at most $k$ mismatches is exponentially decreasing if $k/\lfloor m/q \rfloor < 1 - e/\sigma$ [9]. For our $q = O(m/\log_\sigma m)$, this becomes $k/\log_\sigma m < 1 - e/\sigma$. This condition ensures that the probability of a verification is $\gamma^{\lfloor m/q \rfloor}$, where $\gamma < 1$, and hence the number of verifications is negligible, and the total average time is $O(n \log_\sigma m/m)$, which is again optimal. This is good only for reasonably large alphabets and very small $k$, at most $O(\log_\sigma m)$. For larger $k$ one can choose $q = O(m/(k + \log_\sigma m))$, to get again an optimal $O(n(k + \log_\sigma m)/m)$ average time. Linear worst case time (for short patterns) can be obtained in similar way as in the case of Shift-Or. For long patterns all the bounds must be multiplied by $O(m \log_2(k)/w)$.

# 4 Experimental Results (Preliminary)

We have implemented all the algorithms in C, and compiled with icc 8.1. We ran the experiments in 2.4GHz Pentium4 with 512 Mb RAM, 512 Kb cache, running GNU/Linux 2.4.20-8. We also repeated some experiments with 1.28 GHz UltraSPARC IIIi with 16 Gb RAM, 1 Mb cache, running SunOS 5.9. In this case we compiled with the Sun ONE Studio 8 C compiler.

We performed the experiments using random ASCII ($\sigma = 96$, $n = 10$Mb), and several real texts. These are: the E.coli DNA sequence (4,638,690 characters) from Canterbury Corpus[1], real protein data (5,050,292 characters) from TIGR Database (TDB)[2], and natural language text (the collected works of Charles Dickens, 10,192,446 characters), from Silesia Corpus[3]. The patterns were randomly extracted from the texts, and each test was repeated 100 times. We report the average speed in megabytes per second.

## 4.1 Shift-Or Experiments

We compared our algorithms against BNDM [11] and SBNDM [12], implemented by ourselves. These are in practice the fastest general purpose exact string matching algorithms for $m \leq w$. We also compared against the Boyer-Moore-Horspool algorithm [7], and Boyer-Moore-Horspool-Sunday algorithm [13], but these were not competitive, so we do not report the speeds here.

Table 1 gives the speeds in megabyes per second for all the texts. AOSO denotes our Average-Optimal Shift-Or algorithm, and FAOSO the fast variant of it, using the unrolling trick. Note that the speeds for the plain Shift-Or do not depend on the pattern length. For the fast variants, we used unrolling factor $U = 4$, when the representation fitted into a single computer word, otherwise we were forced to use values $1 \ldots 3$.

As it can be seen, our algorithms are clearly the fastest on DNA in all the cases. Interestingly, the fast variant of the plain Shift-Or algorithm beats our average optimal Shift-Or for $m \leq 8$. The results are quite similar for proteins, but for long patterns BNDM variants have equal performance to our algorithms. Note also that for all cases SBNDM is consistently slightly faster than BNDM. Our approach is faster also in natural language text, while on random ASCII the differences are considerably smaller.

Table 2 repeats the experiments on UltraSPARC IIIi for DNA and natural language. FAOSO is again clearly the fastest alternative, but contrary to the results on Pentium4 the plain AOSO is not competitive.

*Note:* Just before submitting this paper we found a recent work [6] that presents several efficient variants of the BNDM algorithm. We ran some preliminary experiments, comparing the best variants against our algorithms. The variants in [6] are in many case faster than SBNDM, but are competitive against us only

---

[1] http://corpus.canterbury.ac.nz/descriptions/

[2] http://www.tigr.org/tdb

[3] http://sun.iinf.polsl.gliwice.pl/~sdeor/corpus.htm

**Table 1.** Searching speed in megabytes per second for different algorithms on Pentium4. Top left: DNA; Top right: proteins. Bottom left: natural language; Bottom right: random ASCII. Shift-Or processes 131 MB/s, 128 MB/s, 128 MB/s and 132 MB/s, and the fast Shift-Or 776 MB/s, 764 MB/s, 817 MB/s and 820 MB/s for DNA, proteins, natural language and random ASCII, respectively.

| $m, q$ | AOSO | FAOSO | BNDM | SBNDM |
|---|---|---|---|---|
| 4, 2 | 321 | **503** | 181 | 210 |
| 8, 2 | 539 | **763** | 312 | 357 |
| 12, 3 | 702 | **941** | 438 | 492 |
| 16, 3 | 1029 | **1229** | 567 | 598 |
| 20, 4 | 1079 | **1341** | 750 | 804 |
| 24, 4 | 1229 | **1525** | 1106 | 1164 |
| 28, 5 | 1427 | **1638** | 1106 | 1164 |

| $m, q$ | AOSO | FAOSO | BNDM | SBNDM |
|---|---|---|---|---|
| 4, 2 | 580 | **909** | 415 | 512 |
| 8, 4 | 944 | **1267** | 642 | 678 |
| 12, 4 | 1120 | **1376** | 816 | 926 |
| 16, 4 | 1120 | **1459** | 963 | 1025 |
| 20, 4 | 1235 | **1376** | 1175 | 1204 |
| 24, 5 | 1267 | **1338** | 1235 | 1302 |
| 28, 6 | 1302 | 1302 | 1302 | 1302 |

| $m, q$ | AOSO | FAOSO | BNDM | SBNDM |
|---|---|---|---|---|
| 4, 2 | 579 | **884** | 368 | 476 |
| 8, 4 | 1034 | **1262** | 685 | 778 |
| 12, 4 | 1144 | **1279** | 797 | 845 |
| 16, 5 | 1200 | **1389** | 831 | 944 |
| 20, 6 | 1279 | **1389** | 1013 | 1092 |

| $m, q$ | AOSO | FAOSO | BNDM | SBNDM |
|---|---|---|---|---|
| 4, 2 | 599 | 952 | 633 | **1053** |
| 8, 4 | 1124 | **1333** | 1064 | 1220 |
| 12, 4 | 1250 | **1389** | 1299 | 1282 |
| 16, 4 | 1351 | **1389** | 1351 | **1389** |
| 20, 6 | 1449 | **1471** | 1370 | 1429 |

**Table 2.** Searching speed in megabytes per second for different algorithms on Ultra-SPARC IIIi. Left: DNA; right: natural language. Shift-Or processes 91 MB/s and 90 MB/s, and the fast Shift-Or 168 MB/s and 165 MB/s for DNA and natural language, respectively.

| $m, q$ | AOSO | FAOSO | BNDM | SBNDM |
|---|---|---|---|---|
| 4, 2 | 70 | **109** | 103 | 92 |
| 8, 2 | 104 | **193** | 146 | 141 |
| 12, 3 | 132 | **227** | 171 | 164 |
| 16, 3 | 135 | **234** | 194 | 192 |
| 20, 4 | 161 | **256** | 207 | 207 |

| $m, q$ | AOSO | FAOSO | BNDM | SBNDM |
|---|---|---|---|---|
| 4, 2 | 104 | **198** | 142 | 147 |
| 8, 4 | 157 | **250** | 193 | 192 |
| 12, 4 | 160 | **256** | 217 | 220 |
| 16, 4 | 175 | **267** | 232 | 233 |
| 20, 6 | 189 | **275** | 244 | 247 |

**Table 3.** Searching speed in megabytes per second for Average-Optimal Shift-Add on Pentium4. Plain Shift-Add processes 203 NB/s for DNA, 197 MB/s for proteins, 204 MB/s for Natural Language and 208 MB/s for random ASCII.

| Alg | AOSA | AOSA | AOSA | AOSA |
|---|---|---|---|---|
| $m$ | $m = 8$ | $m = 12$ | $m = 16$ | $m = 8$ |
| $k$ | $k = 1$ | $k = 1$ | $k = 1$ | $k = 2$ |
| DNA | 173 | 318 | 333 | 172 |
| Proteins | 342 | 468 | 634 | 259 |
| NL (ASCII) | 347 | 552 | 600 | 291 |
| Rnd (ASCII) | 413 | 633 | 741 | 347 |

for large alphabets and reasonably long patterns, while our algorithms still seem to dominate the cases for small alphabets for small to moderate pattern lengths. Their algorithms are entirely different from ours, except that they also apply a form of loop unrolling, but the method is less useful as applied in BNDM (and variants). The reason is that, assuming that they unroll $U$ times, they can shift only after reading $U$ characters, and the maximum shift is reduced to $m - U + 1$. Our algorithms do not have such limitations.

### 4.2   Shift-Add Experiments

Table 3 gives the speeds for the average-optimal Shift-Add. Our character skipping technique clearly speeds-up Shift-Add as well, the exception being short patterns or large $k$ on DNA alphabet. For the lack of time, we compared only against the plain Shift-Add algorithm.

## 5   Conclusions and Future Work

We have presented new bit-parallel filtering algorithms for exact and approximate (under Hamming distance) string matching algorithms. The algorithms have optimal running times on average, and have extremely simple implementations. This makes the algorithms very fast in practice. The simplicity comes from a novel forward matching technique (as opposed to backward matching as in most competing algorithms) and from the fact that the pattern shifts are constant. This also leads to simple unrolling trick that boosts the search in modern hardware. This trick cannot be applied so successfully to more complex backward matching algorithms.

Finally, we note that the techniques presented in this article can be adapted for some other algorithms as well. An example is the $(\delta, \gamma)$-matching algorithm in [3], which runs in $O(n\lceil m(1 + \log_2(\gamma + 1))/w\rceil)$ time. Using our techniques gives us an $O(n\lceil m(1 + \log_2(\gamma + 1))/w\rceil/q)$ time filtering algorithm. Assuming uniform random distribution of characters, we can obtain $O(n \log_2(\gamma) \log_{\sigma/\delta}(m)/w)$ asymptotic average time by selecting $q = O(m/\log_{\sigma/\delta}(m))$.

### Acknowledgements

We thank the anonymous reviewers for many helpful comments.

## References

1. R. A. Baeza-Yates and G. H. Gonnet. A new approach to text searching. *Commun. ACM*, 35(10):74–82, 1992.
2. R. S. Boyer and J. S. Moore. A fast string searching algorithm. *Commun. ACM*, 20(10):762–772, 1977.

3. M. Crochemore, C. Iliopoulos, G. Navarro, Y. Pinzon, and A. Salinger. Bit-parallel $(\delta, \gamma)$-matching suffix automata. *Journal of Discrete Algorithms (JDA)*, 3(2–4):198–214, 2005.
4. M. Crochemore and W. Rytter. *Text algorithms*. Oxford University Press, 1994.
5. L. He and B. Fang. Linear nondeterministic dawg string matching algorithm. In *Proceedings of the 11th International Symposium on String Processing and Information Retrieval (SPIRE'2004)*, LNCS 3246, pages 70–71. Springer–Verlag, 2004.
6. J. Holub and B. Durian. Fast variants of bit parallel approach to suffix automata. Talk given in *The Second Haifa Annual International Stringology Research Workshop of the Israeli Science Foundation*, 2005. http://www.cri.haifa.ac.il/events/2005/string/presentations/Holub.pdf.
7. R. N. Horspool. Practical fast searching in strings. *Softw. Pract. Exp.*, 10(6):501–506, 1980.
8. D. E. Knuth, J. H. Morris, Jr, and V. R. Pratt. Fast pattern matching in strings. *SIAM J. Comput.*, 6(1):323–350, 1977.
9. G. Navarro. A guided tour to approximate string matching. *ACM Computing Surveys*, 33(1):31–88, 2001.
10. G. Navarro. NR-grep: a fast and flexible pattern matching tool. *Softw. Pract. Exp.*, 31:1265–1312, 2001.
11. G. Navarro and M. Raffinot. Fast and flexible string matching by combining bit-parallelism and suffix automata. *ACM Journal of Experimental Algorithmics (JEA)*, 5(4), 2000. http://www.jea.acm.org/2000/NavarroString.
12. H. Peltola and J. Tarhio. Alternative algorithms for bit-parallel string matching. In *Proceedings of the 10th International Symposium on String Processing and Information Retrieval (SPIRE2003)*, LNCS 2857, pages 80–94. Springer–Verlag, 2003.
13. D. M. Sunday. A very fast substring search algorithm. *Commun. ACM*, 33(8):132–142, 1990.
14. E. Sutinen and J. Tarhio. On using $q$-gram locations in approximate string matching. In P. G. Spirakis, editor, *Proceedings 3rd Annual European Symposium*, number 979 in Lecture Notes in Computer Science, pages 327–340, Corfu, Greece, 1995. Springer-Verlag, Berlin.
15. T. Takaoka. Approximate pattern matching with samples. In Ding-Zhu Du and Xiang sun Zhang, editors, *Proceedings of the 5th International Symposium on Algorithms and Computation*, number 834 in Lecture Notes in Computer Science, pages 236–242, Beijing, P.R. China, 1994. Springer-Verlag, Berlin.
16. S. Wu and U. Manber. Fast text searching allowing errors. *Commun. ACM*, 35(10):83–91, 1992.
17. A. C. Yao. The complexity of pattern matching for a random string. *SIAM J. Comput.*, 8(3):368–387, 1979.

# A Bit-Parallel Tree Matching Algorithm for Patterns with Horizontal VLDC's

Hisashi Tsuji[1], Akira Ishino[1], and Masayuki Takeda[1,2]

[1] Department of Informatics, Kyushu University 33, Fukuoka 812-8581, Japan
[2] SORST, Japan Science and Technology Agency (JST)
{h-tsuji, ishino, takeda}@i.kyushu-u.ac.jp

**Abstract.** The tree pattern matching problem is, given two labeled trees $P$ and $T$, respectively called *pattern tree* and *target tree*, to find all occurrences of $P$ within $T$. Many studies have been undertaken on this problem for both the cases of ordered and unordered trees. To realize flexible matching, a kind of variable-length-don't-care's (VLDC's) have been introduced. In particular, the path-VLDC's appear in XPath, a language for addressing parts of an XML document. In this paper, we introduce horizontal VLDC's, each matches a sequence of trees whose root nodes are consecutive siblings in ordered trees. We address the tree pattern matching problem for patterns with horizontal VLDC's. In our setting, the target tree is given as a tagged sequence such as XML data stream. We present an algorithm that solves the problem in $O(mn)$ time using $O(mh)$ space, where $m$ and $n$ are the sizes of $P$ and $T$, respectively, and $h$ is the height of $T$. We adopt the bit-parallel technique to obtain a practically fast algorithm.

## 1 Introduction

Semistructured data, in particular XML documents, has emerged recently and has been widely spread. Tree pattern matching plays a central role in querying such semistructured data.

Let $\mathcal{N}$ and $\Sigma$ be disjoint finite sets of symbols. One abstraction of XML documents would be ordered labeled trees such that the internal nodes are labeled with elements from $\mathcal{N}$ and the leaves are labeled with elements from $\Sigma$.

The problem we addressed in this paper is to find all occurrences of a pattern tree in a target tree, where the target tree is an ordered labeled tree having two kinds of labels as described above, and the pattern tree is almost the same as the target tree except that the labels of leaves are either constant symbols in $\Sigma$ or variable symbols in $\mathcal{V} = \{x_1, x_2, \ldots\}$. Examples of pattern and target trees are displayed in Fig. 1. The pentagons labeled with variables in the pattern tree are "meta" nodes, each of which is replaced with a (possibly empty) sequence of trees.

Since the general case of this problem is NP-hard, we restrict our attention to the class of *regular* pattern trees such that every variable occurs at most once in a pattern tree. We note that the variables in a regular pattern tree act as

M. Consens and G. Navarro (Eds.): SPIRE 2005, LNCS 3772, pp. 388–398, 2005.

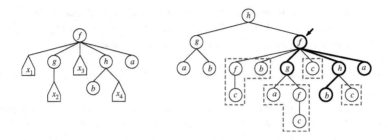

**Fig. 1.** A pattern tree is displayed on the left and a target tree is displayed on the right, where the symbols $f, g, h$ are from $\mathcal{N}$, and the symbols $a, b, c$ are from $\Sigma$. The symbols $x_1, x_2, x_3, x_4$ in the pattern tree are variables from $\mathcal{V}$. The target tree has one occurrence of the pattern tree, the root of which is indicated by an arrow. The variables $x_1, x_2, x_3, x_4$ are respectively replaced with the sequences of trees surrounded with broken lines.

variable-length-don't cares (VLDC's in short), similarly to the string matching case. Zhang, et al. introduced in [20] the notions of *path-VLDC's* and *umbrella-VLDC's*, and the former appears in XPath, a language for addressing parts of an XML document. The path-VLDC's act as VLDC's in the *vertical direction*, from the root to a leaf. On the contrary, the variables of our problem act as VLDC's in the *horizontal direction*, from the left to the right. To our best knowledge, this is the first research dealing with the horizontal VLDC's in tree matching.

It should be stated that Kilpeläinen allowed "logical variables" to be labels of leaves in pattern trees in Chapter 6 of [9]. We note that only a tree is substituted for a variable in his setting, while a sequence of trees is substituted for a variable in our setting. The notion of *internal variables* was introduced by Shoudai, et al. in [18], which are replaced with arbitrary trees.

In this paper we present an online algorithm solving the problem in $O(mn)$ time using $O(mh)$ space, where $m$ and $n$ are the sizes of $P$ and $T$, respectively, and $h$ is the height of $T$. We then adopt the bit-parallel technique to obtain a practically fast algorithm.

## 2   An Overview of Tree Pattern Matching

### 2.1   Various Notions of Occurrence of Pattern Tree

The *ancestor relation* on the nodes of a tree is the reflexive transitive closure of the parent relation. The *left-to-right order* in an ordered tree $T$, denoted by $\preceq_T$, is a partial order on the nodes of $T$ defined as follows: $u \preceq_T v$ if $u = v$ or the lowest common ancestor of $u$ and $v$ has two children $u'$ and $v'$ such that $u'$ and $v'$ are ancestors of $u$ and $v$, respectively, and $u'$ is a left sibling of $v'$.

**Proposition 1.** *For any two distinct nodes $u, v$ of an ordered tree $T$, either of the following statements holds: (1) $u$ is an ancestor of $v$ or vice versa; and (2) $u \preceq_T v$ or vice versa.*

**Definition 1 (occurrence of pattern tree).** *An ordered labeled tree $P$ is said to occur in an ordered labeled tree $T$ if there exists an injection $\varphi$ from the nodes of $P$ to the nodes of $T$ which satisfies the following conditions.*

**(C1)** $\varphi$ *preserves labels*: *For any node $u$ in $P$, the label of $u$ is identical to the label of $\varphi(u)$.*

**(C2)** $\varphi$ *preserves the ancestor relation*: *For any nodes $u, v$ in $P$, $u$ is an ancestor of $v$ in $P$ if and only if $\varphi(u)$ is an ancestor of $\varphi(v)$ in $T$.*

**(C3)** $\varphi$ *preserves the left-to-right order*: *For any nodes $u, v$ in $P$, $u \preceq_P v$ in $P$ if and only if $\varphi(u) \preceq_T \varphi(v)$ in $T$.*

Kilpeläinen and Mannila [10] addresses the tree pattern matching problem with the above notion of pattern occurrence (referred to as the *ordered tree inclusion problem*) and presents an $O(|P||T|)$ time and space algorithm, basing on the dynamic programming technique. The notion might be too general. Replacing the conditions (C2) and/or (C3) with stronger conditions gives restricted notion of occurrence. For example, the condition (C2) can be strengthened as follows.

**(C2')** $\varphi$ *preserves the parent relation*: For any nodes $u, v$ in $P$, $u$ is the parent of $v$ in $P$ if and only if $\varphi(u)$ is the parent of $\varphi(v)$ in $T$.

It is easy to see that (C2') implies (C2). If $\varphi$ satisfies both (C2') and (C3), then siblings in $P$ are mapped to siblings in $T$, and the order of siblings is preserved. That is:

**(C3')** $\varphi$ *preserves the order of siblings*: For any nodes $u, v$ in $P$, $u$ is a left sibling of $v$ in $P$ if and only if $\varphi(u)$ is a left sibling of $\varphi(v)$ in $T$.

We note that (C2') and (C3) hold if and only if (C2') and (C3') hold. In [3] the tree pattern matching for the combination of (C1), (C2'), and (C3') is discussed, and an algorithm is given that runs in $O(|T|\ell(P))$ time after $O(|T| + |P||\Sigma|)$ time and space preprocessing and with $O(|T| + |P||\Sigma|)$ extra space, where $\ell(P)$ denotes the number of leaves in $P$.

The condition (C3') can be strengthened as follows.

**(C3'')** $\varphi$ *preserves the order and adjacency of siblings*: For any nodes $u, v$ in $P$, $u$ is an immediate left sibling of $v$ in $P$ if and only if $\varphi(u)$ is an immediate left sibling of $\varphi(v)$ in $T$.

**(C3''')** $\varphi$ *preserves the numbering of siblings*: For any node $u$ of $P$, $u$ is the $i$-th child of its parent in $P$ if and only if $\varphi(u)$ is the $i$-th child of its parent in $T$.

The notion of occurrence implied by (C1), (C2'), and (C3''') is called *compact occurrence* [3]. In Fig. 2, the pattern occurrence on the left is compact, while the pattern occurrence on the right is not compact. The works [8,11,6,4,5,12,13] are devoted to searching for compact occurrences. Fig. 2 illustrates tree occurrences.

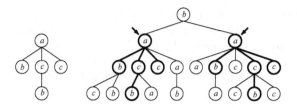

**Fig. 2.** The pattern tree on the left has two occurrences within the target tree on the right, which are indicated by arrows. The left occurrence is compact while the right one is not compact.

## 2.2 VLDC's in Strings and in Trees

For a while we turn to the case of string matching, not tree matching. Let $\star$ be a VLDC that matches any string over an alphabet $\Sigma$. A VLDC pattern is a string over $\Sigma \cup \{\star\}$. For instance, $a\star ba\star c$ is a VLDC pattern that matches any string of the form $aubavc$ with $u, v \in \Sigma^*$. The *substring* pattern matching and the *subsequence* pattern matching are special cases of the VLDC pattern matching in which the patterns are restricted to the form $\star a_1 a_2 \cdots a_k \star$ and to the form $\star a_1 \star a_2 \star \cdots \star a_k \star$, respectively, where $a_i \in \Sigma$ for $i = 1, \ldots, k$ $(k > 0)$.

In the case of tree pattern matching, there can be seen two types of "strings": One is a string of labels spelled out by a path from the root to a leaf (*vertical strings*), and the other is a string of labels spelled out by a left-to-right sequence of siblings (*horizontal strings*). The conditions (C2) and (C2'), respectively, can be regarded as the subsequence matching and the substring matching for "the vertical strings". The conditions (C3'), (C3"), and (C3"'), respectively, can be viewed as the subsequence matching, the substring matching, and the prefix matching for "the horizontal strings". Thus, introducing VLDC's into tree pattern matching in the vertical and in the horizontal directions generalize our problem. As VLDC's for vertical strings, the notion of *path-VLDC's* was introduced in [20]. The path-VLDC's appear in XPath, a language for addressing parts of XML documents, where they are denoted by "//". However, VLDC's for horizontal strings has been not discussed to our best knowledge. In the following section, we introduce horizontal VLDC's into the tree pattern matching.

## 3 Pattern with Horizontal VLDC's

It would be most suitable to define the patterns with horizontal VLDC's as a special case of the *hedges* [17]. The expressions $f(xayg(bc)z)$ and $f(ag(x)y)g(xb)$ $zh(aa)$ are examples of the hedges, where $a, b, c$ are constant symbols, and $x, y, z$ are variable symbols. The hedges resemble the first order terms, but the arities of function symbols are free. The hedges are also called *forests* [19] and *ordered forests* [1] and regarded as data structures suited for representing semistructured data such as XML documents [7,14,15].

From now on, we express hedge $f(axg(b))$ as $[_f x [_g b]_g]_f$. The definition of hedges follows. Let $\Sigma$ be a finite set of constant symbols, and let $\mathcal{V} = \{x_1, x_2, \ldots\}$ be a countable set of variable symbols. Let $\mathcal{N}$ be a set of *names*, and let $B_L = \{ [_f | f \in \mathcal{N}\}$ and $B_R = \{ ]_f | f \in \mathcal{N}\}$, respectively. The elements in $B_L$ (resp. $B_R$) are called the *left brackets* (resp. the *right brackets*). We assume $\Sigma \cap \mathcal{N} = \emptyset$.

**Definition 2.** *The hedges are recursively defined as follows.*

- *The empty string $\varepsilon$ is a hedge.*
- *A constant symbol $c \in \Sigma$ is a hedge.*
- *A variable symbol $x \in \mathcal{V}$ is a hedge.*
- *If $f \in \mathcal{N}$ and $h$ is a hedge, then $[_f h]_f$ is a hedge.*
- *If $h_1$ and $h_2$ are hedges, then the concatenation $h_1 h_2$ is a hedge.*

A hedge is said to be *ground* if it contains no variables. We denote by $\mathcal{H}$ and by $\mathcal{H}_G$ the sets of hedges and ground hedges, respectively. A *substitution* is a mapping from $\mathcal{V}$ to $\mathcal{H}$ specified by

$$x_1 := h_1, \ldots, x_k := h_k \qquad (h_1, \ldots, h_k \in \mathcal{H}).$$

Note that the empty substitution is allowed here. A substitution is naturally extended to the domain $\mathcal{H}$.

**Definition 3** (HEDGEMATCHING). *Given a hedge $p$ and a ground hedge $t$, determine whether there exists a substitution $\theta$ with $p\theta = t$.*

The string version of HEDGEMATCHING in which the input hedges contain no bracket symbols is identical to the membership problem for pattern languages [2], which is known to be NP-complete. Thus HEDGEMATCHING is NP-hard.

A hedge $h$ is *regular* if every variable occurs at most once within $h$. One interesting restriction of HEDGEMATCHING would be REGULARHEDGEMATCHING defined as follows.

**Definition 4** (REGULARHEDGEMATCHING). *Given a regular hedge $p$ and a ground hedge $t$, determine whether there exists a substitution $\theta$ with $p\theta = t$.*

The string version of the above problem is known as the VLDC pattern matching and is solvable in linear time.

We shall consider a variant of the above problem. A hedge $p$ is said to be a *subhedge* of another hedge $t$ if there exist a variable $x$ and a hedge $h$ with $x$ such that $t = h\theta$ for substitution $\theta = \{x := p\}$.

**Definition 5** (REGULARHEDGESEARCHING). *Given a regular hedge $p$ and a ground hedge $t$, find all subhedges $t'$ of $t$ such that $p\theta = t'$ for some substitution $\theta$.*

We concentrate on REGULARHEDGESEARCHING. In the next section we present an algorithm for solving this problem.

## 4    Algorithm for REGULARHEDGESEARCHING

### 4.1    Basic Idea

Consider the regular hedge $P = [_f\, a\, x\, [_g\, b\, ]_g\, ]_f$. If only the strings over $\Sigma$ can be substituted for $x$, the language of $P$ is

$$L = [_f\, a\, \Sigma^*\, [_g\, b\, ]_g\, ]_f.$$

The language $L$ is regular. Fig. 3 shows an NFA accepting the language $\Sigma^* \cdot L$, where the arcs labeled with "$\forall$" denote a state transition by an arbitrary symbol $c \in \Sigma$. We note that in the state-transition diagram the arcs labeled with left brackets are depicted going to the lower-right direction and the arcs labeled with right brackets are depicted going to the upper-right direction. Such hierarchical illustration will be needed in describing our algorithm.

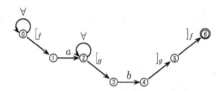

**Fig. 3.** NFA built from the pattern hedge $P = [_f\, a\, x\, [_g\, b\, ]_g\, ]_f$

In reality, arbitrary ground hedges are substituted for variables. We need a mechanism for skipping not only a symbol in $\Sigma$ but also a ground hedge in the form

$$[_n\, h\, ]_n \qquad (n \in \mathcal{N}, h \in \mathcal{H}_G)$$

at a self-loop labeled with "$\forall$" of the NFA.

*Example 1.* Consider the move of the NFA in Fig. 3 running on the ground hedge $T_1 = [_f\, a\, [_g\, a\, ]_g\, [_g\, b\, ]_g\, ]_f$. We want to skip at state 2 the hedge $[_g\, a\, ]_g$, which is a substring beginning at position 3 of $T_1$, so that the NFA accepts $T_1$ after reading the last symbol $]_f$.

*Example 2.* Consider the move of the NFA in Fig. 3 for $T_2 = [_f\, a\, [_f\, [_g\, b\, ]_g\, ]_f\, ]_f$. If we allow the NFA to skip the third symbol $[_f$ of $T_2$ at state 2, then the NFA is in final state 6 just after reading the second symbol $]_f$ from the last. This leads a false detection of the pattern.

### 4.2    The Algorithm

Denote the NFA for a pattern hedge $P$ as mentioned above by

$$M_P = (Q, \Sigma \cup B_L \cup B_R, \delta, Q_0, F)$$

where:

$Q = \{0, 1, \ldots, m\}$ is the set of states;
$\delta : Q \times (\Sigma \cup B_L \cup B_R) \to 2^Q$ is the state transition function;

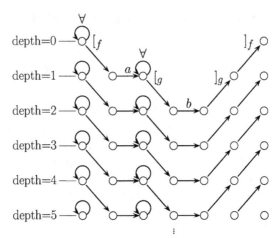

**Fig. 4.** Copies of NFA $M_P$, each finding occurrences of $P = [_f \, a \, x \, [_g \, b \,]_g \,]_f$ at nodes of the corresponding depth in a target tree

$Q_0 = \{0\}$ is the set of initial states; and
$F = \{m\}$ is the set of final states.

Here, $m$ is the number of symbols in the ground hedge obtained from $P$ by removing all variables. For any subset $S$ of $Q$ and any symbol in $\Sigma \cup B_L \cup B_R$, let

$$\delta(S, c) = \bigcup_{q \in S} \delta(q, c).$$

We want to find occurrences of $P$ at *any* position of *any* depth in a target hedge. We create copies of $M_P$ for all possible depths in the target hedge, and simulate the moves of all the copies. See Fig. 4. Naive method would be, for every copy of $M_P$, to store the active states as a set variable and simulate the nondeterministic state transitions by updating the value of the set variable. Since we have to update the set variables for all possible depths for each of the symbols in the target hedge, the simulation is time consuming.

The key idea in overcoming this problem is to parallelize the state transitions of multiple copies of $M_P$ in a *depthwise* manner. Namely, all the states of the copies of $M_P$ are classified into groups according to their depths, and the active states in each group are stored into the corresponding set variable. We use as a stack an array $S$ of sets so that $S[d]$ stores the active states of depth $d$. Initially, set the variable *depth* to 0. The algorithm reads the symbols of the target hedge from the left to the right, and alters the value of *depth* accordingly. That is, it increments *depth* by one when reading a left bracket $[_n$, decrements *depth* by one when reading a right bracket, and do nothing when reading a symbol of $\Sigma$. From the active states in $S[depth]$ for old value of *depth* and from the input symbol, we compute the set of active states for the new value of *depth* by using the state transition function $\delta$ and store them into $S[depth]$ for the new value of *depth*.

---

**Input:**   A ground hedge $T = T[1..n]$ $(T[i] \in \Sigma \cup B_L \cup B_R)$Cand
the NFA $M_P$ for a regular hedge $P$.

**Output:**  All occurrences of $P$ within $T$.

**Method:**

**begin**

    Let *LoopStates* be the set of states having self-loops in $M_P$;

    $depth := 0$; $S[depth] := \emptyset$;

    **for** $j := 1$ **to** $n$ **do**

        $c := T[j]$;

        **if** $c \in B_L$ **then**

            $S[depth + 1] := \delta(S[depth], c) \cup \{0\}$;         $\cdots (1)$

            $depth := depth + 1$;

        **else if** $c \in \Sigma$ **then**

            $S[depth] := \delta(S[depth], c)$;                $\cdots (2)$

        **else** // $c \in B_R$

            $S[depth - 1] := \delta(S[depth], c) \cup (S[depth - 1] \cap LoopStates)$;  $\cdots (3)$

            $depth := depth - 1$;

        **if** $S[depth]$ contains a final state **then**

            Report an occurrence of $P$ at position $j$ of $T$;

**end.**

---

**Fig. 5.** Algorithm for REGULARHEDGESEARCHING

A mechanism for skipping hedges at the states with loops can be realized as follows. Suppose that the current value of *depth* is $d$, and state $s$ has a loop and is active, i.e. $s \in S[d]$. If the next input symbol is a left bracket $[_n$, then the value of $S[d]$ remains without alternation and the algorithm goes to the lower direction with incrementing *depth* by one. The value of $S[d]$ is never changed (and therefore contains $s$) until it returns to the same depth $d$ by reading the corresponding right bracket $]_n$. Among the active states stored in $S[d]$, the ones with loops should remain active.

The algorithm is summarized as in Fig. 5. The move of the algorithm searching $T = [_h [_f a [_g a ]_g [_g b ]_g ]_f ]_h$ for $P = [_f a x [_g b ]_g ]_f$ is displayed in Fig. 6. The trees corresponding to $P$ and $T$ are, respectively, shown on the left and on the right of Fig. 7.

**Theorem 1.** REGULARHEDGESEARCHING *is solved in* $O(mn)$ *time using* $O(mh)$ *space, where* $m$ *and* $n$ *are the sizes of the pattern hedge and the target hedge, respectively, and* $h$ *is the height of the target hedge.*

### 4.3   Efficient Implementation by Bit-Parallel Technique

Now we exploit the bit-parallel technique [16] to obtain an efficient implementation of our algorithm. The set $S[depth] \subseteq Q = \{0, 1, \ldots, m\}$ for every *depth* is

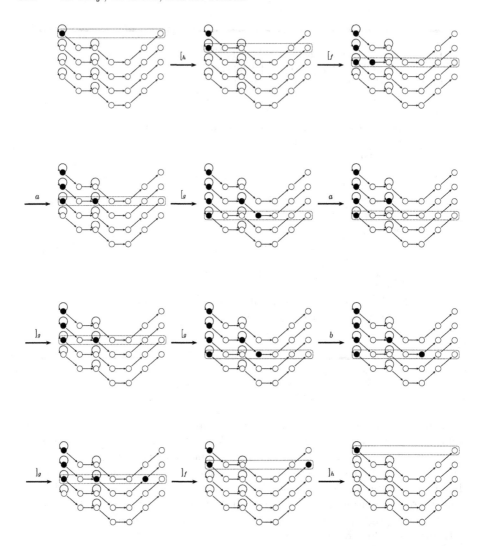

**Fig. 6.** Move of the algorithm, where the pattern hedge is $[_f\, a\, x\, [_g\, b\, ]_g\, ]_f$ and the target hedge is $[_h\, [_f\, a\, [_g\, a\, ]_g\, [_g\, b\, ]_g\, ]_f\, ]_h$. The filled and empty circles mean the active and inactive states, respectively. The rectangle indicates $S[depth]$, the set of active states at the current depth. The states upper than the current depth are sleeping in the stack $S$.

represented with $(m+1)$-bit integer. The state transition function $\delta$ is realized as follows. For each symbol $c$ in $\Sigma \cup B_L \cup B_R$, the mask vector

$$Mask(c) = \{i \mid 1 \le i \le m, \overline{P}[i] = c\}$$

is built, where $\overline{P}$ denotes the ground hedge obtained by removing variables from the given pattern $P$. The set $LoopStates$ is also represented as an $(m+1)$-bit

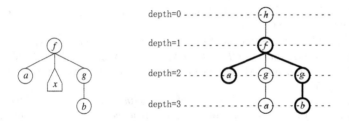

**Fig. 7.** Trees described by the regular hedge $P = [_f\, a\, x\, [_g\, b\,]_g\,]_f$ (on the left) and by the ground hedge $T = [_h\, [_f\, a\, [_g\, a\,]_g\, [_g\, b\,]_g\,]_f\,]_h$ (on the right). An occurrence of $P$ in $T$ is emphasized with thick lines.

integer. Then, for any $S \subseteq Q$ and for any $c \in \Sigma \cup B_L \cup B_R$, The set of the next states $\delta(S, c)$ can be obtained as

$$((S << 1)\&Mask(c))|(S\&LoopStates),$$

where $<<\&$$C\&C|$, respectively, means the bit-shift, the bitwise AND, and the bitwise OR operations. The computation of $\delta(S[depth], c)$ at lines (1),(2),(3) of the algorithm of Fig. 5 can be replaced with a combination of the bit-shift and the bitwise logical operations as mentioned above. The running time of the algorithm is then $O(\lceil m/w \rceil n)$, and is $O(n)$ if $m + 1$ is at most the length $w$ of computer word.

## 5   Conclusion

We addressed the problem of finding a regular hedge $P$ within a ground hedge $T$, and presented an efficient algorithm basing on the bit-parallelism. The regular hedges can be regarded as tree patterns with horizontal VLDC's. We note that Chauve [3] deals with tree pattern matching with a notion of occurrences in which the parent relation and the order and adjacency of siblings are preserved. The problem addressed in this paper is a generalization of the one addressed in [3].

## References

1. P. C. Amoth, T. R. and P. Tadepalli. Exact learning of tree patterns from queries and counterexamples. In *Proceedings of COLT '98*, pages 175–186, 1998.
2. D. Angluin. Finding patterns common to a set of strings. *J. Comput. Sys. Sci.*, 21:46–62, 1980.
3. C. Chauve. Tree pattern matching with a more general notion of occurrence of the pattern. *Inform. Process. Lett.*, 82:197–201, 2002.
4. R. Cole and R. Hariharan. Tree pattern matching and subset matching in randomized $O(n \log^3 n)$-time. In *STOC'97*, pages 66–75, 1997.
5. R. Cole, R. Hariharan, and P. Indyk. Tree pattern matching and subset matching in deterministic $O(n \log n)$-time. In *SODA'99*, pages 245–254, 1999.

6. M. Dubliner, Z. Galil, and E. Magen. Faster tree pattern matching. *J. ACM*, 41(2):205–213, 1994.
7. F. N. Geert Jan Bex, Sebastian Maneth. A formal model for an expressive fragment of xslt. In *Proceedings of CL 2000 (LNAI 1861)*, pages 1137–1151, 2000.
8. C. M. Hoffmann and M. J. O'Donnell. Pattern matching in trees. *J. ACM*, 29(1):68–95, 1982.
9. P. Kilpeläinen. *Tree Matching Problems with Applications to Structured Text Databases*. PhD thesis, Dept. of Computer Science, University of Helsinki, 1992.
10. P. Kilpeläinen and H. Mannila. Ordered and unordered tree inclusion. *SIAM J. Comput.*, 24(2):340–356, 1995.
11. S. R. Kosaraju. Efficient tree pattern matching. In *FOCS'89*, pages 178–183. IEEE Comput. Soc. Press, 1989.
12. F. Luccio and L. Pagli. An efficient algorithm for some tree matching problems. *Inform. Process. Lett.*, 39(1):51–57, 1991.
13. F. Luccio and L. Pagli. Approximate matching for two families of trees. *Information and Computation*, 123(1):111–120, 1995.
14. M. Murata. Transformation of documents and schemas by patterns and contextual conditions. In *Proceedings of Document Processing '96 (LNCS 1293)*, pages 153–169, 1997.
15. M. Murata. Data model for document transformation and assembly (extended abstract). pages 140–152, 1998.
16. G. Navarro and M. Raffinot. *Flexible pattern matching in strings: Practical on-line search algorithms for texts and biological sequences*. Cambridge University Press, Cambridge, 2002.
17. M. Nivat and H. Ait-Kaci. On recognizable sets and tree automata. In *Resolution of Equations in Algebraic Structres*. 1989.
18. T. Shoudai, T. Uchida, and T. Miyahara. Polynomial time algorithms for finding unordered tree patterns with internal variables. In *FCT'01*, pages 335–346, 2001.
19. M. Takahashi. Generalizations of regular sets and their application to a study of context-free languages. *Information and Control*, 27:1–36, 1975.
20. K. Zhang, D. Shasha, and J. T. L. Wang. Approximate tree matching in the presence of variable length don't cares. *J. Algorithms*, 16(1):33–66, 1994.

# A Partition-Based Efficient Algorithm
# for Large Scale Multiple-Strings Matching

Ping Liu, Yan-bing Liu, and Jian-long Tan*

Software Division, Institute of Computing Technology,
Chinese Academy of Sciences, Beijing 100080
{liuping, tjl}@ict.ac.cn, liuyanbing@software.ict.ac.cn

**Abstract.** Filtering plays an important role in the Internet security and
information retrieval fields, and usually employs multiple-strings matching algorithm as its key part. All the classical matching algorithms, however, perform badly when the number of the keywords exceeds a critical
point, which made large scale multiple-strings matching problem a great
challenge. Based on the observation that the speed of the classical algorithms depends mainly on the length of the shortest keyword, a partition
strategy was proposed to decompose the keywords set into a series of subsets on which the classical algorithms was performed. For the optimal
partition, it was proved that the keywords with same length locate in one
subset, and length of keywords in different subsets would not interlace
each other. In this paper, we proposed a shortest-path model for the optimal partition finding problem. Experiments on both random and real
data demonstrate that our algorithms generally has about a 100-300%
speed-up compared with the classical ones.

## 1   Introduction

With the development of Internet, more and more information, including bad
along with good, emerged and congested the network. To secure Internet and
retrieve useful information, filtering systems were designed and deployed on the
gateways to filter out bad things. A filtering system usually employs a string
matching procedure as its key part, and always contains a large scale keywords
set to suit to various focuses. Hence, it is really a great challenge to design an
efficient multi-strings matching algorithm for a large scale keywords set.

String matching problem has been received extensive research, most of which
follow a common procedure, i.e., compare keywords with substring of text within
a fixed length window, and then shift the window from left to right as far as
possible. In [1],according to the way that patterns are compared with the text
in the window, string matching algorithms were categorized into three classes:
prefix searching[2][3][4], suffix searching[5][6][7] and factor searching[8][9][10][11].

The performance of the classical multi-strings matching algorithms are determined mainly by the following three factors: the number of the keywords, the

---

* Corresponding Author.

M. Consens and G. Navarro (Eds.): SPIRE 2005, LNCS 3772, pp. 399–404, 2005.

minimal length of the keywords and the size of the alphabet[1]. In addition, the distribution of keywords in the text would also affect the performance.

However, all of the classical algorithms are inapplicable for a large scale keywords set. Experiment on real data demonstrated that these algorithms perform badly when the number of patterns exceeds a critical point, e.g, 5,000 on a Pentium III CPU (the minimum length of the patterns is 4 bytes). In this paper, we proposed a partition-based strategy to bound the influence of the shortest keywords on the performance, and designed a shortest-path model to work out the optimal partition.

## 2     Properties About Speed of Classical Algorithms

In this section, we analyze the average-case time complexity of three representative multi-strings matching algorithms: SBOM, WuManber and Advanced Aho Corasick. Let use $\Sigma$ to denote the alphabet, $n$ to denote the length of the text, $r$ to denote the number of the pattern, and $b$ to denote the block size in WuManber algorithm. Let us denote the minimum length of a keywords set $S$ as $m(S)$ or $m$, and the maximal one $M(S)$ or $M$.

In [13], the lower bound is given for the average time complexity of the exact multiple strings matching. Assuming that the text and keywords are uniformly and independently chosen from the alphabet $\Sigma$, and $n$ is large enough, a rough estimation of the average-case time complexity of these algorithms is given as follows: **Advanced Aho Corasick algorithm** is $O(n)$, **WuManber algorithm** is $O\left(\frac{n}{(m-b+1)*(1-\frac{(m-b+1)*r}{2*|\Sigma|^b})}\right)$, and **SBOM algorithm** is $O\left(\frac{n*\log_{|\Sigma|} mr}{m-\log_{|\Sigma|} mr}\right)$.

The above analysis implies the following properties about the speed of the classical algorithms, which are confirmed by experimental result on random data:

**Property 1.** The main factors affecting the speed of multi-strings matching algorithms are the size of alphabet, the number of the patterns and the minimum length of the patterns. Hence, the matching time can be denoted as $T(m,r)$ if the size of alphabet is fixed.

**Property 2.** The matching time of a multi-strings matching algorithm increases monotonously when the number of the patterns increase, i.e., $\frac{\partial}{\partial r}T(m,r) > 0$.

**Property 3.** The matching time of a multi-strings matching algorithms decreases monotonously when the minimum length increase, i.e., $\frac{\partial}{\partial m}T(m,r) < 0$.

**Property 4.** The increase rate of the matching time with the number of patterns is independent of the number of the patterns, and decrease when the minimum length increase, i.e., $\frac{\partial}{\partial r}T(m,r) = H(m) > 0$ and $\frac{d}{dm}H(m) < 0$.

## 3     A Partition-Based Matching Algorithm

The shortest keywords, though very small in quantity, have a great negative influence on the matching time. To bound their influence, an intuitional idea is to decompose the keywords set into a series of smaller subsets. According to [1], the most efficient algorithm varies with the pattern number, the minimal length and the size of the alphabet, choosing the most efficient algorithm for each subset would gain another benefit compared with running only the same algorithm on them. Since the influence of the shortest keywords is bounded in smaller subsets rather than the entire one, the sum of matching time on individual subsets is even smaller than the time costed to run on an entire set directly.

For a given keywords set $P = \{p_1, p_2, \cdots, p_n\}$, there are many kinds of feasible partitions, among which the optimal one achieves the minimal matching time. Here, we assumed that the keywords were already sorted according to its length, i.e.,$|p_1| \leq |p_2| \leq \cdots \leq |p_n|$. Then the optimal partition finding problem can be defined as follows:

**Optimal Partition Finding Problem.** Given a sorted keywords multiset $P = \{p_1, p_2, \cdots, p_n\}$, to construct a partition $S_1, S_2, \cdots, S_k$, so that $\bigcup_{i=1}^{k} S_i = P$, $S_i \cap S_j = \emptyset$ ($\forall i, j, 1 \leq i, j \leq k$ and $i \neq j$), and $\sum_{i=1}^{k} T(m(S_i), |S_i|)$ is minimized.

### 3.1   Properties About Optimal Partition

In the following, two properties about the optimal partition are proved, forming a solid foundation to find it.

**Theorem 1.** *There exists an optimal partition, $S_1, S_2, \cdots, S_k$, of the sorted keywords set $P = \{p_1, p_2, \cdots, p_n\}$, and for $\forall i \neq j$, either $\forall a \in S_i, b \in S_j, |a| \leq |b|$; or $\forall a \in S_i, b \in S_j, |a| \geq |b|$.*

*(Proof is omitted for limited space.)*

**Theorem 2.** *In the optimal partition of the sorted keywords set $P = \{p_1, p_2, \cdots p_n\}$, keywords with the same length would not locate in different subsets.*

*(Proof is omitted for limited space.)*

The above two properties imply that the keywords with same length work as a *block*, that is, they would not separate in an optimal partition. Moreover, a subset $S_i$ of an optimal partition contains all the *blocks* with length in the interval $[m(S_i), M(S_i)]$.

### 3.2   Algorithm to Find the Optimal Partition

In this section, we model the optimal partition problem into finding the shortest-path problem in a weighted graph. Given a sorted keywords set $P = \{p_1, p_2, \cdots, p_n\}$, we create a *partition graph* $G$ as follows. For each a *block* with length $i$ in $P$, a node $N_i$ is created to represent it, and an auxiliary node $N_{M(P)+1}$ is created to represent the end of $P$. Let $V = \{N_{m(P)}, N_{m(P)+1}, \cdots, N_{M(P)}, N_{M(P)+1}\}$. The edges of $G$ is specified as follows. For $N_i$ and $N_j \in V$, there is an edge from $N_i$ to $N_j$, denoted as $(N_i, N_j)$ if $i < j$. In fact, an edge $(N_i, N_j)$ is used to represent a

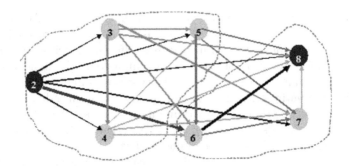

**Fig. 1.** A sketch of finding the optimal partition using the shortest path method

set of *blocks* with length greater than or equal with $i$, but less than $j$. We define $N_m(P)$ as the *source*, and $N_{M(P)+1}$ the *sink*.

For each edge $(N_i, N_j)$, a weight $W(N_i, N_j)$ was assigned to measure the benefit of setting the corresponding *blocks set* as a subset. The minimal time of the three classical algorithms matching a *training text* with the subset was used as an estimation of $W(N_i, N_j)$. Therefore, the optimal partition correspond to the shortest path from *source* to *sink* in the *partition graph G*.

An example is shown in Fig.1. The shortest path in the graph is "$2 \rightarrow 6 \rightarrow 8$", hence the optimal partition has two subset, one containing keywords with length 2,3,4,5, and the other having keywords with length 6,7.

The partition-based multi-strings matching algorithm is given as follows:

---

### PBM: Partition-Based Matching Algorithm

**Input:** A sorted keywords set $P$, a training text $T$, a text to be matched $J$ ;

**Output:** The occurrence of each keyword in $J$;

1. Construct the *partition graph* $G =< V, E >$, here,
    $V = \{N_{m(P)}, N_{m(P)+1}, \cdots, N_{M(P)}, N_{M(P)+1}\}$;
    $E = \{(N_i, N_j)|N_i, N_j \in V, i < j\}$. $W(N_i, N_j)$ is set as above;
2. Finding the shortest path $(e_{i_1}, e_{i_2}, \cdots, e_{i_k})$ from $N_{m(P)}$ to $N_{M(P)+1}$.
    Here, $e_{i_j}$ is an edge in $G$;
3. For each $e_{i_j}$, output a subset containing the corresponding *blocks*, together with the fastest algorithm for this subset.
4. For each subset, running the chosen algorithm in step 3 on it to filter text $J$.

---

## 4   Experimental Result

We implemented the partition-based algorithm, PBM, into a C++ program, and tested it on both random and real data. Here, only experimental studies on real data set are reported.

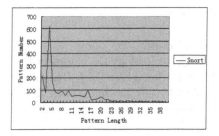

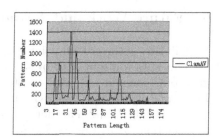

**Fig. 2.** Distribution of pattern lengths in Snort dataset. (pattern number : 2086, pattern length range: 2-40)

**Fig. 3.** Distribution of pattern length in CLAMAV dataset. (pattern number : 26,653, pattern length range : 3-210).

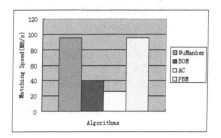

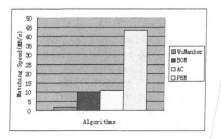

**Fig. 4.** Comparision of matching speed on SNORT dataset

**Fig. 5.** Comparision of matching speed on CLAMAV dataset

Two real data sets were acquired from Internet security field for test, oɪ from Snort, an Intrusion Detection System, the other from the signatures ⟨ ClamAV, an antivirus system. The Snort data contains 2,086 keywords, whi ClamAV has 26,653.

A group in MIT published a dataset for evaluating the performance of IDՏ (Downloadable from http://www.ll.mit.edu/IST/ideval/) We extract part of th dataset, i.e., mit_1999_training_week1_Friday_inside.dat as the *training text* iɪ PBM, and use the full 64MB text as the testing text.

Fig.2 and Fig.3 show the distribution of the length of keywords, and matchinɡ time are showed in Fig.4 and Fig.5, respectively.

On Snort keywords set, only one partition was generated with Wumanber as its most suitable algorithm, which means that PBM would not be faster than the classical algorithms on some special pattern sets, and meanwhile, would not be slower than others.

The advantage of PBM is much more obvious on the ClamAV keywords set. When the length range of keywords is larger and the length distribution is not uniform, partition strategy would be more advantageous, showing its advantage in large scale multi-strings matching.

# 5   Future Work and Acknowledgment

We will survey the critical point deeply in future work.

This work was supported by the National High Technology Development Program of China under Grant No. 2005AA142110, National Key Basic Research & Development Program under Grant No. 2004CB318109. We thank Shen Xingxing, Zhang Ji, Wang Ying for valuable discussion. We also express our deep appreciation to Dr. Dongbo BU for his great help.

## References

1. Gonzalo Navarro and Mathieu Raffinot , "Flexible Pattern Matching in StringsPractical on-line search algorithms for texts and biological sequences", *Camedge University Press,2002*,ISBN 0–521–81307-7. pp15-17,74–76
2. D.E.Knuth,J.H.Morris,V.R.Pratt, "Fast Pattern Matching in Strings",*SIAM Journal on Computing*,Page 323–350,1977
   A.V. Aho and M.J.Corasick, "Efficient string matching:an aid to bibliographic earch", *Communication of the ACM*,18(6):333–340 ,1975
   Wu, U. Manber, "Fast text searching allowing errors",*Communications of the ?M*, 35(10): 83–91 ,1992
   ,Boyer, J.S.Moore, "A fast string searching algorithm",*Communications of the ?*,20(10):762–772 ,1977
   nmentz-Walter, "A string matching algorithm fast on the average", *In Pro-? s of the 6th International Colloquium on Automata, Language and Pro-?ng , number 71 in Lecture Notes in Computer Science*,pages 118–132,1979
   .Manber, "A fast algorithm for multi-pattern searching", *Report TR–94–17 ?ent of Computer Science , University of Arizona,Tucson, AZ ,1994*
   M.Crochemore,    A.Czumaj,    L.Gasienniec,    S.Jarominek,    T.Lecroq, Plandowski, W.Rytter, "Speeding up two string matching algorithms", *hmica*,12(4/5):247–267,1994
   uzen, M.Crochemore, M.Raffinot, "Efficient experimental string matching ak factor recognition", *In proceedings of the 12th Annual Symposium on inatorial Pattern Matching, number 2089 in Lecture Notes in Computer ce*,pages51–72. Springer-Verlag,2001
   lmer, J.Blumer, A.Ehrenfeucht, D.Haussler, R.McConnel. "Complete inverted or efficient text retrieval and analysis", *Jonual of the ACM*,34(3):578–595,1987
   lauzen, M.Raffinot "Factor oracle of a set of words",*Technical report 99–11, tute Gaspard-Monge, University de Marne-la-vallee, 1999*
   ,dong Wang "The Design and Analysis of Computer Algorithms" *Publishing se of Electronic Industry, Beijing. 2001*ISBN7–5053–6391–3P38–81
   lzalo Navarro, Kimmo Fredriksson. "Average Complexity of Exact and Apximate Multiple String Matching",*Theoretical Computer Science (TCS) 321(2-283-290, 2004*

# Author Index

Allali, Julien, 348
Amir, Amihood, 67, 315
Angelov, Stanislav, 167
Ardila, Yoan José Pinzón, 191, 234
Askitis, Nikolas, 91

Backofen, Rolf, 360
Baeza-Yates, Ricardo, 13
Bergroth, Lasse, 301
Bialynicka-Birula, Iwona, 79
Boldi, Paolo, 25
Boughanem, Mohand, 271
Boyer, Frederic, 179
Brini, Asma H., 271

Camacho-Guerrero, José Antonio, 45
Castillo, José Raúl Fernández del, 228
Chauve, Cedric, 335
Chávez, Edgar, 127
Cigarran, Juan M., 49
Clifford, Raphaël, 234
Culpepper, J. Shane, 1

Dubois, Didier, 271
Dupret, Georges, 41

Elkan, Charles, 295

Feng, Yi, 155
Fredriksson, Kimmo, 267, 376

Geva, Shlomo, 29
Gonzalo, Julio, 49
Grabowski, Szymon, 376
Grossi, Roberto, 79
Guignon, Valentin, 335

Hamel, Sylvie, 335
Hermelin, Danny, 360
Hilera, José Ramón, 228
Hyyrö, Heikki, 256

Imafouo, Amélie, 224
Inenaga, Shunsuke, 167
Ishino, Akira, 388

Joy, Mike, 267

Karlgren, Jussi, 151
Kechagias, Dimitrios, 161
Kim, Jin Wook, 315
Kondrak, Grzegorz, 115
Kong, Zhigang, 218
Kopelowitz, Tsvi, 67

Lalmas, Mounia, 218
Landau, Gad M., 315, 360
Lee, Inbok, 191
Lewenstein, Moshe, 67
Lewenstein, Noa, 67
Lipsky, Ohad, 327, 331
Liu, Ping, 399
Liu, Yan-bing, 399
Lloyd, Levon, 161

Macedo, Alessandra Alaniz, 45
Madariaga, Ricardo Sánchez de, 228
Mendoza, Marcelo, 41
Moffat, Alistair, 1
Mohamed, Manal, 234
Mollá, Diego, 139
Moura, Edleno Silva de, 202
Mozgovoy, Maxim, 267

Nwesri, Abdusalam F.A., 206

Oliveira, Arlindo L., 246
Oliveira, Luciene C., 283

Paredes, Rodrigo, 127
Park, Kunsoo, 315
Peñas, Anselmo, 49
Peres, Patrícia Silva, 202
Peterlongo, Pierre, 179
Pimentel, Maria da Graça Campos, 45
Pisanti, Nadia, 179
Pizzato, Luiz Augusto, 139
Porat, Ely, 327, 331

Russo, Luís M.S., 246

Sagot, Marie-France, 179, 348

Sahlgren, Magnus, 151
Salinger, Alejandro, 13
Scholer, Falk, 206
Schürmann, Klaus-Bernd, 55
Silva, Ilmério R., 283
Skala, Matthew, 103
Skiena, Steven, 161
Souza, João N., 283
Stoye, Jens, 55
Sutinen, Erkki, 267

Tahaghoghi, S.M.M., 206
Takeda, Masayuki, 388
Tan, Jian-long, 399

Tannier, Xavier, 29, 224
Tsuji, Hisashi, 388

Valiente, Gabriel, 370
Verdejo, Felisa, 49
Vigna, Sebastiano, 25

Weimann, Oren, 360
White, Daniel, 267
Wu, Zhaohui, 155

Zaanen, Menno van, 139
Zhou, Zhongmei, 155
Zobel, Justin, 91

# Lecture Notes in Computer Science

For information about Vols. 1–3671

please contact your bookseller or Springer

Vol. 3781: S.Z. Li, Z. Sun, T. Tan, S. Pankanti, G. Chollet, D. Zhang (Eds.), Advances in Biometric Person Authentication. XI, 250 pages. 2005.

Vol. 3777: O.B. Lupanov, O.M. Kasim-Zade, A.V. Chaskin, K. Steinhöfel (Eds.), Stochastic Algorithms: Foundations and Applications. VIII, 239 pages. 2005.

Vol. 3775: J. Schoenwaelder, J. Serrat (Eds.), Ambient Networks. XIII, 281 pages. 2005.

Vol. 3772: M. Consens, G. Navarro (Eds.), String Processing and Information Retrieval. XIV, 406 pages. 2005.

Vol. 3770: J. Akoka, S.W. Liddle, I.-Y. Song, M. Bertolotto, I. Comyn-Wattiau, W.-J.v.d. Heuvel, M. Kolp, J. Trujillo, C. Kop, H.C. Mayr (Eds.), Perspectives in Conceptual Modeling. XXII, 476 pages. 2005.

Vol. 3766: N. Sebe, M.S. Lew, T.S. Huang (Eds.), Computer Vision in Human-Computer Interaction. X, 231 pages. 2005.

Vol. 3765: Y. Liu, T. Jiang, C. Zhang (Eds.), Computer Vision for Biomedical Image Applications. X, 563 pages. 2005.

Vol. 3756: J. Cao, W. Nejdl, M. Xu (Eds.), Advanced Parallel Processing Technologies. XIV, 526 pages. 2005.

Vol. 3754: J. Dalmau Royo, G. Hasegawa (Eds.), Management of Multimedia Networks and Services. XII, 384 pages. 2005.

Vol. 3752: N. Paragios, O. Faugeras, T. Chan, C. Schnoerr (Eds.), Variational, Geometric, and Level Set Methods in Computer Vision. XI, 369 pages. 2005.

Vol. 3751: T. Magedanz, E.R. M. Madeira, P. Dini (Eds.), Operations and Management in IP-Based Networks. X, 213 pages. 2005.

Vol. 3750: J.S. Duncan, G. Gerig (Eds.), Medical Image Computing and Computer-Assisted Intervention – MICCAI 2005, Part II. XL, 1018 pages. 2005.

Vol. 3749: J.S. Duncan, G. Gerig (Eds.), Medical Image Computing and Computer-Assisted Intervention – MICCAI 2005, Part I. XXXIX, 942 pages. 2005.

Vol. 3747: C.A. Maziero, J.G. Silva, A.M.S. Andrade, F.M.d. Assis Silva (Eds.), Dependable Computing. XV, 267 pages. 2005.

Vol. 3746: P. Bozanis, E.N. Houstis (Eds.), Advances in Informatics. XIX, 879 pages. 2005.

Vol. 3745: J.L. Oliveira, V. Maojo, F. Martin-Sanchez, A.S. Pereira (Eds.), Biological and Medical Data Analysis. XII, 422 pages. 2005. (Subseries LNBI).

Vol. 3744: T. Magedanz, A. Karmouch, S. Pierre, I. Venieris (Eds.), Mobility Aware Technologies and Applications. XIV, 418 pages. 2005.

Vol. 3740: T. Srikanthan, J. Xue, C.-H. Chang (Eds.), Advances in Computer Systems Architecture. XVII, 833 pages. 2005.

Vol. 3739: W. Fan, Z. Wu, J. Yang (Eds.), Advances in Web-Age Information Management. XXIV, 930 pages. 2005.

Vol. 3738: V.R. Syrotiuk, E. Chávez (Eds.), Ad-Hoc, Mobile, and Wireless Networks. XI, 360 pages. 2005.

Vol. 3735: A. Hoffmann, H. Motoda, T. Scheffer (Eds.), Discovery Science. XVI, 400 pages. 2005. (Subseries LNAI).

Vol. 3734: S. Jain, H.U. Simon, E. Tomita (Eds.), Algorithmic Learning Theory. XII, 490 pages. 2005. (Subseries LNAI).

Vol. 3733: P. Yolum, T. Güngör, F. Gürgen, C. Özturan (Eds.), Computer and Information Sciences - ISCIS 2005. XXI, 973 pages. 2005.

Vol. 3731: F. Wang (Ed.), Formal Techniques for Networked and Distributed Systems - FORTE 2005. XII, 558 pages. 2005.

Vol. 3728: V. Paliouras, J. Vounckx, D. Verkest (Eds.), Integrated Circuit and System Design. XV, 753 pages. 2005.

Vol. 3726: L.T. Yang, O.F. Rana, B. Di Martino, J. Dongarra (Eds.), High Performance Computing and Communcations. XXVI, 1116 pages. 2005.

Vol. 3725: D. Borrione, W. Paul (Eds.), Correct Hardware Design and Verification Methods. XII, 412 pages. 2005.

Vol. 3724: P. Fraigniaud (Ed.), Distributed Computing. XIV, 520 pages. 2005.

Vol. 3723: W. Zhao, S. Gong, X. Tang (Eds.), Analysis and Modelling of Faces and Gestures. XI, 4234 pages. 2005.

Vol. 3722: D. Van Hung, M. Wirsing (Eds.), Theoretical Aspects of Computing – ICTAC 2005. XIV, 614 pages. 2005.

Vol. 3721: A. Jorge, L. Torgo, P. Brazdil, R. Camacho, J. Gama (Eds.), Knowledge Discovery in Databases: PKDD 2005. XXIII, 719 pages. 2005. (Subseries LNAI).

Vol. 3720: J. Gama, R. Camacho, P. Brazdil, A. Jorge, L. Torgo (Eds.), Machine Learning: ECML 2005. XXIII, 769 pages. 2005. (Subseries LNAI).

Vol. 3719: M. Hobbs, A.M. Goscinski, W. Zhou (Eds.), Distributed and Parallel Computing. XI, 448 pages. 2005.

Vol. 3718: V.G. Ganzha, E.W. Mayr, E.V. Vorozhtsov (Eds.), Computer Algebra in Scientific Computing. XII, 502 pages. 2005.

Vol. 3717: B. Gramlich (Ed.), Frontiers of Combining Systems. X, 321 pages. 2005. (Subseries LNAI).

Vol. 3716: L. Delcambre, C. Kop, H.C. Mayr, J. Mylopoulos, O. Pastor (Eds.), Conceptual Modeling – ER 2005. XVI, 498 pages. 2005.

Vol. 3715: E. Dawson, S. Vaudenay (Eds.), Progress in Cryptology – Mycrypt 2005. XI, 329 pages. 2005.

Vol. 3714: H. Obbink, K. Pohl (Eds.), Software Product Lines. XIII, 235 pages. 2005.

Vol. 3713: L. Briand, C. Williams (Eds.), Model Driven Engineering Languages and Systems. XV, 722 pages. 2005.

Vol. 3712: R. Reussner, J. Mayer, J.A. Stafford, S. Overhage, S. Becker, P.J. Schroeder (Eds.), Quality of Software Architectures and Software Quality. XIII, 289 pages. 2005.

Vol. 3711: F. Kishino, Y. Kitamura, H. Kato, N. Nagata (Eds.), Entertainment Computing - ICEC 2005. XXIV, 540 pages. 2005.

Vol. 3710: M. Barni, I. Cox, T. Kalker, H.J. Kim (Eds.), Digital Watermarking. XII, 485 pages. 2005.

Vol. 3709: P. van Beek (Ed.), Principles and Practice of Constraint Programming - CP 2005. XX, 887 pages. 2005.

Vol. 3708: J. Blanc-Talon, W. Philips, D. Popescu, P. Scheunders (Eds.), Advanced Concepts for Intelligent Vision Systems. XXII, 725 pages. 2005.

Vol. 3707: D.A. Peled, Y.-K. Tsay (Eds.), Automated Technology for Verification and Analysis. XII, 506 pages. 2005.

Vol. 3706: H. Fuks, S. Lukosch, A.C. Salgado (Eds.), Groupware: Design, Implementation, and Use. XII, 378 pages. 2005.

Vol. 3704: M. De Gregorio, V. Di Maio, M. Frucci, C. Musio (Eds.), Brain, Vision, and Artificial Intelligence. XV, 556 pages. 2005.

Vol. 3703: F. Fages, S. Soliman (Eds.), Principles and Practice of Semantic Web Reasoning. VIII, 163 pages. 2005.

Vol. 3702: B. Beckert (Ed.), Automated Reasoning with Analytic Tableaux and Related Methods. XIII, 343 pages. 2005. (Subseries LNAI).

Vol. 3701: M. Coppo, E. Lodi, G. M. Pinna (Eds.), Theoretical Computer Science. XI, 411 pages. 2005.

Vol. 3699: C.S. Calude, M.J. Dinneen, G. Păun, M. J. Pérez-Jiménez, G. Rozenberg (Eds.), Unconventional Computation. XI, 267 pages. 2005.

Vol. 3698: U. Furbach (Ed.), KI 2005: Advances in Artificial Intelligence. XIII, 409 pages. 2005. (Subseries LNAI).

Vol. 3697: W. Duch, J. Kacprzyk, E. Oja, S. Zadrożny (Eds.), Artificial Neural Networks: Formal Models and Their Applications – ICANN 2005, Part II. XXXII, 1045 pages. 2005.

Vol. 3696: W. Duch, J. Kacprzyk, E. Oja, S. Zadrożny (Eds.), Artificial Neural Networks: Biological Inspirations – ICANN 2005, Part I. XXXI, 703 pages. 2005.

Vol. 3695: M.R. Berthold, R. Glen, K. Diederichs, O. Kohlbacher, I. Fischer (Eds.), Computational Life Sciences. XI, 277 pages. 2005. (Subseries LNBI).

Vol. 3694: M. Malek, E. Nett, N. Suri (Eds.), Service Availability. VIII, 213 pages. 2005.

Vol. 3693: A.G. Cohn, D.M. Mark (Eds.), Spatial Information Theory. XII, 493 pages. 2005.

Vol. 3692: R. Casadio, G. Myers (Eds.), Algorithms in Bioinformatics. X, 436 pages. 2005. (Subseries LNBI).

Vol. 3691: A. Gagalowicz, W. Philips (Eds.), Computer Analysis of Images and Patterns. XIX, 865 pages. 2005.

Vol. 3690: M. Pěchouček, P. Petta, L.Z. Varga (Eds.), Multi-Agent Systems and Applications IV. XVII, 667 pages. 2005. (Subseries LNAI).

Vol. 3689: G.G. Lee, A. Yamada, H. Meng, S.H. Myaeng (Eds.), Information Retrieval Technology. XVII, 735 pages. 2005.

Vol. 3688: R. Winther, B.A. Gran, G. Dahll (Eds.), Computer Safety, Reliability, and Security. XI, 405 pages. 2005.

Vol. 3687: S. Singh, M. Singh, C. Apte, P. Perner (Eds.), Pattern Recognition and Image Analysis, Part II. XXV, 809 pages. 2005.

Vol. 3686: S. Singh, M. Singh, C. Apte, P. Perner (Eds.), Pattern Recognition and Data Mining, Part I. XXVI, 689 pages. 2005.

Vol. 3685: V. Gorodetsky, I. Kotenko, V. Skormin (Eds.), Computer Network Security. XIV, 480 pages. 2005.

Vol. 3684: R. Khosla, R.J. Howlett, L.C. Jain (Eds.), Knowledge-Based Intelligent Information and Engineering Systems, Part IV. LXXIX, 933 pages. 2005. (Subseries LNAI).

Vol. 3683: R. Khosla, R.J. Howlett, L.C. Jain (Eds.), Knowledge-Based Intelligent Information and Engineering Systems, Part III. LXXX, 1397 pages. 2005. (Subseries LNAI).

Vol. 3682: R. Khosla, R.J. Howlett, L.C. Jain (Eds.), Knowledge-Based Intelligent Information and Engineering Systems, Part II. LXXIX, 1371 pages. 2005. (Subseries LNAI).

Vol. 3681: R. Khosla, R.J. Howlett, L.C. Jain (Eds.), Knowledge-Based Intelligent Information and Engineering Systems, Part I. LXXX, 1319 pages. 2005. (Subseries LNAI).

Vol. 3680: C. Priami, A. Zelikovsky (Eds.), Transactions on Computational Systems Biology II. IX, 153 pages. 2005. (Subseries LNBI).

Vol. 3679: S.d.C. di Vimercati, P. Syverson, D. Gollmann (Eds.), Computer Security – ESORICS 2005. XI, 509 pages. 2005.

Vol. 3678: A. McLysaght, D.H. Huson (Eds.), Comparative Genomics. VIII, 167 pages. 2005. (Subseries LNBI).

Vol. 3677: J. Dittmann, S. Katzenbeisser, A. Uhl (Eds.), Communications and Multimedia Security. XIII, 360 pages. 2005.

Vol. 3676: R. Glück, M. Lowry (Eds.), Generative Programming and Component Engineering. XI, 448 pages. 2005.

Vol. 3675: Y. Luo (Ed.), Cooperative Design, Visualization, and Engineering. XI, 264 pages. 2005.

Vol. 3674: W. Jonker, M. Petković (Eds.), Secure Data Management. X, 241 pages. 2005.

Vol. 3673: S. Bandini, S. Manzoni (Eds.), AI*IA 2005: Advances in Artificial Intelligence. XIV, 614 pages. 2005. (Subseries LNAI).

Vol. 3672: C. Hankin, I. Siveroni (Eds.), Static Analysis. X, 369 pages. 2005.